Corrosion Science and Technology

Third Edition

Corrosion Science and Technology

Third Edition

David E.J. Talbot
James D.R. Talbot

CRC Press is an imprint of the
Taylor & Francis Group, an **informa** business

CRC Press
Taylor & Francis Group
6000 Broken Sound Parkway NW, Suite 300
Boca Raton, FL 33487-2742

First issued in paperback 2020

ISBN-13: 978-1-4987-5241-1 (hbk)
ISBN-13: 978-0-367-73534-0 (pbk)

Visit the Taylor & Francis Web site at
http://www.taylorandfrancis.com

and the CRC Press Web site at
http://www.crcpress.com

Contents

Preface to the Third Edition

In earlier editions of this book, corrosion issues were considered within a coherent framework of chemistry, physics and metallurgy. Guided by this information, corrosion control of selected metals was evaluated in representative technologies according to public safety and economic factors as well as technical feasibility.

In this third edition, the scope is enlarged and diversified. There are chapters on the manufacture, structures, properties and corrosion resistance of aluminium, copper, nickel, titanium, iron, magnesium, zinc, tin, lead, zirconium, hafnium, beryllium, uranium and their alloys and strategies for applying them correctly and economically.

The extended range of technologies now includes aviation, automobile manufacture, food processing and distribution, building construction, marine environments, fossil fuel fired boilers, oilfield operations and nuclear topics. Finally, some of the various ways of corrosion testing and prediction are addressed.

Authors

David E.J. Talbot (1926–2016), PhD, graduated with a BSc and an MSc from the University of Wales and a PhD from Brunel University for research on gas–metal equilibria. From 1949 to 1966, he was employed at the research laboratories of the British Aluminium Company Ltd., contributing to research promoting the development of manufacturing processes and to customer service. From 1966 to 1994, he taught courses on corrosion and other aspects of chemical metallurgy at Brunel University and maintained an active interest in research and development, mainly in collaboration with manufacturing industries in the United Kingdom and the United States. He was a member of the Institute of Materials and had chartered engineer status; he served as a member of the council of the London Metallurgical Society. Dr. Talbot wrote many papers on the chemical aspects of metallurgy, a review on metal–hydrogen systems in *International Metallurgical Reviews*, and a section on gas–metal systems in *Smithells Reference Book.*

James D.R. Talbot, PhD, graduated with a BSc ARCS from Imperial College, London, and earned an MSc from Brunel University. He earned a PhD from the University of Reading for research on the physical chemistry of aqueous solutions and its application to natural waters. Dr. Talbot worked at the River Laboratory of the Institute of Freshwater Ecology, Dorset, United Kingdom, where he assessed and predicted physical chemical changes occurring in river management. He has written papers on the speciation of solutes in natural waters. From 2000 to 2006, he was a lecturer in materials research chemistry at Cranfield University in the United Kingdom, where he specialized in the physicochemical aspects of corrosion, polymer science and process science. He is presently a chemist with interests in species-specific corrosion mechanisms. Dr. Talbot is a current member of the Structure and Properties of Materials Committee of the Institute of Metals, Minerals and Mining. He has published in the fields of corrosion, polymer chemistry, solution chemistry and the chemistry of natural waters.

1

Overview of Corrosion and Protection Strategies

Metals in service often give a superficial impression of permanence, but all except gold are chemically unstable in air and air-saturated water at ambient temperatures, and most are also unstable in air-free water. Hence, almost all of the environments in which metals serve are potentially hostile and their successful use in engineering and commercial applications depends on protective mechanisms. In some metal/environment systems, the metal is protected by passivity, a naturally formed surface condition inhibiting reaction. In other systems, the metal surface remains active and some form of protection must be provided by design. This applies particularly to plain carbon and low alloy irons and steels, which are the most prolific, least expensive and most versatile metallic materials. Corrosion occurs when protective mechanisms have been overlooked, break down or are exhausted, leaving the metal vulnerable to attack.

Practical corrosion-related problems are often discovered in the context of engineering and allied disciplines, where the approach may be hindered by unfamiliarity with the particular blend of electrochemistry, metallurgy and physics which must be brought to bear if satisfactory solutions are to be found. This overview indicates the relevance of these various disciplines and some relationships between them. They are described in detail in subsequent chapters.

1.1 Corrosion in Aqueous Media

1.1.1 Corrosion as a System Characteristic

Some features of the performance expected from metals and metal artefacts in service can be predicted from their intrinsic characteristics assessed from their compositions, structures as viewed in the microscope and past history of thermal and mechanical treatments they may have received. These characteristics control density, thermal and electrical conductivity, ductility, strength under static loads in benign environments and other physical and mechanical properties. These aspects of serviceability are reasonably straightforward and controllable, but there are other aspects of performance which are less obvious and more difficult to control because they depend not only on the intrinsic characteristics of the metals, but also on the particular conditions in which they serve. They embrace susceptibility to corrosion, metal fatigue and wear, which can be responsible for complete premature failure with costly and sometimes dangerous consequences.

Degradation by corrosion, fatigue and wear can be approached only by considering a metal not in isolation but within a wider system with the components, metal, chemical environment, stress and time. Thus a metal selected to serve well in one chemical environment or stress system may be inadequate for another. Corrosion, fatigue and wear can interact synergistically, as illustrated in Chapter 5, but for the most part, it is usually sufficient to consider corrosion processes as a chemical system comprising the metal itself and its environment.

1.1.2 The Electrochemical Origin of Corrosion

From initial encounters with the effects of corrosion processes, it may seem difficult to accept that they can be explained on a rational basis. One example, among many, concerns the role of dissolved oxygen in corrosion. It is well known that unprotected iron rusts in pure neutral waters, but only if it contains dissolved oxygen. Based on this observation, standard methods of controlling corrosion of steel in steam raising boilers include the removal of dissolved oxygen from the water. This appears to be inconsistent with observations that pure copper has good resistance to neutral water whether it contains oxygen or not. Moreover copper can dissolve in acids containing dissolved oxygen but is virtually unattacked if the oxygen is removed, whereas the complete reverse is true for stainless steels. These and many other apparently conflicting observations can be reconciled on the basis of the electrochemical origin of the principles underlying corrosion processes and protection strategies. The concepts are not difficult to follow and it is often the unfamiliar notation and conventions in which the ideas are expressed which deter engineers.

At its simplest, a corroding system is driven by two spontaneous coupled reactions, which take place at the interface between the metal and an aqueous environment. One is a reaction in which chemical species from the aqueous environment remove electrons from the metal; the other is a reaction in which metal surface atoms participate to replenish the electron deficiency. The exchange of electrons between the two reactions constitutes an electronic current at the metal surface and an important effect is to impose an electric potential on the metal surface of such a value that the supply and demand for electrons in the two coupled reactions are balanced.

The potential imposed on the metal is of much greater significance than simply to balance the complementary reactions which produce it because it is one of the principal factors determining what the reactions shall be. At the potential it acquires in neutral aerated water, the favoured reaction for iron is dissolution of the metal as a soluble species which diffuses away into the solution, allowing the reaction to continue, that is, the iron corrodes. If the potential is depressed by removal of dissolved oxygen, the reaction is decelerated or suppressed. Alternatively, if the potential is raised by appropriate additions to the water, the favoured reaction can be changed to produce a solid product on the iron surface, which confers effective corrosion protection. Raising the alkalinity of the water has a similar effect.

1.1.3 Stimulated Local Corrosion

A feature of the process in which oxygen is absorbed has two important effects, one beneficial and the other deleterious. In still water, oxygen used in the process must be resupplied from a distant source, usually the water surface in contact with air, and the rate-controlling factor is diffusion through the low solubility of oxygen in water. The beneficial effect is that the absorption of oxygen controls the overall corrosion rate, which is consequently much slower than might otherwise be expected. The deleterious effect is that difficulty in the resupply of oxygen can lead to differences in oxygen concentration at the metal surface, producing effects which can stimulate intense metal dissolution in oxygen starved regions, especially crevices. This is an example of a local action corrosion cell. There is much more to this phenomenon than this brief description suggests and it is discussed more fully in Chapter 3.

Another example of stimulated corrosion is produced by the bimetallic effect. It comes about because of a hierarchy of metals distinguished by their different tendencies to react with the environment, measured by the free energy changes, formally quantified in electrochemical terms in Chapter 3. Metals such as iron or aluminium with strong tendencies to react are regarded as less noble and those with weaker tendencies, such as copper, are considered more noble. For reasons given later, certain strongly passive metals, such as stainless steels, and some nonmetallic conductors, such as graphite, can simulate noble metals. The effect is to intensify an attack on the less noble of a pair of metals in electrical contact exposed to the same aqueous environment. Conversely, the more noble metal is partially or completely protected. These matters are very involved and are given the attention they merit in Chapters 4 and 20.

1.2 Thermal Oxidation

The components of clean air which are active toward metals are oxygen and water vapour. Atmospheric nitrogen acts primarily as a diluent because although metals such as magnesium and aluminium form nitrides in pure nitrogen gas, the nitrides are unstable with respect to the corresponding oxides in the presence of oxygen.

At ordinary temperatures, very thin oxide films, of the order of 3 to 10 nm thick, protect most engineering metals. These films form very rapidly on contact with atmospheric oxygen but subsequent growth in uncontaminated air with low humidity is usually imperceptible. It is for this reason that aluminium, chromium, zinc, nickel and some other common metals remain bright in unpolluted indoor atmospheres.

1.2.1 Protective Oxides

At higher temperatures, the oxides formed in air on most common engineering metals, including iron, copper, nickel, zinc and many of their alloys, remain coherent and adherent to the metal substrate, but reaction continues because reacting species can penetrate the oxide structure and the oxides grow thicker. These oxides are classed as protective oxides because the rate of oxidation diminishes as they thicken, although the protection is incomplete. The oxide grows by an overall reaction driven by two electrochemical processes, an anodic process converting the metal to cations and generating electrons at the metal/oxide interface, coupled with a cathodic process converting oxygen from the atmosphere to anions and consuming electrons at the oxygen oxide/atmosphere interface. The nature of these ions and the associated electronic conduction mechanisms are quite different from their counterparts in aqueous corrosion. A new unit of oxide is produced when an anion and cation are brought together. To accomplish this, one or the other of the ions must diffuse through the oxide. The ions diffuse through defects on an atomic scale, which are characteristic features of oxide structures. Associated defects in the electronic structure provide the electronic conductivity needed for the transport of electrons from the metal/oxide to the oxide/air interfaces. These structures, reviewed in Chapter 2, differ from oxide to oxide and are crucially important in selecting metals and formulating alloys for oxidation resistance. For example, the oxides of chromium and aluminium have such small defect populations that they are protective at very high temperatures. The oxidation resistance afforded by these oxides can be conferred on other metals by alloying or surface treatment. This is the basis on which oxidation resistance is imparted to stainless steels and to nickel base superalloys for gas turbine blades.

1.2.2 Nonprotective Oxides

For some metals, differences in the relative volumes of an oxide and the metal consumed in its formation impose shear stresses high enough to impair the formation of cohesive and adhesive protective oxide layers. If such metals are used for high temperature service in atmospheres with real or virtual oxygen potentials, they must be protected. An example is the need to can uranium fuel rods in nuclear reactors because of the unprotective nature of the oxide.

1.3 Environmentally Sensitive Cracking

Corrosion processes can interact with a stressed metal to produce fracture at critical stresses of only fractions of its normal fracture stress. These effects can be catastrophic and even life threatening if they occur, for example, in aircraft. There are two different principal failure modes, corrosion fatigue and stress-corrosion cracking (SCC), featured in Chapter 5.

Corrosion fatigue failure can affect any metal. Fatigue failure is fracture at a low stress as the result of cracking propagated by cyclic loading. The failure is delayed, and the effect is accommodated in design by assigning for a given applied cyclic stress, a safe fatigue life, characteristically the elapse of between 10^7 and 10^8 loading cycles. Cracking progresses by a sequence of events through incubation, crack nucleation and propagation. If unqualified, the term fatigue relates to metal exposed to normal air. The distinguishing feature of corrosion fatigue is that failure occurs in some other medium, usually an aqueous medium, in which the events producing fracture are accelerated by local electrochemical effects at the nucleation site and at the crack tip, shortening the fatigue life.

SCC is restricted to particular metals and alloys exposed to highly specific environmental species. An example is the failure of age-hardened aluminium aircraft alloys in the presence of chlorides. A disturbing feature of the effect is that the onset of cracking is delayed for months or years, but when cracks finally appear, fracture is almost imminent. Neither effect is fully understood because they exhibit different critical features for different metals and alloys but, using accumulated experience, both can be controlled by vigilant attention.

1.4 Strategies for Corrosion Control

1.4.1 Passivity

Aluminium is a typical example of a metal endowed with the ability to establish a naturally passive surface in appropriate environments. Paradoxically, aluminium theoretically tends to react with air and water by some of the most energetic chemical reactions known, but provided that these media are neither excessively acidic nor alkaline and are free from certain aggressive contaminants, the initial reaction products form a vanishingly thin impervious barrier separating the metal from its environment. The protection afforded by this condition is so effective that aluminium and some of its alloys are standard materials for cooking utensils, food and beverage containers, architectural use and other applications in which a nominally bare metal surface is continuously exposed to air and water. Similar effects are responsible for the utility of some other metals exploited for their corrosion resistance, including zinc, titanium, cobalt and nickel. In some systems, easy passivating characteristics can also be conferred on an alloy in which the dominant component is an active metal in normal circumstances. This approach is used in the formulation of stainless steels that are alloys based on iron with chromium as the component inducing passivity.

1.4.2 Conditions in the Environment

Unprotected active metals exposed to water or rain are vulnerable, but corrosion can be delayed or even prevented by natural or artificially contrived conditions in the environment. Steels corrode actively in moist air and water containing dissolved air, but the rate of dissolution can be restrained by the slow resupply of oxygen, as described in Section 1.1.3 and by deposition of chalky or other deposits on the metal surface from natural waters. For thick steel sections, such as railroad track, no further protection may be needed.

In critical applications using thinner sections, such as steam raising boilers, nearly complete protection can be provided by chemical scavenging to remove dissolved oxygen from the water completely and by rendering it mildly alkaline to induce passivity at the normally active iron surface. This is an example of protection by deliberately conditioning the environment.

1.4.3 Cathodic Protection

Cathodic protection provides a method of protecting active metals in continuous contact with water, as in ships and pipelines. It depends on opposing the metal dissolution reaction with an electrical potential applied by impressing a cathodic current from a DC generator across the metal/environment interface. An alternative method of producing a similar effect is to couple metal needing protection to a less noble metal, which is sacrificed as explained in Section 1.1.3. The application of these techniques is considered in Chapter 20.

1.4.4 Protective Coatings

When other protective strategies are inappropriate or uneconomic, active metals must be protected by applied coatings. The most familiar coatings are paints, a term covering various organic media, usually based on alkyd and epoxy resins, applied as liquids which subsequently polymerize to hard coatings. They range from the oil based, air-drying paints applied by brush used for civil engineering structures, to thermosetting media dispersed in water for application by electrodeposition to manufactured products, including motor vehicle bodies. Alternatively, a thin coating of an expensive corrosion-resistant metal, usually applied by electrodeposition, can protect a vulnerable but inexpensive metal. One example is the tin coating on steel food cans; another is the nickel/chromium system applied to steel where corrosion resistance combined with aesthetic appeal is required, as in bright trim on motor vehicles and domestic equipment. An important special use of a protective metal coating is the layer of pure aluminium bonded by hot-rolling to aluminium aircraft alloys, which are strong but vulnerable to corrosion.

1.4.5 Corrosion Costs

Estimates of the costs of corrosion draw attention to wasteful depletion of resources, but they should be interpreted with care because they may include avoidable items attributable to the costs of poor design, lack of information or neglect. The true costs of corrosion are the unavoidable costs of dealing with it in the most economical way. Such costs include the prices of resistant metals and the costs of protection, maintenance and planned amortization.

An essential objective in design is to produce structures or manufactured products which fulfill their purposes with the maximum economy in the overall use of resources interpreted in monetary terms. This is not easy to assess and requires an input of the principles applied by accountants. One such principle is the 'present worth' concept of future expenditure, derived by discounting cash flow, which favours deferred costs, such as maintenance, over initial costs, and another is a preference for tax deductible expenditure. The results of such assessments influence technical judgements and may determine, for example, whether it is better to use resources initially for expensive materials with high integrity or to use them later for protecting or replacing less expensive, more vulnerable materials.

1.4.6 Criteria for Corrosion Failure

The economic use of resources is based on planned life expectancies for significant metal structures or products. The limiting factor may be corrosion but more often it is something else, such as wear of moving parts, fatigue failure of cyclically loaded components, failure of associated accessories, obsolescent technology or stock replenishment cycles. The criterion for corrosion failure is therefore premature termination of the useful function of the metal by interaction with its environment, before the planned life has elapsed. Residual life beyond the planned life is a waste of resources. Failure criteria vary according to circumstances and include:

1. Loss of strength inducing failure of stressed metal parts
2. Corrosion product contamination of sensitive material, for example, food or paint
3. Perforation by pitting corrosion, opening leaks in tanks or pipes
4. Fracture by environmentally sensitive cracking
5. Interference with thermal transfer by corrosion products
6. Loss of aesthetic appeal

Strategies for corrosion control must be considered not in isolation but within constraints imposed by cost-effective use of materials and by other properties and characteristics of metallic materials for particular applications. Three examples illustrate different priorities.

1. The life expectancy for metal food and beverage cans is only a few months and during that time, corrosion control must ensure that the contents of the cans are not contaminated. Any surface protection applied must be non-toxic and amenable to consistent application at high speed for a vast market in which there is

intense cost conscious competition between can manufacturers and material suppliers. The metal selected and any protective surface coating applied to it must withstand the very severe deformation experienced in fabricating the can bodies.

2. Aircraft are designed for many years of continuous capital intensive airline operations. Metals used in their construction must be light, strong, stiff, damage tolerant and corrosion resistant. They must be serviceable in environments contaminated with chlorides from marine atmospheres and de-icing salts which can promote environmentally sensitive cracking. Reliable long-term corrosion control and monitoring schedules are essential to meet the imperative of passenger safety and to avoid disruption of schedules through unplanned grounding of aircraft.
3. Canning materials to protect fuel rods in nuclear reactors must resist high temperature attack by the particular heat transfer media in the presence of nuclear radiation, but the choice of material is further constrained by the overriding need for low neutron capture cross section. According to reactor design, materials that have been used include customized alloys based on magnesium and hafnium-free zirconium.

1.4.7 Material Selection

In the initial concept for a metallic product or structure, it is natural to consider using an inexpensive, easily fabricated metal, such as a plain carbon steel. On reflection, it may be clear that unprotected inexpensive materials will not resist the prevailing environment and a decision is required on whether to apply protection, control the environment or to choose a more expensive metal. The choice is influenced by prevailing metal prices.

Metal prices vary substantially from metal to metal and are subject to fluctuations in response to supply and demand as expressed in prices fixed in the metal exchanges through which they are traded. The prices also vary according to purity and form because they include refining and fabricating costs. Table 1.1 gives some recent representative prices.

Table 1.1 illustrates the considerable expense of specifying other metals and alloys in place of steels. This applies especially to a valuable metal such as nickel or tin even if it is used as a protective coating or as an alloy component. For example, the nickel content has an important effect on the prices of stainless steels.

The use of different metals in contact can be a corrosion hazard because in some metal couples, one of the pair is protected and the other is sacrificed, as described earlier in relation to cathodic protection. Examples of adverse metal pairs encountered in unsatisfactory designs are aluminium/brass and carbon steel/stainless steel, threatening intensified attack on the aluminium and carbon steel, respectively. The uncritical mixing of metals is one of the more common corrosion-related design faults, thus it is featured prominently in Chapter 4, where the overt and latent hazards of the practice are explained.

1.4.8 Geometric Factors

When the philosophy of a design is settled and suitable materials are selected, the proposed physical form of the artefact must be scrutinized for corrosion traps. Provided that one or two well-known effects are taken into account, this is a straightforward task. Whether protected or not, the less time the metal spends in contact with water, the less is the chance of corrosion and all that this requires is some obvious precautions, such as angle sections disposed apex upwards, box sections closed off or fitted with drainage

TABLE 1.1

Representative Selection of Metal Prices—June 2017

Metal		Form	Price $/tonne
Pure metals	Aluminium[a]	Primary metal ingot	1900
	Copper[a]	Primary metal ingot	5550
	Lead[a]	Primary metal ingot	2080
	Nickel[a]	Primary metal ingot	8800
	Tin[a]	Primary metal ingot	20,100
	Zinc[a]	Primary metal ingot	2480
Plain low carbon steel[b]		Continuously cast slab	290
		6 mm thick hot-rolled plate, 1 m wide coil	560
		2 mm thick cold-rolled sheet, 1 m wide coil	665
		0.20 mm electrolytic tinplate, 1 m wide coil	1550
Stainless steels[b]	AISI 409	2 mm sheet	1800
	AISI 304	6 mm thick hot-rolled plate, 1 m wide coil	2150
		2 mm thick cold-rolled sheet, 1 m wide coil	2300
	AISI 316	6 mm thick hot-rolled plate, 1 m wide coil	3000
		2 mm thick cold-rolled sheet, 1 m wide coil	3200

Note: Pure metal prices vary with market conditions and prices of wrought products are adjustable by premiums and discounts by negotiation.

[a] Representative Metal Exchange Prices.

[b] Typical price lists.

holes, tank bottoms raised clear of the floor and drainage taps fitted at the lowest points of systems containing fluids. Crevices must be eliminated to avoid local oxygen depletion for the reason given in Section 1.1.3 and explained in Chapter 3. This entails full penetration of butt welds, double sided welding for lap welds, well-fitting gaskets, and so on. If they are unavoidable, adverse mixed metal pairs should be insulated and the direction of any water flow should be from less noble to more noble metals to prevent the indirect effects described in Chapter 4.

1.5 Some Symbols, Conventions and Equations

From the discussion so far, it is apparent that specialized notation is required to express the characteristics of corrosion processes and it is often this notation which inhibits access to the underlying principles. The symbols used in chemical and electrochemical equations are not normal currency in engineering practice and some terms, such as electrode, potential, current and polarization, have meanings which may differ from their meanings in other branches of science and engineering. The reward in acquiring familiarity with the conventions is access to information accumulated in the technical literature with a direct bearing on practical problems.

1.5.1 Ions and Ionic Equations

Certain substances which dissolve in water form electrically conducting solutions, known as electrolytes. The effect is due to their dissociation into electrically charged entities centered on atoms or groups of atoms, known as ions. The charges are due to the unequal distribution of the available electrons between the ions, so that some have a net positive charge and are called cations and some have a net negative charge and are called anions. Faraday demonstrated the existence of ions by the phenomenon of electrolysis in which they are discharged at the positive and negative poles of a potential applied to a solution. Symbols for ions have superscripts showing the polarity of the charge and its value as charge numbers, that is, multiples of the charge on one electron. A subscript (aq), may be added where needed to distinguish ions in aqueous solution from ions encountered in other contexts, such as in ionic solids. The symbols are used to describe the solution of any substances yielding electrolytes on dissolution, for example:

$$\underset{\text{hydrochloric acid gas}}{\mathrm{HCl}} \rightarrow \underset{\text{hydrogen cation}}{\mathrm{H^{+}}_{(aq)}} + \underset{\text{chloride anion}}{\mathrm{Cl^{-}}_{(aq)}} \tag{1.1}$$

$$\underset{\text{solid iron(II) chloride}}{\mathrm{FeCl_2}} \rightarrow \underset{\text{iron cation}}{\mathrm{Fe^{2+}}_{(aq)}} + \underset{\text{chloride anions}}{\mathrm{2Cl^{-}}_{(aq)}} \tag{1.2}$$

The use of these symbols is only a formal convention for convenience in writing equations. The structures and properties of the ions they represent are described in Chapter 2.

1.5.2 Partial Reactions

Equations 1.1 and 1.2 represent complete reactions, but sometimes the anions and cations in solution originate from neutral species by complementary partial reactions, exchanging charge at an electronically conducting surface, usually a metal. To illustrate this process, consider the dissolution of iron in a dilute air-free solution of hydrochloric acid, yielding hydrogen gas and a dilute solution of iron chloride as the products. The overall reaction is:

$$\mathrm{Fe}\ (\text{metal}) + 2\mathrm{HCl}\ (\text{solution}) \rightarrow \mathrm{FeCl_2}\ (\text{solution}) + \mathrm{H_2}\ (\text{gas}) \tag{1.3}$$

Strong acids and their soluble salts are ionized in dilute aqueous solution as illustrated in Equations 1.1 and 1.2 so that the species present in dilute aqueous solutions of hydrochloric acid and iron(II) chloride are not the molecular entities HCl and $FeCl_2$, but their ions ($H^+ + Cl^-$) and ($Fe^{2+} + Cl^-$). Equation 1.1 is therefore equivalent to:

$$\mathrm{Fe} + 2\mathrm{H^+} + 2\mathrm{Cl^-} \rightarrow \mathrm{Fe^{2+}} + 2\mathrm{Cl^-} + \mathrm{H_2} \tag{1.4}$$

The Cl^- ions persist unchanged through the reaction, maintaining electric charge neutrality, that is, they serve as a counter ion. The effective reaction is the transfer of electrons, e^-, from atoms of iron in the metal to hydrogen ions, yielding soluble Fe^{2+} ions and neutral hydrogen atoms which combine to be evolved as hydrogen gas. The electron transfer occurs at the conducting iron surface where the excess of electrons left in the metal by the solution of iron from the metal is available to discharge hydrogen ions supplied by the solution:

$$\mathrm{Fe}\ (\text{metal}) \rightarrow \mathrm{Fe^{2+}}\ (\text{solution}) + 2\mathrm{e^-}\ (\text{in metal}) \tag{1.5}$$

$$2\mathrm{e^-}\ (\text{in metal}) + 2\mathrm{H^+}\ (\text{solution}) \rightarrow 2\mathrm{H}\ (\text{metal surface}) \rightarrow \mathrm{H_2}\ (\text{gas}) \tag{1.6}$$

Processes like those represented by Equations 1.5 and 1.6 are described as electrodes. Electrodes proceeding in a direction generating electrons, as in Equation 1.5, are anodes, and electrodes accepting electrons, as in Equation 1.6, are cathodes. Any particular electrode can be an anode or cathode depending on its context. Thus the zinc electrode:

$$Zn \rightarrow Zn^{2+} + 2e^- \tag{1.7}$$

is an anode when coupled with Equation 1.6 to represent the spontaneous dissolution of zinc metal in an acid, but it is a cathode when driven in the opposite direction by an applied potential to deposit zinc from solution:

$$Zn^{2+} + 2e^- \rightarrow Zn \text{ (metal)} \tag{1.8}$$

1.5.3 Representation of Corrosion Processes

The facility with which the use of electrochemical equations can reveal characteristics of corrosion processes can be illustrated by comparing the behaviuor of iron in neutral and alkaline waters containing dissolved oxygen.

1.5.3.1 Active Dissolution of Iron with Oxygen Absorption

Iron rusts in neutral water containing oxygen dissolved from the atmosphere. The following greatly simplified description illustrates some general features of the process. The concentration (strictly the *activity*, defined later in Section 2.1.3) of hydrogen ions in neutral water is low so that the evolution of hydrogen is replaced by the absorption of dissolved oxygen as the dominant cathodic reaction and the coupled reactions are:

$$\text{Anodic reaction: } Fe \rightarrow Fe^{2+} + 2e^- \tag{1.9}$$

$$\text{Cathodic reaction: } \tfrac{1}{2}O_2 + H_2O + 2e^- \rightarrow 2OH^- \tag{1.10}$$

The notional half quantity of oxygen in Equation 1.10 formally expresses the mass balance for the two equations. The reactions introduce Fe^{2+} and OH^- ions into the solution, which, with simplifying assumptions, co-precipitate as the sparingly soluble compound $Fe(OH)_2$:

$$Fe^{2+} \text{ (solution)} + 2OH^- \text{ (solution)} \rightarrow Fe(OH)_2 \text{ (precipitate)} \tag{1.11}$$

In this system, the transport of Fe^{2+} and OH^- ions in the electrolyte between the anodic and cathodic reactions constitutes an ion current. The example illustrates how a corrosion process is a completed electric circuit with the following component parts:

1. An anodic reaction
2. A cathodic reaction
3. Electron transfer between the anodic and cathodic reactions
4. An ion current in the electrolyte

Control of corrosion is based on inhibiting one or another of the links in the circuit by methods to be described in due course.

The $Fe(OH)_2$ is precipitated from the solution, but it is usually deposited back on the metal surface as a loose defective material which fails to stifle further reaction, allowing rusting to continue. In the presence of the dissolved oxygen it subsequently transforms to a more stable composition in the final rust product. The rusting of iron is not always as straightforward as this simplified approach suggests and is described comprehensively in Chapter 7.

1.5.3.2 Passivity of Iron in Alkaline Water

Iron responds quite differently in mildly alkaline water. The anodic reaction yielding the unprotective soluble ion, Fe^{2+} as the primary anodic product, is not favoured and is replaced by an alternative anodic reaction which converts the iron surface directly into a thin, dense, protective layer of magnetite, Fe_3O_4, so that the partial reactions are:

$$\text{Anodic reaction: } 3Fe + 8OH^- = Fe_3O_4 + 4H_2O + 8e^- \tag{1.12}$$

$$\text{Cathodic reaction: } 2O_2 + 4H_2O + 8e^- = 8OH^- \tag{1.13}$$

Equation 1.13 is Equation 1.10 with notional quantities adjusted to balance Equation 1.12. Information on conditions favouring protective anodic reactions of this kind is important in corrosion control. Pourbaix diagrams, explained in Chapter 3, give such information graphically and within their limitations, they can be useful in interpreting observed effects.

Further Reading

Economic Effects of Metallic Corrosion in the United States, National Bureau of Standards Special Publication, 1978.

2

Structures Participating in Corrosion Processes

2.1 Origins and Characteristics of Structure

Conventional symbols are convenient for use in chemical equations, as illustrated in the last chapter, but they do not indicate the physical forms of the atoms, ions and electrons they represent. This chapter describes these physical forms and the structures in which they exist because they control the course and speed of reactions.

There is an immediate problem in describing and explaining these structures because they are expressed in the conventional language and symbols of chemistry. Atoms can be arranged in close-packed arrays, open networks or as molecules, forming crystalline solids, non-crystalline solids, liquids or gases, all with their own specialized descriptions. The configurations of the electrons within atoms and assemblies of atoms are described in terms and symbols derived from wave mechanics that are foreign to many applied disciplines that need the information.

A preliminary task is to review some of this background as briefly and simply as possible for use later on. At this point, it is natural for an applied scientist to inquire whether an apparently academic digression is really essential to address the practical concerns of corrosion. The answer is that, without this background, explanations of even basic underlying principles can be given only on the basis of postulates that seem arbitrary and unconvincing. With this background, it is possible to give plausible explanations to such questions as why water has a special significance in corrosion processes, why some dissolved substances inhibit corrosion whereas others stimulate it, what features of metal oxides control the protection they afford and how metallurgical structures have a key influence on the development of corrosion damage. Confidence in the validity of fundamentals is an essential first step in exercising positive corrosion control.

2.1.1 Phases

The term *phase* describes any region of material without internal boundaries, solid, liquid or gaseous, composed of atoms, ions or molecules organized in a particular way. The following is a brief survey of various kinds of phases that may be present in a corroding system and applies to metals, environments, corrosion products and protective systems.

2.1.1.1 Crystalline Solids

Many solids of interest in corrosion, such as metals, oxides and salts, are crystalline; the crystalline nature of bulk solids is not always apparent because they are usually agglomerates of microscopic crystals, but under laboratory conditions, single crystals can be produced that reveal many of the features associated with crystals, regular outward geometrical shapes, cleavage along well-defined planes and anisotropic physical and mechanical properties. The characteristics are due to the arrangement of the atoms or ions in regular arrays generating indefinitely repeated patterns throughout the material. This long-range order permits the relative positions of the atoms or ions in a particular phase to be located accurately by standard physical techniques, most conveniently by analysis of the diffraction patterns produced by monochromatic x-rays transmitted through the material. The centers of the atoms form a three-dimensional array known as the *space lattice* of the material.

Space lattices are classified according to the symmetry elements they exhibit. A space lattice is described by its *unit cell,* that is, the smallest part of the infinite array of atoms or ions that completely displays its characteristics and symmetry. The lattice dimensions are specified by quoting the *lattice parameters* that are the lengths of the edges of the unit cell. The complete structure is generated by repetition of the unit cell in three dimensions. Crystallographic descriptions embody a simplifying assumption that atoms and ions are hard spheres with definite radii. Strictly, atomic and ion sizes are influenced by local interactions with other atoms, but the assumption holds for experimental determination of atomic arrangements and lattice parameters in particular solids.

A wide range of space lattices is needed to represent all of the structures of crystalline solids of technical interest. Geometric considerations reveal 14 possible types of lattice, fully described in standard texts*. All that is needed here is a brief review of structures directly concerned with metals and solid ionic corrosion products. These are the close-packed cubic structures and the related hexagonal close-packed structure.

The closest possible packing for atoms or ions of the same radius is produced by stacking layers of atoms so that the whole system occupies the minimum volume, as follows. Spheres arranged in closest packing in a single layer have their centers at the corners of equilateral triangles, illustrated in Figure 2.1. For a stack of such layers to occupy minimum volume, the spheres in every successive layer are laid in natural pockets between

FIGURE 2.1
Assembly of spheres representing the closest-packed arrangement of atoms in two dimensions. The pockets at the centers of the equilateral triangles are sites for atoms in a similar layer superimposed in the closest-packed three-dimensional arrangement.

* For example, Taylor, cited in Further Reading.

contiguous atoms in the underlying layer. Simple geometry shows that the atoms of a layer in a stack are sited in alternate pockets of the underlying layer, that is, at the centers of *either* the upright triangles *or* the inverted triangles in Figure 2.1, yielding either of two different stacking sequences. In one, the positions of the atoms are in register at every second layer in the sequence *ABAB*..., generating the hexagonal close-packed lattice and in the other, they are in register at every third layer in the sequence *ABCABC*..., generating the face-centered cubic lattice.

2.1.1.1.1 The Hexagonal Close-Packed (HCP) Lattice

The hexagonal symmetry is derived from the fact that an atom in a close-packed plane is coordinated with six other atoms, whose centers are the corners of a regular hexagon. In three dimensions, every atom in the HCP lattice is in contact with 12 equidistant neighbors. Geometric considerations show that the axial ratio of the unit cell, that is, the ratio of the lattice parameters normal to the hexagonal basal plane and parallel to it, is 1.633.

2.1.1.1.2 The Face-Centered Cubic (FCC) Lattice

The *ABCABC*... stacking sequence confers cubic symmetry that is apparent in the unit cell, illustrated in Figure 2.2a, taken at an appropriate angle to the layers. Although the atoms are actually in contact, unit cells are conventionally drawn with small spheres indicating the lattice points to reveal the geometry. As the name suggests, the unit cell has one atom at every one of the eight corners of a cube and another atom at the center of every one of the six cube faces. Every atom is in contact with 12 equidistant neighbors, that is, its *coordination number* is 12. Every one of the eight corner atoms in the FCC unit cell is shared with seven adjacent unit cells and every face atom is shared with one other cell, so that the cell contains the equivalent of four atoms, that is [(8 corner atoms) × 1/8] + [(6 face atoms) × 1/2].

The FCC lattice completely represents the structures of many metals and alloys, but its use is extended to provide convenient crystallographic descriptions of some complex structures, using its characteristic that the spaces between the atoms, the *interstitial sites*, have the geometry of regular polyhedra. The concept is to envision atoms or ions of one species arranged on FCC lattice sites with other species occupying the interstices. Considerable use is made of this device later, especially in the context of oxidation, where a class of metal oxides collectively known as spinels have significant roles in the oxidation-resistance of alloys. This application depends on the geometry and number of interstices.

Inspection of the FCC unit cell illustrated in Figure 2.2a reveals that there are two kinds of interstitial sites, *tetrahedral* and *octahedral*. A tetrahedral site exists between a corner atom of the cell and the three adjacent face atoms and there are eight of them wholly contained within every cell. Octahedral sites exist both at the center of the cell between the 6 face atoms and at the middles of the 12 edges, every one of which is shared with 3 adjacent unit cells, so that the number of octahedral interstices per cell is 1 + (12 × ¼) = 4. Since the cell contains the equivalent of four atoms, the FCC structure contains two tetrahedral and one octahedral spaces per atom. The ratios of the radii of spheres that can be inscribed on the interstitial sites to the radius of atoms on the lattice is 0.414 for tetrahedral sites and 0.732 for the octahedral sites. These radius ratios indicate the sizes of interstitial atoms or ions that can be accommodated.

2.1.1.1.3 The Body-Centered Cubic (BCC) Lattice

The BCC structure is less closely packed than the HCP and FCC structures. The unit cell, illustrated in Figure 2.2b, has atoms at the corners of a cube and another at the center.

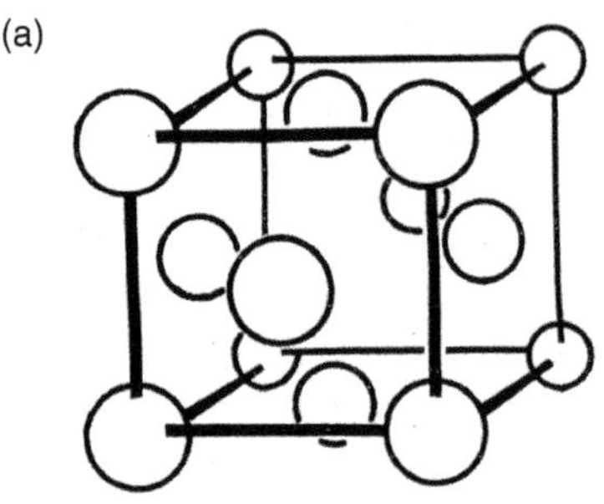

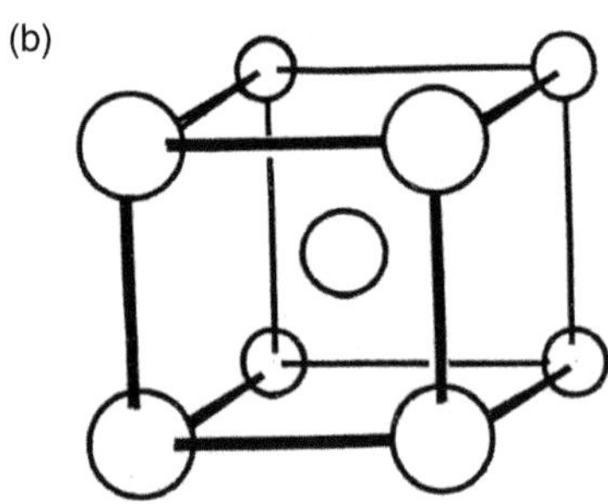

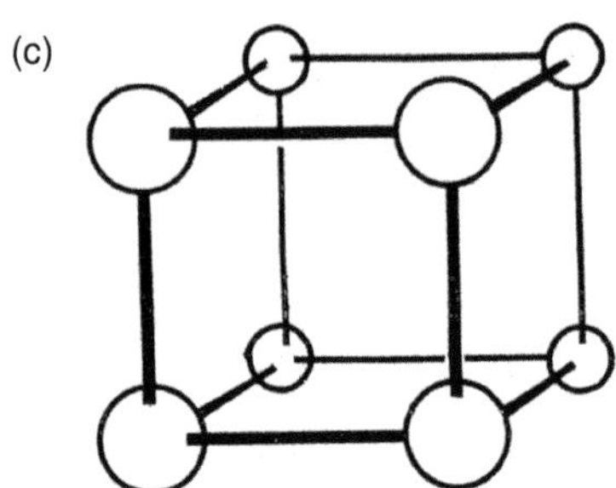

FIGURE 2.2
Unit cell: (a) Face-centered cubic, (b) body-centered cubic and (c) simple cubic.

It has the equivalent of 2 atoms and there are 12 tetrahedral and 3 octahedral interstitial sites with the geometries of irregular polyhedra. Corner atoms at opposite ends of the cell diagonals are contiguous with the atom at the center so that the coordination number is 8.

2.1.1.1.4 The Simple Cubic Lattice

The simple cubic lattice, illustrated in Figure 2.2c, can be formed from equal numbers of two different atoms or ions if the ratio of their radii is between 0.414 and 0.732, as explained later in describing oxides. The structure is geometrically equivalent to an FCC lattice of the larger atoms or ions with the smaller ones in octahedral interstitial sites.

Some characteristic metal structures are summarized in Table 2.1.

TABLE 2.1

Crystal Structures of Some Commercially Important Pure Metals

Crystal Structure	Metals
Face-centered cubic	Aluminium, nickel, α-cobalt, copper, silver, gold, platinum, lead, iron (T > 910°C and <1400°C)
Body-centered cubic	Lithium, chromium, tungsten, titanium (>900°C), iron (T < 910°C and >1400°C)
Hexagonal close pack	Magnesium, zinc, cadmium, titanium (<900°C)
Complex structures	Tin (tetragonal), manganese (complex), uranium (complex)

EXAMPLE 1 DESCRIPTION OF PEROVSKITE STRUCTURE

The structure of perovskite, with the formula $CaTiO_3$, can be described in either of two ways:

1. A body-centered cubic (BCC) lattice with calcium ions at the corners of the unit cell, a titanium ion at the center and all octahedral vacancies occupied by oxygen ions or
2. A face-centered cubic (FCC) lattice with calcium ions at the corners of the unit cell, oxygen ions at the centers of the faces and every fourth octahedral vacancy (the one entirely between oxygen ions) occupied by a titanium ion.

Show that these descriptions are compatible with the formula.

Solution

BCC Description – There are two ions in the BCC unit cell. The eight corner sites, all shared eight-fold, together contribute one calcium atom and the unshared center site contributes one titanium atom. The oxygen ions occupy the octahedral vacancies at the centers of the six faces, all shared two-fold, contributing three oxygen ions. Hence, the description yields the same relative numbers of calcium, titanium and oxygen ions in the structure, that is, 1:1:3, as atoms in the formula. Incidentally, the basic BCC structure can also be envisioned as two interpenetrating simple cubic lattices, one of calcium ions and the other of titanium ions.

FCC Description – There are four ions in the FCC unit cell. The eight shared corner sites together contribute one calcium atom and the six shared face sites together contribute three oxygen ions. One-quarter of the octahedral sites are occupied by titanium ions and since there are four such sites per unit cell, they contribute one titanium atom. This description also yields the same relative numbers of the ions in the structure as atoms in the formula.

2.1.1.2 Liquids

Liquid phases are arrays of atoms with short-range structural order and the atoms or groups of atoms can move relatively without losing cohesion, conferring fluidity. Liquid structures are less amenable to direct empirical study than solid structures. X-ray diffraction studies reveal the average distribution of nearest neighbor atoms around any particular atom, but other evidence of structure can be acquired, particularly for solutions. Liquid metal solutions can show discontinuities in properties at compositions that correspond to changes in the underlying solid phases. The most familiar liquid, water, is highly structured because hydrogen–oxygen bonds have a directional character. Its bulk physical properties, excellent solvent powers and behaviour at surfaces are striking manifestations of its structure, as explained later in this chapter.

2.1.1.3 Non-Crystalline Solids

Certain solid materials, including glasses and some polymeric materials, have short-range order, but the atoms or groups of atoms lack the easy relative mobility characteristic of liquids.

2.1.1.4 Gases

In gaseous phases, attractions between atoms or small groups of atoms (molecules) are minimal, so that they behave independently as random entities in constant rapid translation. The useful approximation of the hypothetical ideal gas assumes that the atoms or molecules are dimensionless points with no attraction between them. It is often convenient to use this approximation for the 'permanent' real gases, oxygen, nitrogen and hydrogen and for mixtures of them. Certain other gases of interest in corrosion, for example, carbon dioxide, sulfur dioxide and chlorine, deviate from the approximation and require different treatment.

2.1.2 The Role of Electrons in Bonding

A phase adopts the structure that minimizes its internal energy within constraints imposed by the characteristics of the atoms that are present. In general, this is achieved by redistributing electrons contributed by the individual atoms. The resulting attractive forces set up between the individual atoms are said to constitute *bonds* if they are sufficient to stabilize a structure. A description of the distributions of electrons among groups or aggregates of atoms that constitute bonds between them is based on a description of the electron configurations in isolated atoms.

An isolated atom comprises a positively charged nucleus surrounded by a sufficient number of electrons to balance the nuclear charge. The order of the elements in the Periodic Table, reproduced in Table 2.2, is also the order of increasing positive charge on the atomic nucleus in increments, e^+, equal but opposite to the charge, e^- on a single electron. The positive charges on the nuclei are balanced by the equivalent numbers of electrons that adopt configurations according to the energies they possess.

TABLE 2.2

Simplified Periodic Table of the Elements

1	2		3	4	5	6	7	0
H								He
Li	Be		B	C	N	O	F	Ne
Na	Mg		Al	Si	P	S	Cl	Ar
K	Ca	Sc Ti V Cr Mn Fe Co Ni Cu Zn First transition series (development of 3*d* shell)	Ga	Ge	As	Se	Br	Kr
Rb	Sr	Y Zr Nb Mo Tc Ru Rh Pd Ag Cd Second transition series (development of 4*d* shell)	In	Sn	Sb	Te	I	Xe
Cs	Ba	La[a] Hf Ta W Re Os Ir Pt Au Hg Third transition series (development of 5*d* shell)	Tl	Pb	Bi	Po	At	Rn
Fr	Ra	Ac[b]						

[a] Lanthanide elements (La, Ce, Pr Nd..., etc.) developing the 4*f* shell.
[b] Actinide elements (Ac, Th, Pa, U..., etc.) developing the 5*f* shell.

Classical mechanics breaks down when applied to determine the energies of electrons moving within the very small dimensions of potential fields around atomic nuclei. The source of the problem was identified by the recognition that electrons have a wave character with wavelengths comparable to the small dimensions associated with atomic phenomena and an alternative approach, *wave mechanics*, pioneered by de Broglie and Schrodinger, was developed to deal with it. This approach abandons any attempt to follow the path of an electron with a given total energy moving in the potential field of an atomic nucleus and addresses the conservation of energy using a time-independent wave function as a replacement for classical momentum. It turns out that the probability that the electron is present at any particular point is proportional to the square of the wave amplitude and this introduces the energy and symmetry considerations with far-reaching consequences described below. The validation of the wave mechanical approach is that it delivers results that account with outstanding success for observations that cannot be explained otherwise. The formulation, solution and interpretation of wave equations is a severe, specialized discipline, beyond the scope of this book, but the conclusions that emerge have far-reaching consequences. Classic monographs by Coulson, Pauling and Hume-Rothery* give reader-friendly explanations. For now, a qualitative description of the approach together with enough terminology to explain structures is all that is needed. It is important to appreciate that the information acquired is not a collection of arbitrary assumptions designed to *explain* observations, but the inevitable result of using the wave equations that *predict* them.

2.1.2.1 Atomic Orbitals

When continuity and other constraints are applied, it is found that only certain solutions to wave equations have physical significance, yielding the following interpretations for isolated atoms:

1. The energies allowed for electrons do not vary continuously but can have only discrete values, referred to as *energy levels*.
2. The allowed energy levels are associated with characteristic symmetries for the electron probability distributions. This geometric feature is important in the formation of structures because it can determine whether bonds between atoms have directional character.

Standard notation used to classify the energy levels and any electrons that might occupy them is derived from the sequence of allowed energies and symmetries associated with them. The allowed energy levels are arranged in *shells*, numbered outward from the nucleus by the *principal quantum number*, $n = 1, 2, 3\ldots$, and so on. The types of symmetry correspond with an historic letter code, *s*, *p*, *d*, *f*, originally devised to describe visible light spectra exited from atoms.

These symmetries have the following characteristics:

s – Spherically symmetrical distribution around an atomic nucleus.

p – Distribution in two diametrically opposite lobes, about an atomic nucleus. By symmetry, there are three mutually perpendicular distributions with the same energy values, designated p_x, p_y and p_z.

* Cited in Further Reading.

d – Distribution in four lobes centered on the nucleus. By symmetry, there are five independent distributions with the same energy values, designated d_{xy}, d_{yz}, d_{xz}, $d_{x^2-y^2}$, d_{z^2}.

f – *f* orbitals are occupied only in the heavier elements and need not be considered here.

Solutions to the wave equations show that the first shell can contain only *s* electrons, the second shell can contain *s* and *p* electrons and the third shell can contain *s*, *p* and *d* electrons. It is usual to refer to the allowed energy levels and their associated electron density probability distributions in an isolated atom as *atomic orbitals* although there is no question of any kind of orbital motion associated with them. The term persists from early attempts to apply classical mechanics to electron energies. The existence of an orbital does not imply that it is necessarily occupied by an electron but describes an allowed discrete energy that an electron *may* occupy if it is present. Any particular orbital that an electron occupies corresponds to its *quantum state*. A further constraint applied is the *Pauli exclusion principle*, explained, for example by Coulson and by Pauling, that limits the occupation of any orbital to two electrons, that must differ in a further quality, called *spin*.

The nominal sequence of allowed energies is (1*s*) (2*s*, 2*p*) (3*s*, 3*p*, 3*d*) (4*s*, 4*p*, 4*d*, 4*f*) (5*s*, 5*p*…, and so on. However, there is a difference in the actual sequence because solutions to the wave equation yield energy values that require occupation of the 3*d* orbital to be deferred until the 4*s* orbital is occupied and there is a similar reversal in sequence for the 4*d*, 4*f* and 5*s* orbitals. Therefore, the actual sequence is (1*s*) (2*s*, 2*p*) (3*s*, 3*p*) (4*s*, 3*d*, 4*p*) (5*s*, 4*d*, 5*p*…, and so on. These changes in the sequence have far-reaching consequences because two series of elements in the Periodic Table, the first and second *transition series*, including most of the commercially important strong metals, are created as the 3*d* and 4*d* orbitals are filled progressively underneath the 4*s* and 5*s* orbitals, respectively. The underlying partly filled *d* orbitals in these metals confer special characteristics on them that govern their interactions with water, the structures of their oxides, their mechanical and physical properties and their alloying behaviour, all of which are of crucial importance in determining their resistance to corrosion.

Applying the Pauli exclusion principle and taking account the three- and five-fold multiplicities of *p* and *d* orbitals, the total number of electrons that can be accommodated in the first four shells are:

First shell ($n = 1$): $2_{(1s)} = 2$
Second shell ($n = 2$): $2_{(2s)} + (3 \times 2)_{(2p)} = 8$
Third shell ($n = 3$): $2_{(3s)} + (3 \times 2)_{(3p)} = 8$
Fourth shell ($n = 4$): $2_{(4s)} + (5 \times 2)_{(3d)} + (3 \times 2)_{(4p)} = 18$

A shell with its full complement of electrons is said to be closed, even when it is provisional, as for the third and fourth shells, where the next shell is started before the *d* orbitals are occupied. The significance of a closed shell is apparent when electron configurations for the elements, given in Table 2.3, are compared with their order in the Periodic Table.

The elements with all shells closed are the noble gases, helium, neon, argon, and so on, that are remarkably stable, as illustrated by their existence as monatomic gases and by their chemical inertness. In contrast, the other elements, that have partly filled outer shells, react readily. It is, therefore, apparent that stability is associated with closed shells.

TABLE 2.3

Electron Configurations of the Lighter Elements

	Electron Shells						
	1	2		3			4
Element	*s*	*s*	*p*	*s*	*p*	*d*	*s*
H	1						
He	2						
Li	2	1					
Be	2	2					
B	2	2	1				
C	2	2	2				
N	2	2	3				
O	2	2	4				
F	2	2	5				
Ne	2	2	6				
Na	2	2	6	1			
Mg	2	2	6	2			
Al	2	2	6	2	1		
Si	2	2	6	2	2		
P	2	2	6	2	3		
S	2	2	6	2	4		
Cl	2	2	6	2	5		
Ar	2	2	6	2	6		
K	2	2	6	2	6		1
Ca	2	2	6	2	6		2
Sc	2	2	6	2	6	1	2
Ti	2	2	6	2	6	2	2
V	2	2	6	2	6	3	2
Cr	2	2	6	2	6	4	2
Mn	2	2	6	2	6	5	2
Fe	2	2	6	2	6	6	2
Co	2	2	6	2	6	7	2
Ni	2	2	6	2	6	8	2
Cu	2	2	6	2	6	10	1
Zn	2	2	6	2	6	10	2

2.1.2.2 Molecular Orbitals and Bonding of Atoms

Stable assemblies of atoms form when the energy of the system is reduced by the combination of atomic orbitals to yield *molecular orbitals.* This can be explained by one or the other of two main approaches, the *molecular orbital* (MO) and *valence bond* (VB) theories, that are regarded as equivalent, except in their treatment of the small electron–electron interactions, as explained, for example by Coulson. The molecular orbital theory is easier to apply to simple molecules and inorganic complexes, including those in which metal atoms bind to other discrete entities such as water, hydroxide or chloride ions. Bonding in metallic phases is easier to envision using valence bond theory. In this section we shall deal mainly with molecular orbital theory.

The fundamental principle of molecular orbital theory is that a bonding orbital between two atoms is derived by a constructive linear combination of atomic wave functions for the component atoms, yielding a *molecular orbital*, in which electrons contributed by the atoms are accommodated with reduced energy. The geometric aspect of atomic orbitals, referred to earlier, is carried over into molecular orbitals derived from them and can impart directionality in the interaction of an atom with its neighbors. The tendency for the molecule to minimize its energy favours interactions yielding orbitals that electrons can populate at the lowest energy.

Criteria contributing to a minimum energy associated with a molecular orbital are:

1. Maximum constructive overlap of the atomic orbital wave functions
2. Minimum electrostatic repulsions between adjacent filled orbitals

The second of these criteria arises because electron-filled orbitals experience mutual electrostatic repulsion and assume orientations with maximum separation.

Sometimes these criteria are best met by transforming some of the dissimilar $2s$, $2p$ and $3d$ atomic orbitals into a corresponding set of mutually equivalent atomic orbitals called *hybrid* orbitals. For example, the $2s$, $2p_x$, $2p_y$ and $2p_z$ orbitals in the free oxygen atom hybridize to form four sp^3 orbitals, with their axes directed towards the corners of a tetrahedron, when it combines with hydrogen to form a water molecule. This phenomenon makes a major contribution to the unique structure and character of liquid water described in Section 2.2.3.

2.1.3 The Concept of Activity

In describing the structures of liquid water, oxides and metals and the interactions within and between them, a physical quantity, *activity*, is applied to the chemical species that are present. Activity is a general concept in chemistry and is rigorously defined*. It is used extensively to determine the energies of chemical reactions and the balance that is struck between the reactants and the products (the *equilibrium*), as explained later in Chapter 3. The concept is now provisionally described in a non-rigorous way to assist in describing interactions that atoms and ions experience in solution hence to explain structural features that they introduce.

When a substance is dissolved in a suitable solvent, its ability to react chemically is attenuated. The property of the substance that is thereby diminished is its activity. The diminished activity is not necessarily, or even generally, proportional to the degree of dilution, because the substance almost always interacts with the solvent or with other solutes that may be present.

The symbol for activity is italic a with a subscript to denote the component to which it refers, for example $a_{C_{12}H_{22}O_{11}}$ and a_{H_2O} represent the activities of sucrose and of water in an aqueous sugar solution. The activity of a species in a solution is a dimensionless quantity referred to its activity in an appropriate *standard state*, in which it is conventionally assigned the value unity. The standard selected is arbitrary, but two approaches are in common use.

In one approach, every component of a solution is treated in the same way and the standard states selected are the pure unmixed substances. There is a hypothetical concept, the *ideal solution* (or *Raoultian solution*, named for its originator) in which it is

* See Lewis, Randall, Pitzer and Brewer cited in Further Reading.

assumed that there is no interaction between the components and the activity of any particular component, *i*, is proportional to its *mole fraction*, N_i, which expresses the fraction present as the ratio of the number of individual entities of the component to the total number of molecular entities in the solution. Few real solutions approach this ideal. More generally, interactions between the components raise or lower their activities from the ideal values. The ratio of the real to the ideal activity is called the *activity coefficient*, *f*, and hence:

$$a_i = f_i \cdot N_i \tag{2.1}$$

The alternative approach, that is often more convenient for dilute solutions, treats the solutes in a different way. Whereas the standard state of the *solvent* is the pure material, the standard state for every solute is some definite composition, usually unit *molality*, a quantity equal to the relative molar mass of the solute (formerly called by the obsolete terms the atomic or molecular weight in grams) dissolved in 1 kg of solution. For species such as ions in aqueous solution that cannot exist as 'pure' substances, such a standard state is not only convenient, but unavoidable.

Many real solutions of interest depart from ideality in reasonably regular ways and various models and devices are available to address them, as summarized, for example by Bodsworth*. For example, although the activity of a species is not generally equal to the mole fraction, it is always a near-linear function of it for a very dilute solution, and in these circumstances the solute is said to exhibit *Henrian activity*. This can be exploited to produce a linear scale, *solution*, in which the numerical values of activities and molalities are interchangeable. In this scale, the activity coefficient, the *Henry's law coefficient*, is constant and is denoted by the symbol, γ.

2.2 The Structure of Water and Aqueous Solutions

2.2.1 The Nature of Water

Liquid water is by far the most efficient medium to sustain metallic corrosion at ordinary temperatures because it is endowed with the following properties and characteristics:

1. Physical stability and low viscosity over a wide temperature range
2. An ability to dissolve and disperse foreign ionic species
3. An ability to dissolve small quantities of oxygen and some other gases
4. Versatility in sustaining neutral, acidic or alkaline environments
5. Some intrinsic electrical conductivity

The value of this information in explaining corrosion processes and their control is clarified by relating it to the structure of liquid water. Contrary to common perceptions, liquid water is a highly organized substance as described in detail by Franks, Davies and Bockris and Reddy, cited in Further Reading. It is essentially a mobile three-dimensional

* Cited in Further Reading.

network of coherent hydrogen and oxygen atoms based on the geometric arrangement and distribution of electric charge within the basic structural unit, the water molecule.

2.2.2 The Water Molecule

The water molecule, formally written, H_2O, contains three atomic nuclei, one of oxygen with a charge, $8e^+$, two of hydrogen each with a charge, e^+, all balanced by the collective charge of the associated electrons, $10e^-$. Its essential characteristics can be formulated using wave mechanics and confirmed by experimental observations. They include molecular geometry, the strengths of the oxygen–hydrogen bonds, the vibrational modes and the distribution of internal charge that is particularly useful in explaining the solvent function of liquid water.

Figure 2.3 is a schematic diagram of the geometrical arrangement that emerges from calculations for the isolated molecule. There are four hybrid orbitals derived from the oxygen $2s$, $2p_x$, $2p_y$ and $2p_z$ atomic orbitals, directed outward from the oxygen nucleus toward the corners of a tetrahedron. Every one of the four orbitals has its full complement of two electrons; the oxygen $1s$ inner orbital, containing the remaining two electrons, persists virtually unchanged in the water molecule. Two of the hybrid orbitals interact with the $1s$ atomic orbitals of the hydrogen atoms to form molecular orbitals and the other two orbitals are uncommitted to bonding and the electrons in them are described as *lone pairs*. This configuration produces an asymmetric distribution of electric charge in the molecule characterizing it as a *polar* molecule.

The hydrogen $1s$ orbital and the single charge on the hydrogen nucleus produces only a small effect on the shape of the molecular orbital, so that the bonding and non-bonding orbitals are roughly similar, but the differential electrostatic repulsion between them yields 104° 27′ and 114° 29′ for the angles between the axes of the two bonding and the two non-bonding orbitals, respectively, instead of 109° 28′ as expected for a regular tetrahedral configuration.

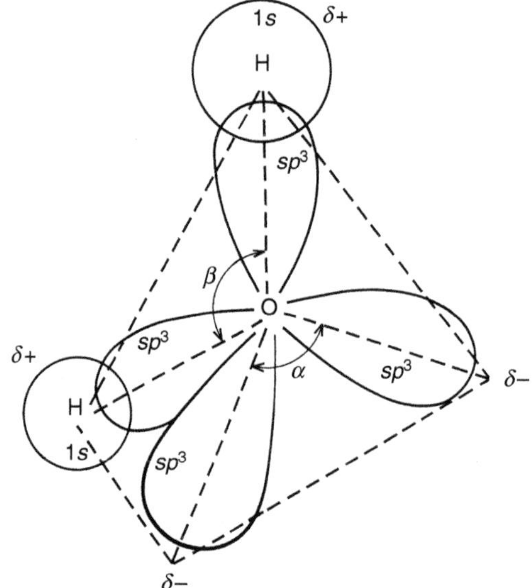

FIGURE 2.3
The tetrahedral configuration of the water molecule. $\alpha = 114° 29'$ $\beta = 104° 27'$.

TABLE 2.4
Physical Stability of Water

Liquid	Liquid Range °C	Boiling Point °C
Water	100	100
Methane	18	–164
Neon	2.8	–245.9

2.2.3 Liquid Water

The *polar* character of the water molecule is responsible for the strong bonding between water molecules in the liquid phase. The magnitude of the effect is strikingly illustrated in Table 2.4, where the physical stability of water is compared with that of liquids with comparable molar masses formed from non-polar molecules that cohere only by weak *Van der Waals* forces derived from attraction between induced dipoles. Such liquids have low boiling points and exist only within narrow temperature ranges.

2.2.3.1 Hydrogen Bonding

The strong coherence of water molecules is due to *hydrogen bonding*, which is a special mechanism exploiting the polar character of the water molecules. The small mass of the hydrogen nucleus allows it to oscillate and interact with the negative charge on the lone pair of an adjacent water molecule, producing a bond with about a tenth the energy of the molecular O–H bond, without losing cohesion in its own molecule. It is true that some other small polar molecules, for example hydrofluoric acid and ammonia, can also link by hydrogen bonding, but this is usually restricted. The unique feature of water molecules is that they can form hydrogen bonds at each of *two* tetrahedral corners, producing a coherent *three-dimensional* network. The detailed arrangement of molecules in liquid water is uncertain, but it is based on the hexagonal ice structure in which every water molecule is linked to four neighbors in a very open network.

The liquid is considered to have short-range order derived from the ice structure but with some of the interstitial spaces occupied by extra molecules constantly interchanging with those in the hydrogen-bonded network, as illustrated in Figure 2.4. This increases the average number of nearest neighbors around a water molecule from 4 in ice to 4.4 in the liquid at its freezing point. This underlies the anomalous *increase* in density when ice melts to form water.

2.2.3.2 Dielectric Constant

By virtue of its polarity and hydrogen-bonding interactions, water is a very effective medium for reducing electrostatic forces contained within it. This is a bulk property described by the *dielectric constant*, ε_r, a dimensionless parameter appearing in the Coulombic force equation:

$$F = \frac{q_1 q_2}{4\pi\varepsilon_0\varepsilon_r \mathrm{r}^2} \tag{2.2}$$

where:

F is the force between two charges q_1 and q_2, at a distance, r

and:

ε_0 is the permittivity of a vacuum

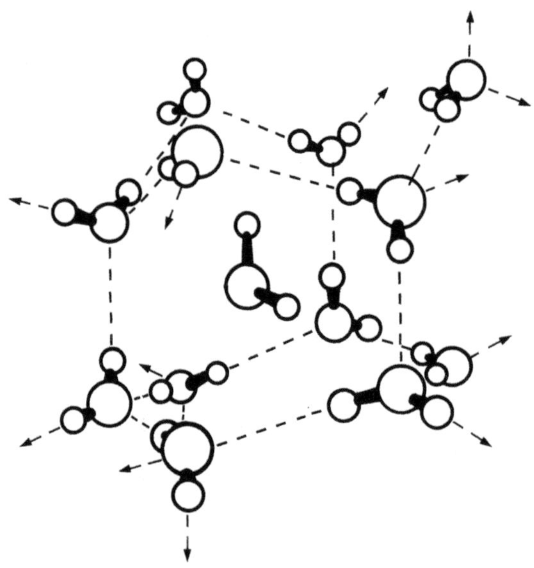

FIGURE 2.4
Schematic diagram of a possible structure of water, envisaged as a dynamic interchange between some molecules in an ice-like hydrogen-bonded network, shown as dotted lines, and others temporarily occupying the spaces.

Water has one of the highest known dielectric constants, that is, 78 at 25°C. The higher the dielectric constant in a medium is, the lower is the energy needed to separate opposite electric charges, and this is one of the factors that makes water such an effective solvent for ionic species.

2.2.3.3 Viscosity

The viscosity of a solvent controls the transport of solutes within it. In liquid water it is related to the *self-diffusion*, that is, the mobility of the molecules through the hydrogen-bonded matrix. Other things being equal, the viscosity of a liquid is an inverse function of the effective molecular size of the solvent, taking account of any coherence between adjacent molecules, yet despite the exceptionally high degree of molecular association in water, its viscosity has a low value, 1.002 Pa s at 20°C. This is attributable to the small size of the unassociated water molecule and the open structure providing easy pathways for diffusing molecules.

2.2.4 Autodissociation and pH of Aqueous Solutions

The autodissociation of water and the concept of pH as an indicator of acidity or alkalinity are not always clearly explained. It is convenient to describe the dissociation by the symbolic equation:

$$H_2O(l) = H^+(aq) + OH^-(aq) \tag{2.3}$$

although some well-intentioned but unhelpful texts use the symbols H_3O^+ or $H_9O_4^+$ for the hydrogen ion, indicating its association with one or more water molecules.

It is much better to describe autodissociation as disorder in the structure of the bulk liquid. In this description, a hydrogen ion is centered on a water molecule with an excess proton, thus carrying a net positive charge e^+. Charge neutrality within the bulk liquid is conserved by the creation of a complementary hydroxyl ion, centered on another water molecule with a proton deficiency. In both cases, the local excess charge is dissipated over a volume of the surrounding hydrogen-bonded water matrix containing several molecules.

The autodissociation is driven by an energy advantage in creating disorder. This is opposed by the energy disadvantage in breaking O–H bonds and separating the charged fragments. The origins of these opposing energy terms are explained in terms of the thermodynamic properties *entropy* and *enthalpy* introduced and defined in Section 2.3.5. For now it is sufficient to anticipate the result that the balance struck between them leads to definite populations of hydrogen and hydroxyl ions related by an *equilibrium constant*, K_w, given by:

$$K_w = a_{H^+} \times a_{OH^-} \tag{2.4}$$

where:

a_{H^+} and a_{OH^-} are the activities of the H^+ and OH^- ions, respectively

Following normal chemical convention, the activity of a pure solvent, in this case hypothetical undissociated water, is equal to unity and is virtually unchanged by the activities of solutes in dilute solution so that its effect is insignificant and is ignored in Equation 2.4. K_w is temperature dependent and examples of its values are given in Table 2.5.

2.2.5 The pH Scale

As Equation 2.3 shows, autodissociation produces equal populations of H^+ and OH^- ions so that their activities are also equal and for pure water at ambient temperatures close to 25°C:

$$a_{H^+} = a_{OH^-} = 10^{-7} \text{ at } 25°\text{C} \tag{2.5}$$

The ratio of hydrogen and hydroxyl ions is altered if acid or alkali is added to water, but the activity product in Equation 2.4 remains equal to K_w, so if, for example a_{H^+} is raised from 10^{-7} to 10^{-4} by adding a small quantity of acid to water at 25°C, a_{OH^-} is reduced to 10^{-10}.

This effect suggests a practical scale for the degree of acidity or alkalinity related to the measured activity of the hydrogen ion but it is difficult to directly measure the activity of a single ion species in isolation because a counter ion species must be present to preserve charge neutrality. The difficulty has been circumvented by defining a scale, the pH scale,

TABLE 2.5

Temperature Dependence of K_w and pH Neutrality

Temperature °C	K_w/mol² dm⁻⁶	pH Neutrality
0	10^{-15}	7.5
25	10^{-14}	7.0
60	10^{-13}	6.5

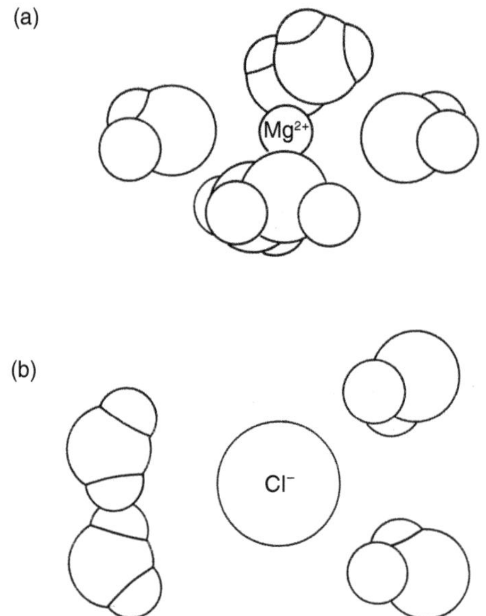

FIGURE 2.5
Typical structures of inner solvation spheres in aqueous solution. (a) Cations – example magnesium. (b) Anions – example chloride.

referring to internationally recommended arbitrary standards based on measurements in solutions that inevitably contain counter ions. The scale is chosen so that for dilute solutions, in which the hydrogen ion activities lie between 10^{-2} and 10^{-12}, the following expression applies without significant error:

$$pH = -\log_{10} a_{H^+} \tag{2.6}$$

Thus for pure water at 25°C and for aqueous solutions in which there is no net effect on autodissociation, $a_{H^+} = 10^{-7}$, yielding a value for pH of 7.0. This value corresponds to equal hydrogen and hydroxyl ion activities and is often referred to as *pH neutrality*. If the pH value is <7.0, the solution has an excess of hydrogen ions and is acidic; if it is >7.0, the solution has an excess of hydroxyl ions and is alkaline. A word of caution is needed because the value of pH varies with temperature as shown in Table 2.5, corresponding to the temperature dependence of K_w. Furthermore, the description a_{H^+} strictly denotes the activity of the entities carrying excess protons in a water matrix. Hence, any factor that disturbs the water matrix, such as close proximity of ions in concentrated solutions, may influence the pH.

2.2.6 Foreign Ions in Solution

Unlike the products of autodissociation, a foreign ion in aqueous solution introduces distinct structural changes in the water. Foreign ions have finite sizes and perturb the structure. A description of the effect is facilitated by the simplifying assumption that ions are spherical or occupy spherical holes in the water matrix.

The charge on a foreign ion imposes a radial electric field in the surrounding water matrix and it re-orientates the adjacent water molecules around the ion to a greater or lesser extent. The molecules around an *anion* are arranged with the hydrogen atoms facing the ion. Those around a *cation* have diametrically opposite configurations, with the non-bonding orbitals containing the lone pair electrons facing the ion. This spherically symmetric reorientation of the water molecules is incompatible with the tetrahedral hydrogen-bonded matrix of undisturbed water. The consequent local modification of the water structure is illustrated for a cation, Mg^{2+}, and an anion, Cl^-, in Figure 2.5 and is known as *solvation*.

Some metal cations with electron-deficient orbitals can use the uncommitted lone pair electrons facing them to increase the cohesion by a tendency to form chemical bonds, so that cations are often more strongly solvated than anions.

Water molecules directly attached to an ion comprise the *primary solvation shell*. Around this shell, there is a region of *secondary solvation* in which the surrounding water molecules are more associated with the primary solvation shell than with the surrounding solvent. Modern theories include a third concentric region in which properties of the solvent such as the dielectric constant are influenced by the electric field centered on the ionic charge.

Foreign ions are classified as structure-forming and structure-breaking according to whether the induced structures around them are more or less ordered than the displaced water matrix. Structure-forming ions have high surface charge densities. They include small singly charged ions such as lithium, Li^+, and fluoride, F^-, and larger multiply charged ions such as calcium, Ca^{2+}, magnesium, Mg^{2+}, aluminium, Al^{3+}, zinc, Zn^{2+}, copper(II), Cu^{2+}, and iron(III), Fe^{3+}.

Structure-breaking ions have low surface charge densities. They include large ions such as potassium, K^+, rubidium, Rb^+, chloride, Cl^-, bromide, Br^-, iodide, I^-, and nitrate, NO_3^-. Salts with structure-breaking ions, for example K^+ or NO_3^-, are very soluble because the cohesive forces in solids composed of ions with low surface charge densities are often not as strong as those for other ions and so that there is little tendency for these solids to precipitate from solution.

In dilute solutions associated with corrosion, the ions are remote from interaction with other charged species and exhibit Henrian activity. Their activities are conveniently referred to the infinitely dilute solution standard. This does not apply to more concentrated ionic solutions because the electrical potentials of neighboring ions in close proximity mutually interfere.

2.2.7 Ion Mobility

Corrosion rates are influenced by the electrical conductivity of aqueous solutions and the diffusion of ionic species within them. Both are related to the limiting velocity of ions in motion, called the ion mobility.

The physical significance of the ion mobility is that an ion in an electrical or chemical potential gradient accelerates to a terminal velocity at which the viscous drag of the solvent balances the accelerating potential. Thus, for most ions, molar conductivity, diffusivity and solvent viscosity are all related. Theoretical equations relating these parameters are derived as explained, for example by Atkins, cited at the end of the chapter, but the significant ones are:

$$\text{Einstein equation:} \quad D_i = u_i kT/ez_i \tag{2.7}$$

$$\text{Nernst–Einstein equation:} \quad \lambda_i = (z_i^2 F^2/RT) D_i \tag{2.8}$$

$$\text{Stokes–Einstein equation:} \quad D_i = kT/6\pi\eta r_i \tag{2.9}$$

where:

D_i is the diffusion coefficient for an ion, i
u_i is the ion mobility
λ_i is the molar conductance
z_i is the charge number for the ion
r_i is the effective (or hydrodynamic) radius
η is the viscosity of the solvent
e is the charge on an electron
R, k and F are the gas, Boltzmann and Faraday constants, respectively

Equations 2.7 and 2.8 give ion mobilities in terms of molar conductances, λ_i, since $R/F = k/e$:

$$u_i = \lambda_i / z_i F \tag{2.10}$$

D_i and λ_i strictly depend on the ionic strength of the medium and the nature of the counter ion. The implied assumption that ions exhibit Stokes behaviour is also a simplification. Nevertheless these equations are useful in comparing ion transport rates. Limiting ion mobilities at 25°C calculated in this way for some common ions in aqueous solution amenable to measurement are given in Table 2.6.

The most striking feature of Table 2.6 is that the values for limiting conductance of ions produced by autodissociation of water, H^+ and OH^- are several times greater than the values for other ions. The exceptionally high mobility of these ions is because they move in a *Grotthus-chain mechanism*, illustrated in Figure 2.6. In this mechanism, the individual ion assemblies do not move bodily through the liquid, but the positions of excess protons (H^+) or proton holes, that form the charge centers of hydrogen (H^+) and hydroxyl (OH^-) ions, are propagated by successive re-orientation of hydrogen bonds along the path

TABLE 2.6

Limiting Ionic Mobilities u_i for Some Ions in Water at 25°C

Ion	Limiting Ion Mobility $u_i/10^{-9}\text{m}^{-2}\text{s}^{-1}\text{V}^{-1}$
H^+	362.5
OH^-	204.8
Na^+	51.9
Ca^{2+}	61.7
Mg^{2+}	55.0
Zn^{2+}	54.7
Mn^{2+}	55.5
Cu^{2+}	56.0
Cl^-	79.1
NO_3^-	74.0
SO_4^{2-}	82.7

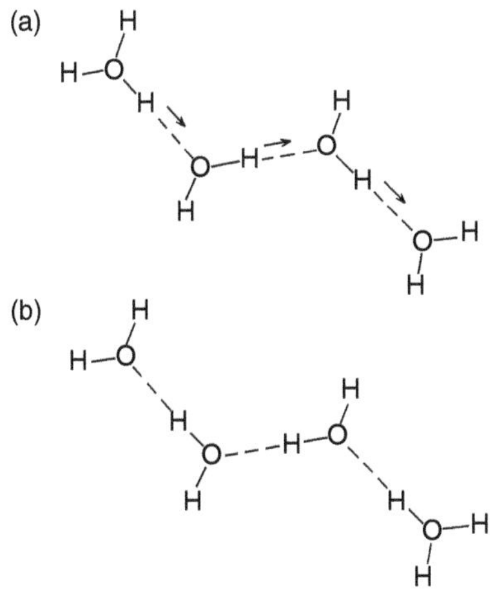

FIGURE 2.6
The Grotthus chain mechanism for proton migration in water. Short full lines – covalent bonds. Longer broken lines – hydrogen bonds. A hydrogen ion moves from the left in (a) to the right in (b) by successive re-orientation of hydrogen bonds between water molecules along the path.

through which the charge centers move, transporting the ions without needing to move atomic species with attached water molecules. As a consequence, mechanisms involving the transport and interaction of protons or hydroxyl ions with other species or surfaces are usually very fast. These include interactions that can contribute to the establishment of the protective passive condition on metal surfaces.

2.2.8 Structures of Water and Ionic Solutions at Metal Surfaces

The structures of water and of aqueous solutions are disturbed in contact with a metal surface. The disturbed structure extends for a only a few water molecules into the bulk liquid but it has a special role in corrosion processes, because it controls the rates of interactions between the metal and solutions of ions and it can be a precursor of a surface passive condition for the metals concerned. Bokris and Reddy, cited in Further Reading, give a comprehensive treatment.

The interfacial structure is due to the polar character of water molecules, and the charge centers in ions that induce electric charges on the metal surface establishing an *electrical double layer,* comprising electrically charged parallel planes separated by distances of the order of magnitude of a few water molecules. The evidence for this state of affairs over such small distances is necessarily indirect and is based mainly on the results of experiments to determine the capacity of the surfaces to hold electric charge. It is difficult to generalize from the accumulated evidence* but Figure 2.7 is a simplified representation of the structure. The figure shows how access of solvated ions to the metal surface is restricted by their inner solvation spheres and by a row of water molecules in contact with the metal, the *hydration sheath,* re-orientated from the bulk water structure under the influence of the surface charge. The locus of the ion centers at their closest approach is known as the *outer Helmholtz plane.* Certain unsolvated ions that are small enough to displace water molecules in the hydration sheath, mainly anions such as Cl^- and Br^-, can be adsorbed directly on the metal surface

* Reviewed for example by Bokris and Reddy, cited in Further Reading.

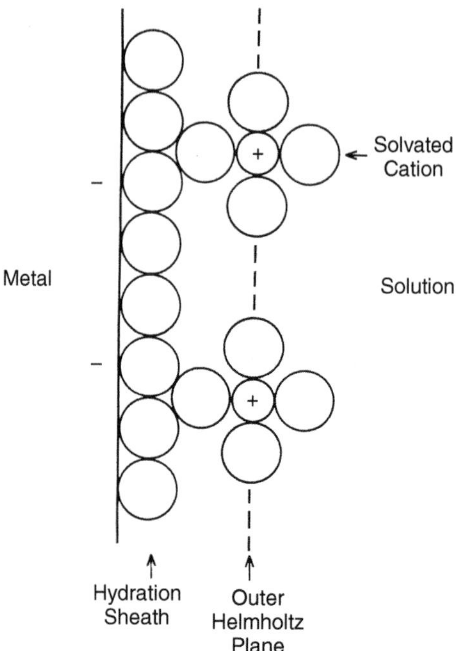

FIGURE 2.7
Schematic diagram representing the arrangement of cations in aqueous solution at a negatively charged metal surface.

by chemical rather than electrostatic interactions and the locus of their centers is known as the *inner Helmholtz plane.* All of these structures are of practical as well as theoretical interest because they have important consequences in surface processes. The activation energy needed to transport an atom from the metal surface to become an ion in the outer Helmholtz plane is one of the rate-controlling factors in the dissolution of a corroding metal, as explained in Section 3.2. Structural differences can also account for differences between the abilities of chloride and fluoride ions to depassivate stainless steels, as explained in Chapters 8 and 30.

2.2.9 Constitutions of Hard and Soft Natural Waters

Natural waters are common environments for corrosion. They exist in such variety that generalizations are difficult, but one frequently encountered aspect of composition is the quality of being 'soft' or 'hard'. The terms are used loosely, but it is generally understood that on evaporation, a hard water leaves a calcareous deposit whereas a soft water does not. It is well-known that corrosion problems are generally less severe in hard than in soft waters implying that calcareous deposits afford some protection and that rainwater can also be acidic and therefore mildly aggressive. These aspects of water quality are influenced by equilibria between the water, atmospheric carbon dioxide and calcium-bearing material with which the water comes into contact. They are considered in detail by Butler, cited in Further Reading.

2.2.9.1 Rainwater

In the absence of pollutants, rainwater is almost pure water that absorbs quantities of oxygen and carbon dioxide in equilibrium with the air as it falls in small droplets to earth. It has an oxygen potential conferred by the dissolved oxygen and a particular pH value derived from the carbon dioxide in solution. It can be calculated as in the following example.

EXAMPLE 2 pH OF RAINWATER

The pH of the water can be determined from the carbon dioxide content, by a *speciation calculation*. The calculation depends on balancing the total electric charges, and so deals both with equilibrium *constants* that relate to activities denoted as usual, that is, a_{X^-} for the activity of an ion, X^-, and with equilibrium *quotients*, that relate to corresponding concentrations, denoted by square brackets, that is, $[X^-]$ for the ion X^-. These quantities are related by:

$$a_{X^-} = (\gamma_{X^-})[X^-]$$

where: γ is the activity coefficient. The symbols for the mean activity coefficients for singly and doubly charged ions are, $\gamma_\pm$, and $\gamma_{2\pm}$, respectively. The activity coefficient for uncharged species in dilute solution, for example CO_2 is approximately 1. An ideal gas is referred to unit activity at atmospheric pressure. For a dilute solution the concentration and activity of the solvent are virtually constant at unity and can be omitted from equilibrium constants and quotients.

The pH of the water is determined by four mutually interacting equilibria:

1. The dissolution of CO_2 in water from the gas phase:

$$CO_2(g) = CO_2(aq) \tag{2.11}$$

$$\text{Equilibrium quotient, } K_H = [CO_2]/p_{CO_2} \tag{2.12}$$

2. The reaction of CO_2 with water, producing bicarbonate and hydrogen ions:

$$CO_2(aq) + H_2O(l) = HCO_3^- + H^+ \tag{2.13}$$

$$\begin{aligned}\text{Equilibrium quotient, } K_1 &= (a_{H^+})(a_{HCO_3^-})/(a_{CO_2}) \\ &= [H^+][HCO_3^-]\gamma_\pm^2/[CO_2]\end{aligned} \tag{2.14}$$

3. The dissociation of the bicarbonate ion into carbonate and hydrogen ions:

$$HCO_3^- = CO_3^{2-} + H^+ \tag{2.15}$$

$$\begin{aligned}\text{Equilibrium constant, } K_2 &= (a_{H^+})(a_{CO_3^{2-}})/(a_{HCO_3^-}) \\ &= [H^+][CO_3^{2-}]\gamma_{2\pm}/[HCO_3^-]\end{aligned} \tag{2.16}$$

4. The autodissociation of water:

$$H_2O = H^+ + OH^- \tag{2.17}$$

$$\begin{aligned}\text{Equilibrium constant, } K_w &= (a_{H^+})(a_{OH^-}) \\ &= [H^+][OH^-]\gamma_\pm^2\end{aligned} \tag{2.18}$$

To preserve electrical neutrality, the sums of the charges on the positive and negative ions must be equal. The charge balance equation is:

$$[H^+] = [HCO_3^-] + [OH^-] + 2[CO_3^{2-}] \tag{2.19}$$

In practice, since K_2 is much smaller than K_1, the concentration of CO_3^{2-} is very small and can be neglected. The $[CO_3^{2-}]$ term can be dropped and $[HCO_3^-]$ and $[OH^-]$ substituted by the following expressions:

From Equations 2.14 and 2.12:

$$[HCO_3^-] = K_1[CO_2] / [H^+]\gamma_\pm^2 = K_H K_1 p_{CO_2} / [H^+]\gamma_\pm^2 \tag{2.20}$$

From Equation 2.18:

$$[OH^-] = K_w/[H^+]\gamma_\pm^2 \tag{2.21}$$

Combining Equations 2.19 through 2.21 and omitting $[CO_3^{2-}]$ gives:

$$[H^+] = (K_H K_1 p_{CO_2} + K_w)/[H^+]\gamma_\pm^2 \tag{2.22}$$

Multiplying both sides of Equation 2.22 by $[H^+]\gamma_\pm^2$, substituting $a_{H+} = [H^+]\gamma_\pm$ and rearranging yields the hydrogen activity:

$$a_{H^+} = [H^+]\gamma_\pm = (K_H K_1 p_{CO_2} + K_w)^{½} \tag{2.23}$$

whence:

$$pH = -\log a_{H^+} = -0.5 \log(K_H K_1 p_{CO_2} + K_w) \tag{2.24}$$

The normal atmosphere contains 0.035% CO_2 (3.5×10^{-4} atm). At a typical ambient temperature of 15°C the values of the equilibrium quotients and constants are:

$K_H = 4.67 \times 10^{-2}$ mol dm^{-3} atm^{-1}
$K_1 = 3.81 \times 10^{-7}$ mol dm^{-3}
$K_2 = 3.71 \times 10^{-11}$ mol dm^{-3}
$K_w = 4.60 \times 10^{-15}$ mol^2 dm^{-6}

Inserting these values into Equation 2.24 yields pH = 5.6 for water in equilibrium with atmospheric carbon dioxide, which is consistent with the results of practical measurements. Use is made of this result in considering corrosion tendencies in Chapter 3. Equation 2.24 also shows that in the absence of carbon dioxide, $p_{CO_2} = 0$ and pH = −0.5 log K_w, in agreement with values given in Table 2.5.

2.2.9.1.1 Oxygen Potential of Rainwater

Oxygen dissolves as molecules in the open network of the water structure. Air contains 21% oxygen and since water that is falling as rain approaches equilibrium with air and oxygen and is nearly ideal at ambient temperatures, the oxygen activity in solution is 0.21, referred to the pure gas as the standard state. This value is assumed in later calculations.

2.2.9.2 Hard Waters

Hard water can often be regarded as a solution in simultaneous equilibrium with both solid limescale (calcite) and atmospheric carbon dioxide, even though other ions such as SO_4^{2-} and Mg^{2+} may be present. Its pH can be calculated as in the following example.

EXAMPLE 3 pH OF A REPRESENTATIVE HARD WATER

The principal solution reaction is:

$$CaCO_3(s) + CO_2(g) + H_2O(l) = Ca^{2+}(aq) + 2HCO_3^-(aq) \tag{2.25}$$

The charge balance equation for ionic species, including ions from dissociation of water is:

$$2[Ca^{2+}] + [H^+] = [HCO_3^-] + 2[CO_3^{2-}] + [OH^-] \tag{2.26}$$

Values of K_H, K_1, K_2 and K_W, given above and K_s, given below, justify the approximation:

$$2[Ca^{2+}] = [HCO_3^-] \tag{2.27}$$

The solubility of $CaCO_3$ is determined by the solubility product:

$$\begin{aligned} K_s &= (a_{Ca^{2+}})(a_{CO_3^{2-}}) \\ &= [Ca^{2+}][CO_3^{2-}]\gamma_{2\pm}^2 \end{aligned} \tag{2.28}$$

and the value of K_s is 3.90×10^{-9} mol^2 dm^{-6}

From Equation 2.16

$$[CO_3^{2-}] = K_2[HCO_3^-]/[H^+]\gamma_{2\pm} \tag{2.29}$$

Substituting $[Ca^{2+}] = ½[HCO_3^-]$ from Equation 2.27 and for $[CO_3^{2-}]$ from Equation 2.29 into Equation 2.28 gives:

$$K_s = ½K_2[HCO_3^-]^2\gamma_{2\pm}/[H^+] \tag{2.30}$$

Substituting for $[HCO_3^-]$ from Equation 2.20 and rearranging yields:

$$[H^+]^3\gamma_\pm^3 = (K_H^2K_1^2K_2\gamma_{2\pm}/2K_s\gamma_\pm) \cdot (pCO_2)^2 \tag{2.31}$$

Since $pH = -\log a_{H^+} = -\log[H^+]\gamma_\pm$, Equation 2.31 gives:

$$pH = -⅓\log(K_H^2K_1^2K_2/2K_s) - ⅓\log \gamma_{2\pm}/\gamma_\pm - ⅔\log p_{CO_2} \tag{2.32}$$

For dilute solutions such as fresh waters, $-⅓ \log \gamma_{2\pm}/\gamma_\pm$ is small and can be ignored. Inserting the values for K_H, K_1, K_2 and K_s at 15°C given above:

$$pH = 5.94 - ⅔\log p_{CO_2} \tag{2.33}$$

Equation 2.33 yields pH = 8.3 for water in equilibrium with calcite and atmospheric carbon dioxide at its normal pressure in the atmosphere, that is, $p_{CO_2} = 3.5 \times 10^{-4}$ atm.

For environments such as artesian pumps, where water in equilibrium with a chalk phase is exposed to high pressures, p_{CO_2} can be higher and the pH of the system is lower.

EXAMPLE 4 PREVENTING LIMESCALE DEPOSITION

Can limescale be prevented from forming from water containing 120 mg dm^{-3} of calcium (i.e., 3.3×10^{-3} mol dm^{-3} of Ca^{2+} ions) by treating it with a waste gas containing 10.5% carbon dioxide (i.e., $p_{CO_2} = 0.105$ atm) at 15°C?

Solution

Applying Equation 2.33 gives the pH at equilibrium:

$$pH = 5.94 - \tfrac{2}{3}\log p_{CO_2} = 5.94 - \tfrac{2}{3}\log(0.105) = 6.59$$

$$a_{H+} = 10^{-(pH)} = 10^{-6.59}$$

Applying Equations 2.16 and 2.20:

$$[CO_3^{\;2-}] = (K_H K_1 K_2 p_{CO_2})/[H^+]^2$$

$$= (4.67 \times 10^{-2})(3.81 \times 10^{-7})(3.71 \times 10^{-11})(0.105)/(10^{-6.59})^2 = 1.05 \times 10^{-6} \text{ mol dm}^{-3}$$

Inserting this value for $\left[CO_3^{\;2-}\right]$ in Equation 2.28:

$$[Ca^{2+}] = K_S/[CO_3^{\;2-}] = (3.55 \times 10^{-9})/(1.05 \times 10^{-6}) = 3.4 \times 10^{-3} \text{ mol dm}^{-3}$$

Hence, the proposition is feasible because the existing concentration of calcium ions, $[Ca^{2+}]$, i.e., 3.3×10^{-3} mol dm^{-3}, is below this value and cannot deposit limescale.

2.3 The Structures of Metal Oxides

Oxides are not only the corrosion products to which metals are converted during oxidation in gaseous environments, but they are also the media through which the reaction usually proceeds. The oxide film initially formed on a metal physically separates it from the gaseous environment, but even when it remains coherent and adherent to the metal surface, it does not necessarily prevent continuing attack because the primary reactants, the metal and/or oxygen, may be able to diffuse through the oxide, allowing the reaction to proceed. The rate at which the reaction continues and the growth pattern of the oxide depend on the oxide structure, especially on its characteristic *lattice defects* that provide the pathways for the diffusing reactants. Lattice defects of certain kinds are accompanied by complementary *electronic defects*, to compensate for charge imbalances that they would otherwise introduce, conferring some electronic conduction on the oxides in which they occur, recognized in the description *semiconductors*. The nature and populations of lattice and corresponding electronic defects vary from oxide to oxide, and the differences are an essential part of explanations for differences in the ability of engineering metals and alloys to resist oxidation at elevated temperatures.

Descriptions of oxide structures relevant to the oxidation of metals are based on assessments of the bonds between ions, the geometry of lattice structures and explanations for various characteristic lattice and electronic defects they contain. Much of this is explained by the ideas of electronegativity, the partial ionic character of oxides and the influence of electron energy bands.

Solid metal oxides are assemblies of positively charged metal cations and negatively-charged oxygen anions formed by transfer of electrons from the metal atoms to the oxygen atoms from which they are composed. The electron transfer is substantial but incomplete so that the bonds between anions and cations have only partial ionic character. Nevertheless electrostatic forces between the oppositely charged ions make a major contribution to the cohesion of an oxide and is one of the dominant factors determining its structural geometry.

The origin of partial ionic character can be explained as follows. In a single bond between two atoms of the same element, the atomic wave functions obviously contribute equally to the molecular wave function, but in a bond between two atoms of different elements, the most energetically favourable molecular orbital may have a greater contribution from the atomic orbital of one atom than from the other. This implies that the electrons in the molecular orbital are associated more with one atom than with the other. In a hypothetical limiting case, the contribution to the molecular orbital is wholly from the atomic wave function of one atom with virtually no contribution from the other. Because the two atoms each formally donate one electron to the molecular orbital, this would be equivalent to the donation of an electron by one atom, A, to the other, B, forming ions:

$$\mathrm{A} + \mathrm{B} = \mathrm{A}^{+} + \mathrm{B}^{-} \tag{2.34}$$

In practice, the limiting case is never completely realized and real bonds between atoms of different elements have aspects of both ionic and covalent character. To consider the dual character of bonds further, a property of atoms called electronegativity is now introduced.

2.3.1 Electronegativity

Electronegativity is generally accepted as a qualitative idea, but a quantitative approach is elusive. It is based on comparisons of the strengths of single bonds between pairs of atoms of the same element, *like atoms,* with the strengths of bonds between pairs of atoms of different elements, *unlike atoms.* The strength of a bond is identified with the energy, E, released when it is formed, and if a pair of unlike atoms, A and B, formed the same kind of bond as pairs of like atoms, the bond energy, $E_{(A-B)}$, would have some mean value, such as the geometric mean, between bond energies, $E_{(A-A)}$ and $E_{(B-B)}$, for like pairs of the constituent atoms:

$$E_{(A-B)} = [E_{(A-A)} \times E_{(B-B)}]^{1/2} \tag{2.35}$$

The actual energies of bonds between unlike atoms, determined experimentally, are found to exceed the predicted bond energies by a quantity, ΔE, which is small for bonds between some unlike atom pairs, for example nitrogen and chlorine, N–Cl or bromine and chlorine, Br–Cl but large for others, for example hydrogen and chlorine, H–Cl or hydrogen and bromine, H–Br. The inference is that the extra energy indicates some characteristic of the bond that varies from one pair of unlike atoms to another. By comparing such results with other chemical evidence, such as the ability of substances to ionize in aqueous solution, it is possible to associate the extra energy of bonds with the extent of

TABLE 2.7

Electronegativity Values on Pauling's Scale for Representative Elements

Element	*Fluorine*	*Oxygen*	*Chlorine*	*Carbon*	*Sulfur*
Electronegativity	4.0	3.5	3.0	2.5	2.5
Element	*Hydrogen*	*Copper*	*Tin*	*Iron*	*Nickel*
Electronegativity	2.1	1.9	1.8	1.8	1.8
Element	*Zinc*	*Titanium*	*Aluminium*	*Magnesium*	*Caesium*
Electronegativity	1.6	1.6	1.5	1.2	0.7

a partial transfer of electrons from one of the atoms in the unlike pair to the other and to attribute it to a difference between them in their attraction for electrons, a concept called *electronegativity*. This property of electronegativity represents the attraction of a neutral atom in a stable molecule for electrons. There are alternative ways of putting the concept on a semi-quantitative basis. Pauling* based a dimensionless scale on the values of excess bond energy found for many pairs of unlike atoms, in which the numbers are obtained by manipulation of the excess energy values to be self-consistent and convenient to use. Mulliken devised an alternative scale based on the averages of measured ionization potentials and electron affinities of the elements. The two scales produce different numbers but agree on the sequence of the elements in order of electronegativity. For illustration, Table 2.7 gives values on Pauling's scale for elements of frequent interest.

Pauling's values range from 0.7 for the least electronegative (or most electropositive) natural element, caesium, to 4.0 for the most electronegative element, fluorine. When two atoms of different elements interact, the more electronegative atom acquires an anionic character and the less electronegative atom acquires a cationic character. In the present context, it is significant that oxygen is the second most electronegative element, with a value of 3.5, which is much greater than the values for metals, which are all <2.4, with most of them <1.9.

2.3.2 Partial Ionic Character of Metal Oxides

The electronegativity concept was developed for single bonds between pairs of unlike atoms in isolation. Its benefits can be fully realized only if it can be used to assess the characters of bonds in real assemblies of atoms. In the particular case of metal oxides, it can be applied, for example, to explain why one kind of lattice defect predominates in some oxides and different kinds predominate in others, as will be seen later. However, this presents two problems that have concerned the originators of the concept. The first problem is to what extent an electronegativity scale derived for single bonds can be transferred to assemblies of ions in a solid with multiple bond contacts. Pauling's inference is that the electronegativity values can be transferred without serious error. The second problem is how to devise a quantitative scale of partial ionic character from the differences in electronegativity between the constituent atom precursors, A and B.

Pauling, cited at the end of the chapter, suggested the empirical equation:

$$\%(\text{ionic character}) = 100\{1 - \exp - (x_A - x_B)^2/4\} \qquad (2.36)$$

where x_A and x_B are the electronegativity values for the atoms A and B.

* Chapter 3 in Pauling's text cited in Further Reading.

TABLE 2.8
Ionic Character of Bonds between Unlike Atoms Due to Electronegativity Difference

Electronegativity Difference	0	0.4	0.8	1.2	1.6	2.0	2.4	3.0
% Ionic character	0	4	15	30	47	63	76	86

TABLE 2.9
Ionic Characters of Some Metal–Oxygen Single Bonds

	Electronegativity			
Bond	**Cation**	**Oxygen**	**Difference**	**Ionic Character %**
Mg–O	1.2	3.5	2.3	74
Cu–O	1.9	3.5	1.6	47
Zn–O	1.6	3.5	1.9	60
Fe–O	1.8	3.5	1.7	50
Ni–O	1.8	3.5	1.7	50
Sn–O	1.9	3.5	1.6	47
Cr–O	1.6	3.5	1.9	60
Al–O	1.5	3.5	2.0	63
Ti–O	1.5	3.5	2.0	63
Si–O	1.8	3.5	1.7	50

An alternative empirical equation, more suitable for large differences in electronegativity is:

$$\%(\text{ionic character}) = 16[x_A - x_B] + 3.5[x_A - x_B]^2 \tag{2.37}$$

Equation 2.36 yields the general relation between ionic character and electronegativity difference given in Table 2.8 and the ionic characters of some particular metal–oxygen single bonds given in Table 2.9. Despite uncertainties in the absolute values of the numbers given in Tables 2.8 and 2.9, they are derived logically and taking them as relative values, they can be used to support other evidence of differences between various oxides to be described.

2.3.3 Oxide Crystal Structures

Atoms in a solid are arranged to minimize the energy of the system. Factors that determine the minimum energy can be complex and may include not only coulombic forces, but also options for mixing wave functions and the strength of possible covalent contributions to the bonding.

If the coulombic forces predominate, as they do for the oxides of most metals of commercial interest, the lowest energy state is realized by closely packing the ions in a space lattice, relaxing the coulombic forces. The close-packed arrangement in any given oxide depends on the relative numbers of anions and cations, the *stoichiometric ratio* and on their relative sizes, the cation/anion *radius ratio*. Table 2.10 gives sizes of some selected ions determined from their closest distance of approach to other ions in appropriate series of compounds.

TABLE 2.10
Molecular Formulae, Cation/Anion Radius Ratios and Structures for Metal Oxides

Metal	Oxide Formula	Ion	Ion Radius/nm (Pauling16)	Cation/Anion Radius Ratio[a]	Oxide Structure
Aluminium	Al_2O_3	Al^{3+}	0.050	0.36	Corundum
Magnesium	MgO	Mg^{2+}	0.065	0.46	Simple cubic
Titanium	TiO_2	Ti^{4+}	0.068	0.48	Rutile
Chromium	Cr_2O_3	Cr^{3+}	0.069	0.49	Corundum
Iron II	FeO	Fe^{2+}	0.076	0.54	Simple cubic
Iron III	Fe_2O_3	Fe^{3+}	0.064	0.45	Corundum
Nickel	NiO	Ni^{2+}	0.072	0.51	Simple cubic
Copper I	Cu_2O	Cu^{+}	0.096	0.69	Cubic
Copper II	CuO	Cu^{2+}	0.072	0.51	Monoclinic
Zinc	ZnO	Zn^{2+}	0.069	0.49	Würtzite
Tin	α-SnO_2	Sn^{4+}	0.071	0.50	Rutile
Magnesium & Aluminium	$MgAl_2O_4$	Mg^{2+} Al^{3+}	0.065 0.050	0.46 0.36	Spinel
Iron(II) & Chromium	$FeCr_2O_4$	Fe^{2+} Cr^{3+}	0.076 0.069	0.54 0.49	Spinel
Iron(II) & Iron(III)	Fe_3O_4	Fe^{2+} Fe^{3+}	0.076 0.064	0.54 0.45	Inverse spinel

[a] Radius of oxygen ion = 0.140 nm.

The close-packed structures with minimum coulombic energies are those in which the smaller metal ions are most efficiently accommodated between the larger oxide ions, and so oxides with the same stoichiometric and radius ratios often exhibit the same structure. These include the simple cubic, rhombohedral–hexagonal, rutile and spinel structures described below. Brucite is included because several metal hydroxides that can form as corrosion products or their intermediaries adopt this structure.

2.3.3.1 Simple Cubic Structures (Rock Salt, NaCl Structure)

The simple cubic structure is favourable for ionic compounds with equal numbers of anions and cations and with cation/anion radius ratios between 0.414 and 0.732, which is the range for octahedral coordination. As Table 2.10 illustrates, the radius ratios for several metal oxides lie within this range. In the simple cubic structure, every ion is surrounded by six contacting ions of opposite charge. Oxides with simple cubic structures include magnesium oxide, wüstite FeO and nickel(II) oxide NiO.

2.3.3.2 Rhombohedral–Hexagonal Structures (Corundum Structure)

Trivalent metals form oxides with the general formula M_2O_3 and different structures are needed to accommodate the ratio of two metal ions to three oxide ions. One of these is the rhombohedral structure, which is best described as a hexagonal close-packed lattice of oxygen ions with metal ions occupying two-thirds of the octahedral spaces. Since the corundum arrangement is based on a hexagonal close-packed lattice, the structure also has hexagonal geometry. Oxides with the corundum structure include corundum itself, α-Al_2O_3, normal haematite Fe_2O_3 and chromium oxide, Cr_2O_3.

2.3.3.3 Rutile Structures

The name *rutile* is derived from its generic member, a natural mineral of TiO_2. The structure is commonly adopted by metal oxides of the form MO_2 where the (metal ion)/(oxygen ion) radius ratio is in the range 0.414–0.732. A ratio within this range favours the cation in octahedral coordination with every metal ion coordinated with six oxygen ions. There are twice as many oxygen ions as metal ions in MO_2, thus every oxygen ion is coordinated with three metal ions in a triangular configuration. The crystal structure is essentially a series of octahedra centered on the metal ions, interconnected by the triangular configurations centered on the oxygen ions and aligned in two sets of mutually perpendicular parallel planes. The overall geometry is tetragonal. Oxides with the rutile structure include titanium oxide, TiO_2, tin(IV) oxide SnO_2, lead(IV) oxide PbO_2 and manganese dioxide, MnO_2.

2.3.3.4 Spinel Structures

A further arrangement is the 2:3 spinel structure in which two metals participate, M^{II} and M^{III}, yielding the divalent and trivalent cations, M^{2+} and M^{3+}, respectively. The name is derived from the generic member, spinel, $MgAl_2O_4$. The structure can be described as a face-centered cubic lattice of oxygen ions in which divalent and trivalent metal ions, M^{2+} and M^{3+}, respectively, occupy tetrahedral and octahedral interstices. Only sufficient interstices are occupied to give the correct stoichiometric correspondence and hence charge balance between the oxygen and metal ions. This is accomplished when half of the octahedral interstices are occupied by ions of the trivalent metal and one-eighth of the tetrahedral sites are occupied by the divalent metal yielding the general formula, $M^{II}M^{III}{}_2O_4$. Members of this group that are important in thermal oxidation are *chromite*, $Fe^{II}Cr^{III}{}_2O_4$ and *spinel*, $Mg^{II}Al^{III}{}_2O_4$, that can form on iron–chromium alloys and aluminium–magnesium alloys respectively. *Magnetite*, $Fe^{II}Fe^{III}{}_2O_4$, one of the oxides that form on pure iron, is an inverse 2:3 spinel, so called because the *divalent* ions, Fe^{2+}, are predisposed to occupy octahedral sites, displacing half of the *trivalent* ions, Fe^{3+}, to the tetrahedral sites. There are other variants of the structure, for example, 1:3 spinel-type oxides in which monovalent and trivalent metals participate, such as *β-alumina*, $Na_2Al_2O_4$.

2.3.3.5 Brucite Structures

The Brucite structure is adopted by several hydroxides, including magnesium hydroxide, nickel hydroxide and iron(II) hydroxide, all important species in aqueous corrosion processes. These hydroxides have the general formula $M(OH)_2$ where M is a divalent metal with a moderate ion size. In this structure, every metal atom is in octahedral coordination with six hydroxyl ions. The octahedra are linked in two dimensions by sharing edges in which every hydroxyl ion is coordinated with three metal ions. The layers so formed are weakly bound together in the third dimension to give a complex layered structure with hexagonal symmetry.

2.3.3.6 Other Structures

The oxides of some engineering metals are exceptional because of peculiarities of their ions. Despite its favourable radius for octahedral, that is, six-fold, coordination with oxygen ions, the zinc ion is predisposed to tetrahedral, that is four-fold, coordination and so zinc oxide adopts a hexagonal structure called *würtzite*.

This brief survey of oxide structures is far from comprehensive but it is sufficient to proceed with a discussion of defect structure.

2.3.4 Conduction and Valence Electron Energy Bands

In approaching the subject of oxide lattice defects, some attention must be paid to the distribution of electrons within oxides because perturbations in this distribution constitute the electronic defects inevitably accompanying certain lattice defects that were referred to earlier. The configurations of electrons in assemblies of ions in solid compounds is a matter of great complexity because they are accommodated in orbitals formed by all of the atoms that constitute the material so that the solid contains an immense number of energy states for possible occupation by the available electrons.

The relevant result for the present discussion is that in contrast with the sharp energy levels in free atoms, the energy states are grouped into ranges or bands and in oxides with distinct ionic character, the bands are separated by an energy gap, the *band gap.* The band below the band gap, the *valence band,* is associated with wave functions contributed by the oxygen atomic orbitals and the band above the gap, the *conduction band,* is associated with wave functions contributed by the metal atomic orbitals. Comparison of the number of energy states in the valence band with the number of available electrons shows that if they were all in the valence band, it would be full. In that event, the oxide would be an electrical insulator because under an applied potential, there would be no unoccupied energy states to allow net transfer of electrons in the valence band and no electrons in the conduction band to act as charge carriers.

This condition is closely approached in oxides with strong ionic character, such as MgO, where the band gap is so wide that the energy required to promote electrons from the valence band to the conduction band is prohibitively high. In oxides with less ionic character, such as ZnO, the band gap is narrower and the conduction band contains a few electrons, conferring a small but significant degree of electronic conductivity, that is, the oxide is a semiconductor. The ability to accept electrons into the conduction band stabilizes the particular kinds of lattice defect that prevail in such oxides, as described in the next section.

2.3.5 The Origins of Lattice Defects in Metal Oxides

The term *lattice defects* in metal oxides or other solids does not mean accidental or sporadic faults that ought not to be present, but has a very specific reference to imperfections in the arrangement of ions on a lattice that are entirely natural features of the material in its normal condition. In metal oxides, the defects can take the form of ions missing from lattice positions, called *lattice vacancies,* additional ions inserted between the ions on their normal lattice sites, called *interstitials* or various combinations of them.

Defects exist because the lowest energy of a phase is associated with a finite degree of disorder in its structure. This principle is quite general and was quoted in Section 2.2.4 in relation to disorder in the structure of water manifesting as autodissociation. The number of defects of any particular kind in an oxide can be quantified in terms of the energy released by a notional process introducing the defects into a hypothetical perfect lattice. The energy released is the sum of two terms. One describes the heat transfer during the process and the other describes internal changes in the state of order of the system. Standard thermodynamic equations are selected to suit the appropriate conditions.

For an isothermal process at constant pressure, for example the ambient atmosphere, the equation is:

$$\Delta G = \Delta H - T\Delta S \tag{2.38}$$

where:

ΔG is the change in *Gibbs function, G*, which is the decrease in the energy of a system when the work done by is expansion against a constant external pressure

ΔH is the change in *enthalpy, H*, which is numerically equal to the heat taken into the system when the pressure on the system is constant

ΔS is the change in *entropy, S*, of the system

T is temperature on the Kelvin (absolute) scale

The quantities G, H and S are *extensive state properties*, that is, properties of a prescribed mass of a given material or system of materials with definite values, irrespective of the previous history of the material.

Entropy, S, is a cornerstone of science and is rigorously defined. The full implications are explained in standard classic texts*. For now, the relevant aspect is that there is a quantitative correlation between the entropy of a system and the statistical distribution of energy and mass within it at the atomic level that includes:

1. The distribution of the total thermal energy within the system among its atoms, molecules or ions and among energy modes available within them, that is translation, rotation and vibration, described as the *degrees of freedom*.
2. The distribution of the atoms, molecules or ions in the three dimensions of the space allocated to the system. Entropy arising from the spatial distribution is described as *configurational entropy*.

2.3.5.1 General Approach

A dimensionless quantity, the *thermodynamic probability, W*, introduced below, expresses the number of ways in which a system in a given state can be implemented. It is related to the entropy of the system by the Boltzmann–Planck equation:

$$S = k \ln W \tag{2.39}$$

where:

k is *Boltzmann's constant*, equal to the universal gas constant, R, divided by Avogadro's number N_A

Equation 2.39 allows S to be calculated statistically from first principles. In the particular case of interest here, it provides a means for calculating the defect populations corresponding to the minimum energy and hence to the most stable state of an oxide.

A statistical method is employed in which the system is represented by a multi-dimensional *phase space* with coordinates representing the three space dimensions and sufficient additional dimensions to represent the momenta of the fundamental particles

* For example Lewis, Randall, Pitzer and Brewer or Glasstone, cited in Further Reading.

in the available degrees of freedom. The phase space is divided into convenient identical small *phase cells* and the number of possible ways of distributing the atoms within the cells is counted. Every possible distribution is a *microscopic state* of the system, but since atoms of any one kind are identical, many microscopic states correspond to the same state in the system as a whole and the number of these microscopic states is the thermodynamic probability of the state. When the statistics are applied to the immense number of atoms, ions or molecules in a real system, the spectrum of probabilities peaks so sharply at one particular state of the system that this state is a virtual certainty. An often quoted illustrative example is a near ideal gas, such as hydrogen at ordinary temperatures and pressures, for which the statistics predict correctly that the gas fills the volume available to it at uniform concentration and temperature.

2.3.5.2 Configurational Entropy of Atoms or Ions on a Lattice

The object is to calculate the increase in entropy, ΔS, produced by creating defects of a particular kind in a hypothetical perfect lattice of N lattice sites, ignoring entropy due to the distribution of thermal energy within the system, which does not enter into the calculation. In this case, the phase cell chosen includes one lattice site. There is only one microscopic state corresponding to a perfect lattice and so the probability, W_1 is one and by Equation 2.39:

$$S_1 = k \ln 1 = 0 \tag{2.40}$$

If n defects are now created and distributed among the N lattice sites, application of standard statistics gives the new probability W_2:

$$W_2 = \frac{N!}{(N-n)!n!} \tag{2.41}$$

For one mole of material, N is Avogadro's number. Using Stirling's approximation, $\ln N! = N \ln N - N$, assuming n/N is small and substituting for W_2 in Equation 2.39, yields:

$$S_2 = -R(n/N) \ln (n/N) = -RX \ln X \tag{2.42}$$

where X is the mole fraction of defects, that is, the ratio of the number of defect sites to the total number of lattice sites and R is the gas constant, 8.314 J K mol^{-1}.

The entropy change due to the creation of the defects is therefore:

$$\Delta S = S_2 - S_1 = -RX \ln X \tag{2.43}$$

2.3.5.3 Equilibrium Number of Defects

The formation of lattice defects is favoured by the release of energy in introducing disorder but constrained by the energy that must be supplied to create the defects. These two energy quantities are represented by the ΔH and $T\Delta S$ terms, respectively, in Equation 2.38. Provided that n/N is small so that the creation of new defects is not influenced by those already present, the energy needed to create defects is a linear function of their mole fraction:

$$\Delta H = \chi NX \tag{2.44}$$

where χ is the energy required to create one defect.

Inserting expressions for ΔS and ΔH from Equations 2.43 and 2.44 in Equation 2.38 gives the decrease in energy of the material by the creation of a mole fraction X of defects:

$$\Delta G = \chi N X + RTX \ln X \tag{2.45}$$

The value of X at the minimum for ΔG, where $d\Delta G/dX = 0$ is:

$$X = \exp(-\chi N/RT - 1) \tag{2.46}$$

Equation 2.46 shows that at a given temperature, there is a definite value of X for the minimum energy of the system. This is the most stable state of the material. Two features of Equation 2.46 are important:

1. The number of defects increases with rising temperature.
2. At a given temperature, the number of defects depends on the energy needed to create them and recalling earlier remarks, this includes energy both to break bonds with adjacent ions and to create complementary electronic defects, as explained shortly.

A schematic illustration of the minimum value for ΔG, given in Figure 2.8, shows how the number of defects depends on temperature and on the enthalpy for creating defects. Raising T displaces the $T\Delta S$ curve to the right, moving the minimum for ΔG to larger values of N. Increasing ΔH raises the slope of ΔH versus N, shifting the minimum to a smaller value of N.

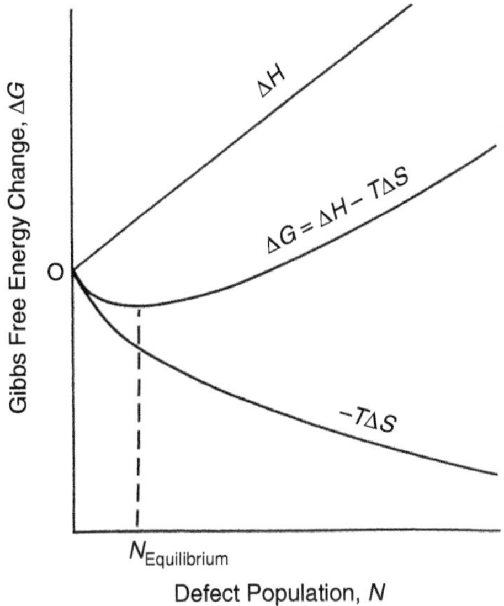

FIGURE 2.8
Gibbs free energy change, ΔG, produced by introducing a defect population, N, into a hypothetical perfect metal oxide. The minimum in the plot of ΔG against N corresponding to the equilibrium value for N, is derived by balancing the enthalpy and entropy terms ΔH and $-T\Delta S$.

2.3.5.4 Nature of Defects in Oxide Lattices

The significant defects in metal oxides are:

1. Schottky disorder (vacant cation and anion sites in equal numbers)
2. Vacant metal cation sites
3. Metal cation interstitials
4. Vacant oxygen anion sites
5. Intrinsic semiconductors

One other type of defect, Frenkel disorder, comprising paired anion interstitials and vacancies, is excluded from metal oxides by the large size and high charge density of the oxygen ion.

Unpaired lattice defects disturb the balance of charge on the lattice and electrical neutrality is preserved by compensating *electronic defects.* Cation interstitials or anion vacancies introduce a net positive charge, which is compensated by excess electrons, described as *interstitial electrons.* Cation vacancies introduce a net negative charge, which is compensated by a deficit of electrons, described individually as *electron holes.*

2.3.5.5 Representation of Defects

The standard symbols for defects in an oxide with a metal cation, M^{z+} are:

Cation vacancies:	$M^{z+}\square$
Cation interstitials:	$M^{z+}\bullet$
Anion vacancies:	$O^{2-}\square$
Interstitial electrons:	$e\bullet$
Electron holes:	$e\square$

Their use can be illustrated by schematic equations describing the formation of defects:

1. Creation of a cation interstitial by inserting a metal atom into the lattice.

$$\underset{\text{metal atom}}{M} = \underset{\text{cation interstitial}}{M^{z+}\bullet} + \underset{\text{interstitial electrons}}{ze\bullet} \tag{2.47}$$

2. Creation of a cation vacancy by removing a metal atom from the lattice.

$$\underset{\text{cation}}{M^{z+}} = \underset{\text{metal atom}}{M} + \underset{\text{cation vacancy}}{M^{z+}\square} + \underset{\text{electron holes}}{ze\square} \tag{2.48}$$

3. Creation of an anion vacancy by removing an oxygen atom from the lattice.

$$\underset{\text{anion}}{O^{2-}} = \underset{\text{½ oxygen molecule}}{\tfrac{1}{2}O_2} + \underset{\text{anion vacancy}}{O^{2-}\square} + \underset{\text{interstitial electrons}}{2e\bullet} \tag{2.49}$$

TABLE 2.11
Classification of Some Metal Oxides by Defect Structure

Defect	Conduction	Oxides
Schottky	Ionic	MgO, Al_2O_3 (<825°C)
Cation vacancies	*p*-type	FeO, NiO, MnO, CoO, Cu_2O, $FeCr_2O_4$, UO_2
Cation excess	*n*-type	ZnO, CdO, BeO, Al_2O_3 (>825°C), $MgAl_2O_4$, UO_3, U_3O_8
Anion vacancies	*n*-type	TiO_2, ZrO_2, Fe_2O_3 (with some cation interstitials)
Mixed	Mixed	Fe_3O_4

2.3.6 Classification of Oxides by Defect Type

An equation in the form of Equation 2.45 applies to every kind of defect in all oxides, but usually one kind predominates in any particular oxide because the energy to create it is less than that for the others and its influence determines much of the character of the oxide. For this reason, it is convenient to classify oxides by their predominant defects, as in Table 2.11.

2.3.6.1 Stoichiometric Oxides

Oxides of strongly electronegative metals, such as magnesium oxide, are characterized by a large band gap, so that the energy needed to insert electrons into the conduction band to compensate for cation interstitials or anion vacancies, as required in Equations 2.47 and 2.49, is very high. The energy for creating electron holes in the valence band to compensate for cation vacancies, as required in Equation 2.48, is also very high, for reasons given earlier. As a consequence, the only significant type of defect is Schottky disorder, which is paired cation and anion vacancies with no requirement for compensating electronic defects. The compositions of these oxides correspond closely to their molecular formula, recognized by the term *stoichiometric*; for example the ratio of metal ions to oxygen ions is 1:1 for MgO and 2:3 for Al_2O_3. Figure 2.9 is a schematic illustration of a Schottky defect pair in magnesium oxide.

2.3.6.2 Cation Excess Oxides

Oxides of less electronegative nontransition metals, such as zinc oxide, are characterized by a small band gap, and the energy needed to insert electrons into the conduction band to compensate for excess cations is not large. Much more energy would be required to remove

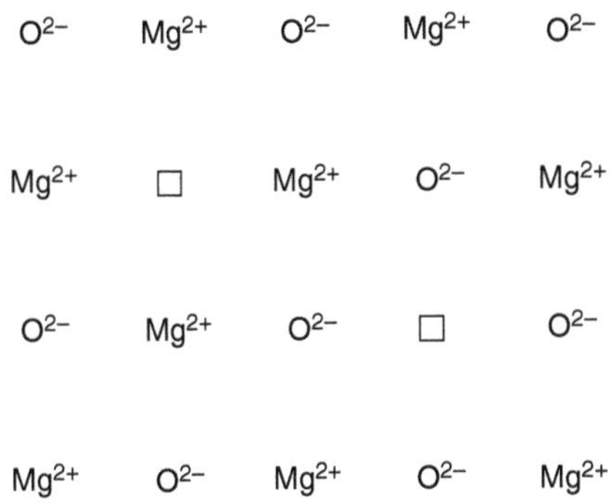

O^{2-} Mg^{2+} O^{2-} Mg^{2+} O^{2-}

Mg^{2+} □ Mg^{2+} O^{2-} Mg^{2+}

O^{2-} Mg^{2+} O^{2-} □ O^{2-}

Mg^{2+} O^{2-} Mg^{2+} O^{2-} Mg^{2+}

FIGURE 2.9
Example of an ionic oxide. Schematic diagram of a Schottky defect pair in magnesium oxide.

Zn2+ O2– Zn2+ O2– Zn2+
e● e●
O2– Zn2+ O2– Zn2+ O2–
Zn2+ ●
Zn2+ O2– Zn2+ O2– Zn2+
O2– Zn2+ O2– Zn2+ O2–

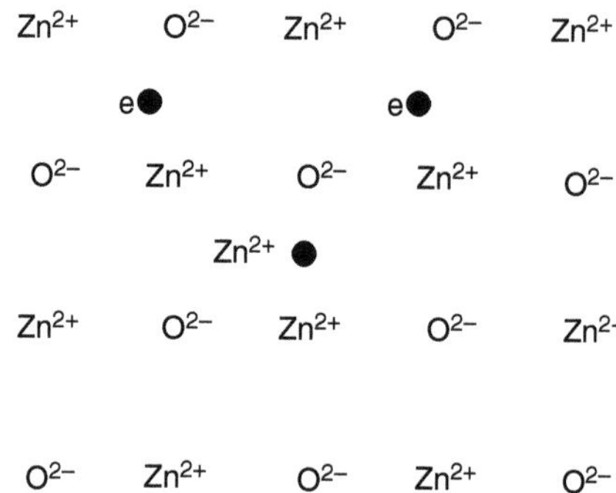

FIGURE 2.10
Example of a cation excess oxide. Schematic diagram of an interstitial cation and electrons in zinc oxide.

electrons from the valence band to compensate for cation vacancies. Consequently, cation interstitials are the predominant defects. Figure 2.10 is a schematic illustration for zinc oxide.

The small population of electrons in an otherwise empty conduction band confers some electron mobility and hence cation excess oxides are semiconductors of electricity. The term *n-type semiconductor* indicates that the conductivity is due to *negative* charge carriers.

2.3.6.3 Cation Deficit Oxides

Recalling the brief discussion in Section 2.1.2.1, the partly filled 3*d* orbitals underlying the nominal 4*s* valency electrons confer special characteristics on metals in the first transition series. For some of these metals, for example manganese, iron, cobalt, nickel and copper, the energies of the *d* electrons are close enough to the energies of the valency electrons to provide a source from which electrons can be withdrawn to compensate for charge imbalance introduced by cation vacancies. For example, it is relatively easy to convert the normal valency of nickel from 2 as in Ni^{2+} to 3 as in Ni^{3+}. The increase in valency corresponds to a deficit of electrons in the *d* orbitals referred to as *electron holes*, e□, hence the predominant defects in the lower oxides of these metals are cation vacancies. The electron deficit applies to the oxide as a whole, but for convenient representation, electron holes are usually indicated by nominally assigning increased valency to arbitrary cations as illustrated for nickel oxide in Figure 2.11.

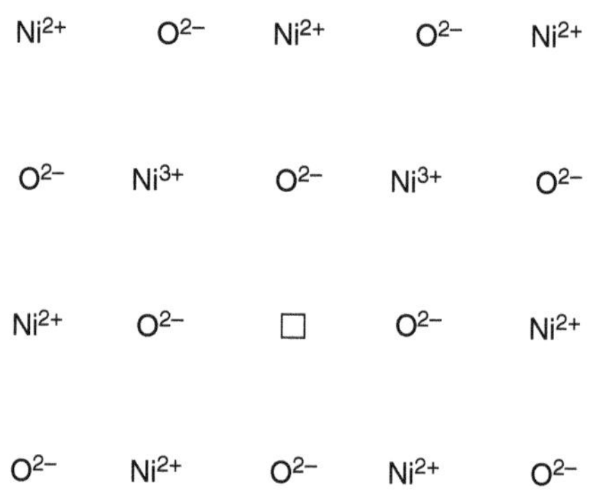

Ni2+ O2– Ni2+ O2– Ni2+
O2– Ni3+ O2– Ni3+ O2–
Ni2+ O2– □ O2– Ni2+
O2– Ni2+ O2– Ni2+ O2–

FIGURE 2.11
Example of a cation vacant oxide. Schematic diagram of a cation vacancy and electron holes (Ni^{3+}) in nickel oxide.

The electron holes confer mobility on the electrons, which in the schematic diagram in Figure 2.11 would be represented by electron transfer between the majority Ni^{2+} ions and the minority Ni^{3+} ions, exchanging their oxidation states, but in reality all of the nickel ions are equivalent with all of them being mainly of Ni^{2+} character with some Ni^{3+} character. Cation deficient oxides are, therefore, also semiconductors. The term *p-type semiconductor* indicates that the conductivity is due to *positive* charge carriers.

2.3.6.4 Anion Deficit Oxides

Some metal oxides with the potential to compensate for excess positive charge exhibit anion vacancies. Most are higher oxides of metals in the transition series, including TiO_2, ZrO_2 and Fe_2O_3. The excess positive charge on the lattice is compensated by absorbing extra electrons into the cation *d*-orbitals. The electron excess can be represented schematically either by assigning *decreased* valency to arbitrary cations or by symbolic interstitial electrons, as illustrated for titanium dioxide in Figure 2.12. Anion deficit oxides are *n*-type semiconductors.

The formation of anion vacancies in an oxide is thus related to the existence of two relatively stable oxidation states for the metal concerned, for example Ti^{4+} and Ti^{3+}, Fe^{3+} and Fe^{2+} and this is also reflected in the ability of these metals to exercise more than one valency in their aqueous chemical behaviour, as considered later with reference to aqueous corrosion.

2.3.6.5 Intrinsic Semiconductors

Intrinsic semiconductors are characterized by equivalent numbers of electrons and electron holes without associated lattice defects. CuO is an example of interest in the present context.

2.3.6.6 Degrees of Non-Stoichiometry

The creation of Schottky defects needs energy to remove both cations and anions from the lattice, so that the enthalpy of formation is high and the defect population in stoichiometric oxides is small but significant.

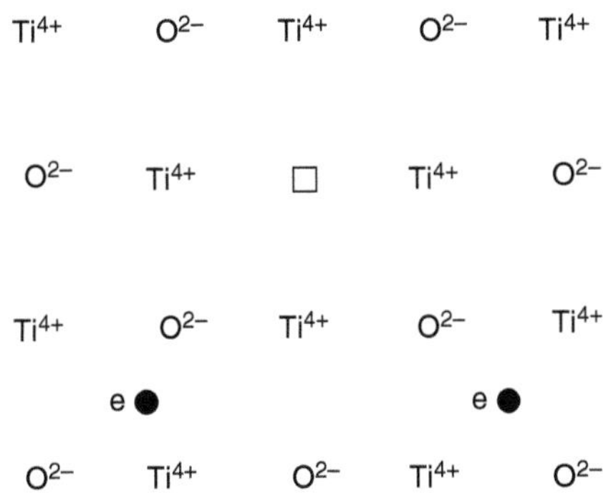

Ti^{4+} O^{2-} Ti^{4+} O^{2-} Ti^{4+}

O^{2-} Ti^{4+} □ Ti^{4+} O^{2-}

Ti^{4+} O^{2-} Ti^{4+} O^{2-} Ti^{4+}

e ● e ●

O^{2-} Ti^{4+} O^{2-} Ti^{4+} O^{2-}

FIGURE 2.12
Example of an anion vacant oxide. Schematic diagram of an anion vacancy and interstitial electrons in titanium oxide.

Cation deficient oxides can exhibit a high degree of non-stoichiometry that is often measurable by chemical analysis. This is because, where the appropriate charge compensation mechanism exists, as for the lower oxides of transition metals, cation vacancies have the lowest enthalpies of formation and are thus the easiest to form. One of the most non-stoichiometric oxides known is nominally FeO, which exists over a composition range $Fe_{0.95}O$ to $Fe_{0.99}O$, that does not even include the stoichiometric composition. Incidentally, the degree of non-stoichiometry of sulfides can be even greater, for example nominal FeS can exist at a chemical composition corresponding to $Fe_{0.92}S$, indicating that about 8% of the cation sites are vacant.

The compositions of cation excess oxides are much closer to the true stoichiometric ratio and although the non-stoichiometry can be detected indirectly, for example by diffusion and electrical conductivity measurements, it is not easy to directly detect by chemical analysis.

2.4 The Structures of Metals

Considerable detail of the electron configurations in bonding was needed to describe the structures of liquid water and of oxides in sufficient depth to explain corrosion phenomena. There is less need to describe the electronic counterpart for metals, but for completeness it is referred to very briefly in Section 2.4.1. Section 2.4.2 describes crystallographic features. Section 2.4.3 summarizes the main features of phase equilibria in alloys to provide a basis for descriptions of particular alloy systems in later chapters. Section 2.4.4 deals with an aspect of structure in its widest sense, which is often overlooked, concerned with characteristics imparted to metals by manufacturing procedures.

2.4.1 The Metallic Bond

Bonding in metals is clearly quite different from that in covalent molecules or in ionic solids. The electron configuration responsible for the cohesion of assemblies of metal atoms must be consistent with the following features of the metallic state:

1. Metals have a unique combination of properties, including *inter alia* strength, ductility and high electrical and thermal conductivities.
2. In general, metallic elements have considerable mutual miscibility even when their chemical valencies differ, as for aluminium and magnesium. They also form intermetallic compounds which appear to disregard valency rules.

The metallic elements are *hypoelectronic* elements, meaning that their outer electron shells are less than half-filled so that there is an insufficient pool of electrons to satisfy either atom-to-atom covalent bonding or the formation of ions. The problem of accounting for the cohesion of metals was solved by Pauling, who conceived that a special orbital, the *metallic orbital*, is set aside by means of which bonding electrons can move from atom to atom, thereby permitting bonds between atoms to switch in turn between all of their neighbors in an unsynchronized way. Without going into detail, this represents *resonance energy* that stabilizes the structure. Metals such as lithium, magnesium and aluminium, which can provide only 1, 2 and 3 valency electrons, respectively, for this kind of bonding are not intrinsically strong and have relatively low melting points. Metals from the transition series can augment their valency electrons by up to 4 extra electrons promoted from the underlying *d* shell and for this reason, metals such as iron, nickel and copper are stronger and have higher melting points.

The resonant bonding accounts for the following metallic characteristics:

1. Metals crystallize in close-packed structures, usually FCC, BCC or hexagonal close pack, characteristic of non-directional bonding for assemblies of atoms of the same radius.
2. Most metals are ductile because the assembly of atoms coheres as a whole and not on an atom-by-atom basis. Thus atoms can move relative to one another, retaining cohesion and planes of atoms can slide past each other by propagation of *dislocations* in the lattice.
3. Metals have high electrical and thermal conductivities because resonant bonding *delocalizes* the bonding electrons, freeing them to move throughout the metal, carrying electric charge and thermal energy.

2.4.2 Crystal Structures and Lattice Defects

The crystal structures of common engineering metals are listed in Table 2.1. Iron is especially interesting and important because the structure is BCC at temperatures <910°C, FCC in the temperature range 910–1400°C, reverting to BCC at temperatures >1400°C. These transformations are manipulated in alloys to produce a wide range of steels with different characteristics. This is illustrated in a later discussion of the formulation of stainless steels to produce the FCC structure for applications where easy formability is required and the BCC structure where resistance to SCC is needed.

Metals contain vacant lattice sites and interstitials, because as explained earlier, the lowest energy state of a lattice is associated with optimum disorder but, since the atoms are neutral and chemically similar, there is no counterpart to the electronic defects and non-stoichiometry in oxides. The defect population density rises with temperature and in particular the population of vacant lattice sites typically approaches 10^{-3}, that is, one vacancy per thousand atoms, near the melting point. The principal effect of the vacancies is to facilitate diffusion of solutes in solid solution by sequential interchange between vacancies and adjacent atoms.

Manufactured metal products are agglomerates of small crystals of various orientations, which can be revealed by etching polished sections of the material. Randomly orientated microscopic crystals are desirable to give uniform isotropic mechanical properties. The *grain boundaries* separate adjacent crystals with different orientations and the lattice structure is distorted in regions within a few atomic diameters of the boundaries to accommodate the changes in orientation. The local disturbance provides preferred sites for occupation by impurities, that is, *segregation* and for the nucleation of minority phases. Such structural features can offer paths for corrosion or SCC, considered in Chapter 5.

2.4.3 Phase Equilibria

Most but not all combinations of metals are miscible in the liquid state. On solidification, a liquid mixture transforms to one or more solid phases and further transformations

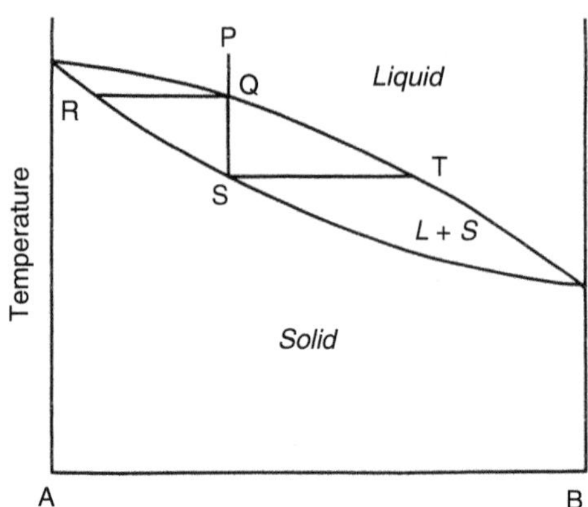

FIGURE 2.13
Solid solutions of two metals A and B completely are miscible in the solid and liquid states.

may occur as the metal cools. Manipulation of compositions and transformations is the basis for alloying.

Five basic kinds of transformation are particularly important, that is, the solidification of solid solutions, eutectic, eutectoid and peritectic transformations and the formation of intermetallic compounds. Phase relationships for a few alloy systems are completely represented by one or another of these transformations, but more often they are derived from combinations of them.

Phase relationships are depicted in equilibrium *phase diagrams,* in which the relative stabilities of phases in an alloy system are represented as functions of two parameters, alloy composition and temperature. Lines representing equilibria between phases are plotted from thermodynamic information, confirmed by experiment. The effect is to divide the diagram into *phase fields,* which enclose the coordinates of composition and temperature over which particular phases are stable. If sufficient time is available, a phase change occurs whenever the temperature or composition of an alloy is changed so that the coordinates representing it on the diagram cross a boundary between adjacent phase fields. The basic phase equilibria are briefly described in Sections 2.4.3.1 to 2.4.3.5 and illustrated in Figures 2.13 through 2.17.

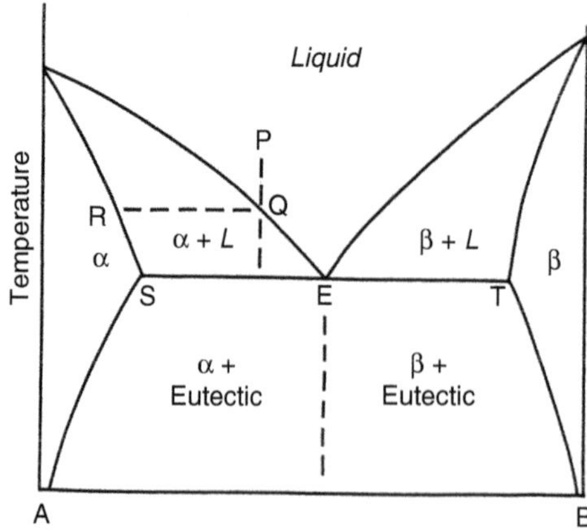

FIGURE 2.14
Eutectic system formed by two metals A and B partially miscible in the solid state.

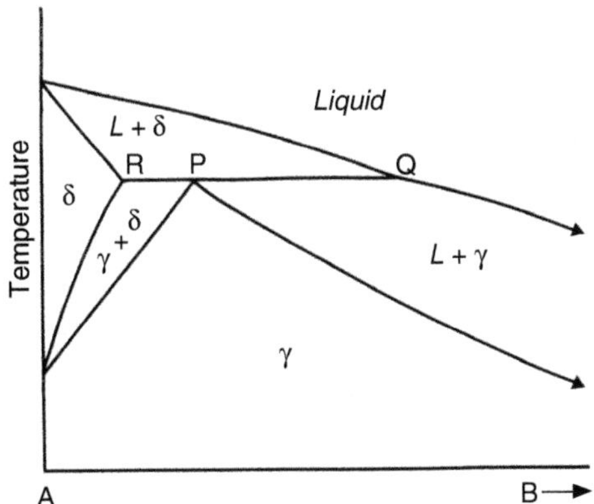

FIGURE 2.15
Part of a phase diagram for metals A and B showing the intervention of a peritectic reaction during solidification.

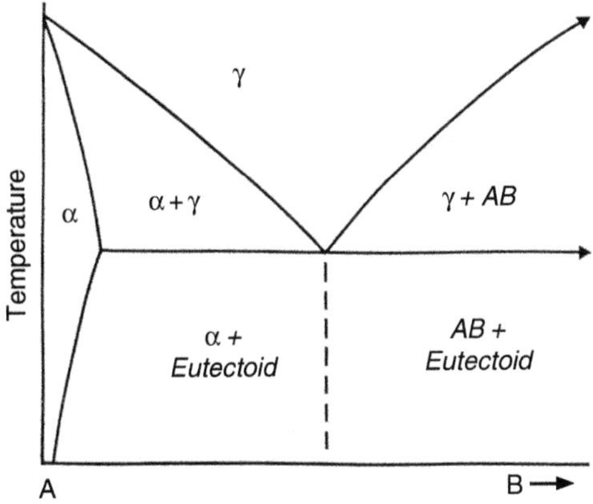

FIGURE 2.16
Part of a phase diagram for metals A and B showing decomposition of the solid phase, γ, into α and a eutectoid of α and an intermetallic compound AB.

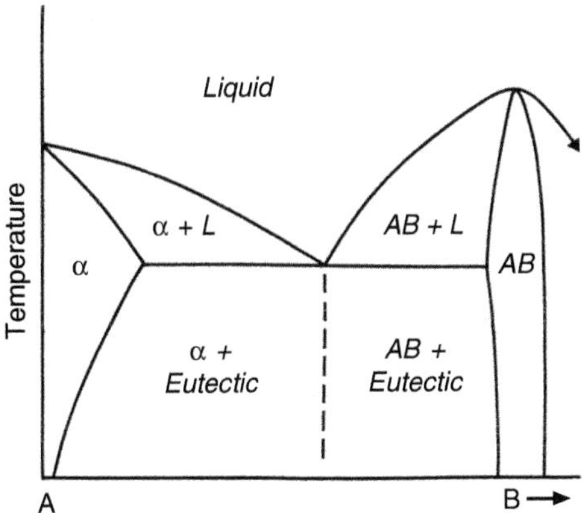

FIGURE 2.17
Part of a phase diagram for metals A and B showing formation of an intermetallic compound.

2.4.3.1 Solid Solutions

Most pairs of metals are partially miscible in the solid state, forming *binary solid solutions.* Some pairs of closely similar metals form complete series of solid solutions and the phase diagram for such a pair of metals, A and B, forming a *binary system,* is given in Figure 2.13. The upper and lower boundaries of the loop are, respectively, the *liquidus* and *solidus.* Above the liquidus, alloys of the two metals are liquid solutions, below the solidus they are solid solutions and in the liquidus/solidus field they are equilibrium mixtures of solid and liquid.

The diagram yields information on the mechanism by which liquid alloys solidify, which is used in Section 2.4.4.1 to explain structural features of castings and ingots. The principle can be appreciated by tracing the progress of solidification when a liquid alloy at a temperature above the liquidus line, represented by the point P in Figure 2.13, is cooled through the liquidus/solidus field. On crossing the liquidus line at Q, a solid solution of composition R is deposited. On further cooling, if freezing were infinitely slow, the instantaneous composition of the growing fraction of solid metal would follow the compositions traced out by the arc RS on the solidus line, leaving a diminishing fraction of liquid with the corresponding compositions traced out by the arc QT on the liquidus line. If the compositions of the corresponding, that is *conjugate,* solid and liquid solutions are assumed to be continuously adjusted to their equilibrium values by interdiffusion of the components, the final solid would be homogeneous with the same composition, Q, as the original liquid. Real casting processes deviate from this idealized mechanism because freezing rates are finite and transfer of the components between the solid and liquid fractions of the metal is incomplete. The phenomenon is called *selective freezing* and one effect is to introduce differential compositions and associated differential electrochemical characteristics in the material that can adversely influence surface-sensitive treatments, such as the brightening and anodizing of aluminium as a decorative finish for vehicle trim or domestic consumer durables.

2.4.3.2 Eutectic Transformations

In a eutectic transformation, a liquid mixture of metals transforms on solidification to form not one but two different phases. In the simple binary system for two metals, A and B, illustrated in Figure 2.14, one is a solution of B in A and the other is a solution of A in B designated by Greek symbols, for example α and β. Liquid alloys with less of metal B than the composition S solidify as the single-phase α solid solution. Alloys with compositions between S and E, such as that represented by the point, P, deposit the α solid solution as they cool through the liquidus/solidus field, labelled $\alpha + \mathrm{L}$. The composition of the deposited solid follows the line RS and that of the remaining liquid follows the line QE. When the composition of the liquid reaches the point E, the remaining liquid solidifies at constant temperature as an intimate mixture of the two phases, α and β, often in a lamellar morphology. The composition and temperature represented at E are the *eutectic composition* and *eutectic temperature.* The α solid solution deposited before the eutectic composition is reached is described as the *primary α phase.* The solidification of alloys with compositions between E and T follows a similar path except that the solid deposited in the corresponding liquidus/solidus field, labelled $\beta + \mathrm{L}$, is the *primary β phase.* Liquid alloys with more metal B than the composition, T, solidify as a single-phase β solid solution. The resultant alloy structures depend on the relative quantities of primary crystals and eutectic. Alloys with compositions close to the eutectic composition exhibit isolated primary phase dendrites distributed in a eutectic matrix. Alloys with compositions remote from the eutectic composition consist mainly of the primary phase, with isolated pockets of eutectic.

2.4.3.3 Peritectic Transformations

In peritectic transformations, the first solid separated from a solidifying liquid alloy reacts with residual liquid at a constant lower temperature to produce a final solid of different structure and composition, illustrated in Figure 2.15. The horizontal line, RQ, marks the peritectic temperature and the point, P, is the peritectic composition. When alloys with compositions between R and Q solidify, they initially deposit δ-solid solution, but when the metal has cooled to the peritectic temperature, the solid δ phase reacts with remaining liquid to form the γ-solid solution. When the reaction is complete, alloys with compositions between R and P leave a residue of δ that subsequently transforms to γ as the alloy cools further. Alloys with compositions between P and Q leave residual liquid that solidifies to γ. Alloys with metal B in excess of the composition Q do not exhibit the peritectic reaction and solidify directly to γ. Some iron alloy systems, notably iron–carbon and iron–nickel (see Brandes in Further Reading) include peritectic reactions.

2.4.3.4 Eutectoid Transformations

Eutectoid transformations have similar characteristics to eutectic transformations but they take place entirely in the solid state. An important example is the transformation of austenite (a solid solution of carbon in γ-iron) to a eutectoid mixture of ferrite (a solid solution of carbon in α-iron) and cementite (an iron carbide, Fe_3C). This is illustrated in the partial phase diagram given in Figure 2.16. The versatility of carbon steels is due mainly to exploitation of this transformation in various ways, as explained in Chapter 7.

2.4.3.5 Intermetallic Compound Formation

Figure 2.17 illustrates a diagram for two metals which form an intermetallic compound, AB. The binding energy of the compound is reflected in its high melting point. Often, the intermetallic compound tolerates a degree of non-stoichiometry and characteristically forms a eutectic system with one of the component metals, as illustrated in the figure.

2.4.3.6 Real Systems

Some real binary alloy systems exhibit simple phase relationships. Examples are copper–nickel and silver–gold (solid solutions), lead–tin, and aluminium–silicon (single eutectic systems). More generally, alloy systems are complex combinations of solid solutions, eutectics, eutectoids, peritectics and compounds. Convenient sources of diagrams for binary systems are Hansen and Anderko* and *Smithells Metals Reference Book**. Some examples are described in Chapters 7 to 12, dealing with corrosion behaviour in specific systems.

* Cited in Further Reading.

2.4.4 Structural Artefacts Introduced during Manufacture

Commercially supplied metals inevitably bear imprints of the manufacturing processes by which they are produced. Artefacts in final products can be acquired from any stage of the production route. They include detritus and gases incorporated in liquid metal prepared for casting, the effects of selective freezing and solidification contraction in castings or ingots and features imparted by mechanical working. The following descriptions summarize various artefacts, which, if uncontrolled, can compromise the expected corrosion resistance. The degree to which such artifacts prevail is described by the semi-quantitative term *metal quality*. It is as much a feature of real metal structures as the intrinsic characteristics summarized above and insistence on good metal quality is an important aspect of corrosion control.

2.4.4.1 Structural Features of Castings and Ingots

2.4.4.1.1 *Effects of Selective Freezing*

Metal crystals grow from the liquid in a characteristic tree-like morphology with orthogonal axes and are called *dendrites*. In real casting processes, there is insufficient time for equilibrium to be maintained and the solidified metal crystals have composition gradients from their centers to their peripheries, called *coring*. Movement of the liquid during freezing usually also produces differences in composition on a macroscopic scale, known as *segregation*. Concentration effects of selective freezing may alter the proportions of phases present in an alloy and even produce phases in the spaces between the dendrites which would not exist in the alloy at equilibrium.

2.4.4.1.2 *Contraction During Solidification*

In poorly designed castings or ingots, the liquid cannot flow to compensate for contraction of the metal accompanying solidification, causing cavities or internal tears. Casting also imposes severe thermal gradients in the metal, causing differential contraction of the solidified metal, which can induce stresses sufficient to crack alloys with insufficient ductility to relax them.

2.4.4.1.3 *Gases Occluded in the Structure*

If, by default, liquid metals or alloys contain dissolved gases or their progenitors, gases evolved in the solidifying structure are trapped in irregular spaces between dendrites, generating *interdendritic porosity*; examples include hydrogen in aluminium alloys, carbon monoxide in some steels and steam in copper.

2.4.4.1.4 *Oxide and Other Extraneous Inclusions*

The surfaces of liquid metals prepared for casting are almost always covered with either a thick oxide dross or a flux or slag cover to prevent oxidation. Unless great care is exercised to prevent the liquid from enveloping its own surface during transfer of the liquid to ingot molds or castings, pieces of oxide, slag or detritus from the surfaces over which the liquid flows become included in the metal structure. If they outcrop through the metal surface, these inclusions can be very potent sites of local corrosion.

Artifacts of the kind described diminish the quality of near net shape castings and if present in ingots for subsequent mechanical working they can persist in modified forms to fabricated products, leading to surface blemishes and internal cracks in sheet, plate or extrusions.

2.4.4.2 Characteristics Imparted by Mechanical Working

2.4.4.2.1 Dispersion of Cast Structure

In products fabricated by extensive unidirectional deformation of cast ingots at high temperatures, such as hot-rolled plates or extruded sections, the metal recrystallizes continuously as hot-working proceeds, replacing the cast dendritic crystals with polygonal crystals which become elongated in the direction of working. At the same time, deformation of the structure by compression across the working direction greatly diminishes the distances over which chemical concentration gradients prevail, facilitating dispersion of heterogeneities introduced during solidification. As a consequence, worked materials have more uniform structures than the ingots from which they are produced.

2.4.4.2.2 Texture, Preferred Orientation and Work Hardening

Thin metal sheet is produced by cold-rolling thicker hot-rolled material. The metal does not recrystallize during cold reduction and becomes harder, that is, it *work hardens*. At the same time, the crystals align with their lattice planes of easiest slip in the working direction, so that the product is anisotropic. It is said to exhibit *preferred orientation* and the degree and direction of the preferred orientation is referred to as the *texture of the material*. Cold-rolled metal can be softened for further fabrication by *annealing*, that is, by reheating it to a moderate elevated temperature. Annealing does not remove texture but may change its orientation referred to the original working direction.

Further Reading

Azaroff, L. V., *Introduction to Solids*, McGraw-Hill, New York, 1960.

Bodsworth, C., *The Extraction and Refining of Metals*, CRC Press, Boca Raton, Florida, 1994, Chap. 2.

Bokris, J. O'M. and Reddy, A. K. N., *Modern Electrochemistry*, Plenum Press, New York, 1970, Chaps. 2. and 7.

Brandes, E. A., *Smithells Metals Reference Book*, Butterworth, London, 1983, Chap. 11.

Butler J. N., *Carbon Dioxide Equilibria and their Applications*, Addison-Wesley, Reading, Massachusetts, 1982.

Cotton, F. A. and Wilkinson, G., *Advanced Inorganic Chemistry*, John Wiley, New York, 1980, Chap. 1.

Coulson, C. A., *Valence*, Oxford University Press, London, 1952.

Davies, C. W., *Electrochemistry*, George Newnes, London, 1967.

Franks, F., *Water*, The Royal Society of Chemistry, London, 1984.

Hansen, M. and Anderko, K., *Constitution of Binary Alloys*, McGraw-Hill, New York, 1957.

Hume-Rothery, W., *Atomic Theory for Students of Metallurgy*, Institute of Materials, London, 1955.

Kubaschewski, O. and Hopkins B. E., *Oxidation of Metals and Alloys*, Butterworths, London, 1962, Chap. 1.

Lewis, G. N., Randall, M., Pitzer, K. S. and Brewer, L., *Thermodynamics*, McGraw-Hill, New York, 1961, Chap. 20.

Pauling, L., *The Nature of the Chemical Bond*, Cornell University Press, Ithaca, New York, 1967.

Smallman, R., *Modern Physical Metallurgy*, Butterworths, London, 1985.

Swalin, R. A., *Thermodynamics of Solids*, John Wiley, New York, 1962, Chaps. 13 and 15.

Taylor, A., *An Introduction to X-Ray Metallography*, Chapman & Hall, London, 1952.

3

Thermodynamics and Kinetics of Corrosion Processes

3.1 Thermodynamics of Aqueous Corrosion

Thermodynamics provides part of the scientific infrastructure needed to evaluate the course and rate of corrosion processes. Its principle value is in yielding information on intermediate products of the complementary anodic and cathodic partial reactions that together constitute a complete process. The structures and characteristics of these intermediate products can control the resistance of a metal surface to attack within such wide limits as to make the difference between premature failure and sufficient life for its practical function.

An illustrative example concerns the behaviour of iron in water containing dissolved oxygen. In neutral water, the anodic reaction produces a soluble ion, Fe^{2+}, as an intermediate product and although the final product is a solid, it is precipitated from solution and does not protect the metal, whereas in alkaline water, the anodic reaction converts the iron surface directly into a thin protective solid layer. Such effects can be explained by thermodynamic assessments.

Classic works develop the principles of chemical thermodynamics *ab initio,* but a concise summary of the background to the basic relations needed is given by Bodsworth in a companion volume of the present CRC series and is recommended to the reader.

3.1.1 Oxidation and Reduction Processes in Aqueous Solution

Corrosion science is frequently concerned with the exchange of electrons between half-reactions, that is, anodic reactions that produce them and complementary cathodic reactions that consume them, and some means of accounting for the electrons is essential. Such an accounting system is based on the concept of oxidation that has wider implications in inorganic chemistry than the name suggests. It is approached through the idea of oxidation states.

3.1.1.1 Oxidation States

The concept of oxidation states expresses the combining power exercised by elements in their compounds and is thus related to their valences. It can be illustrated by comparing the ratios of metal atoms to oxygen atoms in some examples of metal oxides. These ratios, stoichiometric ratios, for Na_2O, MgO, Al_2O_3 and TiO_2, are 2:1, 1:1, 2:3 and 1:2, respectively. The differences in the relative amounts of oxygen with which these metals are combined are expressed on a scale of oxidation states in which the numbers +I, +II, +III and +IV are assigned to sodium, magnesium, aluminium and titanium when present in their oxides. The number applies to the state of combination and not to the element itself and so the oxidation state of an uncombined element is zero. Many elements, including metals in the transition series, can exercise more than one oxidation state. The oxidation state of iron is +II in FeO but +III in Fe_2O_3. For copper it is +I in Cu_2O but +II in CuO.

Metal oxides are formed by transfer of electrons from the electropositive metal atoms to the electronegative oxygen atoms, converting them to cations and anions, respectively. Every increment of one in the oxidation state of a metal represents the loss of an electron and corresponds to oxidation of the metal. The electrons lost by the metal are gained by the oxygen. This constitutes complementary reduction of the oxygen and by counting electrons, it is found that its oxidation state is reduced from zero to −II.

The idea of oxidation and reduction is so useful that the concept of oxidation states is extended to cover other compounds containing elements that differ significantly in electronegativity, whether or not they contain oxygen and whether the transfer of electrons is complete or partial. Thus in the reaction between sodium and chlorine to produce sodium chloride, the sodium is said to be oxidized since it loses its valence electron to chlorine and in the reaction between hydrogen and bromine to form hydrogen bromide, the hydrogen is similarly said to be oxidized although the bond does not have strong ionic character.

The common oxidation states of some commercially important metals and corrosion products in aqueous systems are given in Tables 3.1 and 3.2.

The oxidation state of an element changes when it absorbs or releases electrons by transfer of electrons between chemical species. For example, iron (II) hydroxide in an alkaline solution is oxidized by dissolved oxygen to iron (III) oxide, a component of rust, by the reaction:

$$4Fe(OH)_2 + O_2 = 2Fe_2O_3 \cdot H_2O + 2H_2O \tag{3.1}$$

TABLE 3.1

Oxidation States of Some Selected Metals

Metal	Oxidation State	Oxides	Hydroxides	Aquo Ions	Others
Aluminium	III	Al_2O_3	$Al(OH)_3$	Al^{3+}, AlO_2^-	AlO(OH)
Chromium	III	Cr_2O_3	$Cr(OH)_3$	Cr^{3+}	–
	VI	CrO_3	–	CrO_4^{2-}, $Cr_2O_7^{2-}$	–
Copper	I	Cu_2O	–	Cu^+ unstable	–
	II	CuO	$Cu(OH)_2$	Cu^{2+}	$CuCO_3 \cdot Cu(OH)_2$
Iron	II	FeO	$Fe(OH)_2$	Fe^{2+}	–
	III	Fe_2O_3	$Fe_2O_3 \cdot nH_2O$	Fe^{3+}	FeO(OH)
	II + III	Fe_3O_4	–	–	–
Lead	II	PbO	$Pb(OH)_2$	Pb^{2+}	–
	IV	PbO_2	–	PbO_3^{2-}	–
	II + IV	Pb_3O_4	–	–	–
Magnesium	II	MgO	$Mg(OH)_2$	Mg^{2+}	$MgCO_3$
Nickel	II	NiO	$Ni(OH)_2$	Ni^{2+}	–
Tin	II	SnO	$Sn(OH)_2$	Sn^{2+}	–
	IV	SnO_2	–	SnO_3^{2-}	–
Titanium	II	TiO	$Ti(OH)_2$	Ti^{2+}	–
	III	Ti_2O_3	$Ti(OH)_3$?	Ti^{3+}	$Ti_2O_3 \cdot nH_2O$
	IV	TiO_2	none	TiO^{2+}	$TiO_2 \cdot nH_2O$
Zinc	II	ZnO	$Zn(OH)_2$	Zn^{2+}, ZnO_2^{2-}	–

TABLE 3.2

Oxidation States of Some Nonmetallic Elements of Interest in Corrosion

Element	Oxidation State	Ions
Chlorine	–I	Chloride, Cl^-
Nitrogen	–III	Ammonium, NH_4^+
	+III	Nitrite, NO_2^-
	+V	Nitrate, NO_3^-
Oxygen	–II	Hydroxide, OH^-
Sulfur	–II	Hydrogen sulfide, HS^-
	+IV	Sulfite, SO_3^{2-}
	+VI	Sulfate, SO_4^{2-}

The overall change in oxidation state in a completed reaction must be zero so that in Equation 3.1 as written, 4 iron atoms all increase in oxidation state from +II to +III and this is balanced by a reduction in the oxidation state of the two atoms in molecular oxygen from 0 to –II in the form of the oxide. This illustrates the principle that in any complete system an oxidation reaction must be compensated by an equivalent reduction reaction.

3.1.1.2 Electrodes

The oxidation state of species can also be changed by the intervention of an electrically conducting surface at which electrons can be transferred. Such a system is an electrode. The process proceeds if an electron source or sink, such as another electrode transferring an opposite charge to the same surface, removes the charge transferred.

Electrodes coupled in this way sustain the active corrosion of metals. At one electrode, surface atoms of a metal, for example iron, dissolve as ions, raising the oxidation state of the iron from 0 to II, leaving electrons as an excess charge in the metal:

$$Fe = Fe^{2+} + 2e^- \tag{3.2}$$

At the other electrode, the electrons are absorbed by a complementary reaction on the same surface, where some other species, for example dissolved oxygen, are reduced:

$$½O_2 + H_2O + 2e^- = 2OH^- \tag{3.3}$$

reducing the oxidation state of oxygen from 0 for the element to –II in the hydroxyl ion. The balance of electrons between Equations 3.2 and 3.3 produces Fe^{2+} and OH^- ions in the correct ratio for precipitation of sparingly soluble $Fe(OH)_2$ that continuously removes the ions from solution, allowing the reaction to continue.

$$Fe^{2+} + 2(OH)^- = \downarrow Fe(OH)_2 \tag{3.4}$$

Summation of Reactions 3.2 through 3.4 yields the complete process:

$$Fe + ½O_2 + H_2O = Fe(OH)_2 \tag{3.5}$$

To proceed further, the characteristics of electrode processes must be examined.

3.1.2 Equilibria at Electrodes and the Nernst Equation

In an isolated electrode process, the charge transferred accumulates until equilibrium is established. As an example, consider the dissolution of a metal, M, in an aqueous medium to produce ions, M^{z+}:

$$M \rightarrow M^{z+} + ze^- \tag{3.6}$$

The accumulation of electrons in the metal establishes a negative charge on the metal relative to the solution, creating a potential difference opposing further egress of ions and promoting the reverse process, that is, the discharge of ions and their return to the metal as deposited atoms. A dynamic equilibrium is established when the metal has acquired a characteristic potential relative to the solution. Conditions for equilibrium at a given constant temperature are derived from a form of the Van't Hoff reaction isotherm:

$$\Delta G = \Delta G^{\circ} + RT \ln J \tag{3.7}$$

where J is the activity quotient corresponding to a free energy change, ΔG, and ΔG° is the free energy change for all reactants in their standard states:

$$J = \frac{a(\text{product } 1) \times a(\text{product } 2) \times \text{etc.}}{a(\text{reactant } 1) \times a(\text{reactant } 2) \times \text{etc.}} \tag{3.8}$$

In a chemical change without charge transfer, Equation 3.7 does not specify an equilibrium condition, but in an electrode process, ΔG is balanced by the potential that the electrode acquires. This is expressed in electrical terms by replacing the Gibbs free energy terms, ΔG and ΔG° in Equation 3.7 with the corresponding potential terms, E and E°, given by:

$$\Delta G = -zFE \quad \text{and} \quad \Delta G^{\circ} = -zFE^{\circ} \tag{3.9}$$

where the potential, E°, corresponding to ΔG°, is called the standard electrode potential.

This yields the Nernst equation:

$$E = E^{\circ} - \frac{RT}{zF} \ln J \tag{3.10}$$

The equation is often applied at ambient temperature. Taking this as 298 K (25°C), inserting the values for $R = 8.314$ J mol^{-1} and $F = 96490$ coulombs mol^{-1} and converting to common logarithms:

$$E = E^{\circ} - \frac{0.0591}{z} \log J \tag{3.11}$$

Since any particular electrode process proceeds by the gain or loss of electrons, that is, by reduction or oxidation, it is often convenient to use the term, redox potential, to describe the potential of the system at equilibrium for the prevailing activities of participating species. It is simply an abbreviated form of the expression, reduction/oxidation potential.

3.1.3 Standard State for Activities of Ions in Solution

The matter of selecting standard states for the activities of reacting species to be inserted in the activity quotient, J, must now be addressed. There is no problem in selecting the standard state for the solvent, water, or for pure solid and gaseous phases, for example a metal, a solid reaction product, hydrogen or oxygen, since it is convenient to use the usual definition of the standard state as the pure substance under atmospheric pressure, but two problems must be addressed in selecting a standard state for ions in solution:

1. Ions do not exist in isolation as 'pure substances' so that an arbitrary solution is needed as the standard state. Consistency with standard states often used for nonionic solutions might suggest that this should be based on unit molality, that is, one mole of solute in 1 kg of solvent.
2. Since ions are charged particles with repulsion between like ions and attraction between unlike ions, their activities in solutions are nonlinear functions of the quantities present. Thus selection of the concentrated solutions corresponding to unit molality as the standard state would extend the inconvenience of a composition-dependent activity coefficient to dilute solutions that are frequently of interest.

The problems are resolved by choosing a standard state that is convenient for dilute solutions in which ions are separated by sufficient distances to limit interaction between them to a negligible extent, so that the activities are linear functions of molality. This facilitates treatment of many solutions encountered in electrochemistry, including corrosion processes. The formal rigorous definition is that the standard state corresponds to that solution that has the effect that activity equals molality at infinite dilution.

3.1.4 Electrode Potentials

3.1.4.1 Convention for Representing Electrodes at Equilibrium

The equation for an electrode process is intuitively visualized as written for the direction in which it proceeds. For example, if interest centers on the active corrosion of a metal such as nickel, it could be written in the direction of oxidation, as an anodic reaction with electrons appearing on the right-hand side:

$$Ni \rightarrow Ni^{2+} + 2e^- \tag{3.12a}$$

Alternatively, if interest centers on the electrodeposition of nickel it might seem more appropriate to write it as a cathodic reaction, in the direction of reduction, with electrons appearing on the left-hand side:

$$Ni^{2+} + 2e^- \rightarrow Ni \tag{3.12b}$$

Similarly, the reduction of oxygen complementing the anodic dissolution of metals in aerated water can be written as a cathodic reaction:

$$\tfrac{1}{2}O_2 + H_2O + 2e^- \rightarrow 2OH^- \tag{3.13a}$$

but the reverse process is an anodic reaction generating oxygen in the electrolysis of water:

$$2OH^- \rightarrow ½O_2 + H_2O + 2e^- \quad (3.13b)$$

The matter is important because the formal direction of the reaction determines whether the signs for equilibrium electrode potentials, E', and standard electrode potentials, E°, are positive or negative. It is resolved by a mandatory international convention that an electrode process at equilibrium is formally written in the direction of reduction, that is, with the electrons on the left-hand side, for example:

$$Ni^{2+} + 2e^- = Ni \quad (3.14)$$

Some caution is needed in reading some excellent textbooks that do not observe the convention because they were written before it was established.

3.1.4.2 Choice of a Potential Scale

It is possible to estimate the absolute values of the potentials established at electrodes indirectly by summation of the free energy changes entailed, but such estimates lack the precision needed for both practical and theoretical study. This presents no real difficulty because the important quantities are not absolute potentials but relative potentials between electrodes. These relative potentials are obtained by referring electrodes to a scale in which a standard reference electrode is arbitrarily assigned the value zero.

The requirements of the standard are:

1. It must be internationally agreed.
2. It must be easily reproducible.
3. It must not be susceptible to interference from side reactions.

The standard selected is the standard hydrogen electrode. Reasons for selecting it are deferred pending a discussion of electrode kinetics, but it is defined empirically as follows.

3.1.4.3 The Standard Hydrogen Electrode

The standard hydrogen electrode is the reaction:

$$2H^+ + 2e^- = H_2 \quad (3.15)$$

conducted at 25°C, with reacting species in their standard states, that is, with a_{H_2} and a_{H^+} both equal to unity. It is physically realized by a system in which an inert metal, platinum, is partially immersed in an acid solution containing unit activity of hydrogen ions. The acid is hydrochloric acid at 298 K (25°C), corresponding to a molality of about 1.2 to reproduce the standard state for hydrogen ions. The system is enclosed in a glass vessel and a stream of pure hydrogen at atmospheric pressure is passed through the acid near the platinum. The platinum is 'platinized', that is coated with finely divided platinum powder, giving a very large effective surface on which a layer of hydrogen absorbs. This yields a hydrogen surface exposed to the acid in intimate contact with the platinum that can conduct electrons to or from it.

The standard hydrogen electrode is arbitrarily assigned the potential 0.000 V and an electrode potential referred to it is described as on the *standard hydrogen scale* (SHE). The sign of the potential is determined by the direction of the spontaneous reaction at the electrode when it is coupled to a standard hydrogen electrode. If it is in the direction of reduction, the potential is positive and if it is in the direction of oxidation, it is negative, as in the following examples:

1. $Cu^{2+} + 2e^- = Cu$ proceeds to the right if $a_{Cu^{2+}} = 1$
   $2H^+ + 2e^- = H_2$ proceeds to the left
   so that the sign for $Cu^{2+} \rightarrow Cu$ is positive
2. $Fe^{2+} + 2e^- = Fe$ proceeds to the left if $a_{Fe}{}^{2+} = 1$
   $2H^+ + 2e^- = H_2$ proceeds to the right
   so that the sign for $Fe^{2+} \rightarrow Fe$ is negative

The standard hydrogen electrode cell, though accurate, is tedious in use and substandard scales are based on cells that are more convenient for practical work. These include the saturated calomel (Hg_2Cl_2) scale (SCE) and the silver/silver chloride scale (Ag/AgCl). Substandard electrodes are selected to suit particular applications on grounds of sensitivity, temperature dependence, compactness or nontoxicity and it is important not to omit reference to the scale on which a potential is quoted because the zeros for substandard scales are shifted relative to the SHE scale, for example:

$$0.00 \text{ V (SCE)} \equiv +0.244 \text{ V (SHE) at } 25°\text{C}$$

$$0.00 \text{ V (Ag/AgCl)} \equiv +0.222 \text{ V (SHE) at } 25°\text{C}$$

The potential at an electrode depends not only on its nature, but also on the activities of the participating species and on temperature. Thus a table of potentials for different electrodes is significant only if the potentials are given on the same scale (SHE) and for equal activities. The logical choice is for all reactants in their standard states and the potentials for this condition are described as standard electrode potentials, E°. Table 3.3 gives some standard electrode potentials at 25°C of interest in corrosion science. Note that the change in sign on passing through the hydrogen electrode in the series is purely a result of choosing it as the arbitrary standard. In fact, although it is not known precisely, the absolute value of the standard hydrogen electrode potential is about −0.9 V. Standard electrode potentials, E°, follow the temperature dependence of the standard Gibbs free energies to which they correspond.

Electrodes such as the saturated calomel and silver/silver chloride cells used as reference standards are often based on the electrochemical reaction of a sparingly soluble salt. Historically, electrodes of this kind have been used to measure their very small solubilities.

EXAMPLE 1

Calculate the solubility product, K_s, for silver chloride from the measured potential, 0.222 V (SHE) of the silver/silver chloride standard cell, saturated with AgCl.

Solution

The electrode reaction is:

$$AgCl + e^- = Ag + Cl^- \qquad \text{(Reaction 1)}$$

TABLE 3.3

Selected Standard Electrode Potentials at 25°C

Electrode	Standard Electrode Potential E° V (SHE)
$Au^{3+} + 3e^- = Au$	+1.50
$Cl_2 + 2e^- = 2Cl^-$	+1.360
$\frac{1}{2}O_2 + 2H^+ + 2e^- = H_2O$	+1.228
$Br_2 + 2e^- = 2Br^-$	+1.065
$Ag^+ + e^- = Ag$	+0.799
$Hg_2^{2+} + 2e^- = 2Hg$	+0.789
$Fe^{2+} + e^- = Fe^{3+}$	+0.771
$I_2 + 2e^- = 2I^-$	+0.536
$Cu^+ + e^- = Cu$	+0.520
$Cu^{2+} + 2e^- = Cu$	+0. 337
$2H^+ + 2e^- = H_2$	0.000 (by definition)
$Pb^{2+} + 2e^- = Pb$	−0.126
$Sn^{2+} + 2e^- = Sn$	−0.136
$Ni^{2+} + 2e^- = Ni$	−0.250
$Fe^{2+} + 2e^- = Fe$	−0.440
$Cr^{3+} + 3e^- = Cr$	−0.740
$Zn^{2+} + 2e^- = Zn$	−0.763
$Al^{3+} + 3e^- = Al$	−1.663
$Mg^{2+} + 2e^- = Mg$	−2.370
$Na^+ + e^- = Na$	−2.714

and can be considered as the sum of the reactions:

$$AgCl(s) = Ag^+ + Cl^- \qquad \text{(Reaction 2)}$$

$$Ag^+ + e^- = Ag \qquad \text{(Reaction 3)}$$

Since standard Gibbs free energies, ΔG° are additive:

$$\Delta G_1^{\circ} = \Delta G_2^{\circ} + \Delta G_3^{\circ}$$

Applying the Van't Hoff isobar to Reaction 2 for the saturated solution (that is, $a_{AgCl} = 1$) yields:

$$\Delta G_2^{\circ} = -RT\ln(a_{Ag^+})(a_{Cl^-})/a_{AgCl} = -2.303RT\log(a_{Ag^+})(a_{Cl^-})$$

and replacing ΔG_1° and ΔG_3° by the corresponding standard electrode potentials:

$$zFE_1^{\circ} = 2.303RT\log(a_{Ag^+})(a_{Cl^-}) + zFE_3^{\circ}$$

whence: $\log(a_{Ag^+})(a_{Cl^-})\dfrac{zF}{2.303\,RT} = (E_1^{\circ} - E_3^{\circ}) = 16.92(E_1^{\circ} - E_3^{\circ})$ at 298 K

$E_1^{\circ} = 0.222\text{V (SHE)}$, E_1° is given in Table 3.3 as 0.799 V (SHE) and $z = 1$.

Hence, $\log(a_{Ag^+})(a_{Cl^-}) = (0.222 - 0.799)/0.0591 = -9.763$

and $K_s = (a_{Ag^+})(a_{Cl^-}) = 1.726 \times 10^{-10}$

3.1.5 Pourbaix (Potential-pH) Diagrams

3.1.5.1 Principle and Purpose

Pourbaix (or potential-pH) diagrams, named for the originator, are graphical representations of thermodynamic information appropriate to electrochemical reactions. The presentation of information in this format facilitates its application to practical problems in a wide variety, including corrosion, electrodeposition, geological processes and hydrometallurgical extraction processes. A particular diagram is called 'the Pourbaix diagram for the iron–water system', 'the Pourbaix diagram for the zinc–water system', and so on. They are examples of predominance area diagrams, discussed in a wider context by Bodsworth, cited at the end of the chapter.

The objective is to represent the relative stabilities of solid phases and soluble ions that are produced by reaction between a metal and an aqueous environment as functions of two parameters, the electrode potential, *E*, and the pH of the environment. The information needed to construct a Pourbaix diagram is the standard electrode potentials, E°, or the equilibrium constants, *K*, as appropriate, for all of the possible reactions considered. The purpose and construction of these diagrams is best appreciated by considering a particular system in detail.

3.1.5.2 Example of the Construction of a Diagram – The Iron–Water System

The example selected is the iron–water system both for its technical importance and because it exhibits all of the features to be generally found in the diagrams.

3.1.5.2.1 Selection of Species and Reactions

About 20 known reactions can proceed between various species including water, iron metal, Fe^{2+}, Fe^{3+}, Fe_3O_4, Fe_2O_3, $Fe(OH)_2$, FeO_4^{2-}, $HFeO_2^-$ and others. The first task is to select reactions appropriate to the problem in hand. This is not always straightforward. In considering the corrosion of iron, reactions including Fe, Fe^{2+}, Fe^{3+}, Fe_3O_4 and Fe_2O_3 could be selected on the grounds that they are the stable species in the presence of dissolved oxygen. Alternatively, reactions including Fe, Fe^{2+}, Fe^{3+}, $Fe(OH)_2$ and FeO(OH) or some other combination could be selected because hydroxides can form as primary corrosion products, even though they are unstable in the presence of dissolved oxygen and ultimately become converted to hydrated oxides. For illustration, the species Fe, Fe^{2+}, Fe^{3+}, Fe_3O_4 and Fe_2O_3 that yield the most usual version are selected. The reactions are written conventionally and in a form that may include H^+ ions but not OH^- ions, for example, equilibrium between metallic iron and Fe_3O_4 is written:

$$Fe_3O_4 + 8H^+ + 8e^- = 3Fe + 4H_2O \tag{3.16}$$

and not as the mass/charge balance equivalent equation:

$$Fe_3O_4 + 4H^+ + 8e^- = 3Fe + 4OH^- \tag{3.17}$$

Written in this form, the reactions significant for corrosion are:

Reaction 1. $Fe^{2+} + 2e^- = Fe$

Reaction 2. $Fe_3O_4 + 8H^+ + 8e^- = 3Fe + 4H_2O$

Reaction 3. $Fe_3O_4 + 8H^+ + 2e^- = 3Fe^{2+} + 4H_2O$

Reaction 4. $Fe_2O_3 + 6H^+ + 2e^- = 2Fe^{2+} + 3H_2O$

Reaction 5. $3Fe_2O_3 + 2H^+ + 2e^- = 2Fe_3O_4 + H_2O$

Reaction 6. $Fe^{3+} + e^- = Fe^{2+}$

Reaction 7. $2Fe^{3+} + 3H_2O = Fe_2O_3 + 6H^+$

3.1.5.2.2 The Approach to Calculations

In Reactions 1–7, the solid phases, Fe_3O_4 and Fe_2O_3 and the solvent, water, are present at unit activity. The equilibria are, therefore, governed by the following factors:

1. The charge transferred manifests as an electrode potential, E. This applies only to reactions involving electrons, Reactions 1–6, but not Reaction 7.
2. The activity of hydrogen ions, that is, the pH of the solution.
3. The activities of other soluble ions, for example, Fe^{2+} and Fe^{3+}.

For reactions in which charge is transferred, these quantities are related by the Nernst equation, given as Equation 3.11:

$$E = E^{\circ} - \frac{RT}{zF}\ln J$$

If there is no charge transfer, the activities are related by the normal equilibrium constant, for example for Reaction 7:

$$K = \frac{(a_{H^+})^6}{(a_{Fe^{3+}})^2} \qquad (3.18)$$

3.1.5.2.3 Calculations

Equations 3.11 and 3.18 are applied as appropriate to calculate the required relations as follows:

Reaction 1 $Fe^{2+} + 2e^- = Fe$

Information needed: $E^{\circ} = -0.440\,\text{V (SHE)}$

Apply the Nernst equation:

$$\begin{aligned} E &= -0.440 - \frac{0.0591}{2}\cdot\log\frac{1}{a_{Fe^{2+}}} \\ &= -0.440 + 0.0295\log(a_{Fe^{2+}}) \end{aligned} \qquad (3.19)$$

Reaction 2 $Fe_3O_4 + 8H^+ + 8e^- = 3Fe + 4H_2O$

Information needed: $E^{\circ} = -0.085\,V\,(SHE)$

Apply the Nernst equation:

$$E = -0.085 - \frac{0.0591}{8} \cdot \log \frac{1}{(a_{H^+})^8}$$
$$= -0.085 \; -0.0591\,pH \tag{3.20}$$

Reaction 3 $Fe_3O_4 + 8H^+ + 2e^- = 3Fe^{2+} + 4H_2O$

Information needed: $E^{\circ} = +0.980\,V\,SHE$

Apply the Nernst equation:

$$E = +0.980 - \frac{0.0591}{2} \cdot \log \frac{(a_{Fe^{2+}})^3}{(a_{H^+})^8}$$
$$= 0.980 - 0.2364\,pH - 0.0886\,\log(a_{Fe^{2+}})^3 \tag{3.21}$$

Reaction 4 $Fe_2O_3 + 6H^+ + 2e^- = 2Fe^{2+} + 3H_2O$

Information needed: $E^{\circ} = +0.728\,V\,SHE$

Apply the Nernst equation:

$$E = +0.728 - \frac{0.0591}{2} \cdot \log \frac{(a_{Fe^{2+}})^2}{(a_{H^+})^6}$$
$$= 0.728 - 0.1773\,pH - 0.0591\,\log(a_{Fe^{2+}}) \tag{3.22}$$

Reaction 5 $3Fe_2O_3 + 2H^+ + 2e^- = 2Fe_3O_4 + H_2O$

Information needed: $E^{\circ} = +0.221\,V\,SHE$

Apply the Nernst equation:

$$E = +0.221 - \frac{0.0591}{2} \cdot \log \frac{1}{(a_{H^+})^2}$$
$$= 0.221 - 0.0591\,pH \tag{3.23}$$

Reaction 6 $Fe^{3+} + e^- = Fe^{2+}$

Information needed: $E^{\circ} = +0.771\,V\,SHE$

Apply the Nernst equation:

$$E = +0.771 - \frac{0.0591}{1} \cdot \log \frac{(a_{Fe^{2+}})}{(a_{Fe^{3+}})}$$
$$= 0.771 + 0.0591\,\log(a_{Fe^{3+}}) - 0.0591\,\log(a_{Fe^{2+}}) \tag{3.24}$$

Reaction 7 $2Fe^{3+} + 3H_2O = Fe_2O_3 + 6H^+$

Information needed: $(a_{H^+})^6/(a_{Fe^{3+}})^2 = 10^2$

Taking logarithms: $6\log(a_{H^+}) - 2\log(a_{Fe^{3+}}) = 2$

Hence, rearranging: $pH = \{-\log(a_{Fe^{3+}}) - 1\}/3$ (3.25)

3.1.5.2.4 Plotting Lines on the Diagram

Equations 3.19–3.25 contain, in a suitable form, all of the information needed to plot the diagram. Lines are plotted against coordinates, *E* and pH, representing the conditions for which the activities of soluble ions, for example, Fe^{2+} and Fe^{3+}, are at some specified value.

Consider Reaction 1. If the activity of the soluble ion, Fe^{2+}, is arbitrarily chosen to be unity, that is, 10^0, then inserting $a_{Fe^{2+}} = 1$ into Equation 3.19 gives a value for *E* of −0.440 V SHE. For other values of $a_{Fe^{2+}}$, that is, 10^{-2}, 10^{-4} and 10^{-6}, the corresponding values obtained for *E* are −0.50, −0.56 and −0.62 V (SHE), respectively. Note that these values do not depend on pH, since the hydrogen ion, H^+, does not participate in the reaction and pH, therefore, does not appear as a variable in Equation 3.19. Thus the plots for Reaction 1 appear as a family of horizontal lines, corresponding to the selected values of $a_{Fe^{2+}}$, as illustrated in Figure 3.1. As a convenient shorthand, it is conventional to label the individual lines as 0, −2, −4 and −6 to indicate the values for $a_{Fe^{2+}}$ to which they refer, that is, 10^0, 10^{-2}, 10^{-4} and 10^{-6}.

The interpretation to be placed on the plots is as follows. Points on a selected line represent combinations of *E* and pH for which the soluble ion activity *equals* the value to which the line refers, points above the line correspond to conditions for which the activity is higher and points below the line correspond to conditions for which the activity is lower. Thus the line forms the boundary between two regions, an upper region in which the Fe^{2+} ion is the stable species and a lower one in which metallic iron is the stable species, *with respect to the soluble ion activity for which the line is drawn*. This point is made clearer in Section 3.1.5.4, where application of Pourbaix diagrams to corrosion problems is considered.

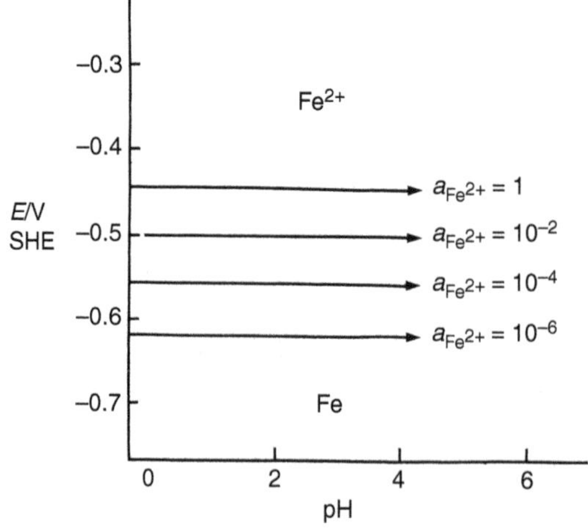

FIGURE 3.1
Potential-pH plots of the equilibrium, $Fe^{2+} + 2e^- = Fe$, for selected values of $a_{Fe^{2+}}$.

3.1.5.2.5 The Complete Diagram

Corresponding series of lines are drawn for Reactions 1–7. Since the reactions are mutually exclusive, they terminate at intersections, as illustrated in the full diagram given in Figure 3.2. The effect of the intersecting lines is to divide the diagram into several *domains*, within which one or another of the species is considered to be stable with respect to the soluble ion activity corresponding to any chosen set of lines.

Figure 3.2 illustrates the following features of Reactions 1–7:

1. Equilibria for Reactions 1 and 6 depend on $a_{Fe^{2+}}$ and E but are independent of pH, because H^+ ions do not participate in the reactions. This yields a family of horizontal lines for the prescribed values of $a_{Fe^{2+}}$.
2. Equilibria for Reaction 7 depend on $a_{Fe^{3+}}$ and pH but are independent of E, because electrons do not participate in the reaction, that is, *it is not an electrode reaction*. This yields a family of vertical lines for the prescribed values of $a_{Fe^{3+}}$.
3. Equilibria for Reactions 3 and 4 depend on all three of the variables, E, pH and $a_{Fe^{2+}}$. This yields families of sloping lines for the prescribed values of $a_{Fe^{2+}}$. The slopes are negative because H^+ ions are on the left side of the reactions as written, leading to negative pHs.
4. Equilibria for Reactions 2 and 5 depend on E and pH but are independent of $a_{Fe^{2+}}$. This yields single sloping lines. The slopes are also negative for the reason given in 3 above.

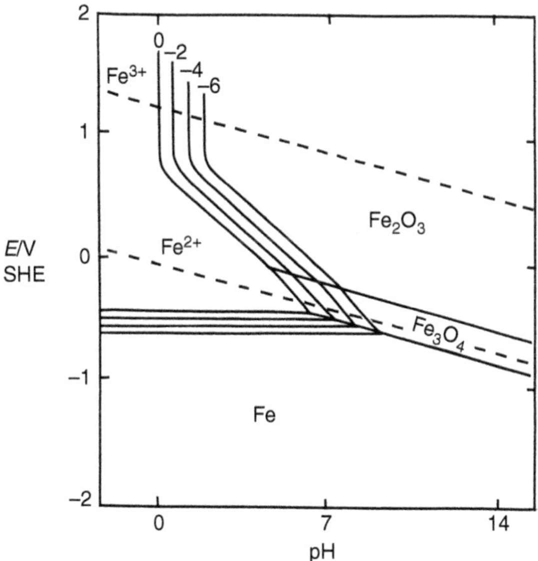

FIGURE 3.2
Pourbaix diagram for the iron–water system at 25°C.
Labels: 0: $a_{Fe^{2+}} = a_{Fe^{3+}} = 1$
−2: $a_{Fe^{2+}} = a_{Fe^{3+}} = 10^{-2}$
−4: $a_{Fe^{2+}} = a_{Fe^{3+}} = 10^{-4}$
−6: $a_{Fe^{2+}} = a_{Fe^{3+}} = 10^{-6}$
Domain for the stability of water shown by dotted lines.

3.1.5.3 The Domain of Stability for Water

Superimposing on a Pourbaix diagram the domain enclosing the combinations of the parameters, E and pH for which water is stable enhances its usefulness. This is defined by lines on the diagram representing the decomposition of water by evolution of hydrogen by Reaction 1 or of oxygen by Reaction 2:

Reaction 1. $2H^+ + 2e^- = H_2 \quad E^\circ = 0.000\,V\,(SHE)$

Reaction 2. $½O_2 + 2H^+ + 2e^- = H_2O \quad E^\circ = +1.228\,V(SHE)$

Note that Reaction 2 is an alternative form of the equation introduced earlier:

$$½O_2 + H_2O + 2e^- = 2OH^-$$

but rewritten in terms of H^+ ions instead of OH^- ions, to permit plotting on the potential pH diagram. The gases are evolved against atmospheric pressure, so that $a_{H_2} = a_{O_2} = 1$.

Application of the Nernst equation yields:

Reaction 1

$$\begin{aligned} E &= 0.000 - \frac{0.0591}{2} \cdot \log \frac{1}{(a_{H^+})^2} \\ &= -0.0591\ \text{pH} \end{aligned} \tag{3.26}$$

Reaction 2

$$\begin{aligned} E &= +1.228 - \frac{0.0591}{2} \cdot \log \frac{1}{(a_{H^+})^2} \\ &= 1.228 - 0.0591\ \text{pH} \end{aligned} \tag{3.27}$$

Equations 3.26 and 3.27 are superimposed as dotted lines on Figure 3.2. The two lines enclose a domain within which water is stable. For combinations of potential and pH above the top line, water is unstable and decomposes evolving oxygen. For combinations below the bottom line, it is also unstable and decomposes evolving hydrogen.

3.1.5.4 Application of Pourbaix Diagrams to Corrosion Problems

The first requirement is to select the appropriate lines from the Pourbaix diagram. By international convention, a metal is considered to be actively corroding if the equilibrium activity of a soluble ion derived from it, for example, $a_{Fe^{2+}}$ or $a_{Fe^{3+}}$ exceeds 10^{-6}. The diagram selected from Figure 3.2 for $a_{Fe^{2+}} = a_{Fe^{3+}} = 10^{-6}$, is given in Figure 3.3. Domains of stability for the species *with respect to* $a_{Fe^{2+}} = a_{Fe^{3+}} = 10^{-6}$, are labelled.

As explained later in discussing kinetics, a metal exposed to an aqueous medium acquires a potential from coupled electrode processes such as those represented in Equations 3.2 and 3.3 that prevail at its surface. This establishes one of the parameters plotted in a Pourbaix diagram. The other parameter is the pH of the aqueous medium. The two parameters define a point in the diagram and the probable response of the metal is indicated by the nature of the stable species within the domain containing the point, as follows:

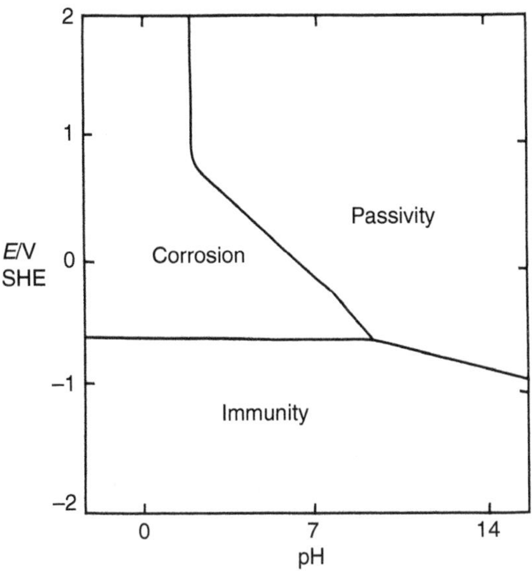

FIGURE 3.3
Pourbaix diagram for the iron–water system at 25°C showing nominal zones of immunity, passivity and corrosion for $a_{Fe^{2+}} = a_{Fe^{3+}} = 10^{-6}$.

1. Domains in which the metal is the stable species so that it is immune to corrosion in the terms for which the diagram is drawn: These domains are *zones of immunity.*
2. Domains in which a soluble ion is the stable species: *Provided that the kinetics are favourable* the metal is expected to corrode. These domains are *zones of corrosion.*
3. Domains in which an insoluble solid, for example, an oxide or hydroxide is the stable species: *If the solid product is formed as an adherent layer, impervious to one or more of the reacting species,* it can protect the metal. These domains are *zones of passivity.*

3.1.5.5 Pourbaix Diagrams for Some Metals of Interest in Corrosion

Figures 3.4 to 3.10 give Pourbaix diagrams for the aluminium, zinc, copper, tin, nickel and titanium water systems, with the domain of stability for water superimposed, constructed using the equations given in the Appendix to this chapter.

Species in the aluminium, zinc, tin and nickel systems relevant to corrosion issues are readily identified, but the selection of appropriate species in copper and titanium must take account of the special features of their aqueous chemistries.

3.1.5.5.1 Species in the Copper–Water System

In the Pourbaix diagram of the copper–water system for $a_{Cu^{2+}} = 10^{-6}$ at 25°C, given in Figure 3.6, the +II oxidation state is represented by both a soluble ion $Cu^{2+}_{(aq)}$ and an oxide, CuO, but the +I state is represented only by the oxide Cu_2O. An associated feature is the retrograde boundary between the domains for Cu^{2+} and Cu_2O. This is because the stability of the Cu^{+} ion is sensitive to its chemical environment as illustrated in the following calculations showing that the activity of the $Cu^{+}_{(aq)}$ ion is unsustainable at the value, 10^{-6}, for which the diagram is prepared, whereas the oxide Cu_2O has a domain of stability.

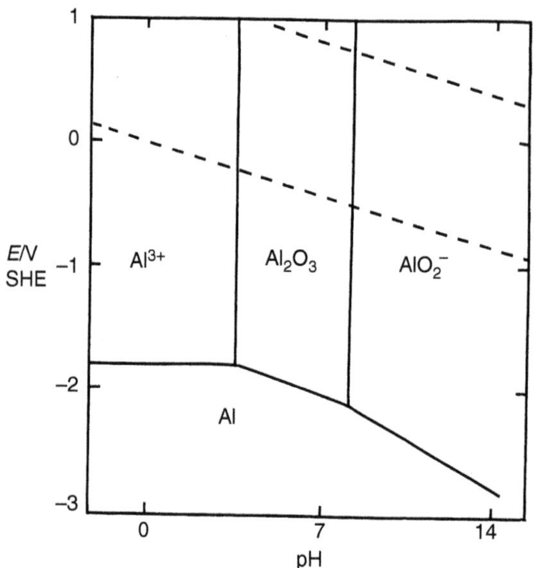

FIGURE 3.4
Pourbaix diagram for the aluminium–water system at 25°C.
$a_{Al^{3+}} = a_{AlO_2^-} = 10^{-6}$.
Domain for the stability of water shown by dotted lines.

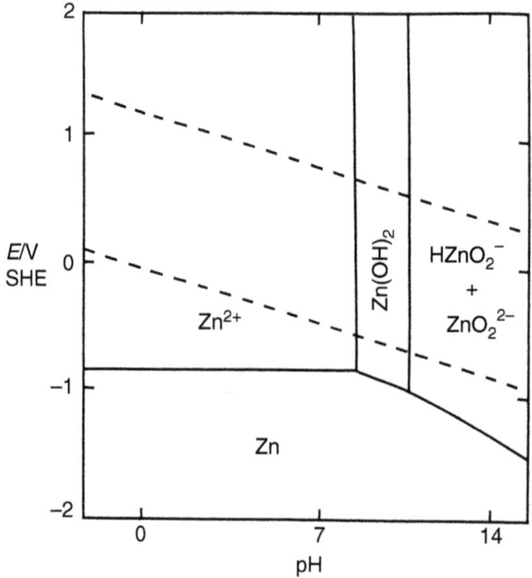

FIGURE 3.5
Pourbaix diagram for the zinc–water system at 25°C.
$a_{Zn^{2+}} = a_{HZnO_2^-} = a_{ZnO_2^{2-}} = 10^{-6}$.
Domain for the stability of water shown by dotted lines.

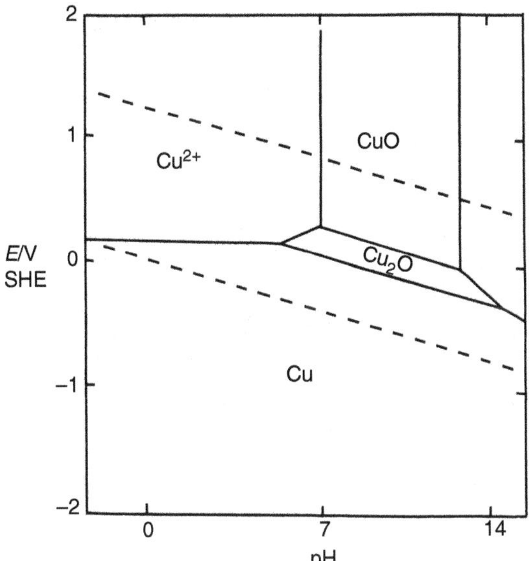

FIGURE 3.6
Pourbaix diagram for the copper–water system at 25°C.
$a_{Cu^{2+}} = 10^{-6}$.
Domain for the stability of water shown by dotted lines.

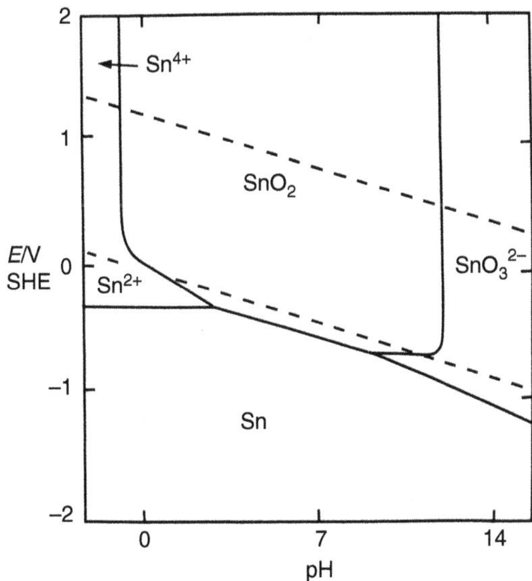

FIGURE 3.7
Pourbaix diagram for the tin–water system at 25°C.
$a_{Sn^{2+}} = a_{Sn^{4+}} = a_{SnO_3^{2-}} = 10^{-6}$.
Domain for the stability of water shown by dotted lines.

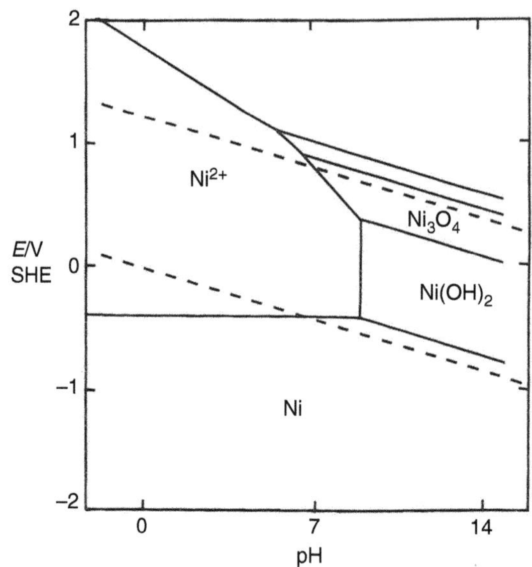

FIGURE 3.8
Pourbaix diagram for the nickel–water system at 25°C.
$a_{Ni^{2+}} = 10^{-6}$.
Domain for the stability of water shown by dotted lines.

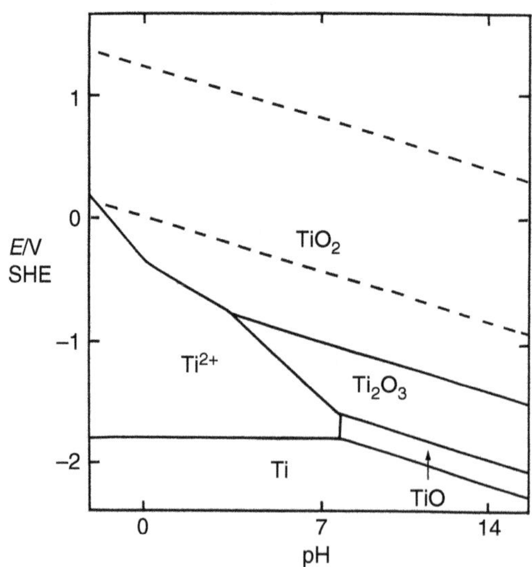

FIGURE 3.9
Pourbaix diagram for the titanium–water system at 25°C considering anhydrous oxides, TiO, TiO_2 and Ti_2O_3.
Notional activities $a_{Ti^{2+}} = a_{Ti^{3+}} = a_{Ti^{4+}} = 10^{-6}$.
Domain for the stability of water shown by dotted lines.

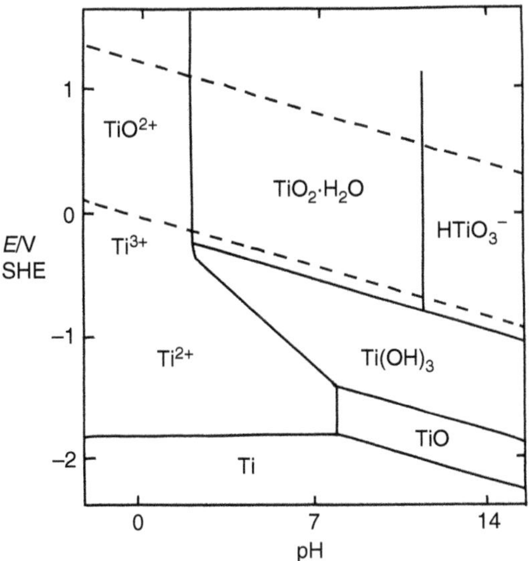

FIGURE 3.10
Pourbaix diagram for the titanium–water system at 25°C considering hydrous oxides, $TiO_2{\cdot}H_2O$ and $Ti(OH)_3$.
Notional activities, $a_{TiO^{2+}} = a_{Ti^{2+}} = a_{Ti^{3+}} = a_{Ti^{4+}} = 10^{-6}$.
Domain for the stability of water shown by dotted lines.

First consider equilibrium between oxidation states I and II for the ions, $Cu^{+}{}_{(aq)}$ and $Cu^{2+}{}_{(aq)}$:

$$2Cu^{+}{}_{(aq)} = Cu^{2+}{}_{(aq)} + Cu_{(s)} \quad \text{(Reaction 1)}$$

The Van't Hoff isobar yields the equilibrium constant from the standard Gibbs free energy:

$$\ln K = \frac{-\Delta G^{\circ}}{RT}$$

The Gibbs free energy for Reaction 1, $(\Delta G^{\circ})_1$, is the weighted sum of the Gibbs free energies for the reactions:

$$Cu^{+} + e^{-} = Cu \quad \text{(Reaction 2)}$$

and:

$$Cu^{2+} + 2e^{-} = Cu \quad \text{(Reaction 3)}$$

that is

$$(\Delta G^{\circ})_1 = 2(\Delta G^{\circ})_2 - (\Delta G^{\circ})_3$$

$(\Delta G^{\circ})_2$ and $(\Delta G^{\circ})_3$ are obtained from standard electrode potentials for Reactions 2 and 3, given in Table 3.3, using the relation, $\Delta E^{\circ} = -zFE^{\circ}$.

Thus:

$$(\Delta G^{\circ})_2 = -(1 \times 96490 \times 0.520) = -50180\,\text{J}$$

and:

$$(\Delta G^{\circ})_3 = -(2 \times 96490 \times 0.337) = -65040\ \text{J}$$

Hence:

$$(\Delta G^{\circ})_1 = 2 \times (-50180) - (-65040) = -35320\ \text{J}$$

The negative value $(\Delta G^{\circ})_1$ indicates that the reaction is spontaneous.

Inserting this value and $R = 8.314$ J mol^{-1} into the Van't Hoff isobar applied to Reaction 1 at $T = 25°\text{C}$ (≡298 K), assuming unit activity for the metallic copper:

$$\ln K = \ln \frac{(a_{Cu^{2+}})}{(a_{Cu^{+}})^2} = \frac{-(-35320)}{8.314 \times 298} = 14.256$$

whence:

$$\frac{(a_{Cu^{2+}})}{(a_{Cu^{+}})^2} = 1.55 \times 10^6$$

Consider a solution in which all of the copper, Cu_T, is assumed to be present initially as Cu^+. On equilibration, every mole of Cu^+ eliminated produces 0.5 mole of Cu^{2+} that is

$$[Cu^{2+}] = 0.5(Cu_T - [Cu^+])$$

where square brackets are concentrations. In dilute solution, $a_{Cu^+} \rightarrow [Cu^+]$ and $a_{Cu^{2+}} \rightarrow [Cu^{2+}]$.

Therefore:

$$0.5(Cu_T - [Cu^+])/[Cu^+]^2 = 1.55 \times 10^6$$

Rearranging yields:

$$3.1 \times 10^6 \times [Cu^+]^2 + [Cu^+] - Cu_T = 0$$

which can be solved to give the concentration of Cu^+ ions in solution for various values of Cu_T. For $Cu_T = 10^{-6}$ mol dm^{-3}, the standard quadratic formula gives:

$$[Cu^+] = \{-1 + (1^2 + 4 \times 3.1 \times 10^6 \times 10^{-6})^{1/2}\}/\{2 \times 3.1 \times 10^6\} = 4.3 \times 10^{-7}\ \text{mol dm}^{-3}$$

Since 57% of the Cu^+ disproportionates at the activity that would be specified in the Pourbaix diagram, it cannot be considered a stable species and no domain can be assigned to it. The disproportionation increases with rising concentrations of copper ions, for example, it is 80% at $Cu_T = 10^{-3}$ mol dm^{-3}, for which $[Cu^+] = 0.02 \times 10^{-3}$ mol dm^{-3}.

Now consider equilibrium between oxidation states I and II for the oxides, Cu_2O and CuO. A hypothetical disproportionation reaction for Cu^+ in Cu_2O, corresponding to Reaction 1 would be:

$$Cu_2O = CuO + Cu \qquad \text{(Reaction 4)}$$

which is half the difference between the reactions:

$$Cu_2O + 2H^+ + 2e^- = 2Cu + H_2O \qquad \text{(Reaction 5)}$$

$$2CuO + 2H^+ + 2e^- = Cu_2O + H_2O \qquad \text{(Reaction 6)}$$

Proceeding as before, using data from calculations for the copper–water system given in the Appendix to this chapter:

$$(E')_5 \text{ (for Reaction 5)} = 0.471 - 0.0591 \text{ pH (SHE)}$$

$$(E')_6 \text{ (for Reaction 6)} = 0.669 - 0.0591 \text{ pH (SHE)}$$

$$(\Delta G)_4 \text{ (for Reaction 4)} = \tfrac{1}{2}\{-zFE'_5\} - \tfrac{1}{2}\{-zFE'_6\}$$

$$= \tfrac{1}{2}\{-2 \times 96490 \times (0.471 - 0.0591 \text{ pH})\} - \tfrac{1}{2}\{-2 \times 96490 \times (0.669 - 0.0591 \text{ pH})\}$$

$$= +19100 \text{ J mol}^{-1}$$

Since all species participating in Reaction 4 are solid substances, assumed pure, all activities are unity and the free energy of the reaction is the standard free energy ΔG°. Since its value is positive, the reaction is not spontaneous so that there is a domain of stability for Cu_2O.

3.1.5.5.2 Species in the Titanium–Water System

Alternative versions of the Pourbaix diagram for the titanium–water system differ in the state of the oxide species considered, that is, *either* the anhydrous oxides TiO_2 and Ti_2O_3 *or* the hydrous oxides derived from them, $TiO_2 \cdot H_2O$ and $Ti(OH)_3$. These versions are illustrated in Figures 3.9 and 3.10, respectively.

The stability of oxidation states II, III and IV for titanium in compounds and complex ions suggests the possible existence of domains for the simple aquo ions, Ti^{2+}, Ti^{3+} and Ti^{4+}, but in practice, only the trivalent ion Ti^{3+} can exist in aqueous solution and even so it is confined to a narrow potential range in highly acidic media. The energy to form a simple tetravalent ion, Ti^{4+}, would be prohibitive due to its high charge density. Cations with tetravalent titanium can exist in oxidizing strongly acidic conditions, but only as oxo ions, for example TiO^{2+}, which is stabilized by the lower charge density due to its greater size and reduced charge. The tetravalent state can be similarly stabilized in complex ions formed in multicomponent aqueous media that by definition do not appear in diagrams for the binary titanium–water system. It is true that there are notional domains for a divalent aquo ion, Ti^{2+}, in the diagrams given in Figures 3.9 and 3.10, but they have no physical reality because they lie well below the domain of stability for water so that there is no aqueous chemistry for this ion.

The principal difference between the diagrams given in Figures 3.9 and 3.10 is that the domain of stability of the passivating anhydrous tetravalent oxide, TiO_2, in Figure 3.9 is much larger than the domain for the corresponding hydrous oxide, $TiO_2 \cdot H_2O$ in Figure 3.10 that is also passivating but which can dissolve in acidic solutions as the oxo ion TiO^{2+}, sometimes considered in association with a water molecule and written $[Ti(OH)_4]^{2+}$. Whichever of the diagrams is most appropriate for any particular application is decided empirically from experience and foreknowledge of the prevailing conditions.

The most impressive feature of titanium is the exceptionally wide range of pH and potential over which it can passivate within the domain of stability for water. These characteristics underlie its outstanding corrosion resistance and its suitability for use in severe conditions that are considered in greater detail together with its metallurgical characteristics in Chapter 12 and which are related to some marine applications in Chapter 25.

3.1.5.6 Limitations of Pourbaix Diagrams

Pourbaix diagrams can be of great utility in guiding consideration of corrosion and other problems, but they apply only for the conditions assumed in their construction and they are not infallibly predictive because they have limitations, as follows:

1. The diagrams are derived from thermodynamic considerations and yield no kinetic information. There are situations in which zones of corrosion suggest that a metal dissolves and yet it does not, due, for example, to the formation of a metastable solid phase or to kinetic difficulties associated with a complementary cathodic reaction. In extremis, some binary systems include species that are ambiguous or so kinetically inert that they can persist almost indefinitely under conditions in which they are thermodynamically very unstable, thereby defeating the purpose of the diagrams. Fortunately, few such systems are important in the present context except for the chromium–water system, which is considered empirically later in relation to electrodeposited coatings and stainless steels.
2. Domains in which solid substances are considered to be stable species relative to arbitrary soluble ion activities $<10^{-6}$ give good indications of conditions in which a metal may be passive. Whether particular metals are actually passivated within these nominal domains and to what extent a useful passive condition can extend beyond their boundaries depends on the nature, adherence and coherence of the solid substance. This reservation is considered further in Sections 9.2.1 and 9.2.2 with respect to the development of passivity on aluminium.
3. The diagrams yield information only on the reactions considered in their construction and take no account of known or unsuspected impurities in the aqueous phase or of alloy components in the metal that may modify the reactions. For example, Cl^- or SO_4^{2-} ions present in solution may attack, modify or replace oxides or hydroxides in domains of passivity, usually diminishing but occasionally enhancing the protective power of these substances. Alloy components can modify surface conditions or introduce microstructural features into the metal that may enhance or destroy passivation according to circumstances.
4. The form and interpretation of a Pourbaix diagram are both temperature dependent, the form because T appears in Equation 3.10 and the interpretation because pH is temperature dependent, as shown in Table 2.5.

3.2 Kinetics of Aqueous Corrosion

In the long term, the degradation of engineering metals and alloys by corrosion is inevitable and so resistance to it is essentially concerned with the rates of corrosion. Reaction rate theory can be quite complicated and is dealt with in specialized texts*, but the following

* For example, Hinshelwood, cited in Further Reading.

brief summary of the essential principles is sufficient to underpin the derivation of some well-known rate equations for electrochemical processes frequently applied in corrosion problems.

The rate of any transformation is controlled by the magnitude of one or more energy barriers that every particulate entity, for example, an atom or an ion, must surmount to transform. These peaks are the energy maxima of intermediate transition states through which the entity must pass in transforming and the energy that must be acquired is the *activation energy*, ΔG^*. The statistical distribution of energy among the particles ensures that at any instant a small but significant fraction of the particles has sufficient energy to surmount the peaks. This fraction and hence the fraction of particles transforming in unit time, the *reaction rate*, r, depends on the value of the activation energy.

An expression for reaction rate that applies to many reactions over moderate temperature ranges is the Arrhenius equation:

$$r = \mathrm{A}\exp\frac{-\Delta G^*}{RT} \tag{3.28}$$

To apply the equation to an electrode process, it must be restated in electrical terms. Ions transported across an electrode carry electric charge, so that the reaction rate, r in Equation 3.28, can be replaced by an equivalent electric current, i. The energy of the process is the product of the charge and the potential drop, E, through which it is carried. Thermodynamic quantities are expressed per mole of substance and because the charge on a mole of singly charged ions is the Faraday, $F = 96490$ coulombs, the Gibbs free energy change, ΔG, of an electrode process is:

$$\Delta G = -zFE \tag{3.29}$$

where z is the charge number on the particular ion species transferred in the process, for example, z is 1 for H^+ or Cl^-, 2 for Fe^{2+}, 3 for Fe^{3+}, and so on.

Replacing r in the Arrhenius equation, Equation 3.28, with an appropriate change of constant:

$$i = k\exp\frac{-\Delta G^*}{RT} = k\exp\frac{zFE^*}{RT} \tag{3.30}$$

3.2.1 Kinetic View of Equilibrium at an Electrode

The equilibrium at an electrode is dynamic and the ionic species are produced and discharged simultaneously at the conducting surface. Taking the dissolution of a metal as a relevant tangible example:

$$\mathrm{M}^{z+} + \mathrm{ze}^- = \mathrm{M} \tag{3.31}$$

the metal dissolves as ions and ions deposit back on the metal surface at equal rates. Since the electrode is at equilibrium, there is no net change in Gibbs free energy, ΔG, in either the forward or reverse process. The chemical free energy change due to the dissolution or deposition of the metal is balanced by an equivalent quantity of electrical work done by the ions in crossing the electric field imposed by the equilibrium electrode potential.

Since they are charged entities, the ion flows constitute two equal and opposite electric currents. The currents leaving and entering the metal, denoted $\vec{i}$ and $\overleftarrow{i}$, respectively, are called the partial currents. Their magnitude at equilibrium is called the exchange current density, i_o:

$$\vec{i} = \overleftarrow{i} = i_o \tag{3.32}$$

The activation energy is the excess energy that must be acquired to transform metal atoms at the metal surface into solvated metal ions. This is because fully solvated ions cannot approach closer to the metal surface than the outer Helmholtz plane because they are obstructed by their own solvation sheaths and the monolayer of water molecules attached to the metal surface, as illustrated in Figure 2.7. Hence, the metal atoms are only partially solvated during the transformation and are, therefore, in a transient higher energy state. The free energy profile for the reaction is shown schematically in Figure 3.11. Applying Equation 3.30 gives the exchange current density as a function of the activation free energy:

$$i_o = \vec{i} = \overleftarrow{i} = k\exp\frac{-\Delta G^*}{RT} \tag{3.33}$$

where:

ΔG^* is the activation energy
T is the temperature
R is the gas constant
k is a constant depending on the process and on the ion activity

3.2.2 Polarization

If equilibrium at an electrode is disturbed, a net current flows across its surface displacing the potential in a direction and to an extent depending on the direction and magnitude of the current. The shift in potential is called polarization and its value, η, is the overpotential. There are three possible components, activation, concentration and resistance polarization.

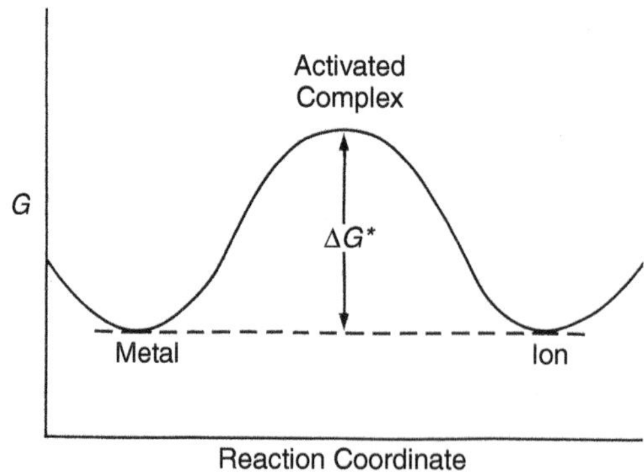

FIGURE 3.11
Schematic energy profile for equilibrium at an electrode.
ΔG^* is the activation Gibbs free energy.

3.2.2.1 Activation Polarization

Activation polarization is a manifestation of the relative changes in the activation energies for dissolution and deposition, when equilibrium is disturbed. It is always a component of the total polarization, whether or not there are also significant contributions from concentration and resistance effects. The polarization is positive, that is anodic, or negative, that is cathodic, according to whether the net current is a dissolution or deposition current. The free energy profile of an electrode subject to activation polarization is shown schematically in Figure 3.12, where the electrode is assumed to be anodically polarized with an overpotential of η. Since $\Delta G = -zFE$, the polarization raises the energy of the metal by $zF\eta$ and that of the activated complex by $\alpha zF\eta$, relative to that of the ions, where α is a symmetry factor defining the position of the maximum in the profile in Figure 3.12, for example, $\alpha = 0.5$ if the maximum is equidistant from the two minima.

At equilibrium, the activation energies for dissolution and deposition both equal ΔG^*. When polarized, the activation energy for dissolution is reduced to $[\Delta G^* - (1 - \alpha)zF\eta]$ and that for deposition is raised to $[\Delta G^* + \alpha zF\eta]$ (from the geometry of the profile in Figure 3.12).

Hence, the partial currents are no longer equal. The dissolution current, $\vec{i}$, is:

$$\vec{i} = k\exp\frac{-\{\Delta G^* - (1-\alpha)zF\eta\}}{RT} \tag{3.34}$$

$$= k\exp\frac{-\Delta G^*}{RT}\exp\frac{(1-\alpha)zF\eta}{RT} \tag{3.35}$$

$$= i_o\exp\frac{(1-\alpha)zF\eta}{RT} \tag{3.36}$$

and by similar reasoning, the deposition current, $\overleftarrow{i}$ is:

$$\overleftarrow{i} = i_o\exp\frac{-\alpha zF\eta}{RT} \tag{3.37}$$

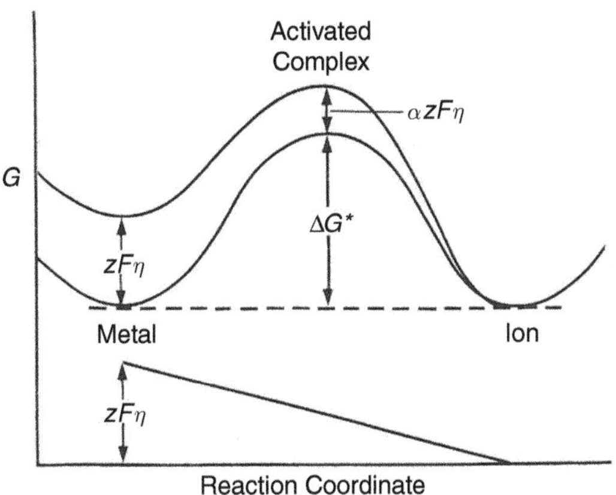

FIGURE 3.12
Schematic energy profile for activation polarization at an electrode.
ΔG^* = Activation Gibbs free energy; η/V = Anodic overpotential.

The net current is the difference between the two partial currents:

$$i_{net} = \vec{i} - \overleftarrow{i} = i_o \exp\frac{(1-\alpha)zF\eta}{RT} - i_o \exp\frac{\alpha zF\eta}{RT} \tag{3.38}$$

This is the *Butler–Volmer* equation expressing the relation between the net anodic current flowing at the electrode, i net and the overpotential, η.

3.2.2.1.1 Tafel or High-Field Approximation

An approximation for Equation 3.38 can be used if the overpotential $\eta > 0.1$ V. For anodic polarization, the value of $\overleftarrow{i}$ is insignificant so that $i_{net} \approx \vec{i}$, and the equation becomes:

$$i_{net} \approx \vec{i} = i_o \exp\frac{(1-\alpha)zF\eta}{RT} \tag{3.39}$$

Since α and z are constant for a given electrode process at constant temperature (usually ambient temperature), this equation can be rearranged as:

$$\eta_{anodic} = \mathrm{b} \log i_o + \mathrm{b} \log i_{anodic} \tag{3.40}$$

For cathodic polarization with $\eta > -0.1$ V, $\vec{i}$ is insignificant, $i_{net} \approx \overleftarrow{i}$, and the equation is:

$$\eta_{cathodic} = \mathrm{b} \log i_o - \mathrm{b} \log i_{cathodic} \tag{3.41}$$

Equations 3.40 and 3.41 are usually combined in the single expression:

$$\eta = \mathrm{b} \log i_o \pm \mathrm{b} \log i \tag{3.42}$$

This *high-field approximation* replicates the equation developed empirically in 1905 by Tafel, and is hence known as Tafel's equation. The constant, b, is the Tafel slope, where:

$$\mathrm{b} = \frac{2.303\,RT}{(1-\alpha)zF} \tag{3.43}$$

3.2.2.1.2 The Symmetry Factor, α

If α is 0.5, Equation 3.38 becomes a hyperbolic sine function:

$$i_{net} = i_o \exp\frac{(1-\alpha)zF\eta}{RT} - i_o \exp\frac{-\alpha zF\eta}{RT} = 2i_o \sinh\frac{zF\eta}{2RT} \tag{3.44}$$

Equation 3.44 can be used to show that the symmetry factor, α, is usually close to 0.5 in the following way. The hyperbolic sine function is symmetrical about the origin. This means that if $\alpha \approx 0.5$, the potential–current relationships are the same for forward and reverse currents, so that the electrode cannot act as a rectifier for AC current. This is the usual situation. However, it is found by experiment that certain electrodes do act as rectifiers and the phenomenon is called Faradaic rectification. This implies that for those particular electrodes the curve is not symmetrical about the origin, and so α is not equal to 0.5.

3.2.2.1.3 Low-Field Approximation

Equation 3.44 can also be used to derive a simple low-field approximation for Equation 3.38. When the overpotential is very small, <0.05 V, the hyperbolic sine function approximates to a linear function of the independent variable, η, so that:

$$i_{net} = 2i_o \sinh \frac{zF\eta}{2RT} \approx \frac{i_o zF\eta}{RT} \tag{3.45}$$

3.2.2.2 Concentration Polarization

As the potential of an electrode is altered further and further away from its equilibrium potential, the net current flowing, whether anodic, i_a, or cathodic, i_c, increases at first according to the Tafel equation, Equation 3.42. However, the current cannot be increased indefinitely because there are limits to the rate at which ions can carry charges through the solution to and from the electrode. This results in excess potential over that predicted by the Tafel equation. The situation is illustrated in Figure 3.13. The effect arises because ions are produced or consumed at the electrode surface faster than they can diffuse to or from the bulk of the solution. In an anodic reaction, the concentration of ions in the immediate vicinity of the electrode is raised above that in the bulk solution. Conversely, in a cathodic reaction, the local concentration is depressed. Thus, the polarization for a given current is greater than that predicted by the Tafel equation. The excess potential is called the *concentration polarization*, η_C. The magnitude of the effect can be examined by applying the Nernst equation, assuming that it is valid in the prevailing dynamic situation.

In the dissolution reaction:

$$M \rightarrow M^{z+} + ze^-$$

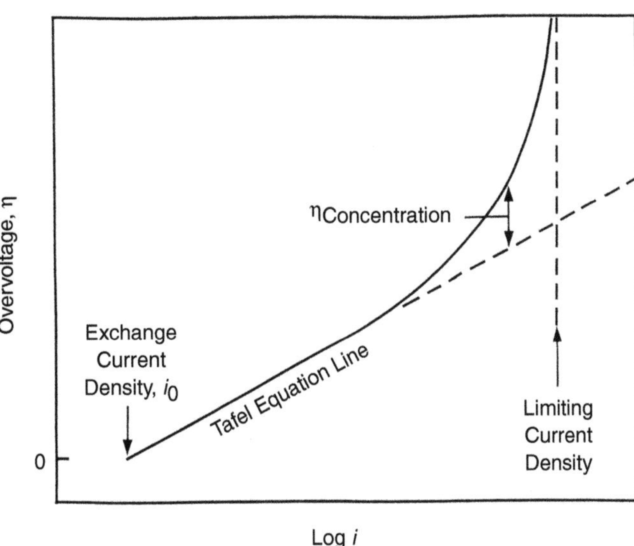

FIGURE 3.13
Logarithmic plot of current density, i, versus overvoltage, η, for a polarized electrode showing a deviation from the Tafel line and limiting current density due to concentration polarization. η is positive for an anodic current, $\vec{i}$, and negative for a cathodic current, $\overleftarrow{i}$.

Let $[a_{M^{z+}}]$ and $[a_{M^{z+}}]_P$, respectively, represent the ion activity in the bulk of the solution and the enhanced activity at the surface of the polarized electrode.

The potential of the unpolarized electrode is the equilibrium potential, E' given by:

$$E' = E^{\ominus} + \frac{0.0591}{z}\log[a_{M^{z+}}] \tag{3.46}$$

When the electrode is polarized, its potential must be referred not to the nominal equilibrium potential but to a higher potential, E'_P, corresponding to equilibrium with the enhanced ion activity at the polarized electrode surface, where:

$$E'_P = E^{\ominus} + \frac{0.0591}{z}\log[a_{M^{z+}}]_P \tag{3.47}$$

For any given current, the concentration polarization, η_C is the difference between the potentials given in Equations 3.47 and 3.46:

$$\eta_c = E'_P - E' = \frac{0.0591}{z}\log\frac{[a_{M^{z+}}]_P}{[a_{M^{z+}}]} \tag{3.48}$$

As the current rises, $[a_{M^{z+}}]_P$ and hence η_C also rises. Eventually, the ion activity at the electrode surface reaches saturation and a limiting current density is reached.

For the reverse (cathodic) reaction:

$$M^{z+} + ze^- \rightarrow M$$

in which metal is deposited, the solution at the electrode surface is depleted of M^{2+} ions and the concentration polarization is in the opposite sense:

$$\eta_c = E' - E_P = \frac{0.0591}{z}\log\frac{[a_{M^{z+}}]}{[a_{M^{z+}}]_P} \tag{3.49}$$

There is a limiting current density because as the deposition current increases, $[a_{M^{z+}}] \rightarrow 0$ and $\eta_C \rightarrow \infty$. The limiting current density for anodic polarization is marked on Figure 3.13.

3.2.2.3 Resistance Polarization

In discussing activation and concentration polarization, ohmic resistances were not considered. For some electrode reactions, ohmic resistances are considerable and especially significant when the reaction itself or a complementary reaction produces films on the electrode surface. The potential drop across such resistance is called *resistance polarization*, η_R.

The total polarization at an electrode is the sum of three components, activation, concentration and resistance polarization:

$$\eta_{TOTAL} = \eta_A + \eta_C + \eta_R \tag{3.50}$$

The effects of these forms of polarization are illustrated by the characteristics of hydrogen evolution and oxygen reduction reactions that feature prominently in corrosion processes.

3.2.2.4 The Hydrogen Evolution Reaction and Hydrogen Overpotential

The hydrogen evolution reaction:

$$H^+ + e^- = ½H_2 \tag{3.51}$$

provides a good subject for exploring characteristics of activation polarization and is a common cathodic reaction supporting the corrosion of metals in acidic aqueous solutions.

In Tafel's equation, Equation 3.42, the constant, b is given by:

$$b = \frac{2.303RT}{(1-\alpha)zF} \tag{3.52}$$

The value of z for the reaction is 1 and provisionally taking the most probable value for α as 0.5, as suggested in Section 3.2.2.1, the theoretical value of b at 298 K is:

$$b = \frac{2.303 \times 8.315 \times 298}{0.5 \times 1 \times 96490} = 0.118 \text{ V/decade} \tag{3.53}$$

The unit, V/decade, appears because η is a logarithmic function of i in the Tafel equation.

EXAMPLE 2

Table 3.4 gives representative results from measurements of potential, *E*, versus current density, *i*, for the evolution of hydrogen on platinum. They apply to 0.1 M hydrochloric acid at 25°C. Do they yield a value for b comparable with the value given in Equation 3.53?

Solution

The equilibrium potential, E', is calculated by applying the Nernst equation. The pH value for 0.1 M hydrochloric acid is close to 1 and at 25°C ($T = 298$ K), the value of the term, $2.303RT/F$ in the Nernst equation is 0.0591. Hence:

$$\begin{aligned} E' &= E^{\circ} - (0.0591/z)\log(1/a_{H^+}) \\ &= 0.00 - 0.591\text{pH} = -0.0591 \times 1 = -0.059 \text{ V (SHE)} \end{aligned} \tag{3.54}$$

The overvoltage, η, is $E - E'$, yielding the values given in the middle row of Table 3.4. These values are plotted as a Tafel plot, η versus log i in Figure 3.14. The plot is linear for $\eta > 0.06$ V, but deviates progressively from linearity as $\eta \to 0$ due to the approximation

TABLE 3.4

Experimental Potential versus Current Relation for Hydrogen Evolution on Platinum

E/V SHE	−0.07	−0.08	−0.12	−0.17	−0.22	−0.27	−0.32	−0.37
η/V	−0.01	−0.02	−0.06	−0.11	−0.16	−0.21	−0.26	−0.31
i/Acm^{-2}	5.0×10^{-4}	1.0×10^{-3}	3.5×10^{-3}	1.0×10^{-2}	3.0×10^{-2}	9.0×10^{-2}	2.5×10^{-1}	6.0×10^{-1}

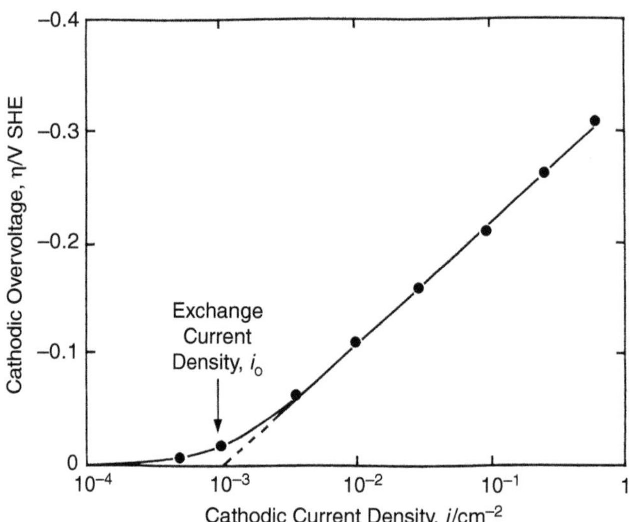

FIGURE 3.14
Tafel plot for hydrogen evolution on platinum from 0.1 M hydrochloric acid at 25°C.

$i_{net} \approx \vec{i}$, implicit in the Tafel equation. This can be exploited to determine the exchange current density, i_o, because extrapolation of the linear part of the plot to $\eta = 0$ disregards the anodic current, $\overleftarrow{i}$, and gives the value of the cathodic current at the equilibrium potential, which by Equation 3.32 is equal to the exchange current density, i_o. From Figure 3.14, this is found to be 1.0×10^{-3} A cm^{-2}. The value of b, found from the slope of the plot is 0.11 V/decade, which is close to the theoretical value given by Equation 3.53, consistent with the assumptions that hydrogen evolution on platinum is a one electron process, that is, $z = 1$, and that the symmetry factor is 0.5. Inserting the values found for b and i_o in Equation 3.41 yields a Tafel equation:

$$\eta_c = -0.33 - 0.11 \log i_c \tag{3.55}$$

This equation applies only to this particular set of results because the value obtained for the exchange current density, i_o is very sensitive to the state of the metal surface.

For metals other than platinum, the hydrogen overpotential is more difficult to assess and its value varies widely from metal to metal. For practical application, the orders of magnitude for different metals are often compared by quoting the overpotential required to evolve hydrogen at an arbitrary current, for example 1 mA cm^{-2}, as in Table 3.5. Hydrogen overpotentials have very important effects in technology. In Chapter 6, it is shown how high hydrogen overpotentials prevent hydrogen evolution during the electrodeposition of metals such as tin and zinc, circumventing practical difficulties that would otherwise arise. Chapter 6 also shows how the high hydrogen overpotential on tin protects it sufficiently to

TABLE 3.5
Hydrogen Overpotentials η/V for Current Density of 1 mA cm^{-2} on Selected Metals

Metal	Platinum	Gold	Nickel	Iron	Copper	Aluminium	Tin	Lead
η/V	0.09	0.15	0.30	0.40	0.45	0.70	0.75	1.0

allow its use as a coating on steel sheets used for cans to conserve mildly acidic foods. The low hydrogen overpotential and high exchange current density on platinum are important factors in the choice and construction of the hydrogen electrode for the standard potential scale, because they contribute to its reproducibility and insensitivity to unwanted side reactions.

3.2.2.5 The Oxygen Reduction Reaction

The reduction of oxygen dissolved in water at metal surfaces exemplifies the influences of concentration and resistance polarization. It is a common cathodic process supporting corrosion of metals because natural waters are constantly replenished with oxygen by recycling through air as rain. In acidic media, the reduction reaction is predominantly:

$$\tfrac{1}{2}O_2 + 2H^+ + 2e^- = H_2O \tag{3.56}$$

for which $E^{\circ} = 1.228$ V SHE, and in neutral and alkaline media it is predominantly:

$$\tfrac{1}{2}O_2 + H_2O + 2e^- = 2OH^- \tag{3.57}$$

for which:

$$E^{\circ} = 0.401\,\text{V SHE}$$

These reactions differ kinetically in the requirement of Reaction 3.56 for a copious supply of hydrogen ions and thermodynamically in the standard states to which activities of the ions are referred. In Reaction 3.56, the standard state is unit activity of hydrogen ions, that is, for pH = 0 at 25°C, but in Reaction 3.57 it is unit activity of hydroxyl ions, that is, for pH = 14 at 25°C, at which temperature $K_W = (a_{H^+}) \times (a_{OH^-}) = 10^{-14}$.

The reduction of oxygen at metal surfaces is difficult to characterize. One problem is its sensitivity to concentration polarization due to the low solubility of oxygen in water. Another is that the conditions of metal surfaces at potentials prevailing during corrosion can differ from those at equilibrium potentials for Reactions 3.56 and 3.57 plotted in the Pourbaix diagrams given in Figures 3.2 to 3.10. These effects are approached quantitatively in Example 3 in Section 3.2.3 after introducing the concept of corrosion potentials.

3.2.3 Polarization Characteristics and Corrosion Velocities

The polarization characteristics for an electrode refer to the relation between current and applied potential, including activation, concentration and resistance contributions. These characteristics differ from electrode to electrode. Some are dominated by activation polarization, others by concentration or resistance polarization as illustrated in the contrasting examples of the evolution of hydrogen on platinum and the reduction of oxygen at an iron surface, just described.

3.2.3.1 Corrosion Velocity Diagrams

Polarization characteristics provide the basis for a convenient graphical method of presenting information on corrosion velocities, introduced by Evans*. The method is to

* Cited in Further Reading.

display on the same diagram the polarization characteristics of all electrode processes that contribute anodic and cathodic reactions in a corroding system. The construction is as follows:

1. The equilibrium potentials, E' for possible reactions are determined, applying the Nernst equation for the prevailing activities. Taking the corrosion of iron in neutral aerated water as an example (neglecting the contribution to the cathodic current from the discharge of hydrogen, that is, found to be small in Example 3 below), the constituent electrochemical reactions, written in the conventional direction, are:

$$Fe^{2+} + 2e^- = Fe \tag{3.58}$$

$$\tfrac{1}{2}O_2 + H_2O + 2e^- = 2OH^- \tag{3.59}$$

2. Assuming that the pH of the water is 7, that it is saturated with oxygen from the air and that by the conventional criterion for corrosion $a_{Fe^{2+}} = 10^{-6}$, application of the Nernst equation yields the values $E'_{Fe^{2+}} = -0.62$ V SHE and $E_{O_2} = 1.0$ V SHE.
3. Axes labelled E and i are drawn as in Figure 3.15, using the symbols, $\vec{i}$ and $\overleftarrow{i}$ to denote anodic and cathodic current densities, respectively.
4. The equilibrium values of E' are marked on the potential axis and the overpotentials are plotted as functions of i for anodic and cathodic polarization.

The corroding metal acquires a steady potential between $E'_{Fe^{2+}}$ and E_{O_2} called the *corrosion potential*, $E_{CORROSION}$. Conservation of electrons determines that the corrosion

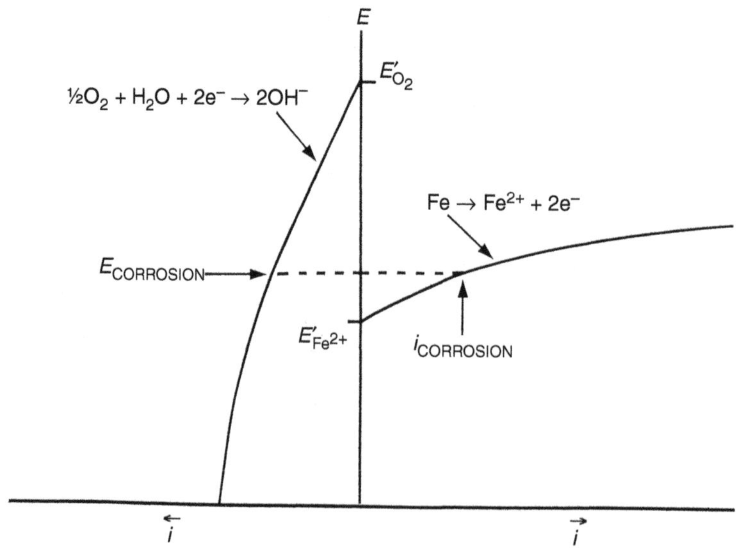

FIGURE 3.15
Schematic corrosion velocity diagram for iron corroding in aerated neutral water. Arbitrary curves indicating high concentration polarization for oxygen absorption. $\vec{i}$ = Current density for anodic reaction; $\overleftarrow{i}$ = Current density for cathodic reaction.

potential has a value at which the total anodic and cathodic currents are equal. The anodic current is called the *corrosion current*, $i_{CORROSION}$, and the metal dissolves at the Faradaically equivalent rate.

Sometimes, two significant cathodic reactions sustain an anodic reaction by which a metal dissolves, for example, in an aerated dilute acid, where both oxygen reduction and hydrogen discharge can contribute to the cathodic current. The corrosion potential then assumes a value at which the anodic current equals the sum of the currents from the cathodic reactions.

EXAMPLE 3

Using the following information, show that the absorption of oxygen is the dominant cathodic reaction sustaining the corrosion of iron in neutral water in equilibrium with air.

1. For the reaction $Fe^{2+} + 2e^- = Fe$:
   $E^{\ominus} = -0.44$ V SHE
   $a_{Fe^{2+}} = 10^{-6}$ (assumed for corrosion by convention)
   Tafel constants: b = 0.06 V/decade, $i_o = 10^{-11}$ A cm^{-2}.
2. For the reaction $2H^+ + 2e^- = H_2$ on iron:
   $E^{\ominus} = -0.00$ V SHE
   Tafel constants: b = 0.12 V/decade, $i_o = 10^{-10}$ A cm^{-2} at pH 7
3. The corrosion potential, $E_{CORROSION}$, on iron in aerated water at pH 7 is −0.45 V SHE

Solution

Anodic Current Density Due to Dissolution of Iron:

Applying the Nernst equation yields the equilibrium potential, $E'_{Fe^{2+}}$:

$$E'_{Fe^{2+}} = -0.44 - \frac{0.0591}{2}\log\frac{1}{10^{-6}} = -0.62\text{ V SHE}$$

Applying the Tafel equation yields the anodic dissolution current for iron:

$$\eta_{ANODIC} = -b\log i_o + b\log i_{ANODIC}$$

Substituting for η_{ANODIC}, b and i_o:

$$E_{CORROSION} - E'_{Fe^{2+}} = -0.06\log 10^{-11} + 0.66\log i_{ANODIC}$$

$$-0.45 - (-0.62) = -0.06\log 10^{-11} + 0.06\log i_{ANODIC}$$

whence,

$$i_{ANODIC} = 6.8\times 10^{-9}\text{ A cm}^2.$$

Cathodic Current Density Due to Discharge of Hydrogen:

Applying the Nernst equation yields the equilibrium potential, E'_{H^+}:

$$E'_{H^+} = 0.00 - \frac{0.0591}{2}\log\frac{1}{(a_{H^+})^2}$$
$$= -0.0591\text{ pH} = -0.0591\times 7 = -0.414\text{ V SHE}$$

Applying the Tafel equation yields the cathodic current due to the discharge of hydrogen:

$$\eta = -b \log i_o + b \log i$$

$$E'_{H_2} - E_{CORROSION} = -0.12 \log 10^{-10} + 0.12 \log i$$

$$-0.414 - (-0.45) = -0.12 \log 10^{-10} + 0.12 \log i$$

whence,

$$i = 2 \times 10^{-10} \text{ A cm}^{-2}$$

Cathodic Current Density Due to Oxygen Reduction:
The anodic current density due to the dissolution of iron equals the sum of the cathodic current densities due to the discharge of hydrogen and the reduction of dissolved oxygen:

$$½O_2 + H_2O + 2e^- = 2OH^-$$

that is, i (dissolution of iron) = i (discharge of hydrogen) + i (reduction of oxygen)

$$6.8 \times 10^{-9} = 2 \times 10^{-10} + i \quad \text{(reduction of oxygen)}$$

whence:

$$i \text{ (reduction of oxygen)} = 6.6 \times 10^{-9} \text{ A cm}^{-2}$$

This is an order of magnitude more than the cathodic current density due to evolution of hydrogen, identifying the reduction of oxygen as the dominant cathodic reaction.

Overpotential for Oxygen Reduction:

For the reaction: $½O_2 + H_2O + 2e^- = 2OH^-$

$$E^\ominus = +0.401 \text{ V SHE}$$

Applying the Nernst equation yields the equilibrium potential:

$$E' = E^\ominus - \frac{0.0591}{z} \log \frac{(a_{OH^-})^2}{(a_{O_2})^{1/2}}$$

$$= +0.401 - \frac{0.0591}{2} \log \frac{(10^{-7})^2}{(0.21)^{1/2}}$$

$$= +0.805 \text{ V SHE}$$

At the corrosion potential the cathodic overpotential, η, for oxygen reduction, is:

$$\eta = E' - E_{CORROSION} = +0.805 - (-0.45) = 1.255 \text{ V}$$

This is a very high overpotential for a cathodic current of only 6.6×10^{-9} A cm^{-2} and implies strong polarization of the reaction. Thus other effects must predominate over activation polarization as predicted by the Tafel equation. These are:

1. Resistance polarization due to a tendency towards film formation and
2. Concentration polarization due to slow diffusion of oxygen, depleting its concentration at the metal surface below the concentration in the bulk of the electrolyte.

These effects conspire to make the corrosion process very sensitive to oxygen concentration.

3.2.3.2 Differential Polarization for Oxygen Reduction and Crevice Corrosion

It is well-known that if oxygen dissolved in neutral water cannot reach part of the surface of an active metal but has unrestricted access to the rest, the oxygen-starved surface can suffer enhanced, sometimes intense, corrosion. The phenomenon is called *differential aeration* and if the oxygen starvation is within crevices, the corrosion is known as *crevice corrosion*.

Crevice corrosion is a common problem but forethought in design and maintenance can eliminate many of its depredations. Crevices can exist through faulty geometric design, lack of full penetration in welds, ill-fitting gaskets and underneath dirt, loose rust or builders' debris. Other differential aeration effects are intensified corrosion at the water line of partly immersed metals and differential corrosion as a function of depth in stagnant water.

Theories of the mechanisms by which differential aeration exerts its effects turn out to be more complex than they first appear and have exercised many minds. The enhanced corrosion of an active metal such as iron in neutral aerated water is evidence that the anodic current density causing dissolution is higher on oxygen-starved surfaces and hence is less heavily polarized than it is on surfaces with open access to oxygen. One possible factor is associated with films produced on an active metal surface by the cathodic oxygen reduction reaction, as illustrated by the calculation given in Section 3.2.3.1. Suppression of film formation in oxygen-starved regions would reduce resistance polarization, thereby locally accelerating anodic dissolution.

Differential aeration can also initiate localized corrosion of passivated metals, but different criteria apply and discussion is deferred to the end of the following general review of passivity after phenomena associated with particular metals and environments have been addressed.

3.2.4 Passivity

In appropriate conditions, some base metals can develop a surface condition that inhibits interaction with aqueous media. The condition is described as passivity and its development is called passivation. The effect is valuable in conferring corrosion resistance on bare metal surfaces even in aggressive environments.

3.2.4.1 Spontaneous Passivation

Some metals passivate spontaneously in water if the pH is within ranges corresponding to potential-independent domains of stability for oxides or hydroxides. These domains appear in the Pourbaix diagrams, for example, within the pH ranges 3.9–8.5 for aluminium illustrated in Figure 3.4, and 8.5–10.5 for zinc illustrated in Figure 3.5. The effect is due to the formation of thin protective oxide or hydroxide films. The existence of a domain of stability for an oxide or hydroxide in the Pourbaix diagram is not the sole condition for passivation. To be protective, the solid product must not only be coherent and adherent to the metal surface, but also its integrity must not be impaired by impurities contributed by the metal or the environment. Foreign ions such as chlorides and sulfates can become incorporated in the passive film, reducing its protective power. Impurities and extraneous inclusions in the metals can contribute minority phases that breach the passive surface. With attention to formulation, metal quality and environmental application, corrosion-resistant alloys based on spontaneous passivation of copper, aluminium, zinc and some other metals are widely applied.

3.2.4.2 Anodic Passivation

For some transition metals and their alloys, passivation is both pH and potential dependent. In these circumstances, the metal may corrode actively at low potentials but can be passivated by raising its potential to a more positive value. The phenomenon is described as *anodic passivation.*

The principle can be illustrated by the behaviour of iron. In aerated neutral water with pH 7, the oxygen absorption reaction polarizes the iron anodically to a potential of approximately −0.45 V SHE, as indicated in Example 3. Reference to the iron–water Pourbaix diagram in Figure 3.2 shows that these coordinates for pH and potential define a point in the domain of stability for Fe^{2+}, so that dissolution of iron is expected as happens in practice. If the anodic polarization is increased to impose a potential more positive than −0.15 V SHE, the new coordinates define a point in the domain of stability for Fe_2O_3, indicating a possible passive condition. This is of limited application for corrosion protection, because the passive condition can be implemented only by imposing the potential needed by adding oxidizing agents to the aqueous environment or by impressing anodic current on iron from an external source. The real value of anodic passivation is realized when the potential at which a metal passivates is low enough to be realized by a cathodic reaction naturally available from the environment in which the metal is required to serve. This is especially true of stainless steels, which are formulated, as described in Chapter 8, to passivate even in moderately strong acids when polarized by the oxygen reduction reaction from oxygen solutions naturally in equilibrium with the atmosphere.

3.2.4.3 Theories of Passivation

Passivation is complex and is difficult to explain. There are two approaches, the film theory, originated by Faraday, and the absorption theory, associated with Kolotyrkin and Uhlig*.

3.2.4.3.1 *Film Theory*

There is experimental evidence for tangible oxide films on some but not all kinds of passivated metal surfaces. When the film is established, its ability to protect the metal depends

* Cited in Further Reading.

on the degree of restraint it imposes on dissolution of the metal. The dissolution is the combined result of three processes:

1. Entry of metal atoms into the film as cations at the metal/film interface
2. Transport of the metal cations or of oxygen anions through the oxide
3. Dissolution of metal cations from the film at the film/environment interface

At ambient temperatures, the processes are driven by the electric field across the film and so the following properties of a film commend it as a passivating agent:

1. Stability over a wide potential range
2. Mechanical integrity
3. Low ionic conductivity
4. Good electron conductivity to reduce the potential difference across the film
5. Low solubility and slow dissolution in the prevailing aqueous medium

3.2.4.3.2 Adsorption Theory

The theory recognizes that strong anodic passivation is a characteristic of alloys of metals in the transition series, notably iron, chromium, nickel, molybdenum and cobalt and it is natural to associate it with the partly filled *3d* or *4d* shells in their electron configurations, described in Chapter 2. An explanation is that singly occupied or unfilled atomic orbitals offer the facility for the highly polarized water molecules and anions in contact with the metal to bind to metal atoms at the surface by sharing electrons, creating the passive condition that protects the metal. A dynamic steady state is envisioned with continuous exchange of molecules between the adsorbed layer and the aqueous environment. Theories of alloying to promote passivating capability by optimizing the bond strength are associated with Uhlig who calculated, for example, that the critical composition for passivation in the iron–chromium binary system is 13.6 at %Cr, which is close to the observed value of 12.7 at %.

The theory can account for passivation, but it must also explain the failure of the metal to passivate at potentials in the active potential range. The difference in behaviour has been attributed to potential-sensitive interaction of susceptible metals with the first row water molecules adsorbed on the surface, described in Section 2.2.8. At the lower potentials in the active range, surface atoms of the metal, M, are assumed to lose electrons and become solvated by first row water molecules, forming soluble cations:

$$M + n(H_2O)_{adsorbed} \rightarrow M(H_2O)_n{}^{z+} + ze^- \qquad (3.60)$$

In the passive range, it is assumed that this interaction is replaced by:

$$M + (H_2O)_{adsorbed} \rightarrow M(O^{2-}) + 2H^+ \qquad (3.61)$$

in which adsorbed water loses protons to become passivating adsorbed anions. The passivating interaction, Equation 3.61, is assumed to begin at an appreciable rate only at the passivating potential, a threshold potential denoting the lower limit of the passive range.

The view that a monolayer of adsorbed anions is sufficient to confer passivity in some circumstances is probably well founded, but ideas of the detailed mechanism are uncertain, because they are undoubtedly complex and difficult to verify by experiment.

3.2.4.3.3 *Compatibility of Film and Adsorption Theories*

The main criticism of the adsorption theory is that although it can explain instantaneous passivation by very thin films, the passive state is more often associated with a film thicker than a monolayer, where the film theory is more appropriate. The two theories are not incompatible and it is best to regard each as revealing part of the truth, the film theory dealing with the nature of the film and the adsorption theory drawing attention to the nature of the metal and to the relation between the kinetics of dissolution and the kinetics of passivation.

3.2.5 Breakdown of Passivity

Metals and alloys that rely on passivity for protection are vulnerable to corrosion failure if the passivity breaks down. The nature and distribution of the ensuing corrosion damage often indicates the cause. If the prevailing conditions cannot maintain passivity, the whole surface becomes active and attack is by uniform dissolution. If isolated sites become active in an otherwise passive surface they are selectively attacked. Local dissolution is often very intense due to the establishment of *active/passive* cells, that is, a galvanic effect with small local anodes at active sites stimulated by an extensive cathode provided by the surrounding passive surface.

These failure modes are now briefly indicated, but they are considered again at greater depth within the contexts of mixed metal systems, SCC and applications of stainless steels, where their effects are more clearly relevant.

3.2.5.1 General Breakdown of Passivity

Failure to establish or maintain the metal in the passive potential range leads to the uniform corrosion of nominally passivating metals. It implies mismatch of the metal or alloy with the environment to which it is exposed. For example, as discussed in Chapter 8, the hydrogen evolution reaction is incapable of imposing a corrosion potential in the passive ranges of stainless steels and the oxygen reduction reaction is essential to establish and maintain passivity. Thus, if a stainless steel is required to resist a nonoxidizing acid it is essential to maintain a sufficient concentration of dissolved oxygen. Assuming correct material selection, general failure by depassivation is often associated with inadvertent failure to replenish the oxygen as it becomes depleted.

For some metals, such as stainless steels, the passive condition is terminated at an upper potential, the *breakdown potential*, and higher potentials are *designated transpassive potentials*. General breakdown of passivity at a transpassive potential is not often a problem because the breakdown potential is usually above corrosion potentials encountered in service.

3.2.5.2 Local Breakdown of Passivity

3.2.5.2.1 *Crevices*

Small quantities of stagnant solutions inside crevices and under shielded areas on a passivated metal surface are selectively depleted of dissolved oxygen because of the difficulty in replenishing it by diffusion from the bulk of the liquid. If, over a period, the oxygen content falls below a critical concentration in such a crevice, the passivity breaks down, establishing an active/passive cell, throwing intense attack on the small local anode thereby created.

3.2.5.2.2 *Pitting*

Pitting corrosion is intense chemical attack at dispersed points on an unshielded passive metal surface, forming pits that can perforate thin gauge metal. They are due to breakdown of passivity at very small isolated sites distributed over the metal surface, creating local active/passive cells. They can be initiated by heterogeneities in the metal surface due to minority phases in the microstructure of the metal as for aluminium alloys or by environmental agents as for stainless steels, notably halide and hypochlorite ions. In particular, the chloride ion imposes severe limitations on the use of passivating metals for service in seawater and in chemical and food processing where the solutions can contain chloride contents in excess of about 0.01 molar, equivalent to a 0.06 mass % solution sodium chloride. Disinfectants containing hypochlorites, ClO^-, are another hazard. If the chloride content is much above 0.1 molar, the tendency to pit may be overtaken by general depassivation. Susceptible metals include not only stainless steels and aluminium, but also some copper-based alloys and mild steels in certain environmental conditions. The highly localized damage can render equipment unserviceable even if attack on the rest of the metal surface is negligible.

3.2.5.3 Mechanical Breakdown of Passivity

Passivity can be broken down by mechanical means to produce damaging effects, by stress-corrosion cracking, corrosion fatigue and erosion–corrosion dealt with in Chapter 5, but mentioned here for completeness. These effects apply to metals generally, whether passivating or not, but the presence of an initial passive condition is an influential factor.

SCC is the result of synergy between sustained mechanical stress and specific agents in the environment, producing cracks causing fracture. For passive metals, applied stress facilitates passivity breakdown at crack sites and prevents repassivation.

Fatigue cracking is caused by application of cyclic stress that first initiates and then propagates a crack. Both initiation and propagation stages are sensitive to environmental intervention, especially in aqueous media, in which event the phenomenon is said to be corrosion fatigue. The mechanical disturbance enhances localized anodic dissolution that concentrates the cyclic stress which in turn amplifies the mechanical disturbance, and so on, setting up a synergistic cyclic mechanism. For passivating metals, the role of mechanical disturbance in stimulating anodic dissolution is to break down passivity during the crack initiation stage and to prevent it reforming at the crack root during propagation.

Erosion–corrosion and impingement describe corrosion damage to passivated surfaces enhanced by the mechanical effects of high velocity or turbulent water flow. These effects are aggravated if the water contains particulates.

3.2.6 Corrosion Inhibitors

Very low concentrations of solutes with particular characteristics can intervene with corrosion kinetics and thereby protect metals from corrosion and are described by the general term, inhibitors. Some occur naturally and others are introduced artificially as a strategy for corrosion control. Examples of their applications include:

1. Preserving existing iron or steel delivery systems for municipal water supplies
2. Maintaining clean thermal transfer surfaces in closed heating and cooling circuits
3. Protecting mixed metal systems
4. Enhancing the protective value of paints

Inhibitors intervene in corrosion kinetics in various ways. Some inhibit cathodic reactions, others inhibit anodic reactions and yet others, *mixed inhibitors*, do both. The detailed mechanisms by which the individual substances produce their effects can be quite complex and are the subject of extensive ongoing research. Nevertheless, information on certain well-established principles is needed to correctly apply inhibitors.

3.2.6.1 Cathodic Inhibitors

Cathodic inhibitors produce deposits on metal surfaces that suppress the cathodic reaction:

$$\tfrac{1}{2}O_2 + H_2O + 2e^- = 2OH^- \quad (3.3)$$

by creating a barrier to oxygen diffusion and preventing transfer of electrons from the metal.

The most familiar cathodic inhibitors are the solutes in hard water supplies that yield scale. In the natural environment, water containing carbon dioxide derived from the atmosphere can dissolve calcium carbonate from calcareous geological material over which it flows, yielding soluble calcium bicarbonate by the reaction described in Section 2.2.9:

$$CaCO_3(s) + CO_2(g) + H_2O(1) = Ca^{2+}(aq) + 2HCO_{3^-}(aq) \quad (2.25)$$

Magnesium carbonate is often associated with calcium carbonate in nature and so some magnesium bicarbonate can be present, produced by an analogous reaction.

In the natural environment, these reactions do not adjust to equilibrium because of difficulty in precipitating the sparingly soluble carbonates, $CaCO_3$ and $MgCO_3$, in open waters. In consequence, hard waters are usually supersaturated by up to an order of magnitude with respect to the carbonates. When the water is abstracted, the surfaces of the systems into which it is transferred can assist nucleation, allowing the reactions to approach equilibrium by precipitating carbonates as a scale on the walls of pipes, tanks and ancillary equipment. The carbonates of some other divalent cations, for example Zn^{2+} and Mn^{2+}, have similar calcite structures so that solutions of their bicarbonates can behave similarly as cathodic inhibitors. Since these cathodic inhibitors depend only on the chemistry of the water, they are applicable to all metals.

Soft waters do not naturally form protective scales but they can be induced to do so by treatment with lime and polyphosphates or silicates that can yield alternative deposits on iron and steel that may include iron oxides, calcium phosphates such as hydroxylapatite, $Ca_5(PO_4)_3OH$ or siliceous material. These materials are mixed inhibitors and are considered again in Section 3.2.6.2 below, in the context of anodic inhibitors.

3.2.6.2 Anodic Inhibitors

Anodic inhibitors suppress anodic reactions by assisting the natural passivation tendencies of metal surfaces or by forming deposits that are impermeable to the metal ions. The attenuation of anodic activity can be confirmed experimentally by a comparison of polarization characteristics for metals in aqueous solutions with and without one of these substances. It is convenient to consider anodic inhibitors in two groups, highly oxidizing and less oxidizing.

3.2.6.2.1 Self-Sufficient Oxidizing Inhibitors

Sodium nitrite and sodium chromate are two of the most effective of all inhibitors. They are oxidizing to iron and are applied to ferrous metals in near neutral solutions. Nitrite, NO_2^- and chromate, CrO_4^{2-}, ions can impose redox potentials on the metal by reactions such as:

$$2NO_2 + 6H^+ + 4e^- = N_2O(g) + 3H_2O \tag{3.62}$$

for which:

$$E^\circ = 1.29\,V\,(SHE)$$

and

$$2Cr^{VI}O_4^{2-} + 10H^+ + 6e^- = Cr^{III}{}_2O_3 + 5H_2O \tag{3.63}$$

for which:

$$E^\circ = 1.33\,V\,(SHE)$$

Even in small concentrations, the available potentials are sufficient to passivate iron by producing a coherent film of γ–Fe_2O_3. This is only a partial explanation for the action of chromates because they are also effective for metals such as aluminium that do not exhibit anodic passivation. Chromium can be identified in the surface film, indicating reinforcement of the existing passive film by the insoluble chromic oxide, $Cr^{III}{}_2O_3$, produced in Reaction 3.63. This topic is considered again in Section 6.4.3.1 in the context of chromate conversion coatings, especially when they are used as underlays to inhibit corrosion by moisture permeating through paints. The additions of sodium nitrite or sodium chromate used depend on water purity and are typically in the range 1 to 10 g dm^{-3} of solution. Contamination with depassivating ions, for example, chlorides increase the quantities needed.

3.2.6.2.2 Inhibitors Assisted by Dissolved Oxygen

Sodium polyphosphates, sodium silicates, sodium tetraborate (borax) and sodium benzoate are nonoxidizing inhibitors used to protect iron and steels in neutral and slightly alkaline waters; they are effective only when the solution contains dissolved oxygen. All of these substances ionize in aqueous solution yielding anions that are the active species at the iron surface with Na^+ counter ions. Polyphosphate ions are polyions containing multiple phosphorus atoms, typically $P_3O_{10}^{5-}$. The tetraborate ion is also a polyion containing four boron atoms, $B_4O_7^{2-}$. The benzoate ion, $(C_6H_5)COO^-$ is a benzene ring with a carboxylic acid group. A feature that these apparently disparate substances have in common is that whereas their sodium salts are very soluble in water, they have the potential to form very insoluble iron (III) phases and hence to lay down deposits on an iron surface. The requirement for dissolved oxygen is probably associated with a need to promote some iron (II) to iron (III) in these phases. Polyphosphates, silicates and tetraborates are used as inhibitors for very large volumes of water such as municipal supplies and concentrations in the range 0.000005 to 0.00002 g dm^{-3} are recommended. Sodium benzoate can be used only for small volumes, as in recirculating systems, because as much as 15 g dm^{-3} may be needed.

Certain other ions, less widely used, for example, vanadates, tungstates and molybdates, require dissolved oxygen to act as inhibitors, even though some of them, especially vanadates, are associated with high redox potentials. They exist in aqueous solution as large polyions and the equilibria between these ions and the insoluble solids that they produce on metal surfaces, such as iron(III) phases, are obscure, not well established and beyond the scope of the present text.

The benefits from many of the anodic inhibitors found to be effective on iron and steel are transferable to other metals such as aluminium, although the mechanisms may differ, but there are some reservations, for example, sodium nitrite can attack lead solders and copper in mixed metal systems, such as automobile cooling systems.

3.2.6.2.3 Safe and Dangerous Inhibitors

The concept of safe and dangerous inhibitors is important in practice. Cathodic inhibitors are safe in the sense that if they are present at insufficient concentration, they simply fail to protect completely and do not stimulate corrosion on unprotected areas. In contrast, anodic inhibitors are dangerous because cathodic reactions are not suppressed on protected areas. If by inadequate initial additions or subsequent poor maintenance the supply of inhibitor fails locally, corrosion is stimulated on the depleted areas because they become anodes in active/passive cells supported by large cathodic currents collected by the passive area.

3.3 Thermodynamics and Kinetics of Dry Oxidation

3.3.1 Factors Promoting the Formation of Protective Oxides

On first exposing a clean metal surface to air at ambient temperature, a thin film of oxide forms within a few minutes, covering the metal and separating it from the hostile gaseous environment. It forms under the influence of an electric field developed by a quantum mechanical effect at the metal surface, explained in the Cabrera–Mott theory introduced briefly in Section 3.3.2, and virtually ceases to grow when it is typically 3–10 nm thick. Although thin, the film is almost impenetrable and the protection it affords explains why many common metals, for example, aluminium, zinc, copper, nickel and even iron are virtually permanent in perfectly dry air at ambient temperature.

At elevated temperatures, reaction continues if the integrity of the film is breached or if metal and/or oxygen atoms diffuse through it at significant rates. In general, metals and alloys can be assigned to one of two classes depending on whether or not their oxide films are protective.

1. *Metals forming protective films*: These metals develop and maintain continuous coherent oxide films adhering to the metal substrate and the oxidation rate is controlled by transport of the metal and/or oxygen through the film itself. The term, protective, is relative because if diffusion is rapid at only moderately elevated temperatures, as it is for metals such as iron and copper, the oxide grows into thick scales, progressively consuming the metal surface. Such metals are unsuitable for high temperature service. Oxidation-resistant metals and alloys, such as stainless steels and alloys based on the nickel-chromium system, are designed *inter alia* for their ability to maintain films sufficiently impermeable to the reacting species to limit the growth of the oxide film in severe high temperature service conditions, such as those prevailing in electric heating elements or gas turbine engines.

2. *Metals forming nonprotective films*: As the oxide thickens, differences between the volumes of oxide and the metal from which it is formed induce tensile or compressive stresses at the metal/oxide interface that can crack or buckle the oxide, so that its protective value is diminished or destroyed. Examples are magnesium, with an oxide/metal volume ratio of 0.8, producing an oxide under tension and uranium, with a volume ratio of >3, producing such a voluminous oxide product that it fails to adhere.

When a metal forms a continuous oxide layer, the oxide protects the metal and the reaction of the metal with air is inhibited. Since the oxide layer separates the reactants, continued growth implies that it is sustained by two synergistic processes:

- Interface reactions by which the metal enters the oxide at the metal/oxide or oxygen enters at the oxide/atmosphere interface
- Diffusion of reactants through the oxide

The rate of oxide growth is controlled by the slower of these two processes.

3.3.2 Thin Films and the Cabrera–Mott Theory

Thin films are characteristic of the almost imperceptible oxidation of engineering metals in dry air at ambient temperatures. Their essential features are:

1. The film growth is initially very rapid, but virtually ceases when the films are only 3–10 nm thick, corresponding to oxide layers of 30–100 atoms.
2. The time-dependence of film thickening, expressed as *growth laws*, is uncharacteristic of diffusion-controlled processes. Examples for different metals include logarithmic, inverse logarithmic and cubic functions of time.

From a practical point of view, these films are vitally important because they confer initial protection on all new surfaces of engineering metals, pending the application of more permanent protection, if required. Theoretical treatments of how the thin films grow are beset by difficulties in acquiring direct information on the structures of such extraordinary thin materials and inconsistencies are found in the growth laws for thin oxides formed on different metals. They are within the remit of physicists, who have exercised great ingenuity in fitting experimentally observed growth laws to theories that are based on short-range electric fields at the metal surface. The basic ideas, due to Cabrera and Mott*, are:

1. Oxygen atoms are adsorbed at the oxide/atmosphere interface.
2. The film is so thin that electrons from the metal pass through it, either by thermionic emission or tunneling (a quantum mechanical effect).
3. The oxygen atoms capture these electrons, becoming anions.

This produces a very strong electric field across the thin film, for example, 10^6 V cm^{-1}, if the film is 10 nm thick, that is mainly responsible for driving metal cations through the

* Cited in Further Reading.

oxide. The short range of electrons available by thermionic emission or tunneling limits the film growth.

3.3.3 Thick Films, Thermal Activation and the Wagner Theory

The growth of thick oxide films at elevated temperatures is more relevant to technological interests in the application and fabrication of metals at high temperatures and it is easier to characterize. Oxidation in these circumstances can be approached by Wagner's theory, which applies to an oxide that is thick enough for equilibrium to be maintained locally at both the oxide/atmosphere and metal/oxide interfaces. According to the theory, the oxide grows by complementary reactions with oxygen at the oxide/atmosphere interface and with the metal at the metal/oxide interface and its rate of growth is controlled by the rate at which reacting species diffuse through the oxide via lattice defects.

The theory must be applied with discretion, because strictly, it represents oxidation processes which fulfill the following implicit assumptions:

1. The oxide remains coherent and adherent to the metal substrate.
2. The growing oxide is a stable isotropic phase, uniform in thickness.
3. Diffusion is via lattice defects in the oxide and is not significantly supplemented by diffusion via alternative paths, for example, crystal boundaries.

With these provisos, the theory can often explain the short-term oxidation behaviour of engineering metals and alloys and it is useful even where it fails to do so, because incompatibility between theory and observation can sometimes identify factors which might otherwise be overlooked.

The particular defects in the oxides determine the reaction mechanisms as explained below.

3.3.3.1 Oxidation of Metals Forming Cation Interstitial (n-type) Oxides

3.3.3.1.1 Atomic Mechanisms

The unit step extending the oxide lattice is adsorption of an oxygen atom from the atmosphere, its ionization to an anion, O^{2-}, and its coordination with a metal cation at the oxide/atmosphere interface. The electrons ionizing the oxygen atom are supplied from the conduction band of the oxide and the cation is extracted from the existing local population of interstitials. Taking zinc as a tangible example, the reaction is:

$$\tfrac{1}{2}O_2 + Zn^{2+}\bullet + 2e\bullet = ZnO \text{ (i.e., } Zn^{2+} + O^{2-}) \tag{3.64}$$

The complementary reaction at the metal/oxide interface is the entry of metal atoms into the oxide as cations and electrons, replenishing the depleted species and consuming the metal:

$$Zn = Zn^{2+}\bullet + 2e\bullet \tag{3.65}$$

Addition of Equations 3.64 and 3.65 yields the overall oxidation reaction:

$$Zn + \tfrac{1}{2}O_2 = ZnO \tag{3.66}$$

3.3.3.1.2 Interface Equilibria

Recalling that the theory assumes local equilibrium at the bounding interfaces of the oxide layer, the equilibrium constant, K, for the reaction given in Equation 3.64, characterizes the defect populations at the oxide/atmosphere and metal/oxide interfaces:

$$K = \frac{1}{[a_{O_2}]^{1/2} \times a_{(Zn^{2+\bullet})} \times [a_{(e\bullet)}]^2} \tag{3.67}$$

Since the defect population is small, $a_{(Zn^{2+\bullet})} \propto n_{(Zn^{2+\bullet})}$ and $a_{(e\bullet)} \propto n_{(e\bullet)}$, where $n_{(Zn^{2+\bullet})}$ and $n_{(e\bullet)}$ are the numbers of interstitial zinc ions and interstitial electrons. Assuming unit activity for ZnO, Henrian activity for the species, $Zn^{2+}\bullet$ and e· and ideal behaviour for oxygen:

$$K = \frac{1}{[p_{O_2}]^{1/2} \times n_{(Zn^{2+\bullet})} \times [n_{(e^\bullet)}]^2} \tag{3.68}$$

From stoichiometric considerations, $n_{(Zn^{2+\bullet})} = \frac{1}{2} n_{(e^\bullet)} = n$. Substituting, rearranging and dropping subscripts, Equation 3.68 becomes:

$$n = p^{-1/6} \cdot k^{-1/3} \tag{3.69}$$

where k includes K and coefficients relating n to $a_{(Zn^{2+\bullet})}$ and $a_{(e\bullet)}$.

The number of interstitials at the oxide/atmosphere interface, n_{atm}, is given by putting p equal to the prevailing oxygen pressure, p_{atm}:

$$n_{atm} = (p_{atm})^{-1/6} \cdot k^{-1/3} \tag{3.70}$$

and the number at the metal/oxide interface, n_o by putting p equal to the dissociation pressure of the oxide, p_o:

$$n_o = (p_o)^{-1/6} \cdot k^{-1/3} \tag{3.71}$$

Oxidation proceeds if the oxygen pressure in the environment exceeds the dissociation pressure of the oxide, that is, if $p_{atm} > p_o$, and since by Equations 3.70 and 3.71, $n_{atm} < n_o$, there is a concentration gradient down which cation interstitials diffuse from the metal/oxide interface, where they enter to the oxide/atmosphere interface where they are consumed.

3.3.3.2 Oxidation of Metals Forming Cation Vacancy (p-type) Oxides

3.3.3.2.1 Atomic Mechanisms

The unit step extending the oxide lattice is also the adsorption of an oxygen atom from the atmosphere, its ionization to an anion, O^{2-}, and its coordination with a metal cation at the oxide/atmosphere interface, but the mechanism is determined by the ability of the oxide to accept cation vacancies. The cation is transferred from a lattice site, adding to the existing local population of lattice vacancies, and the electrons to ionize the oxygen are taken from the metal ion *d* shell, creating electron holes. Taking nickel as an example, the reaction is:

$$\frac{1}{2}O_2 = NiO + Ni^{2+}\square + 2e\square \tag{3.72}$$

The complementary partial reaction at the metal/oxide interface is the entry of atoms from the metal into the oxide, dissociating to cations and electrons, which annihilate cation vacancies and electron holes in the oxide:

$$\text{Ni} + \text{Ni}^{2+}\square + 2\text{e}\square(\text{oxide}) = 0 \quad (3.73)$$

Addition of Equations 3.72 and 3.73 yields the overall oxidation reaction:

$$\text{Ni} + \tfrac{1}{2}\text{O}_2 = \text{NiO} \quad (3.74)$$

3.3.3.2.2 *Interface Equilibria*

The equilibrium constant, *K*, for the reaction given in Equation 3.72 characterizes the defect populations at the oxide/atmosphere and metal/oxide interfaces:

$$K = \frac{a_{(\text{Ni}^{2+}\square)} \times [a_{(\text{e}\square)}]^2}{[a_{\text{O}_2}]^{1/2}} \quad (3.75)$$

whence:

$$n = p^{1/6} \cdot k^{1/3} \quad (3.76)$$

where:

$$n_{(\text{Ni}^{2+}\square)} = \tfrac{1}{2} n_{(\text{e}\square)} = n$$

and *k* includes *K* and coefficients relating *n* to $n_{(\text{Ni}^{2+}\square)}$ and $n_{(\text{e}\square)}$.

The number of cation vacancies at the oxide/atmosphere interface is:

$$n_{\text{atm}} = (p_{\text{atm}})^{1/6} \cdot k^{1/3} \quad (3.77)$$

and the number at the metal/oxide interface is:

$$n_{\text{o}} = (p_{\text{o}})^{1/6} \cdot k^{1/3} \quad (3.78)$$

As before, oxidation proceeds if $p_{\text{atm}} > p_{\text{o}}$, that is, if the oxygen pressure in the environment exceeds the dissociation pressure of the oxide and $n_{\text{atm}} > n_{\text{o}}$, producing a concentration gradient down which cation vacancies diffuse inward from the oxide/atmosphere interface where they are created, to the metal surface where they are annihilated. This is equivalent to the transport of nickel ions outward from the metal through the oxide to the oxide/atmosphere interface, replenishing those consumed in the formation of new oxide.

3.3.3.3 Oxidation of Metals Forming Anion Vacancy (n-type) Oxides

3.3.3.3.1 *Atomic Mechanisms*

Because the lattice defects permit the diffusion of oxygen but not the metal, the unit step extending the oxide lattice takes place at the metal/oxide interface by the entry of a metal atom, its ionization to a cation and its coordination with an oxygen anion. The anion is transferred from a lattice site, adding to the existing local population of lattice vacancies, and the excess electrons are added to the conduction band of the oxide. Taking titanium as an example, the reaction is:

$$\mathrm{Ti} = \mathrm{TiO_2} + 2\mathrm{O}^{2-}\square + 4\mathrm{e}\bullet \tag{3.79}$$

The complementary process at the oxide/atmosphere interface is the entry of adsorbed oxygen atoms, which annihilate anion vacancies and remove excess electrons:

$$\mathrm{TiO_2} + \mathrm{O_2} + 2\mathrm{O}^{2-}\square + 4\mathrm{e}\bullet = \mathrm{TiO_2}\ \text{(with fewer defects)} \tag{3.80}$$

3.3.3.3.2 Interface Equilibria

Changes in such a small defect population are virtually without effect on the activity of the oxide, so that $a_{(\mathrm{TiO_2})} = a_{(\mathrm{TiO_2})}$ (with fewer defects) ≈1 and can be omitted from the activity quotient so that the equilibrium constant for the reaction given in Equation 3.80 is:

$$K = \frac{1}{[a_{\mathrm{O_2}}]\times[a_{(\mathrm{O}^{2-}\square)}]^2\times[a_{(e\bullet)}]^4} \tag{3.81}$$

Proceeding in the same way as for cation interstitial and cation vacancy oxides:

$$n = k^{-1/6} \,.\, p^{-1/6} \tag{3.82}$$

and setting $p = p_o$ and $= p_{atm}$ at the metal/oxide and oxide/atmosphere interfaces, respectively, defines the concentration gradient down which the anion vacancies diffuse.

3.3.3.4 Oxidation of Metals Forming Stoichiometric Ionic Oxides

Two common metals, magnesium and aluminium, form stoichiometric oxides in which the only significant defects are Schottky pairs of cation and anion vacancies. The only sustainable transport mechanism through a homogeneous oxide growing on the hypothetical perfectly pure metals is the slow cooperative diffusion of the pairs by *intrinsic ionic conductivity*.

However, the oxides are so sensitive to impurities that the oxides forming on even high purity commercially produced metals exhibit a degree of *extrinsic conductivity*. This arises because impurity atoms can give electron energy levels introducing some electronic conductivity, which can support independent diffusion of one of the ion species. These impurities are usually traces of other metals, but MgO is also susceptible to the impurity anions OH^- ions injected at the oxide/atmosphere interface by atmospheric water vapour.

Oxygen diffuses via the $O^{2-}\square$ vacancies much more slowly than magnesium does via the $Mg^{2+}\square$ vacancies so that magnesium is the independently diffusing species in the impure oxide. The electron carriers introduced by the impurities permit limited oxidation mechanisms similar to those for cation vacant oxides.

Because the concentrations of impurities introducing electronic conductivity are low, oxidation is very slow. For example, when MgO forms a protective layer, as it can when a very thin layer is formed in dry air on magnesium metal or when it is formed on some aluminium–magnesium alloys for which the oxide/metal volume ratio is favourable the only significant transport of magnesium is through the disorder at the oxide crystal boundaries.

3.3.3.5 Time and Temperature Dependence of Diffusion-Controlled Oxidation

Diffusion control of oxidation implies that the rate at which the oxide thickens is proportional to the rate at which the diffusing species, for example, Zn in ZnO or Ni in NiO, is transported through the oxide, which is inversely proportional to the oxide thickness:

$$\frac{dx}{dt} = \frac{k}{x} \tag{3.83}$$

where x is the instantaneous oxide thickness after the elapse of time, t.

Integration yields a parabolic time law:

$$x = 2kt^{1/2} \tag{3.84}$$

Since it is diffusion-controlled, the dependence of oxidation rate on temperature, T, is derived from that of the diffusion coefficient, Δ, for the diffusing species that, over a limited temperature range, follows an Arrhenius-type relation:

$$\ln \Delta = \frac{-\Delta H^*}{RT} + \text{constant} \tag{3.85}$$

leading to a relation between the parabolic rate constant, k, and T of the form:

$$\log k = \frac{-A}{T} + B \tag{3.86}$$

Experimental verification of Equations 3.84 and 3.86 in the laboratory confirms that the short-term isothermal oxidation of many metals, including iron, nickel and copper, is under diffusion control. Such information is not a reliable guide to long-term oxidation because assumptions implicit in the Wagner theory can break down as the oxide thickens with time. In particular, design for oxidation resistance is best based on pilot tests simulating envisioned applications.

3.3.3.6 Correlations with Other Observations

3.3.3.6.1 *Electrical Conductivities of* n*-type and* p*-type Oxides*

The number, n, in Equations 3.69, 3.76 and 3.82, is a measure not only of the number of lattice defects but also the number of charge carriers (interstitial electrons or electron holes) that confer semiconducting properties on the oxide. These equations predict that the electrical conductivities of semiconducting oxides are functions of the oxygen pressure in the environment. The function is direct for p-type oxides, for example, $n \propto p^{1/6}$ for NiO and inverse for n-type oxides, for example, $n \propto p^{-1/6}$ for ZnO. Experimental evidence for one or the other of these relationships can indicate whether a particular oxide is n-type or p-type.

3.3.3.6.2 *Inert Marker Experiments*

When the metal is the diffusing species, the oxide/atmosphere interface can be identified as the growth interface by observing that experimental inert markers placed on the metal surface become enveloped in the growing oxide. Conversely, when it is oxygen that diffuses so that the metal/oxide interface is the growth interface, the markers remain on the oxide surface.

3.3.3.6.3 Development of Voids at Metal/Oxide Interfaces

With prolonged oxidation, the entry of metal atoms into a cation excess or cation vacant oxide can inject sufficient vacant lattice sites into the metal to condense as voids that are sometimes observed as at the metal/oxide interface, where atomic disorder facilitates nucleation.

3.3.3.7 Effects of Impurities

In real oxidation, impurities entering the oxides from the metal or atmosphere can have pronounced effects on the oxidation rate out of all proportion to the quantities present.

3.3.3.7.1 Valency Effects in Semiconducting Oxides

The introduction into an oxide of impurity cations with an oxidation state different than that of the host metal disturbs the balance of lattice and electronic defects with effects on the oxidation rate that depend on whether the oxide is *n*-type or *p*-type.

1. n-*type oxides – Example ZnO:* The replacement of some host metal cations, Zn^{2+}, by impurity cations in a *lower* oxidation state, for example, Li^{+}, produces a deficit of positive charge on the lattice that is compensated by adjusting the defect structure. The populations of interstitial cations, $n_{(Zn^{2+\bullet})}$ and electrons, $n_{(e\bullet)}$, are interdependent by the equilibrium described in Equation 3.68, so that the adjustment is by both an increase in the population of interstitial cations and a corresponding reduction in the population of interstitial electrons, e•. The additional interstitial cations facilitate the transport of Zn^{2+} ions, enhancing the oxidation rate. Conversely, the replacement of host cations by impurity cations in a *higher* oxidation state, for example, Cr^{3+}, introduces an excess of positive charge that is compensated in the opposite sense, with more interstitial electrons and fewer interstitial cations, diminishing the rate of oxidation.
2. p-*type oxides – Example NiO:* Similar considerations apply to *p*-type oxides, but the consequences are reversed. A deficit of positive charge introduced by impurity cations in a lower oxidation state is compensated by a lower population of cation vacancies, $n(Ni^{2+}\square)$ and a higher population of electron holes, $n(e\square)$, under the control of the equilibrium described in Equation 3.75. This curtails the transport of Ni^{2+} through the oxide, depressing the oxidation rate. Conversely, an excess of positive charge introduced by impurity cations in a higher oxidation state is compensated by a higher population of cation vacancies, $n(Ni^{2+}\square)$ and a lower population of electron holes, $n(e\square)$, enhancing the oxidation rate.

The acceleration or retardation can be striking. An addition of 0.5 mass % of chromium to nickel or of lithium to zinc can increase the oxidation rates by orders of magnitude. These effects are collectively known as *Hauffe's valency rules*, for their originator.

3.3.3.7.2 Catastrophic Oxidation Induced by Impurities

This aptly named effect describes accelerated oxidation associated with the destruction of protection by oxides when liquid phases are introduced by particular impurities contributed from the atmosphere or the metal. The oxides of several metals and other impurities are liquid at relatively low temperatures, especially vanadium, V_2O_5, boron, B_2O_3 and molybdenum, MoO_3, that have melting points of 675°C, 450°C and 795°C, respectively. The melting points of liquid phases may be lowered further by the formation of eutectics with oxides contributed by the metal or with other substances such as sodium sulfate formed from sulfur in oxidizing atmospheres. A potent source of these undesirable impurities is the ash of fossil fuels in products of combustion that impinge on metals. Most coal, fuel oils and gas contain sulfur and the ash of some fuel oils is particularly rich in vanadium oxide. Another source is vanadium and molybdenum from the alloy content of special steels.

3.3.3.8 Loss of Integrity of Protective Oxides

Oxide films formed on some metals may be only temporarily protective, leading to *paralinear oxidation* or to *breakaway oxidation* due to the loss of integrity as they thicken.

3.3.3.8.1 Paralinear Oxidation

As the name suggests, the oxidation rate at first diminishes with time by an apparent parabolic rate law until it reaches a critical value, after which it remains constant. It can be explained as the consequence of two simultaneous processes:

1. The oxide develops as a compact protective *barrier layer* with a diminishing growth rate consistent with the Wagner theory.
2. The outer surface of the barrier layer transforms at a constant rate into an unprotective, for example, cracked or porous form.

When the oxide is thin, the first process dominates, but eventually a steady state is reached at which a linear rate control is exercised by the second process, which restrains growth of the barrier layer to a limiting thickness.

3.3.3.8.2 Breakaway Oxidation

In breakaway oxidation, the oxide begins to grow at a rate diminishing with time, as if to establish a protective layer, but at critical oxide thickness it accelerates. After acceleration, the growth rate may be constant, as for magnesium at 500°C or it may rise to a maximum and then diminish, as for tungsten at 700°C. Among factors that can initiate this behaviour is the development of lateral stresses in an oxide, due to unfavourable metal–substrate/oxide volume ratios, causing it to crack or shear, exposing unprotected metal to the atmosphere.

3.3.4 Selective Oxidation of Components in an Alloy

One or more of the components of some alloy systems can oxidize selectively and the effect can be exploited in alloy formulation to provide protective films.

3.3.4.1 Principles

The expected oxidation product is the oxide of whichever component can reduce the oxygen activity at the alloy surface to the lowest value. This depends on the activities of the components and, therefore, on the alloy composition. Consider two metals, A and B, both of which can oxidize, in a hypothetical alloy system with the following simplifying characteristics:

1. The metal system is a complete series of solid solutions.
2. Only the simple oxides of the metals, AO and BO, can form.
3. The oxides are immiscible.

The oxidation reaction for metal A is:

$$2A + O_2 = 2AO \tag{3.87}$$

and the activities of the reactants are related by the equilibrium constant:

$$K_A = \frac{(a_{AO})^2}{(a_A)^2 \times (a_{O_2})} \tag{3.88}$$

For pure metal, A, in equilibrium with its oxide, $a_A = a_{AO} = 1$, giving a_{O_2} the unique value:

$$[a_{O_2}]_{AO} = 1/K_A \tag{3.89}$$

but when it is diluted by alloying, a_A is reduced to a value depending on composition

$$a_{A\,IN\,ALLOY} = \gamma_A \cdot X_A \tag{3.90}$$

where:
$X_A =$ mole fraction of metal A in the alloy
γ_A = activity coefficient for metal A (note $\gamma_A \rightarrow 1$ as $X_A \rightarrow 1$) and the value of a_{O2} in equilibrium with the metal and its oxide is given by:

$$K_A = \frac{(a_{AO})^2}{(a_{A\,IN\,ALLOY})^2} \times a_{O_2} \tag{3.91}$$

$$= \frac{1}{(\gamma_A \cdot X_A)^2 \times a_{O_2}} \tag{3.92}$$

Substituting for K_A from Equation 3.89 and rearranging:

$$a_{O_2} = \frac{[a_{O_2}]_{AO}}{(\gamma_A \cdot X_A)^2} \tag{3.93}$$

Equation 3.93 gives the minimum oxygen activity to form AO at the metal surface. Applying the same argument to the oxidation reaction for metal B:

$$2B + O_2 = 2BO \tag{3.94}$$

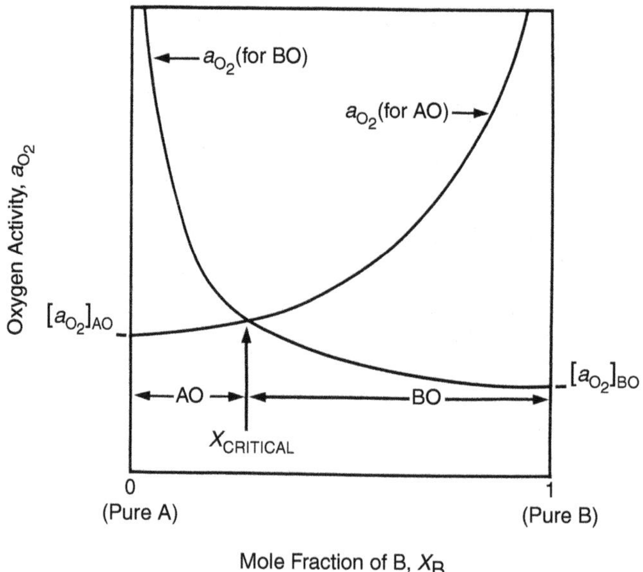

FIGURE 3.16
Oxygen activities in equilibrium with the oxides AO and BO formed on binary alloys in the system A–B, as functions of composition.
$[a_{O_2}]_{AO}$ = Oxygen activity in equilibrium with oxide AO and pure metal A.
$[a_{O_2}]_{BO}$ = Oxygen activity in equilibrium with oxide BO and pure metal B.

$$a_{O_2} = \frac{[a_{O_2}]_{BO}}{(\gamma_B \cdot X_B)^2} \tag{3.95}$$

Equation 3.95 gives the minimum oxygen activity to form BO at the metal surface.

Equations 3.93 and 3.95 plotted in Figure 3.16 show that as the composition of a binary alloy changes progressively from ($X_A = 1$, $X_B = 0$) to ($X_A = 0$, $X_B = 1$):

a_{O_2} for AO formation rises from $[a_{O_2}]_{AO}$ to infinity, and

a_{O_2} for BO formation falls from infinity to $[a_{O_2}]_{BO}$

Theoretically, for compositions between pure metal A and a critical composition, $X_{CRITICAL}$, AO is the exclusive oxidation product because it buffers a_{O_2} at a value below that needed to form BO. Similarly, for compositions between $X_{CRITICAL}$ and pure metal B, BO is the exclusive product because it buffers a_{O_2} at a value below that needed to form AO.

3.3.4.2 Oxidation of Alloys Forming Complex Oxides

The possible oxidation products for some technically important alloy systems include spinel-type oxides containing two metals in addition to the individual oxides of the component metals. This happens when there is a composition range within which the lowest oxygen activity is produced by two metals in combination.

Examples of systems with solid solutions exhibiting this behaviour include:

1. Aluminium–magnesium alloys
2. Nickel–chromium alloys, from which nickel-base superalloys were developed
3. Iron–chromium alloys, which form the basis for stainless steels

3.3.4.2.1 The Aluminium–Magnesium–Oxygen System

Consider the aluminium–magnesium system as a tangible example. The possible oxidation products are MgO, Al_2O_3 and the spinel, $MgAl_2O_4$. The same principles apply as described Section 3.3.4.1, extended to include the spinel. Assuming that the oxides are pure, $a_{Al_2O_3}$, a_{MgO} and $a_{MgAl_2O_4}$ are all unity and can be neglected in expressions for equilibrium constants.

Writing the equations conventionally for one mole of oxygen:

For the reaction: $4/3[Al]_{SOLUTION\ IN\ ALLOY} + O_2 = 2/3Al_2O_3$:

$$K_{Al_2O_3} = \frac{1}{(a_{Al})^{4/3} \times a_{O_2}} \tag{3.96}$$

For the reaction: $2[Mg]_{SOLUTION\ IN\ ALLOY} + O_2 = 2MgO$:

$$K_{MgO} = \frac{1}{(a_{MgO})^2 \times a_{O_2}} \tag{3.97}$$

For the reaction: $½[Mg]_{SOLUTION\ IN\ ALLOY} + [Al]_{SOLUTION\ IN\ ALLOY} + O_2 = ½MgAl_2O_4$:

$$K_{MgAl_2O_4} = \frac{1}{(a_{Mg})^{1/2} \times (a_{Al}) \times a_{O_2}} \tag{3.98}$$

EXAMPLE 4

Show which of the oxides, Al_2O_3, MgO or $MgAl_2O_4$ is most stable for a binary alloy of aluminium with 4 mass % magnesium in solid solution at 500°C.

Solution

The most stable oxide is the one that is in equilibrium with the lowest oxygen activity. The equilibrium oxygen activities are calculated using Equations 3.96, 3.97 and 3.98.

The equilibrium constants, $K_{Al_2O_3}$, K_{MgO} and $K_{MgAl_2O_4}$ are obtained by application of the Van't Hoff isobar to the Gibbs free energies of formation, ΔG° for the oxides:

$$\ln K = \frac{-\Delta G^{\circ}}{RT} \tag{3.99}$$

Standard references, for example, Kubaschewski and Alcock* give Gibbs free energies as functions of temperature. Bhatt and Garg* give values for the activities of

* Kubaschewski, O. and Alcock, C. B., *Metallurgical Thermochemistry*, Pegamon Press, New York, 1979. Bhatt and Garg, *Met. Trans., B*, 7, p. 227, 1976.

aluminium and magnesium in binary alloys. Values for an Al – 4 mass % Mg alloy at 500°C are:

$$a_{Mg} = 0.088 \text{ and } a_{Al} = 0.96$$

The calculations are as follows:

For Al_2O_3 at 500°C (773 K)
$\Delta G^{\circ} = -968529$ J mol^{-1} of oxygen
Applying the Van't Hoff isobar, $\ln K = -(-968529)/(8.314 \times 773) = 150.7$
whence:

$$K = 2.8 \times 10^{65}$$

From Equation 3.96: $K = 1/[(a_{Al})^{4/3} \times a_{O_2}]$
whence:

$$a_{O_2} = 1/[K \times (a_{Al})^{4/3}] = 1/(2.8 \times 10^{65} \times 0.947) = 3.83 \times 10^{-66}$$

For MgO at 500°C (773 K)
$\Delta G^{\circ} = -1043135$ J mol^{-1} of oxygen
Applying the Van't Hoff isobar, $\ln K = -(-1043135)/(8.314 \times 773) = 162.3$
whence:

$$K = 3.1 \times 10^{70}$$

From Equation 3.97: $K = 1/[(a_{Mg})^2 \times a_{O_2}]$
whence:

$$a_{O_2} = 1/[K \times (a_{Mg})^2] = 1/(3.1 \times 10^{70} \times 7.744 \times 10^{-3}) = 4.2 \times 10^{-69}$$

For $MgAl_2O_4$ at 500°C (773 K)
$\Delta G^{\circ} = -1007094$ J mol^{-1} of oxygen
Applying the Van't Hoff isobar, $\ln K = -\ (-1007094)/(8.314 \times 773) = 156.7$
whence:

$$K = 1.14 \times 10^{68}$$

From Equation 3.98: $K = 1/[a_{O_2} \times (a_{Al}) \times (a_{Mg})^{½}]$
whence:

$$a_{O_2} = 1/[K \times (a_{Al}) \times (a_{Mg})^{½}] = 1/(1.14 \times 10^{68} \times 0.96 \times 0.297) = 3.08 \times 10^{-68}$$

Hence, MgO is the most stable oxide for the conditions considered because it is in equilibrium with the lowest oxygen activity.

Figure 3.17 gives plots of oxygen activities at 500°C in equilibrium with the three oxides for alloys with magnesium contents in the range 0 to 10 mass % (equivalent to $X_{Mg} = 0.1$), calculated as in Example 4, using appropriate values of a_{Mg} and a_{Al} given by Bhatt and Garg. The plots for $[a_{O_2}]_{MgAl_2O_4}$, $[a_{O_2}]_{MgO}$ and $[a_{O_2}]_{Al_2O_3}$ intersect, defining *two* critical compositions, 0.16 and 1.05 mass % magnesium, dividing the composition axis into three ranges:

Magnesium mass %	< 0.16	$0.16 - 1.05$	> 1.05
Expected oxidation product	Al_2O_3	$MgAl_2O_4$	MgO

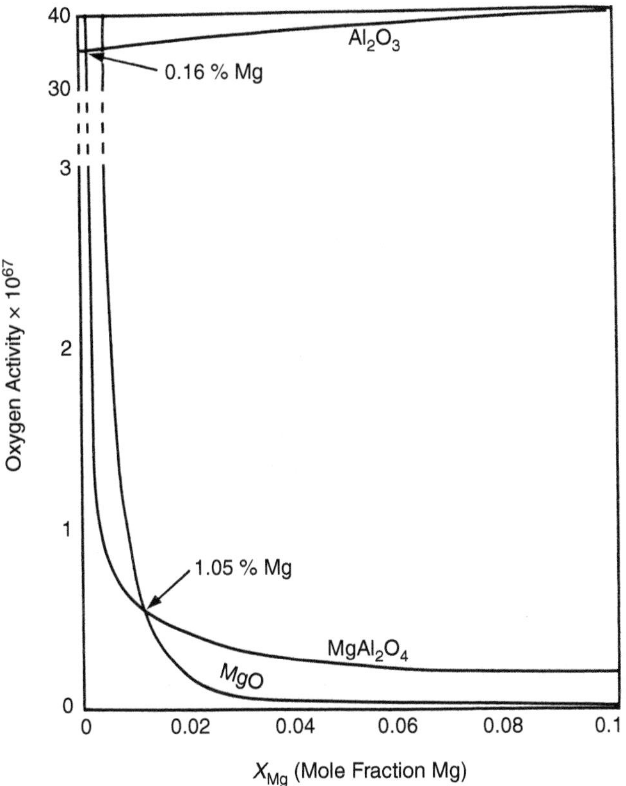

FIGURE 3.17
Oxygen activity in equilibrium with Al_2O_3, MgO and $MgAl_2O_4$ on aluminium–magnesium alloys as functions of magnesium mole fraction. *Note* change of scale for oxygen activity in equilibrium with Al_2O_3. Critical compositions expressed as mass % magnesium.

This result is consistent with industrial experience that magnesium oxide forms in preference to alumina on aluminium–magnesium alloys with high magnesium contents.

3.3.4.2.2 *The Nickel–Chromium–Oxygen System*

The nickel–chromium–oxygen system also includes two single oxides and a spinel, that is, NiO, Cr_2O_3 and $NiCr_2O_4$. The theoretical treatment is essentially similar to that described for the aluminium–magnesium–oxygen system.

3.3.4.2.3 *The Iron–Chromium–Oxygen System*

The iron–chromium–oxygen system resembles the aluminium–magnesium–oxygen and nickel–chromium–oxygen systems, but is complicated by five oxides:

1. A series of three single oxides from the iron–oxygen system, FeO (wüstite), Fe_3O_4 (magnetite) and Fe_2O_3 (haematite), described in Chapter 7
2. The single oxide from the chromium–oxygen system, Cr_2O_3
3. The spinel, $FeCr_2O_4$ (chromite)

If attention is confined to the oxides that form adjacent to the metal surface, that is, FeO, $FeCr_2O_4$ and Cr_2O_3, similar conclusions are reached for the expected oxidation products as for the other systems, but kinetic factors can influence the compositions of the scales.

3.3.4.3 Kinetic Considerations

3.3.4.3.1 Diffusion of Oxidizing Species in Metal

The foregoing considerations apply to the metal composition immediately adjacent to the oxide, that is, the composition at the extreme surface of the metal. However, the selective oxidation of one component of a binary alloy reduces its concentration at the metal surface, shifting the local alloy composition toward the critical composition at which another component oxidizes. Whether or not the critical composition is reached depends on how far the initial composition of the alloy is from the critical value and how rapidly the depleted composition can be restored by diffusion from the bulk of the metal. In practice, there is a range of composition around the critical composition within which both oxides form, giving a two-phase oxide structure.

3.3.4.3.2 Nucleation

The most stable oxide is sometimes difficult to nucleate at the metal surface, allowing the preferential nucleation of a less stable oxide. This can happen for an oxide with a structure that is epitaxially incompatible with the underlying metal or for a spinel with a stoichiometric ratio different from the ratio of different metal atoms in the alloy on which it is favoured. This is considered in Chapter 8 within the context of the practical oxidation resistance of stainless steels. These nonequilibrium structures are often temporary and revert to equilibrium structures, given time. Meanwhile, the effects on oxidation rates can be very significant.

PROBLEMS

Problems for Chapter 3

1. Using the standard approximation, $e^x = (1 + x)$ for small values of x, show that for small overpotentials, η, the Butler–Volmer equation simplifies to $i = i_o F\eta/RT$. If $\beta = 0.5$, $z = 1$ and $T = 298$ K, show that this approximation introduces an error of <1% for η <0.01 V.
2. Standard electrode potentials for the following reactions in alkaline solutions are:

$$2H_2O + 2e^- = H_2 + 2OH^- \qquad E^\circ = -0.827\,V\,(SHE) \tag{1}$$

$$\tfrac{1}{2}O_2 + H_2O + 2e^- = 2OH^- \qquad E^\circ = +0.401\,V\,(SHE) \tag{2}$$

Show that these values are compatible with the standard electrode potentials for the corresponding reactions in acid solutions:

$$2H^+ + 2e^- = H_2 \qquad E^\circ = 0.000\,V\,SHE \tag{3}$$

$$\tfrac{1}{2}O_2 + 2H^+ + 2e^- = H_2O \; E^\circ = +1.228\,V\,SHE \tag{4}$$

3. This problem illustrates that Pourbaix diagrams apply only to the species considered and can be misleading if other information is ignored. Using the following data for 25°C:

$$Cd^{2+} + 2e^{-} = Cd \qquad E^{\circ} = -0.403\,\text{V(SHE)} \qquad \text{(Reaction 1)}$$

$$Cd(OH)_2 + 2H^{+} + 2e^{-} = Cd + 2H_2O \qquad E^{\circ} = +0.017\,\text{V(SHE)} \qquad \text{(Reaction 2)}$$

$$Cd(OH)_2 + 2H^{+} = Cd^{2+} + 2H_2O \qquad (a_{Cd^{2+}})/(a_{H^{+}})^2 = 10^{14} \qquad \text{(Reaction 3)}$$

construct a partial Pourbaix diagram for the cadmium–water system for $a_{Cd^{2+}} = 10^{-6}$. Use it to show that the information it gives is inconsistent with the excellent corrosion resistance of the metal in neutral media. (Note: Information not available from the diagram is that the metal passivates by reaction with atmospheric carbon dioxide, forming insoluble carbonate, $CdCO_3$.)

4. Domestic bleach is produced by treating sodium hydroxide with chlorine gas:

$$2NaOH + Cl_2(g) = Na^{+} + ClO^{-} + NaCl + H_2O \qquad \text{(Reaction 1)}$$

The active agent is the hypochlorite ion, ClO^{-} which exerts its effect by the reaction:

$$ClO^{-} + H_2O + 2e^{-} = Cl^{-} + 2OH^{-} \qquad \text{(Reaction 2)}$$

for which:

$$E^{\circ} = 0.89\,\text{V(SHE)}$$

Calculate the pH of a solution containing 0.01 mol kg^{-1} each of ClO^{-}, Cl^{-} and OH^{-} ions and the potential it imposes on the metal. Referring these values to the diagrams given in Figures 3.2 to 3.10, which of the following metals would you eliminate immediately for use in storage vessels for the solutions (a) iron (or steel) (b) aluminium, (c) zinc, (d) copper, (e) tin, (f) nickel, (g) titanium?

5. Copper oxidizes at 800°C in oxygen at 1 atmosphere pressure, yielding an adherent layer of the metal deficit *p*-type oxide, Cu_2O. Derive expressions for the lattice and electronic defect populations in the oxide at the oxide/oxygen and metal/oxide interfaces. Hence, estimate by how much an increase in the oxygen pressure to 100 atmospheres influences (a) the rate of oxidation of the metal and (b) the electrical conductivity of the oxide.

6. Determine conditions for which iron is theoretically stable in water at 25°C with respect to $a_{Fe^{2+}} = 10^{-6}$. Suggestion: consult the iron–water Pourbaix diagram, given in Figure 3.2, and consider applying a high pressure to the system.

Solutions to Problems for Chapter 3

Solution to Problem 1

If η is small enough, the Butler–Volmer equation:

$$i = i_o \exp\frac{(1-\alpha)zF\eta}{RT} - i_o \exp\frac{-\alpha zF\eta}{RT}$$

can be simplified, applying the standard substitution of the form $e^x = 1 + x$ to both exponentials, yielding the version of the equation known as *the low field approximation*:

$$i \approx i_o \left(1 + \frac{(1-\alpha)zF\eta}{RT} - 1 - \frac{\alpha zF\eta}{RT}\right) = \frac{i_o F\eta}{RT}$$

For $\eta = 0.01$ V, the unmodified Butler–Volmer equation yields:

$$\begin{aligned} i &= i_o \exp\frac{(1-0.5)\times 1\times 96490\times 0.01}{8.314\times 298} - i_o \exp\frac{-(0.5\times 1\times 96490\times 0.01)}{8.314\times 298} \\ &= i_o(\exp 0.195 - \exp -0.195) = 0.392 i_o \end{aligned}$$

For comparison, the low-field approximation yields:

$$i = \frac{i_o \times 96490\times 0.01}{8.314\times 298} = 0.389\, i_o$$

The difference between the two values is $< 1\%$ and diminishes as $\eta \to 0$.

Solution to Problem 2

The standard state for Reaction 1, to which E° applies is $a_{OH^-} = 1$. The standard state for Reaction 3 is $a_{H^+} = 1$, for which $a_{OH^-} = 10^{-14}$, because $K_w = a_{H^+} \times 10^{-14}$ at 298 K.

Applying the Nernst equation to Reaction 1 in the standard state for Reaction 3 yields:

$$\begin{aligned} E &= E^{\circ} - \frac{0.0591}{z}\log J \\ &= -0.827 - \frac{0.0591}{2}\log(10^{-14}) = 0.000\,\text{V}\,(\text{SHE}) \end{aligned}$$

which equals the value of E° for Reaction 3.

Similarly, applying the Nernst Equation to Reaction 2 in the standard state for Reaction 4 yields:

$$\begin{aligned} E &= +0.401 - \frac{0.0591}{2}\log(10^{-14})^2 \\ &= 1.228\,\text{V}\,(\text{SHE}) \end{aligned}$$

which equals the value of E° for Reaction 4. Thus the standard electrode potentials for Reactions 1 and 3 are compatible with the corresponding values for Reactions 2 and 4.

Solution to Problem 3

Proceeding as for the calculations given in the Appendix to this chapter:

1. ***Reaction 1***

$$Cd^{2+} + 2e^- = Cd \qquad E^{\circ} = -0.403\,\text{V SHE}$$

Applying the Nernst equation: $$E = -0.403 - \frac{0.0591}{2}\log\frac{1}{a_{Cd^{2+}}}$$
$$= -0.403 + 0.0296\ \log\ (a_{Cd^{2+}})$$

For:

$$a_{Cd^{2+}} = 10^{-6}: \qquad E = -0.581$$

2. ***Reaction 2***

$$Cd(OH)_2 + 2H^+ + 2e^- = Cd + 2H_2O \qquad E^{\circ} = +0.017\,V\,SHE$$

Applying the Nernst equation: $$E = +0.017 - \frac{0.0591}{2}\cdot\log\frac{1}{(a_{H^+})^2}$$
$$= +0.017 - 0.0591\ pH$$

3. ***Reaction 3***

$$Cd(OH)_2 + 2H^+ = Cd^{2+} + 2H_2O$$

$$(a_{Cd^{2+}})/(a_{H^+})^2 = 10^{14}$$

Taking logarithms:

$$\log(a_{Cd^{2+}}) - 2\log(a_{H^+}) = 14$$

Rearranging:

$$pH = \{-\log(a_{Cd^{2+}}) + 14\}/2$$

For:

$$a_{Cd^{2+}} = 10^{-6}: \ \ pH = 10$$

Plotting lines on a partial Pourbaix diagram using the equations obtained for Reactions 1, 2 and 3 reveal a domain of stability for Cd^{2+} ions for E > 0.558 V (SHE) and pH < 10. Hence, by the conventional criterion for corrosion, that is, $a_{Cd^{2+}} > 10^{-6}\,mol\,dm^{-3}$, Cd^{2+} ions are the stable species with respect to cadmium metal within the domain of stability for water for all pH < 10. The metal might, therefore, be expected to corrode in near neutral waters, but this is inconsistent with experience that cadmium resists humid air and natural waters well enough to serve as a protective coating for steels. An explanation is that in constructing the diagram, no information was included for species derived from carbon dioxide, which is a ubiquitous component of air and natural waters. The good performance of cadmium is usually attributed to passivation by the very insoluble compound, cadmium carbonate.

Solution to Problem 4

0.01 M solutions are fairly dilute and a_{ClO^-}, a_{Cl^-} and a_{OH^-} can all be assumed to be 0.01, that is, 10^{-2}, without serious error.

Since $K_w = 10^{-14}$ at 298 K and $a_{OH^-} = 10^{-2}$, $a_{H^+} = 10^{12}$ and pH = 12.

Applying the Nernst equation to Reaction 2:

$$E = E^{\circ} - \frac{0.0591}{z}\log\frac{a_{Cl^-}\times(a_{OH^-})^2}{a_{ClO^-}}$$
$$= +0.89 - \frac{0.0591}{2}\log\frac{(0.01)\times(0.01)^2}{0.01}$$
$$= +1.01 \text{ V (SHE)}$$

The coordinates, pH = 12, $E = 1.01$ V (SHE), can be referred to the Pourbaix diagrams given in Figures 3.2 and 3.4 to 3.10. They lie well inside the domains of stability for soluble ions for the zinc and aluminium systems, so that these metals can probably be eliminated from consideration. The same coordinates lie in domains of stability of solid phases for the iron, copper, tin, nickel and titanium systems, so that these metals offer prospects for resisting the solution. As always, such conclusions are tentative, pending confirmation.

Solution to Problem 5

The procedure is similar to that used in Section 3.3.3.2 for the formation of NiO on nickel, taking account of the different stoichiometric ratio. The reaction is:

$$\tfrac{1}{2}O_2 = Cu_2O + 2\ Cu^{+}\square + 2e\square$$

The equilibrium constant, K, characterizes the defect populations at the oxide/atmosphere and metal/oxide interfaces:

$$K = \frac{(a_{Cu\square})^2 \times (a_{(e\square)})^2}{(a_{O_2})^{1/2}}$$

whence:

$$n = p^{1/8} \cdot k^{1/4}$$

where:

$n_{(Cu^+\square)} = n_{(e\square)} = n$ and k includes K and coefficients relating n to $n_{(Cu^+\square)}$ and $n_{(e^{\square})}$

The number of cation vacancies at the oxide/atmosphere interface, n_{atm}, is:

$$n_{atm} = (p_{atm})^{1/8} \cdot k^{1/4}$$

where p_{atm} is the external oxygen pressure and the number of vacancies at the metal/oxide interface, n_o, is:

$$n_o = p_o^{1/8} \cdot k^{1/4}$$

where p_o is the oxygen pressure in equilibrium with copper and Cu_2O.

The concentration gradient of vacancies, through the oxide layer, $(n_{atm} - n_o)$, is virtually determined by p_{atm}, because $p_{atm} >> p_o$. According to the Wagner theory, the oxidation rate is a linear function of this concentration gradient, so that if the oxygen pressure is raised from 1 to 100 atmospheres, the oxidation of copper is expected to increase by a factor of $(100)^{1/8} = 1.8$. Since the number of electron holes also equals n_o, the electrical conductivity of Cu_2O in equilibrium with oxygen at the same pressure increases by the same factor.

Solution to Problem 6

Iron is stable in oxygen-free water if there is a common domain of stability. Inspection of Figure 3.2 shows that, by the convention that iron is stable when $a_{Fe^{2+}} < 10^{-6}$, there is only a small gap between the domains of stability of water and of iron at the Fe/Fe_3O_4 boundary. This gap can be closed by increasing the pressure on the system so that $a_{H_2} > 1$, to extend the domain of stability of water to more negative potentials.

For pH > 9, the upper boundary of the domain for iron is determined by the reaction:

$$Fe_3O_4 + 8H^+ + 8e^- = 3Fe + 4H_2O$$

and is described by Equation 3.20 in Section 3.1.5.2:

$$E = -0.085 - 0.0591\ \text{pH}$$

The lower boundary of the domain of stability for water is determined by the reaction:

$$2H^+ + 2e^- = H_2$$

and is given by the Nernst equation:

$$\begin{aligned} E &= 0.000 - \frac{0.0591}{2}\log\frac{p_{H_2}}{(a_{H^+})^2} \\ &= -0.0295\log p_{H_2} - 0.0591\ \text{pH} \end{aligned}$$

The critical condition is when the two boundaries coincide, that is:

$$-0.085 - 0.0591\ \text{pH} = -0.0295\log p_{H_2} - 0.0591\ \text{pH}$$

hence:

$$\log p_{H_2} = 2.876 \text{ and so } p_{H_2} = 752\ \text{atm}$$

and the conditions for a common domain of stability are $p_{H_2} = 752$ atm and pH > 9.

Appendix: Construction of Some Pourbaix Diagrams

1. The Aluminium–Water System

Species considered:

$$Al,\ Al^{3+},\ AlO_{2^-},\ Al_2O_3$$

Reaction 1

$$Al^{3-} + 3e^- = Al$$

Information needed:

$$E^{\circ} = -1.663 \text{ V SHE}$$

Applying the Nernst equation:

$$E = -1.663 - \frac{0.0591}{3} \cdot \log \frac{1}{a_{Al^{3+}}}$$
$$E = -1.663 + 0.0197 \log (a_{Al^{3+}})$$

For:

$a_{Al^{3+}} = 10^{-6}$:	$E = -1.780$	$a_{Al^{3+}} = 10^{-2}$:	$E = -1.701$
$a_{Al^{3+}} = 10^{-4}$:	$E = -1.741$	$a_{Al^{3+}} = 1$:	$E = -1.662$

Reaction 2

$$Al_2O_3 + 6H^+ + 6e^- = 2Al + 3H_2O$$

Information needed:

$$E^{\circ} = -1.550 \text{ V SHE}$$

Applying the Nernst equation:

$$E = -1.550 - \frac{0.0591}{6} \cdot \log \frac{1}{(a_{H^+})^6}$$
$$E = -1.550 - 0.0591 \text{ pH}$$

Reaction 3

$$AlO_{2^-} + 4H^+ + 3e^- = Al + 2H_2O$$

Information needed:

$$E^{\circ} = -1.262\,\text{V SHE}$$

Applying the Nernst equation:

$$E = -1.262 - \frac{0.0591}{3} \cdot \log \frac{1}{(a_{AlO_2^-}) \times (a_{H^+})^4}$$
$$E = -1.262 + 0.0197 \log(a_{AlO_2^-}) - 0.0788\ \text{pH}$$

For:

$$a_{AlO_2^-} = 10^{-6}:\quad E = -1.380 - 0.0788\ \text{pH}$$

$$a_{AlO_2^-} = 10^{-4}:\quad E = -1.341 - 0.0788\ \text{pH}$$

$$a_{AlO_2^-} = 10^{-2}:\quad E = -1.301 - 0.0788\ \text{pH}$$
$$a_{AlO_2^-} = 1:\quad E = -1.262 - 0.0788\ \text{pH}$$

Reaction 4

$$Al_2O_3 + 6H^+ = 2Al^{3+} + 3H_2O$$

Information needed:

$$(a_{Al^{3+}})/(a_{H^+})^3 = 5 \times 10^5$$

Taking logarithms:

$$\log(a_{Al^{3+}}) - 3\log(a_{H^+}) = 5.70$$

Rearranging:

$$\text{pH} = \{-\log(a_{Al^{3+}}) + 5.70\}/3$$

For:

$$a_{Al^{3+}} = 10^{-6}:\quad \text{pH} = 3.9 \qquad a_{Al^{3+}} = 10^{-2}:\quad \text{pH} = 2.6$$

$$a_{Al^{3+}} = 10^{-4}:\quad \text{pH} = 3.2 \qquad a_{Al^{3+}} = 1:\quad \text{pH} = 1.9$$

Reaction 5

$$2AlO_2^- + 2H^+ = Al_2O_3 + H_2O$$

Information needed:

$$(a_{AlO_2^-}) \times (a_{H^+}) = 2.51 \times 10^{-15}$$

Taking logarithms:

$$\log(a_{AlO_{2^-}}) + \log(a_{H^+}) = -14.6$$

Rearranging:

$$pH = \log(a_{AlO_{2^-}}) + 14.6$$

For:

$a_{AlO_{2^-}} = 10^{-6}$:	$pH = 8.6$	$a_{AlO_{2^-}} = 10^{-2}$:	$pH = 12.6$
$a_{AlO_{2^-}} = 10^{-4}$:	$pH = 10.6$	$a_{AlO_{2^-}} = 1$:	$pH = 14.6$

Diagram: Figure 3.4 gives the diagram for $a_{Al^{3+}} = a_{AlO_2^-} = 10^{-6}$.

2. The Zinc–Water System

Species considered:

$$Zn,\ Zn^{2+},\ HZnO_{2^-},\ ZnO_{2^{2-}},\ Zn(OH)_2$$

Reaction 1

$$Zn^{2+} + 2e^- = Zn$$

Information needed:

$$E^{\circ} = -0.763\ \text{V SHE}$$

Applying the Nernst equation:

$$E = -0.763 - \frac{0.0591}{2}\log\frac{1}{a_{Zn^{2+}}}$$
$$= -0.763 + 0.0296\ \log\ (a_{Zn^{2+}})$$

For:

$a_{Zn^{2+}} = 10^{-6}$:	$E = -0.940$	$a_{Zn^{2+}} = 10^{-2}$:	$E = -0.822$
$a_{Zn^{2+}} = 10^{-4}$:	$E = -0.881$	$a_{Zn^{2+}} = 1$:	$E = -0.763$

Reaction 2

$$Zn(OH)_2 + 2H^+ + 2e^- = Zn + 2H_2O$$

Information needed:

$$E^{\circ} = -0.439\ \text{V SHE}$$

Applying the Nernst equation:

$$E = -0.439 - \frac{0.0591}{2} \log \frac{1}{(a_{H^+})^2}$$
$$= -0.439 - 0.0591 \text{ pH}$$

Reaction 3

$$HZnO_{2^-} + 3H^+ + 2e^- = Zn + 2H_2O$$

Information needed:

$$E^o = +0.054 \text{ V SHE}$$

Applying the Nernst equation:

$$E = +0.054 - \frac{0.0591}{2} \log \frac{1}{(a_{HZnO_{2^-}}) \times (a_{H^+})^3}$$
$$E = 0.054 + 0.0296 \log (a_{HZnO_{2^-}}) - 0.0887 \text{ pH}$$

For:

$$a_{HZnO_{2^-}} = 10^{-6}: \quad E = -0.124 - 0.0887 \text{ pH}$$

$$a_{HZnO_{2^-}} = 10^{-4}: \quad E = -0.064 - 0.0887 \text{ pH}$$

$$a_{HZnO_2^-} = 10^{-2}: \quad E = -0.005 - 0.0887 \text{ pH}$$

$$a_{HZnO_2^-} = 1: \quad E = +0.054 - 0.0887 \text{ pH}$$

Reaction 4

$$ZnO_2^{2-} + 4H^+ + 2e^- = Zn + 2H_2O$$

Information needed:

$$E^o = +0.44 \text{ V SHE}$$

Applying the Nernst equation:

$$E = +0.441 - \frac{0.0591}{2} \log \frac{1}{(a_{ZnO_2^{2-}}) \times (a_{H^+})^4}$$
$$= +0.441 + 0.0296 \log (a_{ZnO_2^{2-}}) - 0.118 \text{ pH}$$

For:

$$a_{ZnO_2^{2-}} = 10^{-6}: \quad E = +0.264 - 0.118 \text{ pH}$$

$$a_{ZnO_2^{2-}} = 10^{-2}: \quad E = +0.323 - 0.118 \text{ pH}$$

$$a_{ZnO_2^{2-}} = 10^{-4}: \quad E = +0.382 - 0.118 \text{ pH}$$

$$a_{ZnO_2^{2-}} = 1: \quad E = +0.441 - 0.118 \text{ pH}$$

Reaction 5

$$Zn(OH)_2 + 2H^+ = Zn^{2+} + 2H_2O$$

Information needed:

$$(a_{Zn^{2+}})/(a_{H^+})^2 = 10^{11}$$

Taking logarithms:

$$\log(a_{Zn^{2+}}) - 2\log(a_{H^+}) = 11$$

Rearranging:

$$\text{pH} = \{-\log(a_{Zn^{2+}}) + 11\}/2$$

For:

$$a_{Zn^{2+}} = 10^{-6}: \quad \text{pH} = 8.5 \qquad a_{Zn^{2+}} = 10^{-2}: \quad \text{pH} = 6.5$$

$$a_{Zn^{2+}} = 10^{-4}: \quad \text{pH} = 7.5 \qquad a_{Zn^{2+}} = 1: \quad \text{pH} = 5.5$$

Reaction 6

$$Zn(OH)_2 = H^+ + HZnO_2^-$$

Information needed:

$$(a_{H^+}) \times (a_{HZnO_2^-}) = 2.09 \times 10^{-17}$$

Taking logarithms:

$$\log(a_{H^+}) + \log(a_{HZnO_2^-}) = -16.7$$

Rearranging:

$$pH = \log(a_{HZnO_2^-}) + 16.7$$

For:

$$a_{HZnO_2^-} = 10^{-6}: \quad pH = 10.7 \qquad a_{HZnO_2^-} = 10^{-2}: \quad pH = 14.7$$

$$a_{HZnO_2^-} = 10^{-4}: \quad pH = 12.7 \qquad a_{HZnO_2^-} = 1: \quad pH = 16.7$$

Reaction 7

$$HZnO_2^- = H^+ + ZnO_2^{2-}$$

Information needed:

$$(a_{H^+}) \times (a_{ZnO_2^{2-}}) / (a_{HZnO_2^-}) = 7.8 \times 10^{-14}$$

Taking logarithms:

$$\log(a_{H^+}) + \log(a_{ZnO_2^{2-}}) - \log(a_{HZnO_2^-}) = -13.1$$

Rearranging:

$$pH = 13.1 \text{ (since } a_{ZnO_2^{2-}} = a_{HZnO_2^-} \text{ by definition)}$$

Diagram: Figure 3.5 gives the diagram for $a_{Zn^{2+}} = a_{HZnO_2^-} = a_{ZnO_2^{2-}} = 10^6$.

3. **The Copper–Water System**
   Species considered: Cu, Cu^{2+}, Cu^+, CuO, Cu_2O
   Reaction 1

$$Cu^{2+} + 2e^- = Cu$$

Information needed:

$$E^\ominus = +0.337 \text{ V SHE}$$

Applying the Nernst equation:

$$E = +0.337 - \frac{0.0591}{2} \log \frac{1}{a_{Cu^{2+}}}$$
$$= +0.337 + 0.0296 \log (a_{Cu^{2+}})$$

For:

$a_{Cu^{2+}} = 10^{-6}$: $E = 0.159$ $a_{Cu^{2+}} = 10^{-2}$: $E = 0.278$

$a_{Cu^{2+}} = 10^{-4}$: $E = 0.219$ $a_{Cu^{2+}} = 1$: $E = 0.337$

Reaction 2

$$Cu^{+} + e^{-} = Cu$$

Information needed:

$$E^{\ominus} = +0.520\ \text{V SHE}$$

Applying the Nernst equation:

$$E = +0.520 - \frac{0.0591}{1} \cdot \log \frac{1}{a_{Cu^{+}}}$$
$$E = +0.520 + 0.0591 \log(a_{Cu^{+}})$$

For:

$a_{Cu^{+}} = 10^{-6}$: $E = +0.165$ $a_{Cu^{+}} = 10^{-2}$: $E = +0.402$

$a_{Cu^{+}} = 10^{-4}$: $E = +0.284$ $a_{Cu^{+}} = 1$: $E = +0.520$

Reaction 3

$$2Cu^{2+} + H_2O + 2e^{-} = Cu_2O + 2H^{+}$$

Information needed:

$$E^{\ominus} = +0.203\ \text{V SHE}$$

Applying the Nernst equation:

$$E = +0.203 - \frac{0.0591}{2} \log \frac{(a_{H^{+}})^2}{(a_{Cu^{2+}})^2}$$
$$= 0.203 + 0.0591\ \text{pH} + 0.0591 \log(a_{Cu^{2+}})$$

For:

$a_{Cu^{2+}} = 10^{-6}$: $E = -0.152 - 0.0591\ \text{pH}$

$a_{Cu^{2+}} = 10^{-2}$: $E = +0.085 - 0.0591\ \text{pH}$

$a_{Cu^{2+}} = 10^{-4}$: $E = -0.033 - 0.0591\ \text{pH}$

$a_{Cu^{2+}} = 1$: $E = +0.203 - 0.0591\ \text{pH}$

Reaction 4

$$Cu^{2+} + H_2O = CuO + 2H^+$$

Information needed:

$$(a_{H^+})^2/(a_{Cu^{2+}}) = 1.29 \times 10^{-8}$$

Taking logarithms:

$$2\log(a_{H^+}) - \log(a_{Cu^{2+}}) = -7.89$$

Rearranging:

$$pH = \{-\log(a_{Cu^{2+}}) + 7.89\}/2$$

For:

$a_{Cu^{2+}} = 10^{-6}$:	$pH = 6.95$	$a_{Cu^{2+}} = 10^{-2}$:	$pH = 4.95$
$a_{Cu^{2+}} = 10^{-4}$:	$pH = 5.95$	$a_{Cu^{2+}} = 1$:	$pH = 3.95$

Reaction 5

$$Cu_2O + 2H^+ + 2e^- = 2Cu + H_2O$$

Information needed:

$$E^{\circ} = +0.471\ \text{V SHE}$$

Applying the Nernst equation:

$$E = +0.471 - \frac{0.0591}{2}\log\frac{1}{(a_{H^+})^2}$$
$$= +0.471 - 0.0591\ pH$$

Reaction 6

$$2CuO + 2H^+ + 2e^- = Cu_2O + H_2O$$

Information needed:

$$E^{\circ} = +0.669\ \text{V SHE}$$

Applying the Nernst equation:

$$E = +0.669 - \frac{0.0591}{2}\log\frac{1}{(a_{H^+})^2}$$
$$= +0.669 - 0.0591\,\text{pH}$$

Diagram: Figure 3.6 gives the diagram for $a_{Cu^{2+}} = a_{Cu^+} = 10^{-6}$.

4. The Tin–Water System

Species considered: Sn, Sn^{2+}, Sn^{4+}, SnO_3^{2-}, SnO_2

Reaction 1

$$Sn^{2+} + 2e^- = Sn$$

Information needed:

$$E^\circ = -0.136\ \text{V SHE}$$

Applying the Nernst equation:

$$E = -0.136 - \frac{0.0591}{2}\cdot\log\frac{1}{a_{Sn^{2+}}}$$
$$E = -\ 0.136\ +\ 0.0296\ \log\ (a_{Sn^{2+}})$$

For:

$a_{Sn^{2+}} = 10^{-6}$:	$E = -0.313$	$a_{Sn^{2+}} = 10^{-2}$:	$E = -0.195$
$a_{Sn^{2+}} = 10^{-4}$:	$E = -0.254$	$a_{Sn^{2+}} = 1$:	$E = -0.136$

Reaction 2

$$SnO_2 + 4H^+ + 4e^- = Sn + 2H_2O$$

Information needed:

$$E^\circ = -0.106\ \text{V SHE}$$

Applying the Nernst equation:

$$E = -0.106 - \frac{0.0591}{4}\log\frac{1}{(a_{H^+})^4}$$
$$E = -0.106 - 0.0591\,\text{pH}$$

Reaction 3

$$SnO_2 + 4H^+ + 2e^- = Sn^{2+} + 2H_2O$$

Information needed:

$$E^{\circ} = -0.077 \text{ V SHE}$$

Applying the Nernst equation:

$$E = -0.077 - \frac{0.0591}{2} . \log \frac{(a_{Sn^{2+}})}{(a_{H^+})^4}$$
$$E = -0.077 - 0.118 \text{ pH} - 0.0296 \log(a_{Sn^{2+}})$$

For:

$$a_{Sn^{2+}} = 10^{-6}: \quad E = +0.100 - 0.118 \text{ pH}$$

$$a_{Sn^{2+}} = 10^{-4}: \quad E = +0.041 - 0.118 \text{ pH}$$

$$a_{Sn^{2+}} = 10^{-2}: \quad E = -0.018 - 0.118 \text{ pH}$$

$$a_{Sn^{2+}} = 1: \quad E = -0.077 - 0.118 \text{ pH}$$

Reaction 4

$$Sn^{4+} + 2H_2O = SnO_2 + 4H^+$$

Information needed:

$$(a_{Sn^{4+}})/(a_{H^+})^4 = 2.08 \times 10^{-8}$$

Taking logarithms:

$$\log(a_{Sn^{4+}}) - 4\log(a_{H^+}) = -7.68$$

Rearranging:

$$\text{pH} = -\{\log(a_{Sn^{4+}}) + 7.68\}/4$$

For:

$$a_{Sn^{4+}} = 10^{-6}: \quad \text{pH} = -0.4 \qquad a_{Sn^{4+}} = 10^{-2}: \quad \text{pH} = -1.4$$

$$a_{Sn^{4+}} = 10^{-4}: \quad \text{pH} = -0.9 \qquad a_{Sn^{4+}} = 1: \quad \text{pH} = -1.9$$

Reaction 5

$$Sn^{4+} + 2e^- = Sn^{2+}$$

Information needed:

$$E^{\ominus} = -0.151\,\text{V SHE}$$

Applying the Nernst equation:

$$E = +0.151 - \frac{0.0591}{2} \cdot \log \frac{a_{Sn^{2+}}}{a_{Sn^{4+}}}$$

$$E = +0.151 - 0.0296 \log \left\{ \frac{(a_{Sn^{2+}})}{(a_{Sn^{4+}})} \right\}$$

$$E = +0.151 \text{ since } \frac{(a_{Sn^{2+}})}{(a_{Sn^{4+}})} \text{ is } 1$$

Reaction 6

$$SnO_3^{2-} + 2H^+ = SnO_2 + H_2O$$

Information needed:

$$(a_{SnO_3^{2-}}) \times (a_{H^+})^2 = 6.92 \times 10^{-32}$$

Taking logarithms:

$$\log\left(a_{SnO_{3^{2-}}}\right) + 2\log(a_{H^+}) = -31.16$$

Rearranging:

$$\text{pH} = \left\{\log\left(a_{SnO_{3^{2-}}}\right) + 31.16\right\}/2$$

For:

$a_{SnO_3^{2-}} = 10^{-6}$: pH = 12.6 $a_{SnO_3^{2-}} = 10^{-2}$: pH = 14.6

$a_{SnO_3^{2-}} = 10^{-4}$: pH = 13.6 $a_{SnO_3^{2-}} = 1$: pH = 15.6

Diagram: Figure 3.7 gives the diagram for $a_{Sn^{2+}} = a_{Sn^{4+}} = a_{SnO_3^{2-}} = 10^{-6}$.

5. The Nickel–Water System

Species considered: Ni, Ni^{2+}, $Ni(OH)_2$

Reaction 1

$$Ni^{2+} + 2e^- = Ni$$

Information needed:

$$E^{\circ} = -0.250\,\text{V SHE}$$

Applying the Nernst equation:

$$\begin{aligned} E &= -0.250 - \frac{0.0591}{2}.\log \frac{1}{a_{Ni^{2+}}} \\ &= -0.250 + 0.0296 \log (a_{Ni^{2+}}) \end{aligned}$$

For:

$$a_{Ni^{2+}} = 10^{-6}{:}\quad E = -0.428 \qquad a_{Ni^{2+}} = 10^{-2}{:}\quad E = -0.309$$

$$a_{Ni^{2+}} = 10^{-4}{:}\quad E = -0.368 \qquad a_{Ni^{2+}} = 1{:}\quad E = -0.250$$

Reaction 2

$$Ni(OH)_2 + H^+ + 2e^- = Ni + 2H_2O$$

Information needed:

$$E^{\circ} = +0.110\,\text{V SHE}$$

Applying the Nernst equation:

$$\begin{aligned} E &= +0.110 - \frac{0.0591}{2}\cdot \log \frac{1}{(a_{H^+})^2} \\ &= +0.110 - 0.0591\,\text{pH} \end{aligned}$$

Reaction 3

$$Ni(OH)_2 + 2H^+ = Ni^{2+} + 2H_2O$$

Information needed:

$$(a_{Ni^{2+}})/(a_{H^+})^2 = 1.58 \times 10^{12}$$

Taking logarithms:

$$\log(a_{Ni^{2+}}) - 2\log(a_{H^+}) = 12.20$$

Rearranging:

$$pH = [-\log(a_{Ni^{2+}}) + 12.2]/2$$

For:

$a_{Ni^{2+}} = 10^{-6}$: pH = 9.1 $a_{Ni^{2+}} = 10^{-2}$: pH = 7.1

$a_{Ni^{2+}} = 10^{-4}$: pH = 8.1 $a_{Ni^{2+}} = 1$: pH = 6.1

Diagram: Figures 3.8 and 6.2 give partial diagrams for $a_{Ni^{2+}} = 10^{-6}$ and $a_{Ni^{2+}} = 1$ required, respectively, to correlate with corrosion and electrodeposition of nickel.

6. The Titanium–Water System

Alternative versions are given, depending on the state of hydration of the oxide species.

Version Considering Anhydrous Oxides

Species considered: Ti, Ti^{2+}, Ti^{3+}, TiO

Reaction 1

$$Ti^{2+} + 2e^- = Ti$$

Information needed:

$$E^\circ = -0.630 \text{ V SHE}$$

Applying the Nernst equation:

$$E = -1.630 - \frac{0.0591}{2} \cdot \log \frac{1}{a_{Ti^{2+}}}$$
$$= -1.630 + 0.0296 \log (a_{Ti^{2+}})$$

For:

$a_{Ti^{2+}} = 10^{-6}$: E = −1.807 $a_{Ti^{2+}} = 10^{-2}$: E = −1.689

$a_{Ti^{2+}} = 10^{-4}$: E = −1.748 $a_{Ti^{2+}} = 1$: E = −1.630

Reaction 2

$$Ti^{3+} + e^- = Ti^{2+}$$

Information needed:

$$E^{\circ} = -0.368\ \text{V SHE}$$

Applying the Nernst equation:

$$\begin{aligned} E &= -0.368 - 0.0591.\log \frac{a_{Ti^{2+}}}{a_{Ti^{3+}}} \\ &= -0.368 + 0.0591 \log (a_{Ti^{3+}}) - 0.0591 \log (a_{Ti^{2+}}) \end{aligned}$$

Reaction 3

$$TiO + 2H^+ = Ti^{2+} + H_2O$$

Information needed:

$$(a_{Ti^{2+}})/(a_{H^+})^2 = 6.33 \times 10^{21}$$

Taking logarithms:

$$2\log(a_{H^+}) - \log(a_{Ti^{2+}}) = 21.8$$

Hence, rearranging:

$$\text{pH} = \{-\log(a_{Ti^{2+}}) + 10.9)\}/2$$

For:

$$a_{Ti^{2+}} = 10^{-6}: \quad \text{pH} = 8.45 \qquad a_{Ti^{2+}} = 10^{-2}: \quad \text{pH} = 6.45$$

$$a_{Ti^{2+}} = 10^{-4}: \quad \text{pH} = 7.45 \qquad a_{Ti^{2+}} = 1: \quad \text{pH} = 5.45$$

Reaction 4

$$TiO + 2H^+ + 2e^- = Ti + H_2O$$

Information needed:

$$E^{\circ} = -1.306\ \text{V SHE}$$

Apply Nernst equation:

$$\begin{aligned} E &= -1.306 - \frac{0.0591}{2} . \log \frac{1}{(a_{H^+})^2} \\ &= -1.306 - 0.0591\ \text{pH} \end{aligned}$$

Reaction 5

$$TiO_2 + 4H^+ + e^- = Ti^{3+} + 2H_2O$$

Information needed:

$$E^{\circ} = -0.666\ \text{V SHE}$$

Apply the Nernst equation:

$$\begin{aligned} E &= -0.666 - 0.0591 \log \frac{(a_{Ti^{3+}})}{(a_{H^+})^4} \\ &= -\ 0.666 - 0.2364\ \text{pH} - 0.0591 \log a_{Ti^{3+}} \end{aligned}$$

Reaction 6

$$Ti_2O_3 + 6H^+ + 2e^- = 2Ti^{2+} + 3H_2O$$

Information needed:

$$E^{\circ} = -0.478\ \text{V SHE}$$

Apply the Nernst equation:

$$\begin{aligned} E &= -0.478 - \frac{0.0591}{2} \cdot \log \frac{(a_{Ti^{2+}})^2}{(a_{H^+})^6} \\ &= -0.478 - 0.1773\ \text{pH} - 0.0591 \log a_{Ti^{2+}} \end{aligned}$$

Reaction 7

$$Ti_2O_3 + 2H^+ + 2e^- = 2TiO + H_2O$$

Information needed:

$$E^{\ominus} = -0.123 \text{ V SHE}$$

Apply Nernst equation:

$$E = -1.123 - \frac{0.0591}{2}.\log\frac{1}{(a_{H^+})^2}$$
$$= -1.123 - 0.0591 \text{ pH}$$

Reaction 8

$$2TiO_2 + 2H^+ + 2e^- = Ti_2O_3 + H_2O$$

Information needed:

$$E^{\ominus} = -0.556 \text{ V SHE}$$

Apply the Nernst equation:

$$E = -0.556 - \frac{0.0591}{2}.\log\frac{1}{(a_{H^+})^2}$$
$$= -0.556 - 0.0591 \text{ pH}$$

Version Considering Hydrous Oxides

Species considered: Ti, Ti^{2+}, Ti^{3+}, TiO^{2+}, TiO, $Ti(OH)_3$ and $TiO_2.H_2O$

Reactions 1 through 4 above apply unchanged. Reactions 5 to 8 are modified to represent hydration of Ti_2O_3 and TiO_2, yielding Reactions 9 to 12. Reactions 13 and 14 are required to define a significant zone of stability for the species, TiO^{2+}, thereby introduced.

Reaction 9

$$TiO_2 \cdot H_2O + 4H^+ + e^- = Ti^{3+} + 3H_2O$$

Information needed:

$$E^{\ominus} = -0.029 \text{ V SHE}$$

Apply the Nernst equation:

$$E = -0.029 - 0.0591 \cdot \log\frac{(a_{Ti^{3+}})}{(a_{H^+})^4}$$
$$= -0.029 - 0.2364 \text{ pH} - 0.0591 \log a_{Ti^{3+}}$$

Reaction 10

$$Ti_2O_3.3H_2O + 6H^+ + 2e^- = 2Ti^{2+} + 6H_2O$$

Information needed:

$$E^{\circ} = -0.248\,\text{V SHE}$$

Apply the Nernst equation:

$$E = -\,0.248 - \frac{0.0591}{2}\cdot\log\frac{(a_{Ti^{2+}})^2}{(a_{H^+})^6}$$
$$= -0.248 - 0.1773\ \text{pH}\ -\ 0.0591\log a_{Ti^{2+}}$$

Reaction 11

$$Ti_2O_3.3H_2O + 2H^+ + 2e^- = 2TiO + 4H_2O$$

Information needed:

$$E^{\circ} = -0.894\,\text{V SHE}$$

Apply the Nernst equation:

$$E = -0.894 - \frac{0.0591}{2}\cdot\ \log\frac{1}{(a_{H^+})^2}$$
$$= -0.894 - 0.0591\ \text{pH}$$

Reaction 12

$$2TiO_2\cdot H_2O + 2H^+ + 2e^- = Ti_2O_3\cdot 3H_2O$$

Information needed:

$$E^{\circ} = -0.091\,\text{V SHE}$$

Apply the Nernst equation:

$$E = -0.091 - \frac{0.0591}{2}\cdot\log\frac{1}{(a_{H^+})^2}$$
$$= -0.091 - 0.0591\ \text{pH}$$

Reaction 13

$$TiO^{2+} + 2H_2O = TiO_2.H_2O + 2H^+$$

Information needed:

$$(a_{TiO^{2+}})/(a_{H^+})^2 = 63$$

Taking logarithms:

$$2\log(a_{H^+}) - \log(a_{TiO^{2+}}) = 1.8$$

Hence, rearranging:

$$pH = \{-\log(a_{TiO^{2+}}) - 1.8)\}/2$$

For:

$a_{TiO^{2+}} = 10^{-6}$: pH = 2.1 $a_{TiO^{2+}} = 10^{-2}$: pH = 0.1

$a_{TiO^{2+}} = 10^{-4}$: pH = 1.1 $a_{TiO^{2+}} = 1$: pH = −0.9

Reaction 14

$$Ti^{3+} + H_2O + e^- = TiO^{2+} + 2H^+$$

Information needed:

$$E^\ominus = -0.100 - 0.1182\,pH$$

Apply the Nernst equation:

$$E = -0.100 - 0.0591\log\frac{1}{(a_{H^+})^2}$$
$$= -0.100 - 0.0591\,pH$$

Diagrams: Figures 3.9 and 3.10 give the diagrams for $a_{Ti^{2+}} = a_{Ti^{3+}} = a_{Ti^{4+}} = a_{TiO^{2+}} = 10^{-6}$.

Further Reading

Betts, A. J. and Boulton, L. H., Crevice corrosion: Review of mechanisms, modeling and mitigation, *British Corrosion Journal*, 28, 279, 1993.

Bodsworth, C., *The Extraction and Refining of Metals*, CRC Press, Boca Raton, Florida, 1994.

Bokris, J. O'M. and Reddy, A. K. N., *Modern Electrochemistry*, Plenum Press, New York, 1970, Chap. 8.

Davies, C. W., *Electrochemistry*, George Newnes, London, 1967, Chap. 12.
Evans, U. R., *The Corrosion and Oxidation of Metals*, Edward Arnold, London, 1971.
Fontana, M. G. and Green, N. D., *Corrosion Engineering*, McGraw-Hill, New York, 1967.
France, W. D., *Crevice Corrosion of Metals, ASTM-STP516x*, American Society for Testing and Materials, p. 164, 1972.
Glasstone, S., *Thermodynamics for Chemists*, Van Nostrand, New York, 1947.
Hatch, J. E. (Ed.), *Aluminum: Properties and Physical Metallurgy*, American Society for Metals, 1984, Chap. 7.
Hauffe, K. and Vierk, A. L., *Zeitschrift fur Physikalische Chemie*, 196, 160, 1950.
Hinshelwood, C. N., *Kinetics of Chemical Change*, Clarendon Press, Oxford, 1940.
Institute of Materials, *Proc. 8th European Syposium on Corrosion Inhibitors*, London, 1996.
Ives, D. J. G. and Janz, G. J., *Reference Electrodes*, Academic Press, London, 1961.
Kolotyrkin, Y. M., Electrochemical behavior of metals during anodic and chemical Uhlig, H. H., Electron configuration and passivity in alloys, *Z. Electrochem*, 62, 700, 1958.
Kubaschewski, O. and Hopkins, B. E., *Oxidation of Metals and Alloys*, Butterworths, London, 1962, Chap. 2.
Lewis, G. N., Randall, M., Pitzer, K. S. and Brewer, L., *Thermodynamics*, McGraw-Hill, New York, 1961.
Pourbaix, M., *An Atlas of Electrochemical Equilibria*, Cebelcor, Brussels, 1965.
Rosenfeld, I. L., *Localized Corrosion*, National Association of Corrosion Engineers, Houston, Texas, 1974, p. 373.
Sendriks, A. J., *Corrosion of Stainless Steels*, John Wiley, New York, 1979, Chaps. 4 and 5.
Wagner, C., *Seminar on Atom Movements*, ASM, Cleveland, Ohio, 153, 1951.

4

Mixed Metal Systems

4.1 Galvanic Stimulation

Contact of dissimilar metals in aqueous media is a frequent cause of premature corrosion yet, with appropriate foreknowledge, it can be controlled and with proper design and materials selection it should seldom be encountered. Nevertheless, there are several reasons why it prevails:

1. In good faith, a designer with insufficient chemical background may combine incompatible metals that separately have good corrosion resistance, unaware that the combination constitutes a corrosion hazard. This can happen, for example, by incorrectly specifying steel or stainless-steel fasteners for joining a relatively soft metal such as aluminium.
2. Contractors may assemble systems empirically, using standardized components made from incompatible metals. Such systems may have combinations from among copper and steel pipe, brass fittings, galvanized steel and aluminium bodies.
3. A component or system purposefully constructed from a single metal or from combinations of compatible metals may yet suffer damage stimulated by traces of incompatible metals introduced by contamination in various ways.
4. In areas with hard water supplies, metals in supply systems are protected naturally by coatings of deposited limescale, as explained in Section 2.2.9.2. Besides slowing corrosion generally, it can mask the effect of incompatible mixed metals so that the possible effects are not appreciated. Such experience leads to problems if transferred to soft water areas.

The following discussion first explains the nature and origin of the effect and then considers some common practical implications.

4.1.1 Bimetallic Couples

When two dissimilar metals in electrical contact are exposed to an aqueous medium, the less noble metal in general suffers more corrosion and the more noble metal suffers less than if they were isolated in the same medium. The increased attack on the less noble metal is called *galvanic stimulation.* The combination of metals producing the effect is a *bimetallic couple.*

The effect can be very intense and is a potent cause of the premature failure of the less noble metal. It does not correlate quantitatively either with differences between the standard electrode potentials of a pair of metals, $(E^{\circ}_{\text{FIRST METAL}}) - (E^{\circ}_{\text{SECOND METAL}})$ or between the corrosion potentials that each would have separately in the same environment. Hence, other factors modify the relationship between the metals in a bimetallic couple and the phenomenon is more complex than might appear at first sight.

4.1.2 The Origin of the Bimetallic Effect

The origin of the effect is as follows. In isolation in an environment in which they attract the same cathodic reaction(s), for example, in aerated neutral water:

$$\frac{1}{2}O_2 + H_2O + 2e^- = 2OH^- \tag{4.1}$$

The metals would acquire different corrosion potentials, $E_{\text{CORROSION}}$, but when they are in electrical contact, they must adopt a common potential and hence the potential of the less noble metal is raised and that of the more noble metal is depressed from their respective normal values. This is illustrated in Figure 4.1, which shows the current distribution in a system with two metals, A (more noble) and B (less noble) when brought to a common potential in the same aqueous environment, the *mixed potential*, E_{MIXED}. As a consequence, the corrosion current on metal B is increased from $i_{\text{CORROSION}}$ (B) to $i'_{\text{CORROSION}}$ (B) and that on metal A is reduced from $i_{\text{CORROSION}}$ (A) to $i'_{\text{CORROSION}}$ (A).

The total anodic and cathodic currents are balanced in the system as a whole, but the anodic and cathodic current densities are unequal on both of the two metals. The anodic current exceeds the cathodic current on the less noble metal and the cathodic current exceeds the anodic current on the more noble metal. The proportion of the *total* anodic current each carries and hence the relative intensity of attack depends on their relative

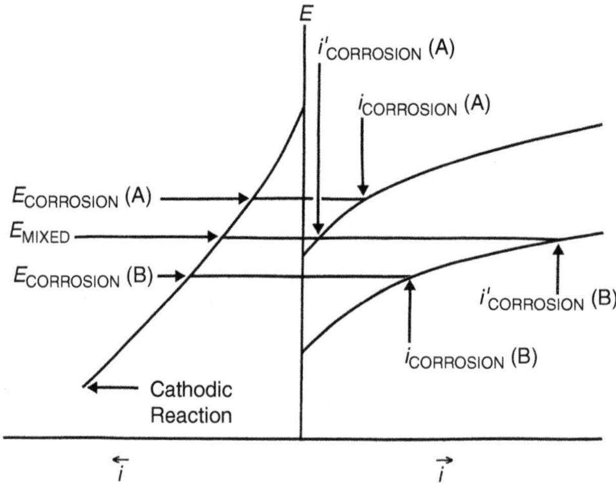

FIGURE 4.1
Current distribution on metals A (more noble) and B (less noble) in a corroding bimetallic system. If the metals are isolated, their corrosion potentials are $E_{\text{CORROSION}}$ (A) and $E_{\text{CORROSION}}$ (B) and the corrosion currents are $i_{\text{CORROSION}}$ (A) and $i_{\text{CORROSION}}$ (B). If they are connected, they acquire a common potential, E_{MIXED}. The anodic current on A falls to $i'_{\text{CORROSION}}$ (A) and that on B rises to $i'_{\text{CORROSION}}$ (B).

areas. If the noble metal area greatly exceeds the base metal area, the attack on the base metal is intense, for example, steel fasteners in copper sheet suffer rapid attack. If the base metal area greatly exceeds the noble metal area, the enhanced attack on the base metal is more widely spread and generally less severe but it may be intense locally, for example, brass fasteners in steel sheet.

4.1.3 Design Implications

4.1.3.1 Active and Weakly Passive Metals

The indiscriminate mixing of metals with large differences in standard electrode potentials in structures or components exposed to aqueous environments, rain or condensation can be a bad design fault. As examples, aluminium or galvanized steel fail rapidly in contact with copper, brass or stainless steels. Contact between such incompatible metals in plumbing, roofing, chemical plant, cooling systems, and so on, must be strictly avoided.

The photograph in Figure 4.2 shows a rather extreme but interesting example. The component illustrated was part of a pressure-sensing device exposed to an external marine environment and used to control a flow of liquid in a large pipe. The photograph shows a painted aluminium cap which was secured to the main body of the device by stainless-steel bolts. The cap was further fitted with brass unions for connection to small-bore tubing forming part of the device. The immediate cause of failure was the production of such a pressure from aluminium corrosion product formed locally within the screw threads that pieces of the aluminium cap cracked away. General corrosion of the component was insignificant. Chemical analysis revealed significant chloride contamination in the corrosion product. Recalling the discussion of the integrity of passivity in Section 3.2.5, it is probable that although stainless steel and aluminium are inherently incompatible, exposure to chloride from the marine environment exacerbates

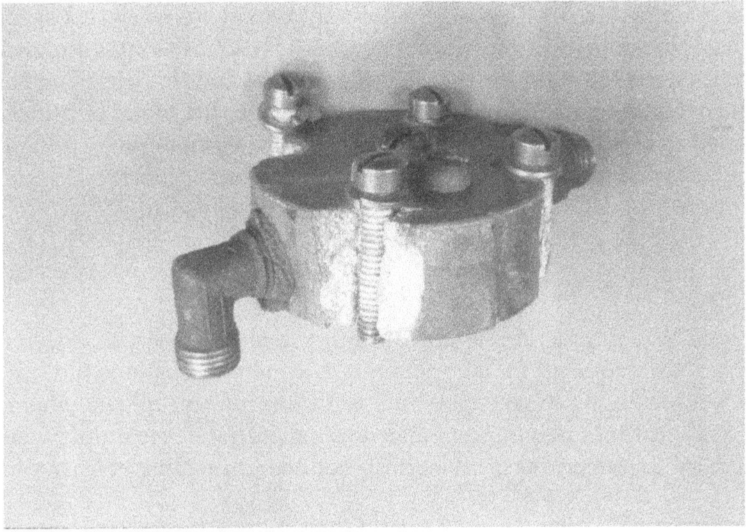

FIGURE 4.2
Aluminium cap cracked by pressure of the product of corrosion galvanically stimulated in the threads of stainless-steel bolts.

the effect by breaking down the passivity of aluminium more readily than the stronger passivity of the stainless steel.

The remedy was to respecify the materials of construction. Since it would have been inconvenient to replace aluminium as the material for the cap, the bolt material was changed to heavily cadmium-plated plain carbon steel. This alone would have been an incomplete remedy because the stainless-steel/aluminium bimetallic couple had masked the latent bimetallic couple presented by the brass union screwed into the cap. It was expected that eliminating the stainless-steel/aluminium couple would activate the brass/aluminium couple so that an essential part of the remedy was to also replace this by an equivalent cadmium-plated item.

4.1.3.2 Active/Passive Couples

Certain combinations of metals and alloys with nominally similar electrode potentials and which might on first consideration seem acceptable can in practice be electrochemically dissimilar and form very active couples. This can happen if one of a pair of metals is strongly passivated in the prevailing environment. Passivation can promote the nobility of base metals, so that in appropriate circumstances, a strongly passivating metal such as titanium, chromium or an alloy in which it is a component can act as the more noble partner in a bimetallic pair with an active base metal such as iron, or a less strongly passivating metal such as aluminium.

An example is the combination of a plain carbon steel with a stainless steel which contains a high chromium content in a mildly oxidizing environment, for example, aerated water. A bimetallic couple forms between the active surface of the carbon steel and the passive surface of the stainless steel, throwing attack on the carbon steel.

4.1.3.3 Cathodic Collectors

An electrically conducting material which does not itself support an anodic dissolution current can accelerate the corrosion of a metal with which it is in electrical contact. It does so because it supplements the area on which cathodic reactions can occur and the complementary anodic current is superimposed on the normal anodic current on the metal with which it is in contact. Such a material thus simulates a noble metal in bimetallic couples, and it is referred to as a *cathodic collector*. Examples of such materials are graphite, certain interstitial metal carbides and borides, for example, titanium diboride TiB_2 and sufficiently conducting *p*-type metal oxides, for example, copper(I) oxide, Cu_2O.

4.1.3.4 Compatibility Groups

In many designs, it is impracticable to avoid mixing metals because different properties or characteristics are required for components. A simple example is the use of mechanical fasteners, such as bolts in constructions with aluminium alloys, where the stresses imposed on screw threads may require a stronger material, for example, a steel. If metals must be mixed, the arrangement must be designed to eliminate or at least minimize bimetallic effects.

One option is to insulate incompatible metals from each other, but this is not always effective because of transport through the aqueous medium as described later in Section 4.1.3.5. It is better to select metal combinations which experience has shown to be acceptable. To aid selection, metals and alloys are assigned to the following *compatibility groups*:

Group 1: Strongly electronegative metals – magnesium and its alloys

Group 2: Base metals – aluminium, cadmium, zinc and their alloys

Group 3: Intermediate metals – lead, tin, iron and their alloys, *except stainless steels*

Group 4: Noble metals – copper, silver, gold, platinum and their alloys

Group 5: Strongly passivating metals – titanium, chromium, nickel, cobalt, stainless steels

Group 6: Cathodic collectors – graphite and electrically conducting carbides, borides and oxides

Cathodic collectors are included because although they cannot sustain *anodic* reactions, they extend the electrical conducting surface available for *cathodic* reactions supporting the anodic reaction of metals with which they are in contact.

Care is needed in applying this information because the compatibility groups do not imply that it is always safe to use metals together from within the same compatibility group. It should be used rather to warn of potentially disastrous situations if metals in different groups are mixed. Generally, the greater the separation of the groups in the above sequence from which metals are combined, the greater is the risk of galvanic stimulation. The groups would, for example, indicate that cadmium-plated or zinc-plated steel bolts may be safe to use for an aluminium alloy structure exposed outside, but that stainless-steel bolts could promote catastrophic corrosion of the adjacent aluminium. Nevertheless, the predictive value of the compatibility grouping is sometimes uncertain, as the following example illustrates. Contact with a small area of graphite may have only a small effect on copper, as the compatibility grouping suggests, but there are circumstances in which a carbon/copper couple is dangerous. If a copper surface is *nearly completely* covered with copper oxide or with a thin film of carbon from carbonization of lubricant residues during the heat treatment of fabricated products which is normally carried out in reducing atmospheres, small areas of copper exposed, for example, at abrasions, can suffer intense attack, leading to perforation by pitting. This is because a high anodic current density on the very small copper anodes is sustained by the total cathodic current collected from the very large area of the oxide or carbon-covered surface. An example of corrosion initiated in this way is illustrated by the photograph of the inside of a copper water main, given in Figure 4.3.

4.1.3.5 Indirect Stimulation

Even when two incompatible metals are not in direct contact, it is still possible for metals to be damaged by bimetallic cells set up indirectly. A common example is perforation of aluminium utensils regularly filled through copper piping with soft water containing a small quantity of dissolved carbon dioxide, such as that supplied from acidic strata or that delivered by domestic water softeners. Since the water is soft, the pipework and utensils are unprotected by chalky deposits, and traces of copper are dissolved from the piping as copper bicarbonate, from which copper metal is subsequently deposited very thinly on the aluminium at weak points in the aluminium oxide film by a replacement reaction:

$$3Cu(HCO_3)_2 + 2Al = 3Cu + 2Al(OH)_3 + 6CO_2 \tag{4.2}$$

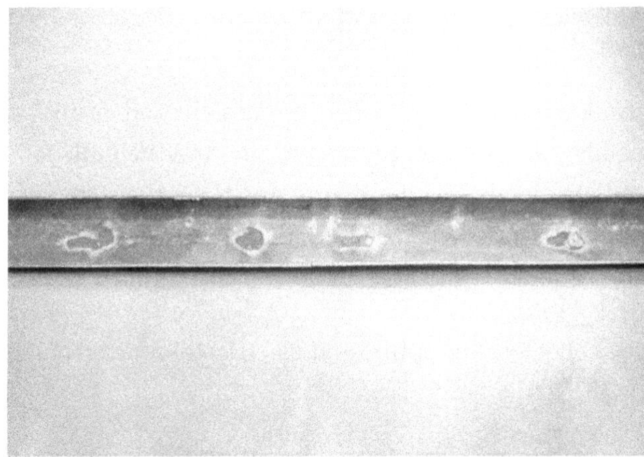

FIGURE 4.3
Corrosion of copper stimulated galvanically by carbonized lubricant residues on the inside of a copper water main.

In this way, multiple small aluminium/copper bimetallic cells are set up *in situ* on the aluminium surface. To illustrate how catastrophic the effect can be, domestic utensils fabricated from commercial pure aluminium, which can have life expectancy of ten or more years, can fail by perforation with a few weeks' use if subjected to indirect stimulation as just described. The photograph of part of the inside base of an aluminium electric kettle in Figure 4.4 illustrates the effect. It was one of several similar items in use in a large institution that had not suffered in this way. On investigation, it was found that most of the copper plumbing was protected by chalky scale deposited from the local municipal hard water supply, but the particular item that failed had been regularly filled from an unprotected spur from a water softener.

Many other examples are found in domestic and industrial plumbing, where incompatible components bought separately are assembled indiscriminately. They can include

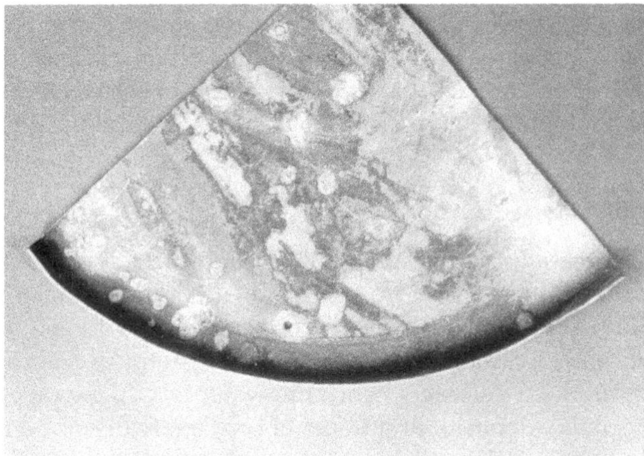

FIGURE 4.4
Part of the inside base of an aluminium kettle showing white corrosion product around multiple galvanic cells stimulated by copper deposition from soft cuprosolvent water. One cell has perforated the aluminium sheet.

copper pipe, steel pipe, galvanized (zinc-coated) steel tanks, brass compression fittings and tin/lead solders.

Systems using incompatible metals can give an illusion of security in particular situations, for example:

1. If the metals are isolated and water flow is from base to noble metals.
2. In hard water where cathodic inhibition by limescale deposits mask bimetallic couples.
3. In recirculating systems, for example, central heating systems or automobile cooling systems, where dissolved oxygen is consumed and not replenished, curtailing the cathodic reactions.

Nevertheless, such situations are unsatisfactory because a change in conditions, for example, a change in the source of water, may activate latent bimetallic effects.

4.1.3.6 Variable Polarity

The complexities of the bimetallic effect were noted in Section 4.1.1. In some systems, the intervention of environmental factors can determine which metal of a bimetallic couple is anodic and which is cathodic. The phenomenon is called *polarity reversal.* An example is the tin coating on food cans described in Section 23.1.2.3.

EXAMPLE

Following rationalization of the water supply industry, the hard water supply to a certain city is replaced by soft water. Within a short time, there is an alarming increase in the corrosion of metals in contact with the water. One particularly prevalent problem is rapid perforation of pure aluminium domestic cooking utensils. Using electron probe microanalysis, a local laboratory has identified traces of copper in rings of corrosion product surrounding the perforations. Assess the probable cause and recommend remedies that could be implemented by (1) householders, (2) plumbers and (3) the water supply company.

Solution

Domestic water supply systems are often installed by builders on the basis of local experience, as implied later in Chapter 24. The change of water supply has probably revealed mixed metal plumbing systems whose potential deleterious effects had previously been suppressed by limescale deposits from hard water. The problem described is a common example of indirect stimulation due to cuprosolvent soft water conveyed by domestic copper plumbing, producing copper/aluminium bimetallic cells *in situ* by a replacement reaction, depositing copper at weak points in the oxide film protecting aluminium surfaces.

1. Householders could counter the effect by replacing aluminium cooking utensils with copper, stainless steel or vitreous enamelled steel alternatives. The strong passive condition of stainless-steel surfaces, explained further in Chapter 8, resists the replacement reactions.
2. Plumbers could remove the problem by replacing all of the copper plumbing with an innocuous material, for example, polyethylene.
3. The water supply company could inhibit the water, for example, by polyphosphate treatment.

4.2 Galvanic Protection

4.2.1 Principle

Bimetallic effects can serve a useful purpose by using one metal to protect another. The protecting metal is the less noble of the pair and is selected for its ability to depress the mixed potential of the system to a value below the equilibrium potential for the protected metal. The principle is implemented within two contexts:

1. Application of galvanically protecting metals as coatings on vulnerable metal substrates.
2. Use of *sacrificial anodes* to protect engineering structures and systems, for example, pipelines and marine equipment. This aspect of galvanic protection is more appropriately considered in Chapters 20 and 25, within contexts which consider the application of sacrificial anodes and impressed currents as alternative means of applying cathodic protection.

4.2.2 Galvanic Protection by Coatings

4.2.2.1 Zinc Coatings on Steel

One of the most effective and inexpensive methods of protecting steel sheet used in ambient environments is to coat it with a thin layer of zinc, either by hot-dipping, that is, *galvanizing*, or more usually by electrodeposition, as described in Chapter 6. In near-neutral aqueous media, the zinc coating resists corrosion by forming a passive surface of $Zn(OH)_2$. At defects in the coating caused by abrasion and at cut edges, the exposed iron is cathodically protected by corrosion of newly exposed zinc, covering it with $Zn(OH)_2$, which stifles further attack. The development of a passive film free from contamination in urban or marine environments can be assured by prepassivating the zinc coating by chromate treatment as described in Chapter 6.

Zinc coatings can be applied to steel sheets with the dual purpose of galvanic protection and providing a substrate for paint systems. The coated material may be heat treated to convert the pure zinc coating to an iron/zinc intermetallic compound that is a better key for the paint system. An example of its use is for parts of automobile body shells described in Chapter 22.

These and other examples of galvanic protection are re-examined within wider contexts in subsequent chapters.

5

The Intervention of Stress

Environment-Sensitive Cracking: Stresses applied to a metal structure can interact synergistically with environmentally induced electrochemical effects in aqueous media causing fracture at loads far below those indicated by the nominal mechanical properties. The effects are confined to the crack paths and there is usually no significant surface damage. Two principal forms of environment-sensitive cracking are recognized:

1. SCC that describes delayed failure under static loading
2. Corrosion-fatigue that describes failure induced by cyclic loading

Embrittlement by hydrogen generated electrochemically is sometimes included with stress-corrosion cracking (SCC) and sometimes treated as a separate effect.

Failure by environment-sensitive cracking is unpremeditated and it is sometimes catastrophic. It can inflict substantial economic loss and at worst it can be life threatening. For example, it can damage the integrity of aluminium alloy airframes in flight, cause explosions of pressurized stainless-steel in chemical plants and initiate collapse of steel civil engineering structures. Naturally, such hazards attract close attention and susceptibility to cracking is often the limiting factor in a design rather than the ultimate strength of metals.

Enhanced Corrosion in Flowing or Turbulent Aqueous Media: Flowing or turbulent aqueous media can apply shear or impact forces to metal surfaces that interact synergistically with electrochemical mechanisms causing failure by severe local surface corrosion. Recognized effects include erosion–corrosion, impingement and cavitation.

5.1 Stress-Corrosion Cracking

5.1.1 Characteristic Features

SCC is a system property influenced by factors contributed by the metal and by the environment. The following features are characteristic:

Conjoint Action: Cracking is caused by the synergistic combination of stress and a specific environmental agent, usually in aqueous solution. Separate or alternate application of stress and exposure to the agent is insufficient.

Stress: Constant stress intensity in a crack-opening mode, K_1, K_2 or K_3, is sufficient. It may be applied externally or by internal strains imparted by fabrication, contraction after welding or mechanically corrected mismatch.

Environment: Conditions for SCC are highly specific and for a given metal or alloy, cracking is induced only if one or another of a particular species, the *specific agents*, is present. General corrosion or the presence of other environmental species is insufficient.

Crack Morphology: The cracks appear brittle with no deformation of adjacent metal, even if the mode of failure under stress alone is ductile.

Life-to-Failure: The life decreases with increasing stress and is the sum of two parts:

1. An induction period which determines most of the life, for example, weeks or years
2. A rapid crack propagation period, typically hours or minutes

Crack Path: The crack path is a characteristic of particular metals or alloys. For some it is *intergranular,* that is, along the grain boundaries between the metal crystals; for others it is *transgranular,* that is, through the crystals, avoiding grain boundaries; for yet others it is indiscriminate.

Features of some important alloy systems exhibiting SCC are summarized in Table 5.1, which illustrates some of the factors which influence susceptibility. They include:

1. Nature and the composition of the metal
2. Crystal structure of the metal
3. Thermal and mechanical treatments given to the metal
4. Species present in the environment
5. Temperature
6. Magnitude and state of stress

The relative importance of the factors listed above differs from metal to metal, adding to the complexity of the phenomenon and for practical reasons, separate bodies of evidence are built up by interests in the safe use of particular metal product groups, such as aluminium alloys, stainless steels or plain carbon steels. For example, users of high-strength aluminium alloys pay particular attention to the sensitivity of the alloys to heat-treatments with special reference to the structure near grain boundaries which provide the crack paths. In contrast, users of stainless steels are less concerned with such aspects, but take account *inter alia* of the fact that the cracking is especially associated with the face-centered cubic austenite phase, in which it is transgranular. Hence, to gain an impression of the problems encountered, it is more useful to consider some selected important alloy groups rather than to attempt to generalize.

TABLE 5.1

Characteristics of SCC in Some Alloy Systems

System	Specific Agent	Crack Path	Remarks
Aluminium alloys	Cl^- ions	Grain boundaries (intergranular)	Age-hardened alloys and alloys with >3% magnesium
Stainless steels	Cl^- ions	Through grains (transgranular)	Serious for austenitic steels at high temperatures
Plain carbon steels	OH^- or NO_3^- ions, hydrogen sulfide	Grain boundaries (intergranular)	The terms caustic and nitrate cracking are sometimes used
Brasses	Ammonia	Indiscriminate	Formerly called season cracking
Magnesium alloys	Cl^- and $Cr_2O_7^{2-}$ ions	Through grains (transgranular)	Mainly wrought alloys in the Mg/Al/Zn system

5.1.2 SCC in Aluminium Alloys

5.1.2.1 Susceptible Alloys

SCC afflicts some but not all aluminium alloys. Alloys at risk include those for critical applications such as airframes, especially since runway deicing materials and marine atmospheres are sources of the chloride ion that is the specific agent of corrosion. The susceptible alloys, specified by the Aluminium Association codes given in Table 9.2, Chapter 9, are:

1. Alloys that develop strength by quenching from high temperature and reheating at lower temperature to produce a dispersion of submicroscopic precipitate particles within their structures, a procedure called age hardening, explained in Chapter 9:

    Aluminium–copper–magnesium–silicon alloys in the AA 2000 series

    Aluminium–zinc–magnesium–copper alloys in the AA 7000 series

    Aluminium–magnesium–silicon alloys in the AA 6000 series

    Aluminium alloys containing lithium

2. Alloys that develop strength by mechanically working them:

    Aluminium–magnesium-based alloys in the AA 5000 series

The age-hardening alloys suffer SCC only when hardened. The aluminium–magnesium alloys are susceptible only for magnesium contents over 3%. Commercial grades of pure aluminium, aluminium–manganese alloys and aluminium–silicon alloys are immune. The crack path is intergranular and is associated with the characteristics of the grain boundary area.

5.1.2.2 Probable Causes

Aluminium alloys, in general, derive protection by passivation within the pH range 3.9 to 8.6, consistent with the Pourbaix diagram in Figure 3.4. and effects related to local corrosion such as SCC are associated with passivity breakdown. Most theories are based on this assumption.

Crack propagation during SCC is discontinuous and is generally attributed to alternating mechanical and electrochemical actions. Attention is drawn to the capacity of the depassivating chloride ions to stimulate localized corrosion at the grain boundaries, producing fissures which act as stress raisers. Stress concentrated at the fissures opens them, exposing new surfaces sensitive to further corrosion, extending the cracks and increasing the stress concentration. The alternating electrochemical and mechanical effects constitute an iterative synergistic process. The following section considers how the grain boundaries become selective sites for the electrochemical aspect of SCC.

5.1.2.3 Mechanisms

The intergranular crack path naturally focusses attention on the metal structure in the immediate vicinity of the grain boundaries and explanations are sought by two different approaches that are not necessarily mutually exclusive. One is based on anodic activity due to selective precipitation of intermetallic phases at the boundaries and the other is based on the entry of cathodically produced hydrogen along grain boundaries.

5.1.2.3.1 Grain Boundary Precipitation

A feature common to aluminium alloys susceptible to SCC is susceptibility to precipitation of intermetallic compounds at the grain boundaries.

The age-hardening alloys are formulated to promote, by a suitable sequence of heat-treatments, a submicroscopic precipitate of intermetallic compound particles highly dispersed within the crystals. Its function is to obstruct the movement of dislocations in the lattice and thereby harden and strengthen the metal. There is, however, a predisposition for some larger precipitate particles to nucleate and grow at the grain boundaries, with the result that precipitate particles are aligned along the grain boundaries and surrounded on both sides by very narrow bands of metal that are depleted in the alloy components contributing to the precipitate, known as the *grain boundary denuded zone*. There is thus established a continuous network of three closely adjacent bands of metal with different compositions, the precipitate, the denuded zone and the matrix, between which there are possibilities for electrochemical interactions. The role of the specific agent, the chloride ion, is to initiate these interactions by assisting in local depassivation.

In aluminium–copper–magnesium–silicon alloys of the AA 2000 series, the grain boundary precipitate includes the compound, $CuAl_2$, and the denuded zone is anodic to both the precipitate and the matrix, creating possibilities for local action electrochemical cells, somewhat similar in action to the bimetallic cells described in Chapter 4. In high-strength aluminium–zinc–magnesium–copper alloys of the AA 7000 series, the precipitate, $MgZn_2$, is anodic to the matrix, creating local action cells of reverse polarity. Other age-hardening alloys can be considered similarly, but with differences, in detail.

Heat-treatment programmes for the age-hardening alloys have a pronounced effect on the susceptibility to SCC and must be carefully designed and controlled to minimize it. In practice, the alloys are strengthened by two sequential heat treatments. The first is a high temperature *solution treatment*, at a temperature in the range 500–510°C, according to the alloy, in which the alloy components responsible for the hardening are dissolved and the metal is quenched in water to retain them in solid solution. The second is a low temperature *artificial aging treatment* at temperatures of the order of 170°C, according to the alloy, in which the alloy components in enforced solution go through a complex sequence of clustering, coalescence and precipitation. During the aging treatment, the strength and hardness of the metal rise to a maximum *peak hardness* and then as the precipitation process nears completion, they decline and the metal is then described as *overaged*. From a purely mechanical viewpoint, the best combination of properties is reached at peak hardness but unfortunately, this coincides with maximum susceptibility to SCC. In critical applications, reduction of sensitivity to SCC is more important than fully exploiting the mechanical properties available, and a suitable compromise is attained by controlled overaging. There are alternative approaches, including two-stage aging and modified solution treatment or quenching. Alloys containing copper can be *naturally aged* by holding them at ambient temperature after quenching, but materials hardened in this way are very susceptible to SCC.

Plain aluminium–magnesium alloys derive their strength from work hardening and not age hardening, but because the solubility of magnesium falls rapidly with falling temperature, alloys with more than 3 mass % magnesium tend to precipitate the compound Mg_5Al_8 at the grain boundaries. This compound is anodic to the matrix, creating an internal bimetallic system. If the magnesium content is much higher than this, there is a risk of slow precipitation continuing at ambient temperature and greater susceptibility to SCC can develop over the long term. Magnesium contents of alloys for critical applications do not normally exceed 5 mass %.

5.1.2.3.2 Hydrogen Embrittlement

An alternative or supplementary theory of SCC is based on experimental observations that although hydrogen is virtually immobile in the matrices of aluminium alloys at ambient temperature, it can permeate the grain boundaries in quenched aluminium alloys containing magnesium. The idea is that hydrogen generated by a complementary cathodic reaction at the metal surface enters the metal via the grain boundaries, and migrates to stressed locations, where it reduces the bond strength between metal atoms so that they separate by *decohesion* under the applied stress, an effect well-known in other contexts.

5.1.2.3.3 Structural Factors

Intergranular SCC is expected to be sensitive to the metal structure. In rolled sheet and plate and sections fabricated by extrusion, the metal crystals or grains are elongated in the direction of working. Therefore, a much greater grain boundary area is presented to a stress applied normal to the working direction, the *short transverse* direction, than to stresses parallel to or across it, the *longitudinal* and *transverse* directions, and the sensitivity to SCC is correspondingly greatest in the short transverse direction. Problems due to sensitivity in this direction are confined mainly to thick sections, such as rolled plate and forgings. In thinner sheet, where the stresses are longitudinal or transverse, the manganese and chromium contents, often present in alloy formulations, are said to assist resistance to SCC by promoting elongated grain structures. The chromium content is also credited with improving resistance to SCC by inducing general precipitation of aging precipitates throughout the grains in preference to grain boundaries.

5.1.3 SCC in Stainless Steels

5.1.3.1 Brief Description of the Phenomena

SCC in stainless steels exemplifies the uncertainties and controversy which surround the topic in general. For a long time it was assumed that SCC was confined to austenitic stainless steels formulated to be composed almost entirely of the face-centered *austenite* phase, as described in Chapter 8, and that cracking was always transgranular, that is, through the grains. Ferritic stainless steels, composed of the body-centered *ferrite* phase, also described in Chapter 8, were assumed to be immune, introducing a crystallographic consideration. It is now known that SCC can afflict ferritic steels and that the crack paths can be intergranular in both ferritic and austenitic steels in special circumstances, but the detail is uncertain, due mainly to less experience of the effects in ferritic steels because, in general, they are not used for such critical applications as austenitic steels. Further complexity is sometimes introduced by including cracking in stainless steels in which carbides are present at the grain boundaries, due to incorrect heat treatment or to unsuitable selection of steels for welding, as described in Chapter 8. The following discussion is restricted to transgranular cracking in correctly heat-treated austenitic stainless steels which constitute the main body of the practical problems encountered.

5.1.3.2 Environmental Influences

The specific agent is chloride and the problem is serious only at elevated temperatures, especially for environments of boiling or superheated aqueous solutions. The time-to-failure is reduced as the temperature is raised and as the chloride content in the environment increases. The effective chloride content may be much higher than the nominal content if there is a concentration mechanism, by for example, cyclic evaporation of condensates or evaporation of water from dilute chloride solutions leaking from or dripping on to a heated pressure vessel. The susceptibility to cracking is not particularly sensitive to pH, but in general, it is more severe in more acidic solutions. In principle, the dissolved oxygen content is an important factor because it provides the oxygen absorption cathodic reaction to complement the electrochemical contribution to SCC, but it is difficult to divorce from the temperature-sensitivity of the effect because the solubility of oxygen in aqueous media diminishes rapidly as the temperature is raised and, in any case, since SCC is a phenomenon associated with solutions at high temperatures, the oxygen content is inevitably low.

5.1.3.3 Sensitivity to Steel Structure and Composition

The relationships between the susceptibility to SCC and steel compositions, condition and microstructures are complex, often known only empirically and sometimes not reproducible.

The cracks are transgranular and can exhibit multiple branching. The presence of some ferrite in nominally austenitic stainless steels can block the cracks suggesting that ferrite is more resistant to SCC. Worked austenitic steels do not have sufficient ferrite for the effect to have any practical value, but castings of equivalent steels have non-equilibrium structures as described in Chapter 8 and can contain as much as 13% ferrite, persisting from an uncompleted peritectic reaction, which affords a useful degree of protection. For the same reason, *duplex steels* that are formulated to contain both austenite and ferrite whether cast or wrought are less susceptible than austenitic steels.

The influence of the composition of an austenitic stainless steel on its susceptibility to SCC is not well defined. High nickel and chromium contents improve resistance to SCC but, judging from published information, the influences of other alloy components, including molybdenum and carbon, which are often components of austenitic stainless steels, do not seem to be reproducible, perhaps because of uncertainties in the interactions between alloy components.

5.1.3.4 Mechanisms

The electrochemical contribution to SCC is confirmed by the academic observation that it can be suppressed by cathodic protection and enhanced by galvanic stimulation, but there is no consensus on how stress and environmental attack interact. Two stages need explanation, that is, induction and crack propagation.

5.1.3.4.1 Induction

Induction must be associated with local breakdown of the passive film, establishing active/passive cells that initiate the electrochemical contribution. Since cracking is transgranular, intrinsic features of the austenite grains or their surfaces must be invoked and not features of their boundaries, as for aluminium alloys. One idea is that local perforation of the passive layer occurs by penetration of microscopic slip steps; this is supported by a relationship between susceptibility to SCC and the energy of faults in the sequence of atom layers, the *stacking fault energy*. Another is that chloride-induced pits can act both as stress raisers and sites for active/passive cells. Dissolution is driven by the potential difference between the large passive metal surface and the active metal at the crack tips.

5.1.3.4.2 Propagation

The cracks can follow crystallographic planes but do not do so consistently and this must be reflected in any proposed explanation. There are two approaches, depending on the view taken on whether the electrochemical or mechanical contribution propagates the cracks. One view asserts that cracks are propagated by electrochemical dissolution of a narrow strip of metal and the role assigned to the mechanical contribution is to maintain the metal at the crack tip in an active condition by causing it to yield, preventing re-passivation. The other view considers that the metal is separated in tension and the electrochemical contribution is to resharpen the crack tip when it is blunted by yielding. There is probably some truth in both ideas.

5.1.4 Stress Corrosion Cracking in Plain Carbon Steels

5.1.4.1 Brief General Description of the Phenomena

Plain carbon steels are susceptible to SCC in the presence of one or another of specific agents that include nitrates, bicarbonates and alkalis and also hydrogen sulfide. Cracking in alkalis is sometimes called *caustic cracking*. All of these agents have technological significance. Additions of nitrates and *pH correction*, which means raising the pH, are sometimes used in moderation to inhibit general corrosion of steel and hydrogen sulfide is encountered in oilfield sour liquors. The crack paths are characteristically intergranular.

SCC promoted by nitrates can occur at ambient temperatures, but it is more often experienced with superheated solutions under pressure where the temperature exceeds 100°C. Caustic cracking is associated with hot solutions with high pH values, but mildly alkaline solutions also pose risks because the pH can be raised locally, for example, by evaporation of water from leaks in heated vessels such as boilers.

5.1.4.2 Mechanisms

Unlike aluminium alloys and stainless steels, plain carbon steels are active in neutral aqueous solutions, but increasing the pH or raising the potential by nitrate additions induces passivity due to the formation of a surface film of magnetite, Fe_3O_4, as suggested by the Pourbaix diagram in Figure 3.3. At any faults in the film, the exposed iron is the anodic partner in an active/passive cell and consequent intense local attack stimulated by the large surrounding passive surface can provide the electrochemical component for SCC. Explanations for the synergism between electrochemical and mechanical components of SCC can be sought along the same lines as for the aluminium and stainless-steel systems described above. The intergranular crack path suggests that structural features at the grain boundaries, such as the presence of carbides, facilitate the depassivation that can initiate and sustain cracking.

Cracking in hydrogen sulfide solutions is a form of hydrogen embrittlement in which the sulfide stimulates catalytic activity assisting the entry of cathodically produced hydrogen into the metal where it can promote decohesion. Steels with exceptionally low sulfur content <0.002% are produced for application to structures at risk because they are more resistant.

5.2 Corrosion Fatigue

5.2.1 Characteristic Features

Corrosion fatigue cracking differs from SCC in two respects:

1. *Stress* – The cracking is induced by a *cyclic* applied stress in a crack-opening mode.
2. *Environment* – Combinations of environmental conditions and metal compositions for corrosion fatigue are not specific.

Fatigue failure of a metal is characterized by delayed fracture associated with cracking induced by cyclic stresses well below the maximum constant stress that the material can bear. A good general account is given in Reed–Hill's standard text on physical metallurgy*. The life expectancy depends not only on the properties of the metal and on the stress system, but also on the nature of the environment. All environments influence the fatigue life but the distinction of most concern is the marked reduction in life expectancy when an air environment is replaced by aqueous media, for which the term *corrosion fatigue* implies interaction between mechanical and electrochemical factors. The overall effect is usually quantified on a comparative basis by laboratory fatigue tests, usually by rotating bending or reverse bending tests using internationally standardized samples. By convention, the results are presented graphically as *S–N* curves, in which the number of stress cycles needed to cause failure, *N*, is expressed as a function of the stress amplitude, *S*. A logarithmic plot of cycles-to-failure is used only for convenient presentation. The life-to-failure becomes progressively shorter as the cyclic stress amplitude is raised. A typical example of effect of an aqueous environment is illustrated in Figure 5.1, which compares

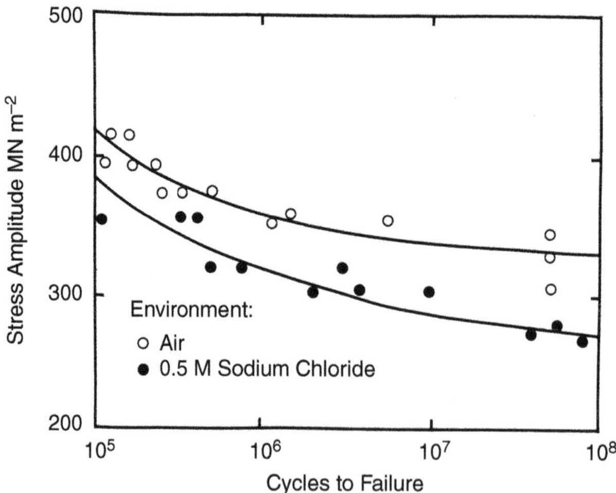

FIGURE 5.1
Comparison of fatigue lives of AISI 316 stainless steel in air and in 0.5 M sodium chloride solution. Results of tests on 3 mm thick plate in reverse bending at 24 Hz. Nominal composition of steel 17% Cr, 12% Ni, 2.5% Mo, 0.06% C. (Reproduced with permission of the Institute of Materials Minerals and Mining, London).

* Cited in Further Reading.

the fatigue life of an austenitic stainless steel in 0.5 M sodium chloride solution with its fatigue life in air. For steels cyclically loaded in air, there is an *endurance limit*, that is, a safe limiting stress below which failure does not occur, but for nonferrous metals and for metals cyclically loaded in aqueous media, there is no safe limit and on the basis of empirical information and experience, stresses are restricted to values ensuring survival for a prescribed design life, typically 10^7 or 10^8 cycles.

5.2.2 Mechanisms

5.2.2.1 Mechanical Events Causing Failure

Fatigue failure of a metal is the culmination of a complex sequence of events, summarized briefly as follows:

Crack initiation – An incubation period that can constitute a significant part of the fatigue life precedes crack initiation. In this period, the characteristic event is the formation of *persistent slip bands*, PSBs, a term describing highly localized regions of cyclically yielding metal at the surface which can be revealed by etching, that are the precursors of cracks.

Crack propagation – Cracks grow initially along crystallographic planes within the PSBs, a process called *Stage 1 cracking*, but usually then change direction to follow planes normal to the maximum applied stress, a process called *Stage 2 cracking*, often leaving microscopic striations on the fracture surface, marking successive increments of crack advance.

Overload fracture – Ultimately, the load-bearing section is reduced to an area which is unable to sustain the maximum stress and the metal separates by fast fracture.

5.2.2.2 Intervention of the Environment

The events listed above provide considerable scope for intervention by the environment to assist the surface events associated with the PSBs. Several environment/stress interactions are possible in principle, as follows:

1. Local corrosion damage, for example pitting, intensifying the surface stress
2. Stimulated dissolution of active metal by the yielding within PSBs
3. Dissolution of metal exposed by the rupture of passive films over PSBs
4. Localized embrittlement by hydrogen absorbed from cathodic reactions

According to circumstances, all are possible contributing factors, founded on well-established concepts. Local corrosion damage can act in the same way as other surface imperfections in raising the local surface stress above the nominal applied stress, thereby reducing fatigue life, but in many other examples of environment-accelerated failure, there is no local corrosion damage and one of the other factors must be invoked. These imply various forms of electrochemical activity stimulated by the applied cyclic stress. They have been identified experimentally by monitoring corrosion current transients synchronized with stress cycles. For a passivating metal, such as a stainless steel, their amplitudes increase if depassivating ions are present and for an active metal, such as steel, their development depends on the accumulation of plastic strain in the PSBs. These observations provide evidence for the electrochemical activity envisioned.

5.3 Enhanced Corrosion in Flowing or Turbulent Aqueous Media

5.3.1 Erosion–Corrosion

Erosion–corrosion refers to accelerated corrosion induced by rapid relative motion or turbulence of an aqueous environment at the surface of a metal, which disperses the protective mechanisms that apply in quiescent media. As the name implies, one aspect of the effect is to scour the metal surface, interfering with the formation of films that would otherwise offer protection. It applies both to easy passivating metals and to active metals that derive some protection from resistance polarization from surface films as deduced from the calculation given in Section 3.2.3. If the moving liquid carries solid particles in suspension, the scouring effect is so much the greater. The relative movement also tends to sweep away the boundary layer of static liquid present at the metal/liquid interface. This further stimulates corrosion by dispersing concentration polarization, especially for the oxygen reduction reaction and for anodic reactions yielding soluble products.

5.3.2 Cavitation

Cavitation is produced by the impingement of a liquid on a metal at high velocity and is particularly associated with rapidly moving metal parts in water, such as propellers and pump impellers. The relative movement induces a hydrodynamic condition that creates streams of small cavities in the liquid which collapse, delivering multiple sharp blows at the metal surface. The disturbance disrupts protective films, leading to a corroded surface with a characteristic rough and pitted appearance.

Material properties which militate against cavitation and also erosion–corrosion are good general corrosion resistance, strong passivating characteristics and hardness of the metal.

5.4 Precautions against Stress-Induced Failures

Irrespective of the particular characteristics of systems susceptible to SCC, corrosion-fatigue or other stress/environment interactions, it is prudent to take obvious precautions related to materials, specific agents and the magnitude of applied stresses.

5.4.1 Materials

Probably the greatest hazard is failure to appreciate the problems. It is essential to be aware of the metal/environment systems that are susceptible to SCC and take the available information into account when specifying materials. A few examples suffice to make the point. If a stainless steel must be used in an application in which there is a possibility of exposure to chlorides, it is wise to consider one of the steels with lower susceptibility to stress-corrosion cracking such as a duplex steel, even if there is a cost or technical incentive to consider a more susceptible steel, such as a fully austenitic steel. As already noted, it is prudent to forego the maximum mechanical properties available for age-hardening aluminium alloys by controlled overaging to reduce susceptibility to stress-corrosion cracking.

5.4.2 Environments

In all load-bearing applications of metals susceptible to stress-corrosion, it is essential to be sure that neither obvious nor latent sources of the specific agent(s) are present. Examples of obvious sources are process liquors, marine atmospheres and deicing salts containing chlorides and oil installations contributing sour liquors. Examples of less obvious sources are spills, leaks, condensates and salts leached from insulation by drips from above.

It is not always appreciated that fatigue is a system property and is environmentally sensitive. Information given in the context of material characterization alone is usually based on fatigue tests conducted in air, but it is essential that even comparative data is evaluated and matched to the actual stress system and environment of the application envisioned.

5.4.3 Stresses

If metals must be used in a system in which environmentally sensitive cracking or corrosion-fatigue is a possible hazard, viable designs are still possible if stresses in all parts of a structure are kept below known safe limits, using information from experience or reliable sources. A designer would not deliberately exceed nominal safe stresses, discounted by the usual factors of safety but, in practice, design stresses may be inadvertently supplemented by other stresses. Stress-corrosion cracking and corrosion-fatigue are surface-sensitive and stress-raisers can markedly increase nominal surface stresses. For this reason, abrupt changes in section and artefacts due to poor surface finishing should be eliminated. Metals can, and frequently do, carry internal stresses which are additive to the external loads. Some of these stresses are imparted by the metal supplier, especially stresses locked in by mechanical working and by differential contraction following heat treatments. Others are imparted by contraction after welding and by careless assembly. These stresses can and should be minimized by careful working and heat-treatment practices and if possible, eliminated by stress-relieving heat treatments. Surface compressive stress is considered beneficial and shot peening is sometimes advocated to introduce it.

5.4.4 Geometry

Suitable geometric design to control stresses and stress-raisers is important to reduce the risks of environmentally sensitive cracking and also to curtail damage by erosion–corrosion and cavitation corrosion. Systems containing flowing liquids should be designed for minimum relative movement and avoid profiles that contribute to turbulence.

5.4.5 Monitoring

In critical situations where environmentally sensitive cracking is possible, it may be advisable to regularly monitor the condition of the metal and of the environment during service. The information can be used to forestall impending damage and adverse environmental changes.

Further Reading

American Society for Metals, *Hydrogen Embrittlement and Stress-Corrosion Cracking*, Metals Park, Ohio, 1984.

Gangloff, R. P. and Ives, M. B. (Eds.), *Environment-Induced Cracking of Metals*, National Association of Corrosion Engineers, Houston, Texas, 1990.

Hatch, J. E. (Ed.), *Aluminum Properties and Physical Metallurgy*, American Society for Metals, Metals Park, Ohio, 1984, Chapter 7.

Newman, R. C. and Procter, R. P. M., Stress corrosion cracking, 1965–1990, *Br. Corros. J.*, 25, 259, 1990.

Parkins, R. N., Predictive approaches to stress-corrosion cracking failure, *Corrosion Sci.*, 20, 147, 1980.

Proc. Symp. on Effect of Hydrogen Sulfide on Steel, Canadian Institute of Mining and Metallurgy, Edmonton, Canada, 1983.

Reed-Hill, R. E., *Physical Metallurgy Principles*, Van Nostrand, Princeton, New Jersey, 1964, p. 559.

Scamens, G. M., *Hydrogen-Induced Fracture of Aluminum Alloys*, Symposium on Hydrogen Effects in Metals, TMS AIME, Warrendale, Pennsylvania, 1980, p. 467 and *Aluminum*, 59, 332, 1982.

Staehle, R. W. (Ed.), *Stress-Corrosion Cracking and Hydrogen Embrittlement of Iron Base Alloys*, National Association of Corrosion Engineers, Houston, Texas, 1977.

Staehle, R. W., Forty, A. J. and Van Rooyen, D. (Eds.), *Fundamental Aspects of Stress-Corrosion Cracking*, National Association of Corrosion Engineers, Houston, Texas, 1969.

Talbot, D. E. J., Martin, J. W., Chandler, C., and Sanderson, M. I., Assessment of crack initiation in corrosion fatigue by oscilloscope display of corrosion current transients, *Metals Technology*, 9, 130, 1982.

6

Protective Coatings

Often, the best strategy to control corrosion of an active metal is to apply a protective surface coating. A further advantage is that it is usually possible to combine the protective function with aesthetic appeal that is valuable for vehicles, domestic and business consumer durables, architecture and even for ephemeral packaging where sales potential is enhanced by attractive appearance. The principal applications of surface coatings on metals is for the protection of iron and steel products and structures because of the sheer volume of production, estimated at 7.5×10^8 tonnes per annum worldwide. Equivalent coatings are applied to other metals, but sometimes as much for appearance as for protection.

Casual observation shows that most steel products and structures are coated either with other metals or with paints. Most metal coats are produced by electrodeposition but the older practice of *hot-dipping*, in which the substrate is passed through the liquified coating metal, is still important for producing zinc-coated, *galvanized*, steel.

The term, *paints*, describes filled polymer binding media, including air-drying and stove-drying formulations, irrespective of trade descriptions. Other less conspicuous but important coatings are generated by reactions in which the metal surfaces themselves participate. Three are of particular interest, phosphate coatings that provide a key for paints on steel, anodic films that extend the applications of aluminium and chromate coatings that can protect or provide a key for paints on nonferrous metals.

6.1 Surface Preparation

6.1.1 Surface Conditions of Manufactured Metal Forms

The surfaces of manufactured metal forms, including sheet, plate and sections, are seldom suitable for the application of coatings. Poor surface quality, detritus and contamination undermine the adhesion of electrodeposits and detract from the protection afforded by paints. They must be prepared before coatings are applied. The preparation needed depends on the metal and the manufacturing and fabricating processes by which they are produced.

6.1.1.1 Rolled Surfaces

The surfaces of rolled products, plates, sheets and sections can be and usually are heavily contaminated. The kind of contamination depends on the metal, as described below for steels and aluminium alloys.

Steels are hot-rolled at high temperatures, of the order of 1100°C, and the steel surface oxidizes to form thick *mill scale* that spalls away from plain carbon steels under the roll pressure. It is more difficult to detach from some alloy steels and patches may remain on the finished hot-rolled product. When hot-rolling is finished, the metal temperature is still high enough to form a final scale as it cools. If the hot-rolled product is to be cold-rolled, the scale must first be removed. During cold-rolling, the roll gap is lubricated with mineral oils and traces remain on the surface of the finished metal sheet. Any intermediate heat-treatments in the schedule provide further opportunities for scale formation.

Different criteria apply to aluminium alloys that are hot-rolled at temperatures of the order of 500°C. They do not form thick scales like steels, but are subject to a different surface effect. The deformation in the roll gap introduces differential velocities between the periphery of the roll and the metal being rolled, so that the metal slips backwards on the roll face on entry and forwards on exit. The relative motion induces transfer of aluminium particles to the roll face that are carried around and transferred back to the metal as the roll revolves. As successive ingots are rolled, a steady state is established in which the roll surface carries a constant thin coating of the metal and the extreme surface of the emerging stock is a layer of compacted flake, a few μm thick. The roll is lubricated and cooled with a recirculated soluble oil/water emulsion that interacts with the metal surface yielding metal soaps, degrading and contaminating the lubricant. The flake is compacted on the metal surface in this unsavory environment, producing a surface condition illustrated for pure aluminium in Figure 6.1. Subsequent cold-rolling further compacts and burnishes the flake surface and also impresses its own characteristic artifact on the surface, that is, reticulation, a network of very fine, shallow surface cracks, offering traps for further contamination.

6.1.1.2 Extruded Surfaces

Extruded aluminium sections may be scored, but they are reasonably clean because they are abraded against the hard dies through which they are forced.

6.1.1.3 Surfaces of Press-Formed Products

Aluminium alloys, plain carbon steels and austenitic stainless steels have extensive applications for hollowware produced by press-forming, usually by deep-drawing in which rolled sheet is deformed by a punch into a shaped cavity. The sheet is restrained by pressure pads to prevent wrinkling and lubricated with mineral oil to control friction. The metal is

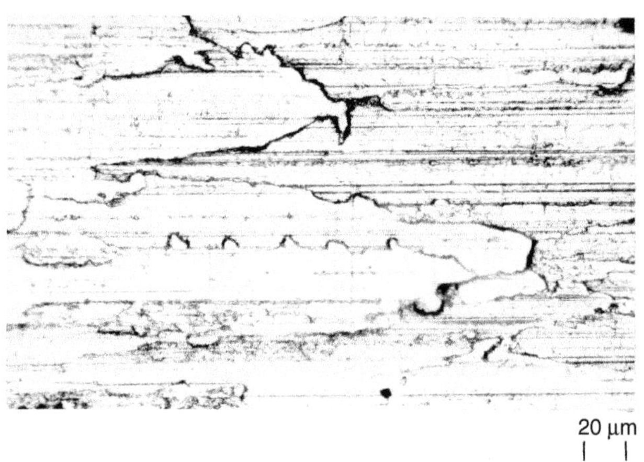

FIGURE 6.1
Compacted flake on hot-rolled pure aluminium (AA 1100) surface.

heated adiabatically by the heavy deformation and the transient surface temperature can be high enough to carbonize some of the lubricant. The resulting contamination is loose to the touch but very resistant to removal by physically and chemically active solutions.

6.1.2 The Cleaning and Preparation of Metal Surfaces

6.1.2.1 Descaling Steels

6.1.2.1.1 Manual Methods

Manual methods such as grinding, shot and sand blasting and flame cleaning are slow, subjective and costly. Recourse to them is justified only if chemical descaling or weathering is inappropriate. Shot and sand blasting remove scale and roughen the surface providing a useful key for paint. Flame cleaning detaches scale by differential expansion and burns off oil and grease, leaving a dry surface that can be painted.

6.1.2.1.2 Weathering

Exposing steel to the weather for about six months exploits atmospheric corrosion to undermine mill scale so that it is easily removed. Weathering is often an economic way of preparing structural steelwork for painting. Subsequent cleaning must be conscientious, because residual scale stimulates corrosion under paint. Regions shielded from rain may not be adequately cleaned and need remedial attention. When correctly practiced, weathering and subsequent phosphate treatment yields substrates suitable for painting.

6.1.2.1.3 Acid Pickling

Mill scale on hot-rolled steel products is removed by pickling in sulfuric acid. It is effective, economical and adaptable to online use in steel mills.

Dissolution of isolated iron oxides in acids is very slow, but oxides in contact with the metal can be attacked by *reductive dissolution*, meaning that the oxide is dissolved by a cathodic reaction. This exploits the structure of scale formed on iron and steel, explained more fully in Chapter 7. Scale formed at temperatures <575°C comprises an inner layer of magnetite, Fe_3O_4 against the metal, overlaid by an outer layer of haematite, Fe_2O_3. Scale on hot-worked steel is cracked due to differential contraction on cooling. Iron is anodic to magnetite and acid percolating through the cracks to the metal establishes local cells in which cathodic dissolution of magnetite:

$$Fe_3O_4 + 2e^- + 8H^+ \rightarrow 3Fe^{2+} + 4H_2O \tag{6.1}$$

is stimulated by the anodic dissolution of iron:

$$Fe \rightarrow Fe^{2+} + 2e^- \tag{6.2}$$

The dissolution reactions proceed along the magnetite/iron interface, loosening the scale so that it falls away.

Scale formed at temperatures >575°C has a third oxide layer, wüstite, FeO, interposed between the metal and the magnetite layer, that is particularly amenable to reductive dissolution, because it is unstable at temperatures <575°C and on cooling it tends to decompose yielding a eutectoid mixture of iron and magnetite:

$$4FeO \rightarrow Fe + Fe_3O_4 \tag{6.3}$$

In pickling, multiple dissolution cells rapidly consume the former wüstite layer, detaching the overlying scale. Hot-working finishing temperatures above 575°C are, therefore, conducive to subsequent easy descaling.

The simplest procedure is to immerse the material as pieces or in coils in static 0.05–0.1 molar sulfuric acid at 60–80°C, treated with restrainers that are organic additives to inhibit dissolution of the descaled metal. The acid is replenished and replaced as necessary. High-speed pickling is needed to descale a moving continuous rolled steel strip. The strip is flexed to crack the scale and passed through 0.25 molar acid at 95°C flowing countercurrent to it in a sequence of tanks, after which the metal is washed and dried online. Looping pits are strategically placed to enable the end of a strip to stop momentarily for the following strip to be joined by welding, permitting continuous operation.

A hazard in pickling is cathodic hydrogen generated on descaled metal:

$$2H^+ + 2e^- = H_2$$

At ambient and slightly elevated temperatures, hydrogen can diffuse into iron and steels, raising blisters at subcutaneous flaws. In thick sections, it can be occluded and cause embrittlement by decohesion, as referred to in Chapter 5 in the context of stress-corrosion cracking. These problems can be averted by attention to the pickling programme and to the metal quality.

There are various modifications to the basic pickling process, including the application of externally impressed cathodic current to protect exposed descaled metal or impressed anodic current to suppress hydrogen embrittlement.

6.1.2.2 Cleaning Aluminium Surfaces

Many applications of aluminium depend on a combination of corrosion resistance and attractive appearance that can be imparted by metal surface finishing treatments. Whatever treatment is applied, the cleanliness of the metal surface is critically important in determining both the initial quality of the finish and its durability.

The surface of the metal as it emerges from casting or fabrication operations is both physically and chemically contaminated. Physical contamination can comprise lubricant residues from working operations, adherent dirt and particles embedded from mechanical polishing. Chemical contamination includes oxidation and corrosion products.

The principles used to remove contamination are straightforward, but acceptable procedures are more difficult to devise. The constraints are:

1. Costs of cleaning
2. Ability of cleaning agents to wet the surfaces
3. Dispersion and trapping detritus to prevent redeposition
4. High quality requirements for cleaned surfaces
5. Toxicity and environmental impact of cleaning agents

6.1.2.2.1 *Solvent Degreasing*

Physical detritus and oily residues are loosened by organic solvents. The choice and application of a solvent is restricted by cost, flammability, low flash points and the toxicity of chlorinated hydrocarbons and of benzene. Kerosene is reasonably safe, but because its volatility is low it evaporates slowly and it can leave oily residues.

The most effective application of solvents is in vapour degreasing in which articles are first immersed in hot solvent to remove the worst contamination and then cooled and held in its vapour to attract condensate that washes off oils and detritus. There are many variants of the process in both the selection of solvent and physical arrangement of the plant. The choice of solvent is not easy. Trichloroethane, $C_2H_3Cl_3$ is effective, but although safe in many respects, it is a candidate for depleting the ozone layer. Trichloroethylene, C_2HCl_3 is less likely to cause ozone depletion, but it is more toxic and there is a small risk of dangerous exothermic polymerization catalysed by traces of aluminium chloride formed from hydrogen chloride produced by photolytic decomposition of trichloroethylene. To counter this risk, commercial trichloroethylene is *stabilized* by adding a soluble amine as a scavenger for hydrochloric acid. For this reason, the more stable solvent, tetrachloroethylene is preferable.

A limitation of degreasing with organic solvents is that they are inefficient in removing water-soluble contaminants such as soaps and soluble oil residues from lubricants used in machining and metal forming. This can be countered by using a two-phase cleaning fluid in which an aqueous phase and an organic solvent are used together and are made compatible by incorporating an emulsifying agent, such as triethanolamine with oleic acid that produces a soap with both hydrophilic and oleophilic properties. Such a system dissolves or loosens both oil- and water-soluble contaminants so that they can both be washed away by water. Arrangements can be made to apply the fluids by immersion in tanks or by spray.

6.1.2.2.2 *Alkaline Cleaning*

Physical cleaning by solvent degreasing is followed by alkaline cleaning, dissolving a surface layer of metal to release firmly adherent oxidation products and subsurface contamination, for example, metal soaps compacted under flake as illustrated in Figure 6.1.

Reference to the aluminium Pourbaix diagram in Figure 3.4 shows that it is easier to attack aluminium in alkaline than in acidic solutions. A pH value in the range 9 to 11 is sufficient. Sodium hydroxide attacks the metal too readily and damages the surface by etching it. Cleaning solutions are, therefore, based on sodium carbonate, Na_2CO_3, and sodium tertiary phosphate, Na_3PO_4, whose etching action can be suppressed by the addition of sodium metasilicate, Na_2SiO_3, assumed to be due to the deposition of a very thin layer of hydrated silica on the freshly exposed aluminium surface. In practice, solutions containing all three salts together with surfactants to promote wetting are used. The concentrations and relative proportions of the components vary within the range 10 to 60 g dm^{-3} and are determined by experience. Operation is by immersion of the metal for a few minutes in the solution at a temperature in the range 75°C to 95°C.

The suitability of nominally cleaned metal for further surface treatments cannot be judged by appearance alone, but may be assessed by spraying it with water. A tendency for the water falling on the metal to break into globules indicates inadequate cleaning.

6.1.2.3 *Preparation of Aluminium Substrates for Electrodeposits*

Electrodeposits do not adhere well to aluminium and aluminium alloy substrates unless they are specially prepared because of interference from the air-formed oxide film and the reactivity of the metal. Although alternative, more complicated techniques are sometimes used, the most widely applied surface preparation is to deposit a thin layer of zinc on the aluminium surface by chemical replacement. The process is reliable, inexpensive and suitable for most aluminium alloys.

Cleaned aluminium alloy articles are immersed in an alkaline solution of sodium zincate, Na_2ZnO_2, which dissolves the oxide film and replaces it with the zinc coating.

Solutions are prepared by dissolving zinc oxide in sodium hydroxide solution. The relative proportions are fairly critical to produce adherent zinc deposits. The replacement reaction is represented by coupled anodic and cathodic reactions, such as:

$$2Al + 8OH^- \rightarrow 2AlO_2^- + 4H_2O + 6e^- \tag{6.4a}$$

$$3ZnO_2^{2-} + 12H^+ + 6e^- \rightarrow 3Zn + 6H_2O \tag{6.4b}$$

Typical solutions contain 40 to 50 g dm^{-3} of zinc oxide and 400 to 450 g dm^{-3} of sodium hydroxide but there are many proprietary variants. A representative treatment is 1 to 3 minutes immersion at ambient temperature, but it may be varied to suit alloy compositions and pretreatments given to the metal.

6.1.2.4 Chemical and Electrochemical Polishing of Aluminium Alloys

Several metals can be brightened by controlled chemical or electrochemical dissolution in specially formulated solutions. There are niche applications for other metals, but the principal industrial use is for aluminium alloys for the following reasons:

1. For many decorative uses, value added to the product justifies the costs of treatment.
2. The bright surface can be protected by anodizing, as described in Section 6.4.2.
3. The colour of the bright surface matches nickel/chromium coating on steel or plastics and mixtures of them in, for example, automobile trim, presenting an integrated appearance.
4. Chemically and electrochemically polished aluminium can be not only bright but specular, that is, yielding undistorted image clarity, a quality needed for reflectors.

6.1.2.4.1 Chemical Polishing

Whether a solution etches, polishes or passivates an aluminium surface depends on which of the following responses it elicits:

1. Direct anodic dissolution of the metal

$$Al \rightarrow Al^{3+} + 3e^- \tag{6.5}$$

2. Formation of a very thin oxide or hydrated oxide film on the metal surface

$$2Al + 3H_2O \rightarrow Al_2O_3 + 6H^+ + 6e^- \tag{6.6}$$

   that does not suppress Reaction 6.5 but exercises diffusion control over it.
3. Formation of a thicker or qualitatively different film that suppresses Reaction 6.5, passivating the metal.

Polishing corresponds to the second of these responses because it is well established that brightening is associated with a tangible film on the metal, qualitatively similar to but thinner than that formed in anodizing, *qv.* The role of the film is to control dissolution, overriding differential dissolution due to the effects of crystallographic features of the metal surface which would otherwise be responsible for etching.

Several proprietary solutions produce the required brightening. They are mainly, but not exclusively, based on mixtures of concentrated phosphoric, sulfuric and nitric acids. Phosphoric acid is the preferred acid for dissolution because it attacks aluminium reasonably uniformly. The nitric acid is the oxidizing agent that forms the film. The less expensive sulfuric acid is sometimes used as a partial replacement for phosphoric acid as an economy. A representative solution is 80.5% phosphoric acid + 3.5% nitric acid + 16% water by volume. An addition of 0.01 to 0.2 mass % of copper improves the brightness obtained. The solution is used hot, for example, at 90°C, and the work is immersed in it for a time between 15 seconds and 5 minutes. The quality of the finish depends on skill and experience in working the solution since the best results are obtained when optimum concentrations of aluminium phosphate (6 to 12 mass %) and free phosphoric acid (65 to 75 mass %) have been established during use. Some patented brand names of solutions in widespread use are ALCOA's *R5*™ and Albright and Wilson's *Phosbrite 159*™.

As concerns with the environment and safety at work increase, polishing baths containing nitric acid that evolve nitric oxides may give way to solutions free from nitric acid and based on phosphoric acid and sulfuric acid, such as Albright and Wilson's *Phosbrite 156*™, that yields bright but nonspecular finishes, suitable for less critical decorative applications such as anodized domestic and bathroom fittings.

6.1.2.4.2 Electrochemical Polishing

The very best specular finishes are given by electrochemical polishing. One of the more successful is the British Aluminium Company's process used under the registered trademark *Brytal*. The process is expensive and works well only for high purity aluminium, that is, >99.8%, preferably 99.99%, and aluminium–magnesium alloys based on these purities, but the specular reflectivity of Brytal treated and anodized reflectors for use in, for example, laser cavities, is unrivalled except by silver, which tarnishes easily. The electrolyte is a solution of 20 mass % sodium carbonate, Na_2CO_3, and 6 mass % sodium tertiary phosphate, Na_3PO_4, both of high purity, in deionized water yielding a pH of 11.0 to 11.6. The bath is heated by steam coils to 90°C and operated with an anodic potential of +7 to 12 V applied to the work against an aluminium cathode. It is agitated to disperse concentration polarization, with care to avoid disturbing a sludge deposited in the tank. Manipulation is expensive through the jigging and provision of electrical connections to the work.

Like chemical polishing, electrochemical polishing is associated with diffusion control of the dissolution process by a film on the metal surface. It is assisted by an overlying viscous liquid phosphate complex of unknown constitution. The condition correlates with a *polishing range* in the anodic polarization characteristics for the process in which current density is independent of potential.

6.2 Electrodeposition

6.2.1 Application and Principles

Electrodeposition provides a convenient means of applying a protective coating of one metal on another. A particular attraction is the ability to control the thickness and uniformity of the coatings fairly accurately through the Faradaic equivalent of the total electric charge passed, allowing for any inefficiencies. This permits the surface qualities of expensive metals such as nickel and tin to be imparted to metals of lower value, such as steels, by applying them as very thin coatings. Often a coating is required to have both protective value and aesthetic appeal and in that case there is a further economy in using electrodeposition because it is usually possible to produce bright decorative finishes that require no subsequent treatment. Furthermore, electrodeposition is versatile, commending it for applications ranging from batch plating of small parts to continuous plating of strip emerging at high speed from steel mills.

6.2.1.1 Cathodic and Anodic Reactions

Provided that hydrogen evolution is suppressed, metals can be deposited from aqueous solution by application of a sufficient cathodic potential. The species from which the metal is extracted can be cations, for example:

$$Zn^{2+} + 2e^- \rightarrow Zn \tag{6.7}$$

oxyanions, for example:

$$Cr_2O_7^{2-} + 14\,H^+ + 12e^- = 2Cr + 7H_2O \tag{6.8}$$

or complex cyanide anions, for example:

$$[Cu(CN)_2]^- + e^- = Cu + 2CN^- \tag{6.9}$$

Deposition from cations is relatively simple and is usually the option with lowest cost, but deposition from anions is sometimes essential to obtain deposits that match particular applications. The cathodic potential needed is the sum of the equilibrium potential for the reaction and the total polarization for deposition at the required rate. Activation, concentration and resistance polarization can all affect the nature of the deposit.

The return current is led into the plating bath by anodes that are preferably made of the same metal as that deposited, because the depletion of metal ions at the cathode is continuously replenished by matching dissolution at the anode, for example:

$$Zn(anode) \rightarrow Zn^{2+} = 2e^- \tag{6.10}$$

For some metals, notably chromium, it is impracticable to use anodes of the metal deposited and inert anodes must be used, in which case the anodic reaction by which the return current enters the bath is oxygen evolution:

$$H_2O \rightarrow ½O_2 + 2H^+ + 2e^- \tag{6.11}$$

The depleted metal ion content of the bath is replenished by additions of metal-bearing salts.

6.2.1.2 Hydrogen Discharge

Metal deposition reactions compete with the discharge of hydrogen:

$$2H^+ + 2e^- = H_2(gas) \tag{6.12}$$

for which, as derived in Chapter 3:

$$E_{HYDROGEN} = -0.0591\ pH \tag{6.13}$$

Plating solutions are concentrated and the metal activity is of the order of unity so that the potential at which the metal is deposited, $E_{DEPOSITION}$, is close to the standard potential for the electrode process, E°. A particular metal can be deposited from aqueous solution only if the cathodic reaction by which it is deposited is thermodynamically favoured or kinetically easier than Reaction 6.12. This depends on which of the following circumstances applies:

1. $E_{DEPOSITION}$ is less negative than $E_{HYDROGEN}$ and the discharge of hydrogen is not thermodynamically favoured. An example is copper deposited from Cu^{2+}.
2. $E_{DEPOSITION}$ is more negative than $E_{HYDROGEN}$ but the hydrogen overpotential on the metal is so high that the discharge of hydrogen is either suppressed as for zinc deposited from Zn^{2+} and tin deposited from Sn^{2+} or limited to a tolerable degree as for nickel deposited from Ni^{2+}.
3. $E_{DEPOSITION}$ is so much less negative than $E_{HYDROGEN}$ that the hydrogen discharge reaction cannot be polarized sufficiently to reach the potential needed to deposit the metal. This is why aluminium and magnesium cannot be electrodeposited from aqueous solution.

6.2.1.3 Throwing Power

Except for materials with the simplest geometry, such as flat sheets, it is seldom possible for all parts of the substrate to be equidistant from anode surfaces. Therefore, the current density and hence the thickness of the deposit is not uniform but varies from place to place due to differences in the ohmic resistance of the path through the electrolyte to an anode. The distribution of deposition current due to this factor is the *primary current distribution*. Anodes are shaped and disposed around the substrate so that it is as uniform as possible.

The actual current distribution is usually more uniform than the primary current distribution suggests because of the effects of polarization. The total potential across the electrodeposition cell is the sum of three components:

1. Any difference between the equilibrium potentials for reactions at the anode and cathode
2. The potential drop across the ohmic resistance of the electrolyte
3. Polarization at the cathode and at the anode

The potential applied across the cell is constant so that variations in the potential drop due to variations in ohmic resistance are offset by opposite variations in the potential associated with polarization. The effect is to deflect deposition current away from places where the current density is high, so that the deposit is more uniform than would be predicted from the primary current distribution. The quality of a plating bath that promotes this uniformity is called its *throwing power*. A high throwing power is essential for deposits applied to complex shapes.

6.2.1.4 Illustrative Selection of Deposition Processes

Processes yielding useful metal deposits share essential principles and a comprehensive review would involve unnecessary repetition. Detailed empirical descriptions are readily available in handbooks and trade journals. Sections 6.2.2 through 6.2.6 describe the deposition of selected metals chosen for their wide commercial applications and because as a group they sufficiently illustrate the electrochemical background and operating procedures to generally characterize electrodeposition. The selection comprises nickel, copper and chromium which are the components of a common protective and decorative coating system, tin which is used extensively in contact with food stuffs and zinc that is applied to steels for protection in external atmospheres. Typical formulations for solutions used in depositing these metals are collected for reference in Table 6.1.

TABLE 6.1

Repesentative Solutions for Electrodeposition

Metal	Solution	Solutes	g dm^{-3}	Temperature °C	Cathode Current Density A dm^{-2}
1. Nickel	Watts solution and derivatives	$Ni_2SO_4 \cdot 6H2O$	240	0–50	0.5–2.0
		$NiCl \cdot 6H_2O$	45		
		H_3BO_3	30		
		Additives in derivatives			
2. Copper	Cyanide strike solution	CuCN	23	40	0.5
		NaCN	34		
		Na_2CO_3	15		
3. Copper	Concentrated cyanide solution	CuCN	55	80	3
		KCN	100		
		Na_2CO_3	7.5		
		Brightening agents			
4. Chromium	Chromate solution	CrO_3	400	45	20
		H_2SO_4	4		
5. Tin	Acid sulfate solution	$SnSO_4$	55	45	30
		H_2SO_4	100		
		$C_6H_5(OH)SO_3H$	100		
		β-naphthol	1 + gelatin		
6. Tin	Acid fluoborate solution	$SnSO_4$	54	35	10–50
		HBF_4	120		
		Proprietary additives			
7. Tin	Halide solution	$SnCl_2$	75		
		NaF	25		
		KF	50		
		NaCl	45		
		β-naphthol			
8. Tin	Alkaline stannate solution	$Na_2SnO_3 \cdot 3H_2O$	90	80	1.0–2.5
		NaOH	7.5		
		$NaC_2H_3O_2 \cdot 3H_2O$	15		
9. Zinc	Acid sulfate solution	$ZnSO_4 \cdot 7H_2O$	350	30	3.0
		NH_4Cl	30		
		$NaC_2H_3O_2 \cdot 3H_2O$	15		
		H_2SO_4 to pH 4			
		Proprietary additives			
10. Zinc	Alkaline cyanide solution	$Zn(CN)_2$	60	50	2.0
		NaCN	25		
		NaOH	50		

Pourbaix diagrams are sometimes useful in interpreting the chemistry of electrodeposition of particular metals and the appropriate versions are those for ion activities of unity, because they apply to metal concentrations close to those used in practice.

6.2.2 Electrodeposition of Nickel

6.2.2.1 General Considerations

Electrodeposited nickel is widely used and one of the main applications is as part of a corrosion and tarnish-resistant coating system for steel in which it is deposited on a copper undercoat and overlaid by a bright chromium deposit, a finish colloquially known as 'bright chromium plating'.

Nickel cannot be plated successfully from complex ions and commercial baths are based on simple nickel salts, using consumable anodes. The character of the deposit is sensitive to the pH of the plating solution and serviceable coatings can be produced only if the pH is controlled at a value close to pH 4. The object is to avoid codepositing nickel hydroxide which occurs at higher values without risking embrittlement by hydrogen which is occluded at lower values. This can be explained conveniently using the nickel/water Pourbaix diagram for $a_{Ni^{2+}} = 1$, given in Figure 6.2.

The diagram indicates that sufficient nickel is soluble for viable electrodeposition only if pH < 6. The metal is electrodeposited by depressing the potential of the substrate to lie in the domain of stability for nickel. The line representing the equilibrium potential for hydrogen evolution is below the equilibrium potential for nickel deposition if pH > 4.2, suggesting that it might be possible to deposit nickel without discharging hydrogen in the pH range 4.2 to 6. However, practical processes operate with a cathodic overpotential and

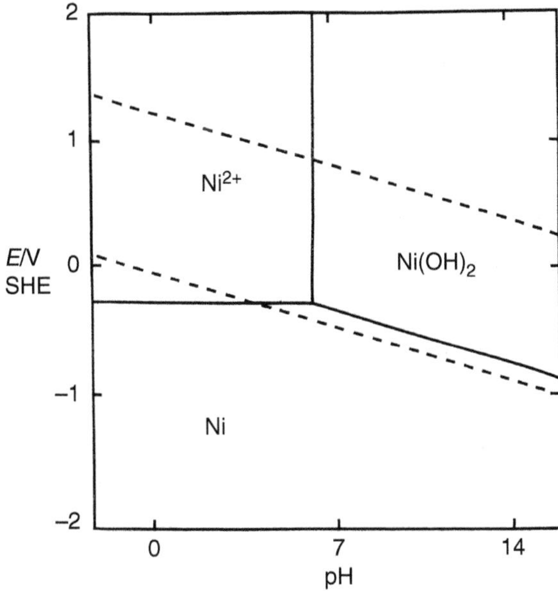

FIGURE 6.2
Partial Pourbaix diagram for the nickel/water system at 25°C, $a_{Ni^{2+}} = 1$. Domain for the stability of water shown by dotted lines.

some hydrogen is always evolved. Without countermeasures, hydrogen evolution poses the following problems:

1. If the initial pH > 5, the depletion of H^+ ions at the cathode causes the local pH to drift into the $Ni(OH)_2$ domain and nickel hydroxide is incorporated in the deposit.
2. If the pH < 3, hydrogen is evolved so rapidly that some is occluded in the deposit.

To solve the problem, boric acid is added to the solution as a pH buffer to maintain a constant characteristic hydrogen ion activity, corresponding to pH 4, by the incomplete dissociation:

$$H_3BO_3 + H_2O \leftrightarrows [B(OH)_4]^- + H^+ \tag{6.14}$$

Some practical problems must also be addressed in formulating a plating bath, as follows:

1. Hydrogen bubbles nucleating at preferred sites on the cathode divert the deposition current, leaving local depressions in the deposit known as *pits*. This is countered with anti-pitting agents, which are wetting agents to detach bubbles and oxidizing agents, such as hydrogen peroxide, that modify the cathodic reaction to produce water instead of hydrogen:

$$2H^+ + H_2O_2 + 2e^- = 2H_2O \tag{6.15}$$

2. Impurities embrittle and discolour the plate and interfere with subsequent chromium plating and both the salts and the consumable anodes must be pure. Unfortunately, pure nickel anodes passivate easily, interrupting the current and this is averted by adding depassivating Cl^- ions to the solution.

A formulation meeting these requirements is the *Watts solution*, named for its originator, using nickel sulfate as the main source of nickel with a typical composition given in Table 6.1. It is suitable for industrial applications but the deposits are soft, coarse grained and dull. It must be modified for bright decorative deposits, hard deposits and high-speed plating.

6.2.2.2 Bright Nickel Plating

Bright plating solutions are Watts solutions with organic additives in empirical proprietary formulations. They contain brighteners that poison easy nucleation sites on the substrate refining the grain and levellers that even out undulations. The structure of a bright deposit is featureless except for alternating light and dark bands, parallel to the surface, visible in microsections.

The buffered value of pH can be adjusted in the range pH 3 to pH 5 as required by additions of sulfuric acid or nickel carbonate. The high-purity nickel anodes may be dosed with oxygen or sulfur to assist dissolution and are enclosed in fabric bags to contain the residues.

Steel, iron, aluminium and zinc alloys are attacked by the acidic nickel plating solution and are prepared to receive nickel by depositing an acid-resistant copper undercoat from alkaline solution. Aluminium and its alloys are first depassivated by sodium zincate treatment.

6.2.2.3 *Other Nickel Plating Processes*

Raising the proportion of nickel chloride in the Watts solution confers the following benefits:

1. The greater ionic mobility of Cl^- ions allows higher current densities.
2. The deposits are harder.

The concentration of boric acid must be raised and the pH reduced to allow for the greater pH drift accompanying the high current density. These formulations are more expensive.

There are other nickel electrodeposition processes that are used less often, but are based on the principles described, for example, using nickel fluoborate and nickel sulfamate as the nickel source.

6.2.2.4 *Electroless Nickel/Phosphorus Plating*

It is convenient to include a complementary process, electroless nickel/phosphorus plating, with corresponding electrodeposition. It is used on a much smaller scale but it has special characteristics that recommend it for some applications. It is based on the reduction of nickel chloride to metallic nickel with sodium hypophosphite, catalysed by the metal substrate on which a nickel/phosphorus alloy is deposited.

A basic composition for an acidic electroless nickel plating bath is:

Nickel chloride, $NiCl \cdot 6H_2O$	30 g dm^{-3}
Sodium hypophosphite, $NaH_2PO_2 \cdot H_2O$	10 g dm^{-3}
Sodium hydroxyacetate, $NaC_2H_3O_3$	50 g dm^{-3}

Nickel chloride is the source of nickel, the hypophosphite is the reducing agent and the sodium hydroxyacetate functions as a buffer and complexing agent.

The solution is operated at a temperature >90°C with pH in the range 4–6. It can yield a semi-bright deposit at a rate of ~0.015 mm per hour. The reaction path is not clear, but the overall reaction is:

$$Ni^{2+} + (H_2PO_2)^- + H_2O \rightarrow Ni + 2H^+ + (H_2PO_3)^- \tag{6.16}$$

Side reactions produce elemental phosphorus which is incorporated into the deposit and hydrogen gas which must be prevented from adhering to the deposit by agitating the bath.

The drift to higher pH values in the plating bath due to the production of hydrogen ions must be corrected by regular additions of sodium hydroxide to maintain a value in the correct range. If the pH is too low, deposition is unacceptably slow and if it is too high, the bath is unstable and may deposit nickel spontaneously. Nickel chloride and the hypophosphite are continuously replenished as they become depleted. Iron, steel, aluminium and nickel can be coated directly by immersion, but cathodic polarization is required to initiate the reaction on certain other metals, including copper and copper alloys.

A coating contains 6–10 mass % of phosphorus and, as deposited, it is described as 'amorphous', that is, it is unresponsive to x-ray and electron diffraction, but if it is heat treated at ~400°C for half an hour it yields a structure with equal quantities of metallic nickel and nickel phosphide, Ni_3P.

The special features of electroless deposits that accrue from the method of production and the incorporation of phosphorus include:

1. Uniform thickness, with no concept of 'throwing power' as for electrodeposits
2. Freedom from porosity
3. Greater hardness than for electrodeposited nickel, that may be further increased by heat treatment to produce a structure containing nickel phosphide if required
4. Resistance to tarnishing in ambient atmospheres

Some special applications are derived from these characteristics. The uniformity of thickness is exploited to extend the application of nickel deposits to complex shapes and to the interiors of tubes that are difficult to coat in other ways. The freedom from porosity permits very thin deposits to be used for protective coatings. The hardness available together with resistance to tarnishing can be exploited in wear-resistant applications including critical parts, such as electrical contacts and instrument parts.

Electroless nickel deposition has certain disadvantages in comparison with corresponding electrodeposition. It is an expensive process due to the high price of sodium hypophosphite, slow deposition and the close control required for temperature and pH. Variants of the process have been proposed to accelerate the deposition rate and improve the stability of the bath.

6.2.3 Electrodeposition of Copper

6.2.3.1 Acid Sulfate Baths

Acceptable deposits of copper can be obtained from acidic copper(II) sulfate solution, using organic additives to refine the structure that is otherwise coarse grained. These deposits have little application as protective coatings because copper is a galvanic stimulant to base substrates such as steel exposed at defects in the coating. However, copper deposition from acidic baths has applications in electroforming, because it is easy to build up thick deposits as illustrated by its former use for typeface in printing.

6.2.3.2 Alkaline Cyanide Baths

The most important current use of electrodeposited copper is as an undercoat plated from alkaline copper cyanide solutions on steel, forming a suitable substrate for comprehensive protective coating systems. Alkaline solutions have the advantage over acidic solutions that copper can be plated directly on base metals including steel, zinc and aluminium with much less risk of undermining the adhesion of the coating by chemical interaction with the substrate. Even so, the solution from which copper is first applied to steel, the copper *strike solution*, is relatively dilute, to discourage deposition of copper by chemical displacement. A typical strike solution, containing CuCN and NaCN, using copper plates as consumable anodes is given in Table 6.1. The balance of solutes is to obtain a pH in the range 11.5–12.5. The coat is struck with a low current density at moderate temperature.

The two cynanides interact and dissociate, yielding cuprocyanide anions:

$$CuCN + NaCN = Na^+ + [Cu(CN)_2]^- \tag{6.17a}$$

and the cathodic reaction depositing copper is:

$$[Cu(CN)_2]^- + e^- = Cu + 2CN^- \tag{6.17b}$$

Deposition of the complex, negatively charged anion at a cathode leads to high activation polarization, diminishing the effect of easy nucleation sites and yielding a bright deposit. The coating is built up by depositing more copper on the strike coat with a higher anode current density, from a more concentrated solution, with a typical composition given in Table 6.1.

6.2.4 Electrodeposition of Chromium

6.2.4.1 Applications

Electrodeposited chromium is used extensively to give a brilliant, decorative and tarnish-resistant finish. The deposits are very thin, usually <1 μm, and by themselves they offer little protection because they are passive, but porous and galvanic corrosion stimulated at a substrate such as steel exposed by the pores can undermine and detach the coat. For this reason, chromium is deposited over an initial thicker nickel deposit that provides the main protection. The nickel must be bright to benefit from the brilliance imparted by the chromium.

Industrial applications exploit the hardness of chromium deposits that can approach 900 Brinell, and for this purpose, thick deposits, for example 25 μm, are applied directly to steel for wear resistance. Examples include bearing surfaces and surfaces mating to close tolerances such as the contacts of measuring gauges.

6.2.4.2 Principles of Deposition

It is difficult to produce metallic electrodeposits of chromium from aqueous solution and its successful development owes as much to empirical craft as to electrochemical science. Chromium deposited from Cr^{3+} cations is unsuitable for decorative or industrial use because it contains entrained oxide and a viable electrodeposition process is based on solutions of chromate ions that yield deposits with the required characteristics.

The chromium source is >99% pure chromic acid anhydride, CrO_3 that yields several chromium-bearing species in mutual equilibrium in aqueous solution:

$$CrO_3 + H_2O = H_2CrO_4 \tag{6.18a}$$

$$CrO_3 + H_2O = 2H^+ + CrO_4^{2-} \tag{6.18b}$$

$$CrO_3 + H_2O = H^+ + HCrO_4^- \tag{6.18c}$$

$$2CrO_3 + H_2O = 2H^+ + Cr_2O_7^{2-} \tag{6.18d}$$

in all of which the chromium is hexavalent. The relative abundance of these species is pH dependent and even in undisturbed solution, it is complex, but in the dynamic situation at a cathode with the high current density used in electrodeposition it is uncertain. Selecting one of these species, $Cr_2O_7^{2-}$ to characterize cathodic reactions that deposit the metal:

$$Cr_2O_7^{2-} + 14\,H^+ + 12e^- = 2Cr + 7H_2O \tag{6.19}$$

To add to the complexity, the process does not work properly unless a small quantity of one of certain other acids is added to the solution. In practice, sulfuric acid is added in sufficient quantity to give a CrO_3/H_2SO_4 ratio between 50 and 150.

6.2.4.3 Operation of Chromic Acid Baths

It is impracticable to use consumable chromium anodes because they are a more expensive source of the metal than chromic acid anhydride. A further disadvantage is that anodic dissolution of chromium is so much more efficient than cathodic deposition that it produces a surplus of chromium in the bath, partly in the form of the unwanted Cr^{3+} ion. Insoluble anodes of a lead–tin or lead–antimony alloy are used and the depletion of the chromium content is replenished by additions to the solution as required.

The concentration of chromic acid anhydride in the bath is 250 to 400 g dm^{-3}. The cathode efficiency is very low, about 15%, and this, together with the high valency state of the chromium, imposes a heavy consumption of electric power for deposition. Raising the chromium content of the electrolyte raises its conductivity, but this is offset by reduced cathode efficiency. From experience, the concentration and current densities are optimized to secure the most economic operation. This turns out to be 400 g dm^{-3} with lower current densities for thin bright deposits and 250 g dm^{-3} with high current densities for industrial hard deposits.

Some hexavalent chromium ions are reduced at the cathode to the trivalent ions, Cr^{3+}. These are reoxidized at the anode, catalyzed by lead dioxide, PbO_2 that forms on the anode surface and the proportion of chromium in the form of Cr^{3+} is kept in check by using anodes with a large surface area. A passive layer of lead chromate formed on anode surfaces during idle periods must be cleaned off before resuming deposition.

6.2.4.4 Quality of the Deposit

To produce a bright deposit, the temperature and current density must be optimized to lie within a restricted *bright plating range* as indicated in Figure 6.3. Outside of this range, the coating has an unacceptable appearance.

The throwing power of the chromate bath is poor and sometimes negative because of the inverse relation between cathode efficiency and current density. However, for decorative coatings, it is less important than the ability of the system to yield bright deposits over the whole substrate. These effects cause difficulty in controlling the process and its success is due to careful jigging, the use of auxiliary anodes, dummy cathodes and shields to supplement or deflect local currents and strict temperature control, for example 45°C ± 1°C.

The existence of the bright plating range, the role of SO_4^{2-} in assisting deposition and the very low cathode efficiency are all difficult to explain. It is known that a thin membrane is formed on the cathode during deposition, perhaps by a supplementary reaction, yielding an oxide species, for example:

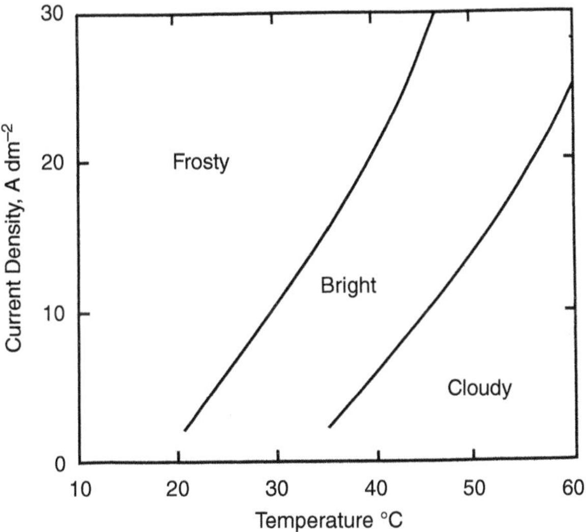

FIGURE 6.3
Influences of temperature and current density on the appearance of chromium electrodeposited on nickel substrates.

$$Cr_2O_7^{2-} + 8H^+ + 6e^- = Cr_2O_3 + 4H_2O \quad (6.20)$$

Resistance polarization for metal deposition reactions introduced by the film is probably one factor contributing to the brightness of the deposit.

6.2.5 Electrodeposition of Tin

6.2.5.1 *General Principles*

The Pourbaix diagram for $a_{Sn^{2+}} = a_{SnO_2^{2-}} = a_{HSnO_2^-} = 1$, given in Figure 6.4, indicates that tin is sufficiently soluble for electrodeposition if the pH of the solution is either <0.5 or >14. This applies to ions formed by interaction of tin with water, that is, Sn^{2+}, Sn^{4+}, SnO_3^{2-} and $HSnO_2^-$. The diagram cannot give the further information that the halide ions, Cl^- and F^- extend the solubility of tin in the intervening pH range by forming complex anions, $SnCl_3^-$ and SnF_3^-.

The minimum potentials for tin deposition lie significantly below the corresponding potentials for hydrogen at all values of pH, but hydrogen evolution is suppressed, because the hydrogen overpotential on tin is very high.

Three types of commercial electroplating baths are in use, strongly acidic solutions, strongly alkaline solutions and halide solutions. Acidic and halide solutions contain divalent tin as Sn^{2+}, $SnCl_3^-$ or SnF_3^-, whereas alkaline solutions contain quadrivalent tin as, SnO_3^{2-}, so that, theoretically, the same quantity of electricity yields only half as much tin from alkaline as from acid or halide baths. The inferior power economy of alkaline baths is compounded by a lower cathode efficiency, that is about 75%, compared with >95% for acid baths and by lower conductivity of the electrolyte. The tin deposits all have a matt finish. It is technically possible but not commercially viable to deposit

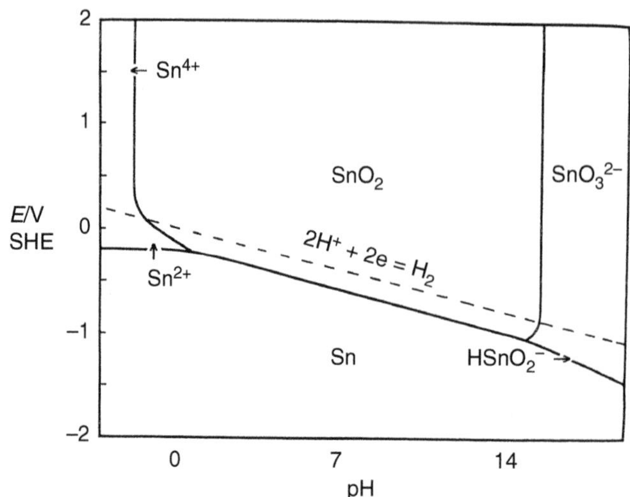

FIGURE 6.4
Pourbaix diagram for the tin/water system at 25°C. $a^{2+}_{Sn} = a_{SnO_2^{2-}} = a_{HSnO_2^-} = 1$.

bright tin. Matte coatings are brightened by flash heating the substrate to above 232°C to melt the tin.

6.2.5.2 *Acid Tin Baths*

6.2.5.2.1 Acid Sulfate Baths

Acid sulfate solutions are often used for high-speed deposition in tinplate manufacture. As expected, deposits from the simple Sn^{2+} ions:

$$Sn^{2+} + 2e^- = Sn \tag{6.21}$$

are coarse grained, but the structure is refined by deactivating easy nuclei with phenol or cresol, solubilized as the corresponding sulfonic acid derivatives:

$$C_6H_5OH(\textit{phenol}) + H_2SO_4 = C_6H_5(OH)SO_3H(\textit{phenol sulfonic acid}) + H_2O \tag{6.22}$$

supplemented by organic additives such as β-naphthol and gelatine. Table 6.1 gives a typical formulation. For high-speed deposition, as in tinplate lines, the bath is operated at 45°C with a current density of 30 A dm^{-2} and it is agitated to disperse concentration polarization. For batch plating, it is operated at a lower temperature, 25°C, and lower current density. Pure tin consumable anodes are used to replenish the depleted tin content:

$$Sn = Sn^{2+} + 2e^- \tag{6.23}$$

The anode current efficiency exceeds that of the cathode and to prevent upward drift of the tin content, the consumable tin anodes are supplemented by inert anodes.

6.2.5.2.2 *Fluoborate Baths*

Fluoborate baths also use tin(II) sulfate as the source of tin, but the low pH is maintained by fluoboric acid instead of sulfuric acid as in the formulation given in Table 6.1. Proprietary additives are used instead of phenol or cresol.

6.2.5.3 Alkaline Stannate Baths

The expense of alkaline stannate baths is justified when high throwing power is essential for complex shapes. A typical formulation is given in Table 6.1. The increased throwing power is due to the high cathodic overpotential for deposition from the complex ion SnO_3^{2-}, in which the tin is quadrivalent, by the cathodic reaction:

$$SnO_3^{2-} + 6H^+ + 4e^- = Sn + 3H_2O \tag{6.24}$$

The depletion of quadrivalent tin ions is replenished by dissolution of consumable tin anodes:

$$Sn + 3H_2O = SnO_3^{2-} + 6H^+ + 4e^- \tag{6.25}$$

but it is accompanied by a minority reaction producing some divalent ions: SnO_2^{2-}:

$$Sn + 2H_2O = SnO_2^{2-} + 4H^+ + 2e^- \tag{6.26}$$

that are subsequently oxidized to the quadrivalent ions by dissolved oxygen:

$$2SnO_2^{2-} + O_2 = 2SnO_3^{2-} \tag{6.27}$$

unbalancing the anodic and cathodic reactions and causing the tin content to drift upwards. This is dealt with using supplementary inert anodes, controlling the anode current density and adjusting the composition of the solution as required.

6.2.5.4 Halide Baths

Halide baths are also used for tinplate manufacture. They are solutions of tin(II) chloride, $SnCl_2$, with alkali metal fluorides and chlorides, and β-naphthol as an additive as in the formulation given in Table 6.1. The components interact and ionize yielding the anions $SnCl_3^-$ and SnF_3^-:

$$SnCl_2 + NaCl = Na^+ + SnCl_3^- \tag{6.28}$$

$$3SnCl_2 + 3NaF = 3Na^+ + 2SnCl_3^- + SnF_3^- \tag{6.29}$$

The structure of the deposit is refined due to the high polarization associated with deposition from complex ions:

$$SnCl_3^- + 2e^- = Sn + 3Cl^- \tag{6.30}$$

$$SnF_3^- + 2e^- = Sn + 3F^- \tag{6.31}$$

6.2.6 Electrodeposition of Zinc

Zinc coatings are used extensively to protect steel. They are well suited for the purpose because zinc is inexpensive, the coating processes are easy and the protection is galvanic and very effective. Electrodeposition is used as an alternative to hot-dip galvanizing where the higher cost is justified by close dimensional tolerances, as in screw threads or by improved appearance. Bright deposits are technically feasible, but the cost is not often justified for the utilitarian applications for which zinc coatings are used.

As with tin, zinc can be electrodeposited from both acidic and alkaline baths, using consumable anodes. The former is simple to operate but the latter has higher throwing power for complex shapes.

6.2.6.1 Acid Sulfate Baths

The standard electrode potential for zinc is −0.76 V SHE, which is very negative with respect to potentials for hydrogen evolution, but the hydrogen overvoltage on pure zinc is high and hydrogen evolution can be suppressed. This is favoured by a formulation with a high $a_{Zn^{2+}}/a_{H^+}$ ratio, implying a high concentration of zinc sulfate, $ZnSO_4$, with the least acidity needed to keep it in solution. For this reason the pH is buffered in the range 3 to 4. Organic additives from among β-naphthol, glucose, dextrin and glycerol are added to refine the deposit. The anodes and solutes must be free from impurities like nickel and cobalt that codeposit with the zinc, inviting hydrogen evolution by lowering the hydrogen overvoltage. A typical formulation using sodium acetate as the pH buffer is given in Table 6.1.

6.2.6.2 Alkaline Cyanide Baths

Alkaline baths are solutions of zinc cyanide, sodium cyanide and sufficient sodium hydroxide to raise the pH > 13 to maintain a high zinc content in solution. A typical solution used without brighteners or other additives is given in Table 6.1.

The improved throwing power over acid baths is probably due to high activation polarization associated with the deposition of zinc from anions. The ions present include, Na^+, ZnO_2^{2-}, $HZnO_2^-$, $Zn(CN)_4^{2-}$, CN^- and OH^-. Pourbaix diagrams are clearly of little value for solutions with so many ionic species.

6.3 Hot-Dip Coatings

In hot-dipping, a solid metal is immersed in a liquid bath of another metal and withdrawn with an adherent film of the liquid that solidifies on cooling. It is the simplest method of applying protective metal coatings to base metals, but there are limitations. The melting point of the protecting metal must be substantially lower than that of the substrate metal and preferably low enough to allow the substrate to retain work hardening, if required. There must be sufficient interdiffusion between the metals to enable a bond to form, but not so much as to consume the thin layer of protecting metal. In practice, the process is most useful for applying coatings of zinc, tin or aluminium to steel. The processes for the three metals differ in detail but follow the same essential sequence of operations:

1. The steel is cleaned by pickling as described in Section 6.1.2.1.
2. It is fluxed with chlorides to promote wetting by the liquid metal.
3. It is immersed for a very short time in the liquid metal.
4. Excess liquid is drained off or wiped off by appropriate means, for example, exit rolls for sheet or dies for wire.
5. Post-dipping treatments are applied as required.

6.3.1 Zinc Coatings (Galvanizing)

Galvanizing to give sacrificial protection to steel is the most common application of hot-dipping and provides a major market for zinc. It is used to coat familiar fabricated articles, for example, water tanks and buckets, but its main use is for semi-finished fabricated steel products including sheet and pipes, where it is an alternative to electrodeposition.

Zinc melts at 419.5°C and dipping is carried out in the temperature range 430 to 470°C. The flux is a mixture of zinc and ammonium chlorides applied to the steel before dipping, either by immersion in the liquid salt mixture or by pre-fluxing with an aqueous solution of the salts dried on the surface.

If pure zinc is used, an *alloy layer* is interposed between the zinc coating and the steel substrate. It comprises a sequence of brittle iron/zinc intermetallic compounds consuming some of the zinc and its thickness is manipulated to suit the application. If the zinc coating is intended as the sole protection and the material is to be fabricated, the brittle alloy layer can be virtually eliminated by adding 0.15 to 0.25% of aluminium to the liquid zinc, provided that the immersion time is short and the temperature is not too high. For some purposes, for example, for some automobile body panels, a zinc coating is used to supplement protection by paint, in which case the formation of the alloy coating is promoted to produce the so-called iron/zinc (IZ) sheet that is better suited to phosphating. This is done by reheating the coated steel to allow all of the zinc to diffuse further into the iron surface.

A characteristic feature often observed on galvanized steel surfaces is the *spangle*, a two dimensional macroscopic grain clearly visible to the unaided eye. It has no disadvantage if appearance is unimportant.

6.3.2 Tin Coatings

Hot-dip tinning was the original process for the production of tinplate, but it has been superseded by electrodeposition to apply the very thin coatings now required. It is still useful for smaller scale work especially to protect pieces of awkward shape, for example, for equipment used in the food industry. Tin melts at 232°C and dipping is carried out in the temperature range 240°C to 300°C using zinc chloride as the flux. The steel is inserted into the tin through a bath of the liquid flux confined in a frame on the tin surface, and withdrawn through a bath of palm oil also floating on the surface. The solidified tin coating has only a thin alloy layer of the compound $FeSn_2$ interposed between it and the steel.

6.3.3 Aluminium Coatings

Hot-dip aluminizing is the most difficult of the three processes because the melting point of aluminium, 660.1°C, is high and the oxide film on aluminium is tenacious, so that it is not easy to produce a smooth finish. The steel is fluxed in a fused salt bath at a temperature approaching the temperature of the liquid aluminium to avoid chilling it. A thick brittle alloy layer, based on the intermetallic compound, $FeAl_3$, is always present between the aluminium and the steel. A disadvantage of the alloy layer is that it limits the degree of deformation possible in any subsequent fabrication, but for applications such as automobile exhaust systems, advantage can be taken of the high melting point of $FeAl_3$, 1160°C, to produce heat-resisting coatings by allowing all of the aluminium to be absorbed into the alloy layer by subsequent heating.

6.4 Conversion Coatings

Conversion coatings bear the name because the metal substrate participates in forming its own coating. Three processes are especially valuable, phosphate coatings on steels and zinc, anodic coatings on aluminium alloys and chromate coatings on aluminium and zinc.

Phosphate coatings are an essential part of protective paint systems applied to manufactured steel products such as automobiles, bicycles, domestic and office equipment. Without pretreatment of the metal substrate, paints adhere only by weak Van der Waals forces and by interference keying to surface roughness. Phosphate coatings are chemically bound to the metal and their physical nature offers a firm mechanical key for paints. For this vital but hidden role, they are among the most indispensable coatings in use.

Anodic coatings can greatly enhance the corrosion resistance and aesthetic appearance of aluminium and some of its alloys and they also have other useful attributes that extend the commercial application of the metal and its alloys in several directions. Moreover, in practice, the treatment is simple and, as a result, it is extensively applied.

Chromate coatings have miscellaneous applications; they are best known for enhancing the corrosion resistance of tin or zinc coatings applied to steel and for providing a key for paint on aluminium alloys, but they can also serve as corrosion-resistant coatings in their own right.

6.4.1 Phosphating

Phosphating is a commercial process, producing coatings of stable, insoluble phosphates of iron(II), zinc(II), manganese(II), nickel(II) cations or more often mixtures of them, bonded to a metal substrate, usually steels or zinc-coated steels but sometimes aluminium alloys.

6.4.1.1 Mechanism of Phosphating

Phosphoric acid is tribasic and forms a series of three iron(II) salts, primary, secondary and tertiary phosphates, $Fe(H_2PO_4)_2$, $FeHPO_4$ and $Fe_3(PO_4)_2$. The primary phosphate is appreciably soluble in water, but the secondary phosphate is sparingly soluble and the tertiary phosphate is virtually insoluble. Using the symbol, ↓, to indicate the insoluble and sparingly soluble phosphates, equilibria between the various phosphates and free phosphoric acid are:

$$Fe(H_2PO_4)_2 = \downarrow FeHPO_4 + H_3PO_4 \tag{6.32}$$

$$\text{for which: } K_1 = \frac{a_{H_3PO_4}}{a_{Fe(H_2PO_4)_2}} \tag{6.33}$$

and

$$3Fe(H_2PO_4)_2 = \downarrow Fe_3(PO_4)_2 + 4H_3PO_4 \tag{6.34}$$

$$\text{for which: } K_2 = \frac{(a_{H_3PO_4})^4}{(a_{Fe_3HPO_4})^3} \tag{6.35}$$

The activity quotients in Equations 6.33 and 6.35 give critical ratios of free phosphoric acid to primary phosphate needed to precipitate insoluble phosphates.

A phosphate coating is produced by treating the metal with an aqueous solution containing the soluble primary phosphate and free phosphoric acid, either by immersion or by spray. A pregnant solution is prepared, with just sufficient free phosphoric acid to prevent spontaneous precipitation. An iron or steel article immersed in or sprayed with the solution destabilizes it by consuming free phosphoric acid:

$$Fe + 2H_3PO_4 = Fe(H_2PO_4)_2 + H_2 \tag{6.36}$$

and presents a surface on which the insoluble phosphates nucleate. The ratio of free phosphoric acid to the total phosphate content is critical and a basic phosphating solution is most efficient if the ratio is 0.12–0.15 and it is close to boiling. Too much free acid impairs the growth of the coating and too little initiates precipitation within the solution. Zinc and manganese form equivalent series of phosphates and there are advantages in formulating phosphating solutions with combinations of phosphates of the three metals. Simple phosphate processes are very slow, even when hot, and treatments of 30–60 minutes are required; longer treatments are needed for iron or manganese solutions than for zinc solutions.

6.4.1.2 Accelerated Phosphating Processes

The criterion for the speed of a phosphating process is the time to cover the metal surface completely, not the time to precipitate a given coating thickness. Written in ionic form, the dissolution of iron, which drives the process, is:

$$\text{Anodic reaction: } Fe \rightarrow Fe^{2+} + 2e^- \tag{6.37}$$

$$\text{Cathodic reaction: } 6H^+ + 2PO_4^{3-} + 2e^- \rightarrow H_2 + 2H_2PO_4^- \tag{6.38}$$

The reaction creates a mosaic of anodic and cathodic sites over the metal surface. Since it is the cathodic reaction that consumes phosphoric acid, deposition of secondary and tertiary phosphates is at the local cathodic sites, where one of the reaction products is hydrogen gas that forms a surface film inhibiting the spread of the coating over the metal surface. The reaction is polarized by the hydrogen and it can be depolarized by adding oxidizing agents, for example, hydrogen peroxide or sodium nitrite.

Using hydrogen peroxide, the depolarizing reaction is:

$$H_2O_2 + H_2 = 2H_2O \tag{6.39a}$$

Using sodium nitrite in acid solution, the reaction is more complex and depends ultimately on the weakness of nitrous acid relative to phosphoric acid, producing the overall result:

$$2NaNO_2 + 2H_3PO_4 + H_2(\text{adsorbed}) \rightarrow 2NaH_2PO_4 + 2H_2O + 2NO(\text{gas}) \tag{6.39b}$$

Depolarizing the cathodic reaction also stimulates reaction at the anodic areas and the cathodic areas spread at their expense. Most phosphate processes are accelerated, reducing treatment times to 1–5 minutes at 40–70°C, which is suitable for spray application on production lines. Coatings produced in accelerated processes are thinner, smoother and yield better paint finishes. The coating thickness is expressed by custom in mass per unit area and is typically 7.5 g m^{-2}. The condition of the metal surface influences the quality of the coating. Coatings on heavily strained metal can be uneven. Selective nuclei introduced by surface detritus or deep etching yield coarse-grained deposits, but multiple nuclei derived, for example, from precipitation within zinc-bearing solutions, refine the grain.

There are options for producing phosphate coatings in situations where regular phosphating is impracticable, such as on site preparation or *ad hoc* use, but the coatings are inferior. Phosphoric acid can be brushed on steel but residual free acid can cause blistering under paint. Proprietary alkali metal phosphate solutions are available that give very thin coatings suitable as a key for paint provided that subsequent exposure is benign.

6.4.2 Anodizing

Anodizing is the formation of thick oxide films on metal substrates, driven by an anodic potential applied to the metal in a suitable electrolyte. The process is important commercially for protective coatings on aluminium products, but it also has niche applications for tantalum and titanium, notably for high integrity electrolytic condensers.

Films on aluminium and its alloys have the following general characteristics that extend the applications of the metal:

1. There is a thin dense adherent film adjacent to the metal surface called the *barrier layer.*
2. The barrier layer is overlaid by a thicker layer of microporous material that can be sealed by a post-anodizing treatment, yielding an impermeable hard film.
3. Values for the electrical resistance and dielectric constant of the film material are both very high, that is, $4 \times 10^{15}\ \Omega\ cm^{-1}$ and 5 to 8, respectively, at 20°C.
4. With suitable metal compositions and anodizing treatments, the films can be transparent.

6.4.2.1 Mechanism of Anodizing

The electrolyte in which the anodic potential is applied to the metal determines which of two kinds of film is formed:

1. Non-solvent electrolytes, for example boric acid or ammonium phosphate that are indifferent to the anodic film.
2. Solvent electrolytes, for example sulfuric, oxalic and chromic acids that attack the film.

Non-solvent electrolytes produce barrier films with thicknesses corresponding to 1.4 nm/V of applied potential. They are used as dielectrics in electrolytic condensers, but they have no applications as protective coatings because very high potentials, 400 V or more, are needed to produce films that are thick enough.

Solvent electrolytes provide a solution to this difficulty. They attack the barrier layer at discrete points. The barrier layer reforms continuously under the points of attack to restore the thickness corresponding to the applied potential, leaving a porous overlay and developing the film structure represented schematically in Figure 6.5. The pore structure is regular, imposing a cellular pattern on the outer part of the film that, in plan, appears as an array of close-packed hexagons with cell walls of about the same thickness as the barrier layer. Thus the film thickness is not potential-dependent as for non-solvent electrolytes, but time-dependent and films thick enough for use as protective coatings can be grown in less than an hour at potentials in the range 12–60 V.

Diffuse electron diffraction patterns indicate that the structure of the freshly formed oxide lacks long-range order, but it spontaneously transforms slowly to γ-Al_2O_3. After anodizing, the porous film can be sealed in boiling water or steam, converting the γ-Al_2O_3 to a monohydrate, $Al_2O_3.H_2O$, which occupies sufficient extra volume to eliminate the porosity.

6.4.2.2 Practice

General anodizing is carried out in sulfuric acid and less often in chromic acid. The procedures are not difficult and they are adaptable to continuous operation for simple shapes. The acid is contained in rubber or lead-lined tanks and the workpieces are supported in jigs, with care to ensure good electrical contact. Since an anodic potential is applied to them, the jigs must be made from strongly passivating metal. Aluminium and titanium are used in sulfuric acid, but titanium is unsuitable for chromic acid. The anodic potential is applied between the workpieces and aluminium or lead cathodes. Since the coating is

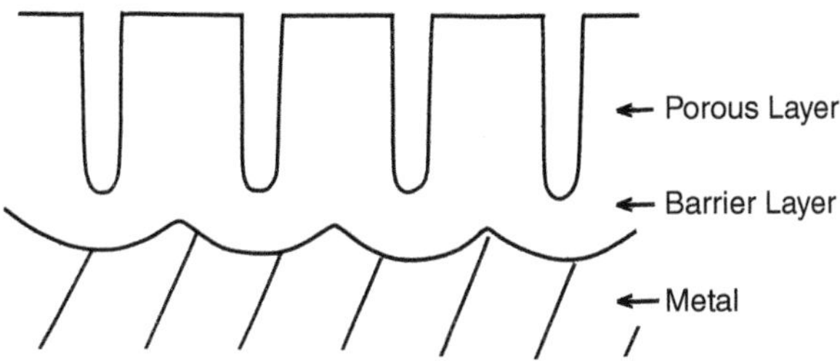

FIGURE 6.5
Schematic section through anodic film on aluminium, showing barrier layer and overlying porous layer.

TABLE 6.2
Typical Operating Conditions for Producing Anodic Coatings

Electrolyte	Temperature °C	Potential *E*/V	Current Density A dm^{-2}	Coat Thickness μm	Appearance
Sulfuric acid 10–15 mass %	15–24	10–22	1–3	3–50	Transparent or white
Chromic acid 3–10 mass %CrO_3	30–40	30–50	0.3–0.4	2–8	Opaque gray
Oxalic acid 3–8 mass %	20–40	30–60	1–3	10–60	Transparent yellow

Note: Typical treatment times are 30 to 60 minutes.

insulating, the film thickness is self-regulating and concepts of primary current density and throwing power do not apply.

Some qualities of the film make conflicting demands on the anodizing conditions and the practice is adapted to suit the application. Higher current densities, less concentrated acid and lower temperatures are needed for hard, abrasion-resistant films than for transparent films. Close control of operating parameters is essential to determine the character of the product. The current density is monitored using a dummy workpiece as a sensor and temperature is controlled by heating and cooling coils in the tank. Table 6.2 gives typical operating conditions.

The metal surface is prepared for anodizing by etching, chemical brightening as described in Section 6.1.2.4 or brushing as required and it must be clean, free from grease detritus and crevices. After anodizing, the work is rinsed in cold water to remove electrolyte. At this stage, the porous film is absorptive and must be handled with care to preserve its appearance. It is sealed by boiling in water or exposure to steam for about an hour.

6.4.2.3 Applications

6.4.2.3.1 Coatings for Corrosion Protection

Anodic films markedly improve the corrosion resistance of aluminium and its alloys. If a yellow appearance is acceptable, it is enhanced by sealing in a solution of an inhibitor, for example, sodium dichromate. Recommended film thicknesses for corrosion protection are:

Permanent outdoor exposure with no maintenance: 25 μm
Permanent outdoor exposure with good maintenance: 15 μm
Indoor exposure with good maintenance: 5 μm

A possible hazard in relying on anodic films for corrosion protection is intense attack by active/passive cells established at imperfections in the film overlying porosity and inclusions. This should not be a problem with material from reputable suppliers, but it is prudent to specify that the material must be of adequate quality for anodizing. Nevertheless, for some applications, such as for thin-walled vessels holding liquids, it may be preferable to accept general corrosion of untreated metal rather than to risk perforation by anodizing it.

6.4.2.3.2 Decorative Protective Coatings

Decorative applications depend on the transparency of the film that is produced in sulfuric acid and the facility to seal dyes in the outer porous layer of the film before sealing.

Transparent films preserve the brilliance of aluminium and customized alloys brightened by mechanical and chemical polishing. Acceptance standards are high, imposing a requirement for special alloys based on high purity aluminium. Alloy components and impurities present as intermetallic compounds in the metal structure resist anodizing and are incorporated in the film, rendering it dull. Elements in solid solution are less bothersome. Iron is a particularly undesirable impurity and for the highest quality anodized finish it is limited to 0.006 mass %. This implies the use of 99.99% purity aluminium, which is too soft even for decorative use, but may be hardened by a small magnesium content in solid solution without detriment to the brightening treatments or the transparency of the anodic film. One such alloy has the composition 1 mass % magnesium, 0.25 mass % copper, balance 99.99% pure aluminium. For applications in which higher strength is required, or where the metal is machined, an age-hardening alloy is available with the composition 0.7 mass % magnesium, 0.3 mass % silicon, 0.25 mass % copper, balance 99.99% pure aluminium. For less demanding mass markets, a less expensive range of alloys is available, based on lower purity, 99.8% aluminium. They are given special heat treatment during manufacture to keep iron in solid solution.

6.4.2.3.3 Miscellaneous Applications

Some specialized applications of anodic films are worth noting. Films from non-solvent electrolytes are used as dielectrics in electrolytic condensers. They are formed *in situ* on coils of >99.8% pure aluminium foil sealed in the condensers after etching to increase the true surface area. Aluminium wire can be electrically insulated with regular anodic films for transformer and motor windings to exploit the heat-resistance of the film. Films 100 μm thick can be used as hard facings for moving aluminium parts and surfaces subject to hard wear, for example, fuel pumps, camera parts, bobbins and pulleys. Polished alloys based on high purity aluminium protected by thin anodic films are used in maintenance-free reflectors for heat, lasers and UV radiation.

6.4.3 Chromating

Chromating is a term covering several different treatments that produce firmly adherent coatings with a small but effective reserve of chromium in a high oxidation state immediately available to metal surfaces. The coatings can serve one or both of two purposes:

1. To enhance the corrosion resistance of passive metals
2. To act as a key for paint systems

The coatings are applied by immersing the metal or article to be coated in an appropriate solution containing a soluble chromate, usually sodium dichromate, $Na_2Cr_2O_7$, for a few minutes. Simple neutral chromate solutions do not produce substantial deposits because they passivate the metal, inhibiting further reaction. Effective solutions, therefore, contain depassivating agents that may include sulfates, chlorides, fluorides and nitrates with or without pH control to establish conditions in which passivity is broken down to allow the reaction to proceed. The pH and the nature and concentrations of depassivating agents depend on the metal to be treated. This provides the scope for the development of proprietary treatments, developed for particular metals. The coatings can be produced on aluminium alloys, zinc, cadmium and magnesium, but not on iron and steel.

The composition of the coating can be quite complex and varies both from metal to metal and according to the way it is produced. The predominant active species is usually a slightly soluble hydrated chromium chromate, $Cr_2^{III}(Cr^{VI}O_4)_3$, that is the source of the oxidizing ion, $Cr^{VI}O_4^{2-}$. Other components can include the oxide of the metal treated and other species introduced from the particular solutions used to produce the coatings.

6.4.3.1 Coatings on Aluminium

Chromate coatings on aluminium alloys have important functions as keys for paint systems, as corrosion inhibitors and as aesthetically pleasing colour finishes. The term *chromating* is usually applied to chromate/phosphate as well as plain chromate coatings. They fulfill these roles economically and well, except for the most critical corrosion-resistant and decorative applications, for which anodizing is preferred where its extra cost is justified. They are applied by treating the work with suitable aqueous solutions of chromate, $Cr^{VI}O_4^{2-}$ and phosphate, PO_4^{3-} ions together with an activator such as fluoride ions to prevent passivation while the coatings are forming. The simplicity of processing gives chromating an advantage over anodizing for coating large and awkward workpieces. Chromating competes to some extent with alternative less effective simpler metal finishing processes, such as hydrated oxide (boehmite) coatings produced by controlled chemical oxidation of the metal in water, reviewed by Wernick, Pinner and Sheasby cited at the end of the chapter.

Chromate and chromate/phosphate coatings have broadly similar applications, but chromate/phosphate coatings are the more versatile and more widely applied. Proprietary treatments introduced by the American Chemical Paint Company are known under the name *Alodine* in the United States and *Alochrome* in Europe.

The structures of chromate/phosphate coatings are indeterminate and the compositions vary through the thickness with the highest chromium and phosphate contents towards the outside. Surface analysis indicates that they are based on hydrated chromium phosphate, $CrPO_4 \cdot 4H_2O$ and boehmite, AlO(OH) containing hydrated chromium oxide, Cr_2O_3 and chromium chromate, $Cr_2(CrO_4)_3$. Consequently, it is difficult to determine the reaction mechanisms, but they must comprise balanced anodic and cathodic reactions. In principle, the dissolution of aluminium by the anodic reaction:

$$Al \rightarrow Al^{3+} + 3e^- \tag{6.40}$$

can support the formation of chromic oxide, chromium chromate and chromium phosphate by notional overall cathodic reactions such as:

$$2Cr^{VI}O_4^{2-} + 10H^+ + 6e^- = Cr^{III}{}_2O_3 + 5H_2O \tag{6.41}$$

$$5Cr^{VI}O_4^{2-} + 16H^+ + 6e^- = Cr^{III}{}_2(Cr^{VI}O_4)_3 + 8H_2O \tag{6.42}$$

$$2Cr^{VI}O_4^{2-} + 8H^+ + PO_4^{3-} + 3e^- = Cr^{III}PO_4 + 4H_2O \tag{6.43}$$

Equations such as these show that the composition of the film deposited is critically dependent on the activities of $Cr^{VI}O_4^{2-}$ and PO_4^{3-} ions in the solution which are related to their concentrations, and on the activity of hydrogen ions, that is on the pH. The ratio of the activator to chromate activity is also critical. Satisfactory coatings are obtained only within optimum composition ranges. Outside of these ranges, the coating either does not form or has an unsatisfactory physical form. Effective solutions are obtained either by mixing chromic acid anhydride, CrO_3, phosphoric acid, H_3PO_4 and an activator source such as sodium fluoride, NaF, or by mixing sodium secondary phosphate, Na_2HPO_4, potassium dichromate, $K_2Cr_2O_7$, and sulfuric or hydrochloric acid. Two examples among several patented compositions are:

1. H_3PO_4 (75 volume %) 64 g dm^{-3} + CrO_3 10 g dm^{-3} + NaF 5 g dm^{-3}
2. $NaH_2PO_4 \cdot H_2O$67 g dm^{-3} + $K_2Cr_2O_7$15 g dm^{-3} + $NaHF_2$4 g dm^{-3} + $H_2SO_4$5 g dm^{-3}

The treatment is given by immersing workpieces in the solution contained in stainless-steel or aluminium vessels at a controlled temperature below 50°C. The immersion time is in the range 2 to 15 minutes, depending on the quality of coating required. After removal from the chromating solution, the work is rinsed in clean water, briefly immersed in dilute chromic or phosphoric acid and dried. The solution is maintained by removing sludge and replenishing the solutes consumed. For continuous operation, concentrated solutions can be applied hot, for example, at 50°C by spray.

The coatings are hard and the mass deposited is typically 2×10^{-4} g dm^{-2}, but it is adjusted to suit particular applications by varying the time of contact between metal and solution. They exhibit attractive pale straw yellow to bluish green colours that can be exploited for decorative effect. The particular colour depends on the alloy, the condition of the metal, the solution composition and the time of immersion, so that maintaining colour match is not easy. Wrought alloys in the AA 3000, AA 5000 and AA 3000 series yield coatings of good appearance. Castings are more difficult because the colour responds to segregation of alloy components, grain size effects and defects such as porosity and included oxides occasionally present. Alloys containing silicon do not respond well because they contain virtually elemental silicon as a eutectic component. Good corrosion resistance is associated with strong colour probably because it indicates a high concentration of $Cr^{VI}O_4^{2-}$ ions.

Plain chromate coatings are produced from proprietary solutions containing chromic acid, sodium dichromate and sodium fluoride but without a phosphate component and often with potassium ferricyanide described as an accelerator. These coatings are also indeterminate in composition and structure, but seem to be based on mixtures of hydrated aluminium and chromium oxides. They are thinner and less aesthetically pleasing as coloured finishes than chromium/phosphate coatings, but they are reputed to provide a better base for paint. The process is also more difficult to control.

Arrangements can be made to apply thin chromate coatings to the cleaned surfaces of aluminium alloy strip as a preliminary treatment in continuous painting or lacquering lines. In one example of such a system, a concentrated solution of chromic acid is partially reduced, yielding a mixture of Cr^{III} and Cr^{VI} ions, and treated with colloidal silica to adjust the consistency for application to the metal strip by rubber or plastic rolls.

6.4.3.2 Coatings on Zinc

Zinc is used extensively as a cathodic protective coating for steel. Although the natural corrosion resistance of zinc in neutral environments is good it can be enhanced, especially during its initial exposure by pre-passivation with a chromate coating. The coating also prevents the formation of *white rust*, a disfiguring white bloom of zinc carbonate formed by the action of atmospheric carbon dioxide on the surface oxide film. The treatments are based on the New Jersey Zinc Company's Cronak™ process using an aqueous solution of 182 g dm^{-3} of sodium dichromate and 11 g dm^{-3} of sulfuric acid. Various equivalent proprietary solutions are also available. The metal is immersed in the solution for 10 to 20 seconds at ambient temperature, yielding a coating mass of about 1.5 g m^{-2}.

6.4.3.3 Coatings on Magnesium Alloys

The application of chromate coatings to magnesium alloys is important as a preparation for protective paint systems not only to provide a key for the paint, but also to clear away any natural corrosion product formed on the metal during storage and prevent its reformation because it is alkaline and can degrade paint applied over it. Successful chromate coatings can be applied by proprietary treatments of such diversity that it would be tedious and unrewarding to enumerate them. The source of chromate is sodium or potassium dichromate and the solutions differ in pH, in the activator used and whether the activator is applied to the metal surface before or during treatment. In a representative example, the metal is first treated in an aqueous solution of hydrofluoric acid to activate the surface with a film of magnesium fluoride, MgF_2, and then the chromate coating is formed by immersing the metal in a 10%–15% aqueous solution of sodium or potassium dichromate with pH maintained in the range 4.0 to 5.5. All of the treatments are slow and immersion times of as much as 30 minutes are required at ambient temperatures.

6.5 Paint Coatings for Metals

Paint is the generic term for filled organic media applied to surfaces as fluids that subsequently harden into protective coatings. When applied to a metal, the hardened coating has the following functions:

1. To provide an environment at the metal surface incompatible with the mechanisms that sustain corrosion
2. To maintain its own integrity against chemical degradation, mechanical forces and natural ultraviolet radiation
3. To give an aesthetically pleasing, glossy coloured appearance; this is especially important for manufactured domestic consumer durables and automobiles

The material must be protective, amenable to the method of application and be available at the lowest cost consistent with the service envisioned. These are conflicting requirements that cannot all be satisfied by a single formulation and paint coatings are multilayered systems comprising primers, intermediate coats and finish coats.

6.5.1 Paint Components

Paints have three principal kinds of components:

1. Organic binding media that determine the drying characteristics of the paint and control the chemical and physical properties of the hardened coating

2. Pigments that are insoluble fillers finely ground and mixed with the binding media
3. Solvents to impart physical characteristics, for example, viscosity, needed for application to the metal surface

Other components are added to modify the rheology of the liquid, to assist dispersion of pigments, to accelerate drying, to impart plasticity in the hardened paint and to reduce skinning of air-drying paints.

6.5.1.1 Binding Media

The product of a binding medium for a paint is a coherent continuous network structure.

The precursors of this network are discrete structural units large enough to determine the form of the final structure yet small enough to be dispersed in a solvent for application to the substrate as a liquid. An essential requirement is that the structural units have *functional groups*, that is, chemically active groups of atoms that enable them to *cross-polymerize*, that is link together to create a three-dimensional structure. The means by which the linking is activated is the first distinction between the various binding media available. They are of two kinds:

1. Air-drying oils that polymerize spontaneously by absorbing oxygen from the air
2. Synthetic resins that do not polymerize spontaneously but do so when heated

Air-drying oils are the natural products linseed, soya bean and tung oils. The rate of drying can be increased if the oils are pretreated by heating them in the presence of catalysts to produce *boiled oils* or *stand oils*.

Synthetic resins are of several kinds, distinguished both by the organization of the basic structural units and by the functional groups responsible for cross-polymerization. There is a wide variety of functional groups that can be employed and they can be attached to various kinds of basic structural units. Detailed descriptions of the production and polymerization of these resins are available in specialized texts. For now, it is sufficient to identify them by name and to associate them with the particular chemical and physical attributes they can confer on paints. They are classified according to their chemical constitutions. The following list gives the most widely applied materials, but is by no means exhaustive.

6.5.1.1.1 Alkyd Resins

Alkyd resins are *polyesters*, produced as basic structural units by the reaction of polyhydric alcohols with monobasic and dibasic organic acids. Their formation is exemplified in Figure 6.6 for the production and subsequent cross-linking of the polyhydric alcohol, glycerol and the anhydride of dibasic phthalic acid. The links are formed between the hydroxy groups, –OH, on the alcohols and the characteristic carboxylate groups, –COOH, on the acid anhydride. These are both common precursors of alkyd resins. Various alcohols and acids can be used and blended to produce a range of resins to suit particular applications.

6.5.1.1.2 Amino Resins

Amino resins are characterized by the use of nitrogen atoms in the structural unit. Two of them, urea formaldehyde and melamine formaldehyde, are common constituents of some of the more durable binding media. The structural unit is made by

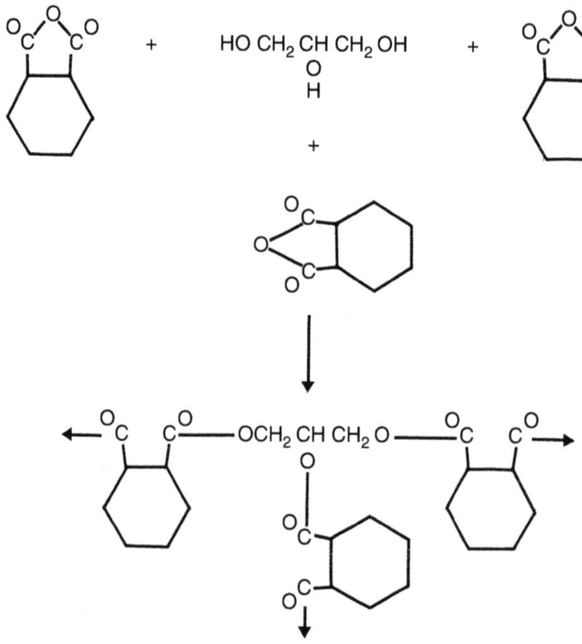

FIGURE 6.6
Cross-linking of phthalic anhydride with glycerol. The C–O–C ring in phthalic anhydride breaks to give two links, forming chains, but glycerol has three OH linking groups to cross-link the chains yielding a three-dimensional structure.

reaction of urea or of melamine with formaldehyde to add functional groups needed for cross-polymerization.

6.5.1.1.3 *Epoxy Resins*

Epoxy resins are complex molecules that have two kinds of groups available for cross-linking, hydroxyl groups and epoxide groups, one at either end. They can be cross-linked with other resins, including melamine formaldehyde, phenols, urea and others.

6.5.1.1.4 *Blends for Air-Drying Paints*

Drying oils are essential components of binding media for air-drying paints used on site, where the option to harden them by thermal treatment is impracticable. However, paints based solely on natural drying oils lack abrasion resistance and gloss and do not weather well. The oils are, therefore, blended with sufficient quantities of synthetic resins to improve their properties without losing the ability to dry spontaneously in air after application.

Air-drying paints are not used for manufactured durable items like automobiles, domestic equipment and office equipment because air-drying is incompatible with production line assembly, it is difficult to maintain control of the paints, the finish is not good enough and the properties of the hardened paint are usually unsuitable.

6.5.1.1.5 *Blends for Stoving Paints*

Binding media in stoving paints have a greater proportion of synthetic resins and are hardened by a thermal treatment, *stoving* or *baking*, typically for 30 minutes at 130°C to 180°C, to suit the composition. Blends of resins are selected to suit applications.

6.5.1.2 Pigments and Extenders

Pigments and extenders are fillers with the following functions:

1. To give the paint *body*, that is, the ability to form a thick coherent film
2. To impart colour either by self-colour or by accepting *stainers*
3. To protect binding media by absorbing ultraviolet light
4. To act as corrosion inhibitors in primer paint coatings

Pigments are classed as corrosion inhibiting pigments and colouring pigments. Extenders are used to replace part of the pigment content to reduce costs. Some common examples are:

Inhibiting pigments: Red lead, Pb_3O_4, calcium plumbate, $CaPbO_3$, zinc chromate, $ZnCrO_4$ and metallic zinc dust

White pigments: Rutile and anastase, both forms of titanium dioxide, TiO_2

Black pigment: Carbon black

Red pigment: Red oxide, a naturally occurring Fe(III) oxide

Yellow pigment: Yellow oxide, a natural hydrated Fe(III) oxide

Organic pigments: Stable insoluble organic substances with various colours

Extenders: Barytes ($BaSO_4$), Paris white ($CaCO_3$), dolomite ($CaCO_3 \cdot MgCO_3$), Woolastonite ($CaSiO_4$), talc and mica

Pigments and extenders are ground to particle sizes in the range 0.1 to 50 μm, as appropriate, for intimate mixing with the binding media.

6.5.1.3 Solvents

Solvents disperse the binding media and pigments, reduce viscosity and impart characteristics needed for applying the paint to metal surfaces.

6.5.1.3.1 *Organic Solvents*

Traditional solvents for paints are volatile organic liquids that reduce the viscosity of the unpolymerized binding media, facilitating application. They subsequently evaporate, depositing the binding media/pigment mixture. They are selected from among hydrocarbons, for example, white spirit, xylene and toluene, alcohols, for example, methanol and ethanol, ketones and esters.

6.5.1.3.2 *Water*

Binding media can be modified by processes called *solubilization* to disperse them in water as ionic species that can be applied to metal surfaces by electrodeposition. There are two options, the binding media can be given either anionic or cationic character that can be deposited at metal anodes or cathodes, respectively. Cationic character is the better option because deposition at a metal cathode does not damage phosphate coatings and gives paint coatings that have superior performance in service. Paints formulated with these modified resins are known as *waterborne* paints. Cationic character is imparted by treating suitable insoluble resins with an organic acid yielding the resin anion and a counter ion:

$$\underset{\text{(insoluble resin)}}{\mathrm{R}-\underset{\mathrm{R}}{\overset{\mathrm{R}}{\overset{|}{\underset{|}{\mathrm{N}}}}}:} \; + \; \underset{\text{(acid)}}{\mathrm{R'COOH}} \; \rightarrow \; \underset{\text{(soluble resin)}}{\mathrm{R}-\underset{\mathrm{R}}{\overset{\mathrm{R}}{\overset{|}{\underset{|}{\mathrm{N}}}}}-\mathrm{H}^{+}} \; + \; \underset{\text{(counter ion)}}{\mathrm{R'COO}^{-}} \tag{6.44}$$

where R represents groups of atoms based on carbon that complete the structure of the unpolymerized binding medium unit and R′ is the structure characterizing the acid.

Resins that are not amenable to solubilization can be linked to those that are, so that a wide range of binding media is available for electropaints. The paint is prepared by dispersing solubilized binding media in water with finely ground pigments in suspension and other components as required.

6.5.2 Applications

6.5.2.1 *Traditional Paints*

Paints dispersed in traditional solvents are applied by all of the obvious methods. Brush application is for onsite work, but it is labour intensive and the quality of the result depends on the painter's skill. Manual spraying also depends on skill and wastes paint in overspray. A modern development of spraying, more economical in the use of paint and independent of operator skill, is to deliver the paint through spinning nozzles into a strong electrostatic field to charge the paint particles so that they are attracted to earthed workpieces. Immersion and flow coating both require control of large volumes of paint to maintain its quality, allowing for solvent evaporation, selective drain off from workpieces and contamination.

6.5.2.2 *Waterborne Paints*

Electropainted waterborne paints are extensively applied on production lines as primer coats for automobiles and domestic equipment and for lacquers on food cans. Using paints with cationic solubilized resins, a cathodic, that is negative, potential of 100 to 200 V is applied to a metal workpiece against an inert counter electrode. The resin is discharged and deposited on the metal by a complex cathodic process with the overall result:

$$\underset{\text{solubilized resin}}{\mathrm{R}-\underset{\mathrm{R}}{\overset{\mathrm{R}}{\overset{|}{\underset{|}{\mathrm{N}}}}}:-\mathrm{H}^{+}} \; + \; 3\mathrm{H}^{+} \; + \; 4\mathrm{e}^{-} \; \rightarrow \; \underset{\text{deposited resin}}{\mathrm{R}-\underset{\mathrm{R}}{\overset{\mathrm{R}}{\overset{|}{\underset{|}{\mathrm{N}}}}}} \; + \; 2\mathrm{H}_2 \tag{6.45}$$

complemented by an overall reaction at the anode of the form:

$$\underset{\text{counter ion}}{\mathrm{R'COO}^{-}} + 3\mathrm{OH}^{-} \rightarrow \underset{\text{organic acid}}{\mathrm{R'COOH}} + \mathrm{O}_2 + \mathrm{H}_2\mathrm{O} + 4\mathrm{e}^{-} \tag{6.46}$$

Pigments are entrained in the coating in the same proportions as in the paint, yielding a coating about 12 μm thick. The paint is circulated through filters and heat exchangers to keep pigments in suspension, remove detritus and remove heat generated by the flow of electric current. Drag-out losses are recovered by filtering rinse water from the coated workpieces.

6.5.3 Paint Formulations

Endless combinations of binding media blends, pigments and solvents are available to devise paint formulations to match expected service conditions, methods of application and acceptable material costs. Paint formulation is a highly specialized activity based on scientific principles but also relying on experience and empiricism. Detailed formulations are difficult for the uninitiated to interpret. It is a professional skill best left to experts.

Binding media blends are customized to meet the requirements of the applications within cost constraints. If required for waterborne paints, they must possess chemical functional groups needed for solubilization. Alkyd resins are relatively inexpensive and are among the most widely used general purpose resins. Melamine formaldehyde is an expensive amino resin, but it confers outdoor durability and impact resistance and is a component of stoving automobile paints. Urea formaldehyde is less costly but less durable. Epoxy and polyurethane resins confer chemical and solvent resistance needed for refrigerators and washing machines, but are also expensive. Binding media for lacquers are selected for durability and clarity to add gloss to underlying paint coats and, of course, they are not pigmented.

Pigments are selected according to requirements for colour and corrosion resistance and are often mixed with extenders for economy. Concern over toxicity now severely restricts the use of lead-bearing pigments and they are not used for domestic products or automobiles.

Factors that determine the choice of organic solvents for traditional paints include cost, volatility, low toxicity and acceptable fire hazard.

6.5.4 Protection of Metals by Paint Systems

Paint films are readily permeable to water and oxygen and when exposed to the weather they are often saturated with both. They protect the metal by blocking electrochemical activity. Outer shell electrons in binding media are mostly localized in covalent bonds so that the paint is a poor conductor of electrons and ions. A significant corrosion circuit cannot be established with the open environment but only through water and oxygen absorbed in the paint structure, where both the anodic and cathodic reactions are inhibited by polarization.

Protection can be supplemented by inhibiting pigments incorporated in the priming paint coats. Chromate pigments are probably slow-release sources of the soluble anodic inhibitor, $Cr^{VI}O_4^{2-}$, at the metal surface, along the lines explained in Section 3.2.6.2. Red lead and calcium plumbate are proven inhibitors, but it is not clear why. The $Pb^{VI}O_2^{2-}$ ion is virtually insoluble in water containing dissolved oxygen so that its potential as an anodic inhibitor cannot be realized. One view attributes the inhibiting properties to lead soaps formed by interaction with the binding media. Zinc dust pigment protects steel sacrificially if sufficient is present to make electrical contact between the particles.

Further Reading

Lambourne, R. (Ed.), *Paint and Surface Coatings*, John Wiley, New York, 1987.
Rowenheim, F. A., *Modern Electroplating*, John Wiley, New York, 1974.
Waldie, J. M., *Paints and Their Applications*, Routledge/Chapman & Hall, New York, 1992.
Wernick, S., Pinner, R. and Sheasby, P. G., *The Surface Treatment of Aluminum and Its Alloys*, ASM International, Ohio, 1990.

7

Corrosion of Iron and Steels

Iron and steels are the most versatile, least expensive and most widely applied of the engineering metals. They are unequaled in the range of mechanical and physical properties with which they can be endowed by alloying, mechanical working and heat treatment. Their main disadvantage is that iron and most alloys based on it have poor resistance to corrosion in even relatively mild service environments and usually need the protection of coatings or environment conditioning. This generalization excludes stainless irons and steels that are formulated with high chromium contents to change their surface chemistry, as explained in Chapter 8.

An essential preliminary is to characterize the structures of steels and irons by briefly reviewing some essential phase relationships and introducing the terminology used to describe them. Table 7.1 gives examples of the materials considered.

7.1 Iron and Steel Microstructures

7.1.1 Solid Solutions in Iron

Pure iron exists as a body-centered cubic (BCC) structure at temperatures below 910°C and above 1400°C, but in the intermediate temperature range, it exists as a face-centered cubic (FCC) structure. Both structures can dissolve elements alloyed with the iron. Solutions in BCC iron are called *ferrite*, denoted by phase symbols, α and δ, identifying solutions formed in the lower and higher temperature ranges, respectively. Solutions in FCC iron are called *austenite*, denoted by the symbol, γ. The solutes stabilize the phases in which they are most soluble. Chromium, molybdenum, silicon, titanium, vanadium and niobium are *ferrite stabilizers*. Nickel, copper, manganese, carbon and nitrogen are *austenite stabilizers*.

In their binary systems with iron, ferrite stabilizers progressively expand the δ and α ferrite fields, usually until they merge into a single field enclosing an austenite field, the *γ-loop*, exemplified by the iron–chromium system illustrated in Figure 8.1 and described in Chapter 8. Austenite stabilizers expand the austenite field, introducing a peritectic reaction by which austenite forms from δ-ferrite and liquid, and a eutectoid reaction in which it decomposes to α-ferrite and a second phase as in the iron–carbon system described in the next section.

7.1.2 The Iron–Carbon System

The iron–carbon system is the basis for the most prolific ferrous materials, carbon steels and cast irons. There are three invariant points in the standard phase diagram, given in Figure 7.1.

TABLE 7.1
Some Typical Compositions of Carbon Steels and Irons

Material	Composition Mass %				Typical Application
	C	Mn	Si	Others	
Ultra low carbon steel	<0.005	0.20	<0.10	$S < 0.003$ $P < 0.015$	Two-piece food cans
Low carbon steel	0.05	0.30	<0.10	$S < 0.003$ $P < 0.015$	Automobile body panels
Low carbon steel	0.08	0.60	<0.10	$S < 0.05$ $P < 0.025$	Three-piece food cans
Hypoeutectoid steel	0.20	0.70	<0.15	$S < 0.05$ $P < 0.025$	General constructional steels and forgings
Hypoeutectoid steel	0.30	0.80	<0.15	$S < 0.05$ $P < 0.025$	General engineering
Hypoeutectoid steel	0.40	0.80	<0.15	$S < 0.05$ $P < 0.025$	Nuts and bolts
Hypoeutectoid steel	0.50	0.80	<0.15	$S < 0.05$ $P < 0.025$	Gears
Eutectoid steel	0.80	0.80	<0.15	$S < 0.05$ $P < 0.025$	Dies, chisels, vice jaws, springs
Hypereutectoid steel	1.20	<0.35	<0.30	$S < 0.05$ $P < 0.025$	Twist drills, files, taps, woodworking tools
Gray cast iron	3.5	0.50	2.00	$S < 0.05$ $P < 0.025$	General purpose castings
Nodular cast iron	3.7	0.40	2.40	1.0 Ni 0.05 Mg	Casings with some ductility
Malleable cast iron	3.0	0.30	0.80	–	Casings with some ductility

1. A peritectic at 1492°C, 0.18 mass % carbon, in which austenite is in equilibrium with δ-ferrite deposited during solidification and the residual liquid:

$$\delta + \text{liquid} = \gamma \tag{7.1}$$

2. A eutectoid at 732°C, 0.80 mass % carbon, in which austenite is in equilibrium with ferrite and *cementite*, Fe_3C:

$$\gamma = \alpha + Fe_3C \tag{7.2}$$

3. A eutectic at 1130°C, 4.3 mass % carbon at 1130°C in which the liquid is in equilibrium with austenite and cementite:

$$\text{liquid} = \gamma + Fe_3C \tag{7.3}$$

The boundaries of the austenite phase field extend to the peritectic and eutectic points. The maximum solubilities of carbon are 1.7 mass % in austenite at the eutectic temperature, 1130°C and 0.035 mass % in ferrite at the eutectoid temperature, 732°C.

The standard diagram represents metastable equilibrium because cementite is unstable at temperatures below 800°C. It separates on cooling in place of the thermodynamically

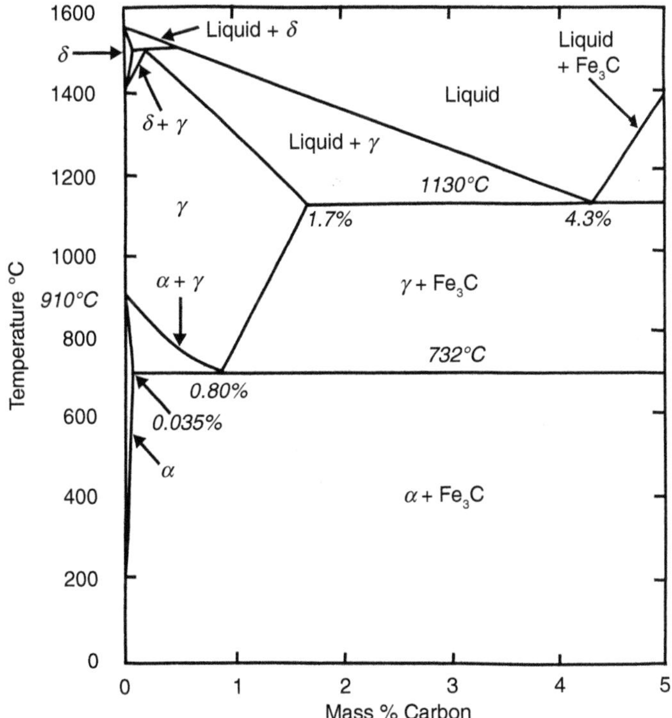

FIGURE 7.1
Equilibrium phase diagram for the iron–carbon system.

stable phase, graphite, which is difficult to nucleate. A technically important exception is that graphite can be induced to form in cast iron eutectics by suitable alloying or heat treatments.

7.1.3 Plain Carbon Steels

The structures of steels have been the subject of an immense amount of observation and experience, commensurate with their economic importance. They are the subjects of standard texts that should be consulted for serious study. The following summary does no more than indicate the origins of the structures encountered.

The compositions of most plain carbon steels lie in the range 0.1–1.0 weight % carbon. Their structures are determined primarily by the effects of their carbon contents although the compositions include other components, inherited from steelmaking processes. Most steels contain 0.5–1.7 mass % of manganese and all contain sulfur and phosphorus, limited to <0.05 mass % each, because higher concentrations embrittle the steel. In some special quality steels such as deep-drawing qualities and qualities to resist stress-corrosion cracking in oil field sour gas, sulfur contents are further limited to 0.002 mass %.

7.1.3.1 Normalized Steels

The production routes for wrought steels include hot deformation and heat treatments at temperatures in the austenite phase field. If the steels cool in air through the eutectoid transformation, *normalizing,* they adopt structures that depend on the carbon content in relation to the eutectoid composition.

7.1.3.1.1 The Eutectoid Composition

As its temperature falls, a steel of the eutectoid composition, 0.80% carbon, passes through the eutectoid point and transforms from austenite to colonies of very fine alternating laminae of ferrite and cementite, called *pearlite* from its pearlescent appearance under the microscope.

7.1.3.1.2 Hypo-Eutectoid Compositions

Steels with lower carbon contents, *hypo-eutectoid steels,* cross the boundary of the austenite field and deposit *pro-eutectoid* ferrite as they cool through the two-phase ferrite/austenite field. Carbon is rejected by the ferrite concentrates in the remaining austenite until it reaches the eutectoid composition at the eutectoid temperature and transforms to pearlite. The final structure comprises crystals of ferrite among colonies of pearlite. The fraction of pearlite in the structure rises approximately linearly with carbon content from zero for a steel with 0.035% carbon to a completely pearlitic structure at the eutectoid composition. In general, steels become stronger, harder and less ductile as the quantity of pearlite increases.

7.1.3.1.3 Hyper-Eutectoid Compositions

Steels with higher carbon contents, *hyper-eutectoid steels,* cross the boundary of the austenite field and deposit *pro-eutectoid* cementite as they cool through the two-phase austenite/cementite field. Carbon consumed in forming the cementite depletes the carbon content of the remaining austenite until it reaches the eutectoid composition at the eutectoid temperature and transforms to pearlite. The final structure comprises grains of cementite among colonies of pearlite. The pro-eutectoid cementite still further hardens the steel.

7.1.3.2 Quenched and Tempered Steels

If steels are reheated to a temperature in the austenite field and solutionized, that is allowed to transform completely to austenite and then cooled rapidly by quenching in water, the eutectoid reaction is suppressed and replaced by an alternative transformation that yields the tetragonal phase, *martensite,* given the phase symbol, α'. The martensite forms as plates in the austenite by realignment of the atoms without rejecting carbon and is thus a diffusionless transformation product. This transformation forms the basis of hardening steels by heat treatment. The hardness of the martensite increases with rising carbon content and steels designed for hardening usually have >0.36 mass % of carbon. In the as-quenched condition, steels have a simple structure of martensite plates set in retained austenite and although they are hard, they are unusable because they are unstable and brittle. To stabilize the structure, alleviate the brittleness and develop useful mechanical properties, quenching is always followed by *tempering,* that is, reheating the steel for a short time at a temperature below the eutectoid temperature. Tempering sets in motion a complicated sequence of events including the precipitation of $Fe_{2.4}C$, *epsilon carbide,* within the martensite plates and decomposition of the retained austenite to *bainite,* an intimate two-phase structure of ferrite with epsilon carbide. If tempering is prolonged, the structure further transforms into a dispersion of fine spherical cementite particles, Fe_3C in ferrite and the steel becomes softer. Thus tempering is adaptable to produce customized structures to suit various applications.

Plain carbon steels can be hardened only in thin sections, typically <2 cm thick, because in thicker sections the structure in the interior does not cool rapidly enough to transform to martensite. This difficulty is surmounted by introducing small quantities of other metals into the composition, producing *low alloy steels*, with martensitic transformations that are more tolerant of slower cooling. They are said to have greater *hardenability*. The metals that promote hardenability most effectively are manganese, chromium and vanadium. With judicious alloy formulation, steels with as little as 3 mass % of alloy additions can be hardened through 15 cm thick sections.

In practice, steels are solutionized at temperatures about 50°C above the lower boundary of the austenite phase field for a period of 0.4 hour per cm of the metal thickness and quenched in cold water or in oil. Subsequent tempering is in the temperature range 400–550°C for periods between a few minutes and an hour or two, as required to produce the prescribed structure.

7.1.4 Cast Irons

Although cast irons are based on the iron–carbon system with compositions near the eutectic composition, they are really multicomponent alloys, usually containing a substantial quantity of silicon, a significant quantity of manganese and sometimes also phosphorus. As the name implies, they have good casting characteristics including fluidity and the ability to take sharp imprints from molds. They are inexpensive because the primary raw materials that supply the iron and most of the alloy content are blast furnace iron, cast iron scrap and steel scrap, together with trimming additions to adjust composition.

According to the metastable equilibrium phase diagram in Figure 7.1, a plain iron–carbon alloy of near eutectic composition is expected to solidify as an austenite/cementite eutectic in which the austenite component subsequently transforms to pearlite on cooling through the eutectoid temperature. These irons are produced if the carbon and silicon contents are low and are called *white irons*. They are brittle and used only for special purposes, such as wear resistance. General purpose irons are formulated to produce graphite in the eutectic transformation and this is accomplished by raising the carbon and silicon contents and sometimes adding phosphorus. Most of the carbon content is present as graphite, except for 0.8% needed to form the cementite component of the pearlite into which the austenite transforms at the eutectoid temperature. Graphitic irons are called *gray irons* from the colour of their fractures.

There are three kinds of gray iron, differentiated by the morphology of the graphite:

1. *Flake irons* are standard gray irons in which the graphite exists as flake-like structures, illustrated in Figure 7.2. They are strong in compression but weak in tension, they cannot be deformed and they are not easy to machine.
2. *Nodular irons* are formulated to alter the morphology of the graphite component of the eutectic so that it is deposited as nodules instead of flakes, generally improving the mechanical properties. The nodular morphology is produced by adding magnesium to the metal immediately before casting. Magnesium reacts with any sulfur present and the nodular structure does not form in iron with >0.02 mass % sulfur.
3. *Malleable irons* yield a favourable graphite morphology by post-casting treatment. They solidify with cementite/pearlite structures and are reheated to temperatures in the range 850–900°C to convert the cementite to graphite. Graphite produced in this way is called *temper carbon*. The thermal cycle can be adjusted to yield either a pearlite or ferrite matrix.

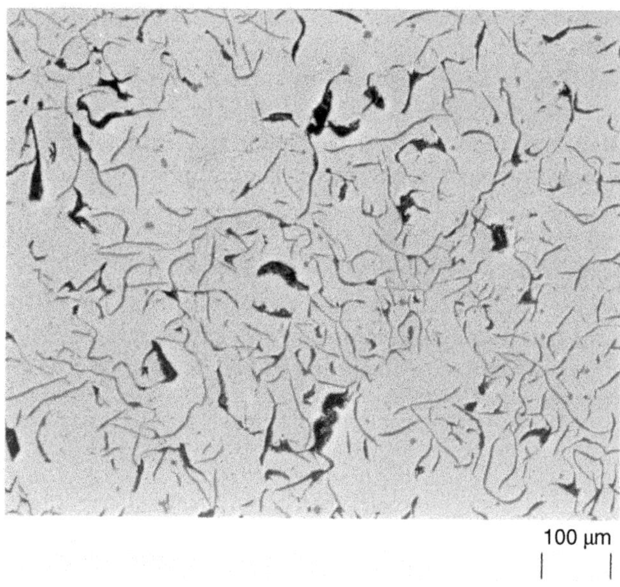

FIGURE 7.2
Graphite flakes in a gray cast iron.

7.2 Rusting

Rust is not a determinate chemical substance, but a complex material that changes continuously as it develops through the precipitation, evolution and transformation of chemical species in the iron–oxygen–water system. These events are conditioned by the prevailing environment.

7.2.1 Species in the Iron–Oxygen–Water System

Iron can exist in oxidation states Fe(II), Fe(III) and Fe(VI). The highest state, Fe(VI), in the ferrate ion, FeO_4^{2-} is stable only in highly oxidizing alkaline media and is seldom significant. The common oxidation states are Fe(II) and Fe(III) formerly called *ferrous* and *ferric iron*, respectively. Both form oxides, hydrous oxide phases, soluble cations and corresponding series of salts. The redox potential of the Fe(III)/Fe(II) couple in dilute aqueous solution, E° is +0.771 V (SHE), but it is sensitive to the chemical environment of the metal atoms and in other contexts the facility with which Fe^{2+} and Fe^{3+} change their oxidation states depends on phase relationships.

7.2.1.1 Iron(III) Oxides, Hydrous Oxides and Derivatives

Fe(III) is the stable oxidation state of simple compounds and aqueous solutions in equilibrium with atmospheric oxygen. There are two Fe(III) oxides, the common red-brown oxide, α-Fe_2O_3 (*haematite*) and γ-Fe_2O_3. α-Fe_2O_3 has the corundum (Al_2O_3) structure with Fe^{3+} ions in the octahedral spaces of a hexagonal close-packed lattice of oxygen ions. γ-Fe_2O_3 has a structure, sometimes incorrectly called 'spinel like', with Fe^{3+} ions distributed randomly in the octahedral and tetrahedral interstices of an FCC lattice of oxygen ions. It is produced by dehydrating lepidocrocite, γ-FeO(OH), or oxidizing the lower oxide, Fe_3O_4.

Fe(III) is precipitated from neutral or alkaline solution as a red-brown gelatinous material loosely described as 'ferric hydroxide' but there is no clear evidence that it contains discrete hydroxyl ions and its constitution is uncertain. Some chemical equations use the notation '$Fe(OH)_3$' but it is probably a hydrous Fe(III) oxide $Fe_2O_3 \cdot nH_2O$, where n approaches 3. It loses water on standing, becoming an amorphous powdery material.

Partially hydrated Fe(III) oxide phases are difficult to characterize unequivocally. There are at least two brown or brown-black hydroxyoxides, α-FeO(OH) (*goethite*) and γ-FeO(OH) (*lepidocrocite*), as well as other phases that can be regarded as oxides containing variable quantities of water, $Fe_2O_3 \cdot nH_2O$. All of these compounds are only very slightly soluble in water unless it is strongly acidic or alkaline. For this reason, the reaction of Fe(III) with anions of weak acids often yields hydrous Fe(III) oxide, $Fe_2O_3 \cdot nH_2O$. One such reaction is with the carbonate ion, usually present in natural waters:

$$2Fe^{3+} + 6CO_3^{2-} + (3+n)H_2O = Fe_2O_3 \cdot nH_2O + 6HCO_3^- \quad (7.4)$$

7.2.1.2 Iron(II) Oxides, Hydroxides and Derivatives

The Fe(II) ion has a lower charge density than the Fe(III) ion and because of this its salts are much less susceptible to hydrolysis. It forms simple salts with most common anions except when they are oxidizing enough to raise the oxidation state from Fe(II) to Fe(III).

Fe(II) forms an oxide, FeO (*wüstite*), with a simple cubic structure, but it is unstable at temperatures below 570°C. Fe(II) does, however, participate in Fe_3O_4 (*magnetite*), an anhydrous mixed Fe(II)/Fe(III) oxide, stable at ambient temperature with the inverse spinel structure, described in Section 2.3.3.4. It can form by oxidation of iron at temperatures below 570°C or by decomposition of FeO formed at higher temperatures:

$$4FeO = Fe + Fe_3O_4 \quad (7.5)$$

Unlike Fe(III), Fe(II) forms a distinct hydroxide, $Fe(OH)_2$ with the $Mg(OH)_2$ structure and is white or pale green when pure but discolours to yellow in contact with air by superficial oxidation to a hydrous Fe(III) oxide:

$$2Fe^{II}(OH)_2 + \tfrac{1}{2}O_2 = Fe^{III}{}_2O_3 \cdot H_2O + H_2O \quad (7.6)$$

In contact with water, a corresponding oxidation reaction occurs more readily:

$$2Fe^{II}(OH)_2 + 2OH^- = Fe^{III}{}_2O_3 \cdot H_2O + 2H_2O + 2e^- \quad (7.7)$$

driven by the reduction of dissolved oxygen:

$$\tfrac{1}{2}O_2(\text{solution}) + H_2O + 2e^- = 2OH^- \quad (7.8)$$

The reaction rate is strongly pH dependent as indicated by the presence of OH^- in Equation 7.7 and occurs most rapidly in alkaline media. Fe(II) hydroxide is amphoteric and can dissolve in both strong acids and alkali, although it has some solubility in neutral solution yielding Fe^{2+} and OH^- ions:

$$Fe(OH)_2 = Fe^{2+} + 2OH^- \quad (7.9)$$

for which the solubility product at 18°C is:

$$K_s = (a_{Fe^{2+}})(a_{OH^-})^2 = 1.64 \times 10^{-14} \quad (7.10)$$

In hard water containing carbon dioxide at significant pressures, Fe(II) can also form Fe(II) carbonate, $FeCO_3$ (*siderite*), which is oxidized by dissolved oxygen to hydrous Fe(III) oxide:

$$4FeCO_3 + 2nH_2O + O_2 = 2[Fe_2O_3 \cdot nH_2O] + 4CO_2 \quad (7.11)$$

7.2.1.3 Protective Value of Solid Species

When formed as coherent, adherent layers on the metal, the close structures of any of the anhydrous Fe(III) oxides, α-Fe_2O_3, γ-Fe_2O_3 and Fe_3O_4, can passivate the metal by inhibiting the transport of reacting species between the metal and its environment. In contrast, the amorphous character of many of the hydrated forms is unsuited to the formation of protective layers.

7.2.2 Rusting in Aerated Water

The conventional Pourbaix diagram for the iron–water system, given in Figure 3.2, and corrosion velocity diagrams as given in Figure 3.13 are useful indicators of corrosion possibilities, but they cannot disclose how the transformations they depict occur.

The diagram in Figure 3.2 shows that, at 25°C for $a_{Fe^{2+}} = 10^{-6}$, there is no common zone of stability for metallic iron and water across the entire pH range, so that iron and steels are unstable in the presence of water. This implies a thermodynamic *tendency* for the metal to transform, but whether or not it actually does so at a significant rate depends on the routes that the transformations must follow and the nature of intermediate products. These in turn depend on pH and on the absence or presence of contaminants, notably chlorides.

7.2.2.1 Fresh Waters

Natural waters including rain, rivers and supplies derived from them are near-neutral aqueous media approaching equilibrium with the atmosphere, in which the absorption of oxygen is the dominant cathodic reaction, as determined by the calculation given in Section 3.2.3.1. Oxygen potentials in these environments are close to the equilibrium line for oxygen absorption superimposed on the iron–water Pourbaix diagram given in Figure 3.2. This line is wholly within the domain of stability of Fe_2O_3 for $a_{Fe^{2+}} = a_{Fe^{3+}} = 1$ in the pH range 2–16, so that given time and opportunity, a phase of this composition must ultimately appear. The transformation is, however, approached by electrochemical partial reactions at the metal surface that place the corrosion potential in the domain of stability for Fe^{2+}, so that the primary products are produced by the anodic dissolution of iron:

$$Fe = Fe^{2+} + 2e^- \quad (7.12)$$

complemented by the cathodic absorption of oxygen:

$$½O_2 + H_2O + 2e^- = 2OH^- \quad (7.13)$$

When the concentrations of Fe^{2+} and OH^- ions close to the metal surface exceed the low solubility product given by Equation 7.10, the first solid product, $Fe(OH)_2$, is produced:

$$2Fe^{2+} + 4OH^- = 2Fe(OH)_2 \tag{7.14}$$

Precipitation of a solid from a liquid phase requires energy to create the solid/liquid interface and this is usually less on any pre-existing surface (*heterogeneous nucleation*) than from the liquid (*homogeneous nucleation*). Thus although the $Fe(OH)_2$ is produced by the interaction of species in solution, it is usually deposited on the iron surface.

The initial deposit of $Fe(OH)_2$ is only the provisional corrosion product and it is slowly oxidized to compositions approaching that of the stable Fe(III) hydrous phase, by reactions based on that given in Equation 7.7. As the rust ages, its constitution changes continuously, with associated changes in colour from yellow through red to brown, and it may contain various species from among those described in Sections 7.2.1.1 and 7.2.1.2. Such an unpromising material, derived from the initial $Fe(OH)_2$ deposit, obstructs diffusion of oxygen to the metal surface but lacks the coherence needed to protect the iron. If the metal is totally submerged and corrosion is uniform, iron is lost from the surface at a steady rate of 0.15 mm per year, after an initial period of a few days during which the rust is first established. In general, the corrosion rate does not vary significantly with the composition or condition of clean iron and steel products, excluding stainless steels and irons.

7.2.2.2 Sea Waters

Ocean waters approximate to a 3.5 mass % solution of sodium chloride, but also contain significant quantities of other solutes. In estuaries and land-locked seas, these solutes can be diluted considerably by fresh waters delivered by rivers but polluted by man-made effluents. In ocean surface water, the oxygen contents are close to equilibrium with the atmosphere, the pH range is 8.0–8.3 and the temperature range is –2–35°C.

The high concentrations of ionic solutes might be expected to influence corrosion rates of iron and steels, but long-term field tests have shown that the corrosion rates for small unshielded test panels are remarkably consistent in all ocean waters irrespective of location and temperature and they are similar to the rates in fresh waters. Despite these similarities, sea waters can cause more problems than fresh waters in some respects. These aspects are considered in Chapter 25 in relation to corrosion in marine environments.

7.2.2.3 Alkaline Waters

Waters are sometimes deliberately made alkaline for corrosion control or acquire alkalinity from environments such as cement. Confining attention to ambient temperatures, information from the iron–water Pourbaix diagram for 25°C in Figure 3.2 shows that if the pH is above about 9, the triangular domain of stability for Fe^{2+} (defined in Section 3.1.5.4 as $a_{Fe^{2+}} > 10^{-6}$) is eliminated and there is a common boundary between the domains of stability for metallic iron and magnetite, Fe_3O_4. Thus corrosion potentials must lie within the domain of stability of magnetite, Fe_3O_4, or haematite, Fe_2O_3, so that one or other of these oxides is the primary anodic product, for example:

$$3Fe + 4H_2O \rightarrow Fe_3O_4 + 8H^+ + 8e^- \tag{7.15}$$

which nucleates directly on the iron surface and grows into an adherent coherent layer that passivates the metal.

Successful passivation depends on forming a layer that is complete and free from attack by aggressive ions in the environment. Thus if the pH is only just high enough to passivate the metal there is a risk of local attack on areas of metal remaining active that become anodes in active/passive cells. Aggressive ions, especially chlorides can attack passive surfaces producing small local anodes that develop into pits, a common form of attack on metals depending on passivated surfaces for protection. This effect is considered in detail in Chapter 8, within the context of the more extensive evidence available for stainless steels.

7.2.2.4 Suppression of Corrosion by Impressed Currents

The information in the iron–water Pourbaix diagram suggests that in near-neutral aqueous media, it should be possible to suppress corrosion either by impressing a cathodic current on the metal to depress the potential of the metal into the domain of stability for metallic iron or by impressing an anodic current to raise it into the domain of stability for haematite, Fe_2O_3 to passivate the metal. Both techniques work, but there are important practical constraints. Impressing a cathodic current is a standard method of cathodic protection, described in Chapter 20. It has the merit that it is fail-safe; if the current does not succeed in depressing the potential into the Fe^{2+} domain, corrosion is reduced even if not completely suppressed. An applied anodic current is not fail-safe. If it does not succeed in raising the potential of the whole metal surface to the domain of stability for magnetite, Fe_3O_4, or haematite, Fe_2O_3, corrosion is increased. For this reason, it is dangerous to rely on an externally impressed anodic current. However, anodic protection is applied chemically by the redox potential of powerful oxidizing agents such as chromates or nitrites, when used as anodic inhibitors, as described in Section 3.2.6.2.

7.2.3 Rusting in Air

Rusting of bare iron and steel surfaces is generally slower in outside air than in water but is much more variable, ranging from near zero to over 0.1 mm per year and it is less predictable. The factors that influence it differ from location to location and vary with time. They include climate, season and weather, atmospheric pollutants, temperature cycles and the initial conditions and orientations of the iron or steel surfaces. There are so many independent variables that empirical comparisons between the results of limited field tests can be unreliable and the effort to acquire statistically significant information is protracted because rates of rusting can take several years to settle to steady values. More progress can be made by applying scientific principles and logic to interpret accumulated local experience.

There are three sources of water to sustain the electrochemical processes in air, atmospheric humidity, precipitation and wind- or wave-driven spray.

7.2.3.1 Rusting Due to Atmospheric Humidity

Rust can form on iron even when it is not wet to the touch. It is less in dry than in humid air and more in industrial or marine locations than in rural areas. These observations are rationalized on the basis of Evans* and his colleagues' classic experiments to clarify the effects of relative humidity and pollutants on rusting. Iron samples do not rust appreciably in pure clean air even when nearly saturated with water vapour, but if as little as 0.01% by volume of sulfur dioxide is present as a pollutant, the metal rusts rapidly when the relative

* Cited in Further Reading.

humidity exceeds a threshold value of about 70%. Dust contaminated with certain ionic salts has a similar effect in initiating rusting. Evans proposed a simple but elegant explanation as follows. An aqueous electrolyte is needed to produce rust and although water cannot condense spontaneously from *unsaturated* air, it can be induced to do so by hygroscopic salts. The role of pollutants is to provide particles of such salts on the metal surface. If the pollutant is sulfur dioxide, as it is in industrial locations, the salt is Fe(II) sulfate, $FeSO_4$ and in marine locations, the salt is sodium chloride, NaCl produced from sea spray. In cities there are other possibilities including the cocktail of pollutants released into the atmosphere by automobiles. As expected, the water vapour pressures in equilibrium with saturated solutions of these various salts correspond closely with the threshold values of relative humidity needed to initiate rusting. Evans observed the following sequence of events in experiments exposing iron to humid air polluted by sulfur dioxide:

1. Deposition of Fe(II) sulfate
2. Formation of a barely visible mist of water
3. Initiation of rust spots
4. Extension of the rust spots to cover the surface

Once rusting has started it is not arrested by removing the pollutant from the atmosphere, so that contamination that initiates rusting continues to sustain it. A corollary of this observation is that the rate of rusting is influenced by the degree of pollution in the initial conditions of exposure and this is borne out by experience that rusting on a clean iron surface first exposed in winter, when pollution is at a maximum, continues at a faster rate than it does if first exposed in summer when pollution is at a minimum.

In atmospheric rusting, the electrolyte is not present in bulk and ideas of the prevailing electrochemical reactions need revision from those that apply in aqueous corrosion. Evans suggests that the orthodox cathodic reaction:

$$\tfrac{1}{2}O_2 + H_2O + 2e^- = 2OH^- \qquad (7.16)$$

is replaced by the reduction of an Fe(III) species in the rust to a species containing Fe(II):

$$\text{Fe(III)} + e^- \rightarrow \text{Fe(II)} \qquad (7.17)$$

followed by its reoxidation back to Fe(III) by atmospheric oxygen:

$$2\text{Fe(II)} + \tfrac{1}{2}O_2 + H_2O = 2\text{Fe(III)} + 2OH^- \qquad (7.18)$$

Various schemes are possible, depending on what species are present in the rust as discussed in Sections 7.2.1 and 7.2.2 provided that there is a conducting path to supply electrons from the metal. Evans' own suggestion is that the reaction is between goethite, $Fe^{III}O(OH)$, and magnetite, $Fe^{II} \cdot Fe_2{}^{III}O_4$.

7.2.3.2 Rusting from Intermittent Wetting

Wetting from liquid water as rain, condensation or spray introduces aspects of aqueous corrosion, described in Section 7.2.2, with the difference that wetting is discontinuous. This contribution to rusting depends on the relative periods for which the metal is wet or dry.

If wetting is by rain or snow, the metal remains wet for the sum of the periods of precipitation and subsequent drying. The drying period depends on humidity, temperature, wind velocity and sunlight and hence on the general climate in regions of interest and on microclimates due to local geography. Melting snow remains for a considerable period and can leave hygroscopic contaminants such as highway deicing salts that promote subsequent rusting after the metal has dried. A familiar example is rusting due to mud poultices accumulated in traps in automobile underbodies. Further factors are the orientation of metal surfaces that allow or prevent drainage and the sponge effect of pre-existing rust.

Wetting from condensation is induced by temperature cycles above and below the dew point. It is most prevalent when there is a wide difference between day and night temperatures in temperate and tropical regions. Heavy metal sections are most vulnerable because their temperature cycles are out of phase with air temperatures.

7.2.4 Rusting of Cast Irons

Despite their vulnerability, cast irons have given remarkably good service when used unprotected. Historically, they were used extensively for buried pipes and structures, tanker fittings and other services, starting when they were one of the few inexpensive materials available for such duties. Some such systems can still survive after a century. Part of the reason for longevity is that cast iron sections are thicker than steel counterparts and in former times they were overdesigned with greater structural reserves to absorb wastage. More recent systems benefit from experience in identifying the most suitable applications and applying protective coatings or cathodic protection.

7.2.4.1 Gray Flake Irons

The general purpose gray flake irons differ from steels in their response to wet environments in several respects. Graphite is an electrical conductor and flakes outcropping at the surface of these irons can act as cathodic collectors on a microscale, as explained in Section 4.1.3.3, stimulating corrosion of the adjacent metal structure. Corrosion of this kind, developing along the flakes, is called *graphitic corrosion.*

The metal is leached out from among the network of graphite flakes, illustrated in Figure 7.2, leaving the surface contour of the metal intact. If the matrix is pearlite, as it usually is, carbonaceous material from dissolution of its cementite component becomes trapped in the mesh of residual graphite together with corrosion products and the composite mass acts as a resistance to diffusion delaying reaction. The value of this residue in protecting the iron depends on its consolidation and that in turn depends on the graphite morphology in the unattacked iron. Fine flakes produce stronger residues and protect the metal better than coarse flakes and residues from irons containing phosphorus are reinforced by iron phosphide.

In near-neutral aerated waters, corrosion is primarily by oxygen absorption, but corrosion by hydrogen evolution is serious when stimulated by sulfate-reducing bacteria in anaerobic waters containing sulfates that can prevail in locations such as polluted estuary waters and sewers. This form of attack can afflict other metals, but it is particularly associated with gray cast irons because it is easy to identify. The bacteria produce an enzyme, *hydrogenase,* that enables them to use hydrogen, accelerating corrosion by depolarizing the cathodic reaction:

$$SO_4^{2-} + 4H_2 \rightarrow S^{2-} + 4H_2O \quad (7.19)$$

Sulfides trapped in the graphite residue provide evidence for the bacteriological activity.

7.2.4.2 Other Irons

Nodular and malleable and white irons lack the graphite network that is responsible for the effects described in Section 7.2.4.1 and, therefore, corrode in a more orthodox fashion.

Normal gray irons have poor resistance to acids and compositions have been developed to meet these conditions. Raising the silicon content substantially yields irons for which initial corrosion leaves a residue of silica, SiO_2 on the surface that confers subsequent resistance to most common acids except hydrofluoric acid, hydrochloric acid and also for some unresolved reason, nominal sulfurous acid, H_2SO_3 (which is actually a solution of sulfur dioxide). For acid resistance, the silicon content must be >11 mass % or preferably 14%. The most useful attribute of these irons with high silicon contents is their resistance to sulfuric acid in all concentrations and all temperatures. But they are hard, brittle and lack tensile strength and their use is justified only when their superior acid resistance is needed.

Cast irons containing high nickel or nickel plus copper contents have been developed for improved corrosion and heat resistance. The structures are essentially graphite in an austenite matrix. Their main disadvantage is the expense of the nickel content, which must be at least 18 mass % or 13.5 mass % if copper replaces some of the nickel. Their acid resistance is better than that of regular gray flake irons, but not as good as that of irons with high silicon contents. They perform well in aggressive waters and are applied in special situations.

7.3 The Oxidation of Iron and Steels

The oxidation of iron at oxygen pressures near atmospheric pressure is worthy of study both because it is technically important and because it is complex and exemplifies many of the principles which underlie the oxidation of metals in general. The oxidation behaviour can be described within three mutually consistent contexts:

1. Crystallographic and defect structures of the oxides
2. Phase equilibria for the iron–oxygen system
3. Wagner's theory of oxidation mechanisms

7.3.1 Crystallographic and Defect Structures of Oxides

There are three stable oxides of iron:

1. FeO (*wüstite*), stable at temperatures >570°C, is a simple cubic cation vacant *p*-type oxide. The only mobile species is Fe^{2+}, but its diffusivity is high.
2. Fe_3O_4 (*magnetite*) is an inverse spinel, in which iron fulfills both divalent and trivalent roles. The mobile species are Fe^{2+}, Fe^{3+} and O^{2-}.
3. Fe_2O_3 (*haematite*) is a rhombohedral *n*-type oxide with both cation interstitials and anion vacancies. The defect populations are much smaller than those in FeO and Fe_3O_4. The mobile species are Fe^{3+} and O^{2-}.

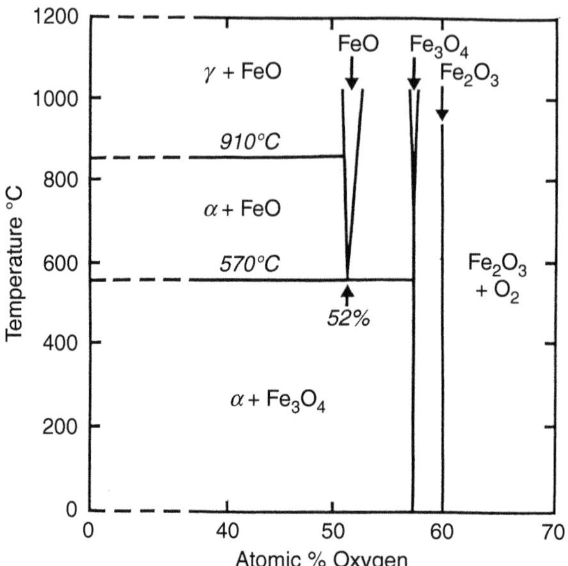

FIGURE 7.3
Equilibrium phase diagram for the iron–oxygen system.

7.3.2 Phase Equilibria in the Iron–Oxygen System

Figure 7.3 gives a part of the phase diagram for the iron–oxygen system relevant to the oxidation of the solid metal. There are four solid phases, pure iron and the oxides FeO, Fe_3O_4 and Fe_2O_3.

The solubility of oxygen in iron is so small that the solid solution phase field is vanishingly narrow and is not indicated in the figure. It has a negligible effect on the α-Fe to γ-Fe transformation which is represented in Figure 7.3 by the horizontal line at 910°C across the iron/FeO phase field. The FeO phase field is relatively wide, corresponding to the ability of the oxide to sustain an exceptionally high population of cation vacancies. The maximum composition range is 0.51–0.54 mole fraction of oxygen at 1370°C, which does not include the stoichiometric mole fraction, 0.50, indicating that the oxide always contains at least 0.02 fraction of vacant cation sites. The composition ranges for Fe_3O_4 and Fe_2O_3 are much narrower, but are detectable at high temperatures.

FeO is unstable below 570°C, at which temperature it decomposes to iron and Fe_3O_4 by the eutectoid reaction:

$$4FeO = Fe + Fe_3O_4 \tag{7.20}$$

The eutectoid reaction is reflected in the plots of the standard Gibbs free energy of formation versus temperature, *oxygen potential diagrams,* illustrated in Figure 7.4. The lines for $FeO \rightarrow Fe_3O_4$ and for $Fe \rightarrow FeO$ intersect at 570°C, corresponding to the eutectoid temperature. At higher temperatures, the free energy of formation of FeO is more negative than that of Fe_3O_4, so that it is the more stable oxide, but at lower temperatures the relative free energies are reversed, so that it becomes unstable, yielding a mixture of iron and Fe_3O_4 by Equation 7.20.

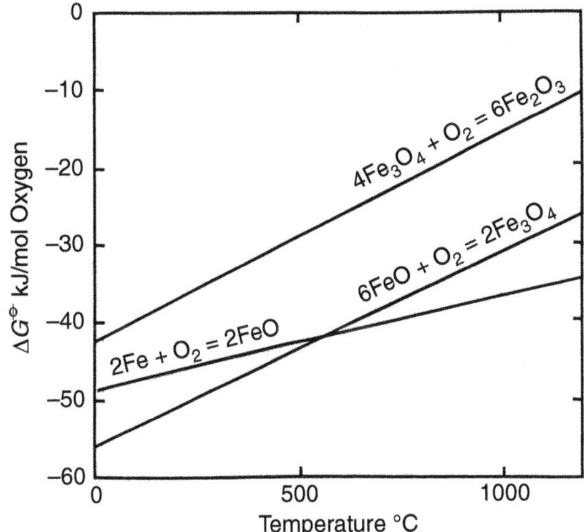

FIGURE 7.4
Gibbs free energy of formation for iron oxides as functions of temperature. Lines in descending order indicate increasing stability. Note the reversal of the relative stabilities of FeO and Fe_3O_4 at 570°C.

7.3.3 Application of Wagner's Oxidation Mechanisms

7.3.3.1 Nature of Scales

Scales formed in ambient air comprise layers of the stable oxides in a sequence of increasing oxygen activity from the metal surface to the atmosphere, illustrated schematically in Figure 7.5. Scales formed at temperatures >570°C have *three* layers but scales formed at temperatures <570°C, where FeO is unstable have only *two* layers.

The Wagner theory of oxidation described in Chapter 3 for a single oxide layer bounded by the metal/oxide and oxide/atmosphere interfaces can be extended to a multilayered oxide system by assuming that local equilibrium is also maintained at internal interfaces. Scales forming on iron have interfaces in the sequences Fe/FeO, FeO/Fe_3O_4, Fe_3O_4/Fe_2O_3, Fe_2O_3/atmosphere and Fe/Fe_3O_4, Fe_3O_4/Fe_2O_3 and Fe_2O_3/atmosphere at temperatures above and below 570°C, respectively.

This implies that there is a composition gradient across every oxide layer corresponding to the width of its phase field shown in Figure 7.3. For example, at 900°C, the mole fractions of oxygen in equilibrium with iron at the Fe/FeO interface and at the FeO/Fe_3O_4 interface are about 0.051 and 0.053, respectively, giving a difference of 0.02 mole fraction of oxygen across the FeO layer. This corresponds with the difference in the population of cation vacancies across the layer that provides the driving force for the diffusion of Fe^{2+} ions. Analogous considerations apply to the Fe_3O_4 and Fe_2O_3 layers.

7.3.3.2 Diffusing Species and Growth Interfaces

Sustained oxidation depends on the mechanisms by which oxygen from the atmosphere and iron from the metal can reach growth interfaces, where new oxide is formed.

The inner oxide layer, cation-vacant FeO, is permeable to iron in the form of Fe^{2+} ions, but impermeable to oxygen because there is no diffusion path for O^{2-} ions; FeO can therefore grow at the FeO/Fe_3O_4 interface but not at the Fe/FeO interface.

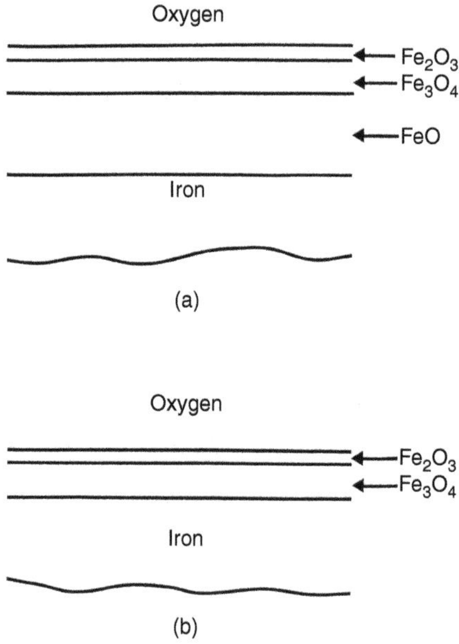

FIGURE 7.5
Sequence of oxide layers formed on iron. (a) >570°C; (b) <570°C.

The middle layer, the spinel, Fe_3O_4, is permeable to both iron and oxygen in the forms of the ions, Fe^{2+}, Fe^{3+} and O^{2-}. Thus it can grow at either interface and also transmit species needed for growth of the FeO and Fe_2O_3 layers on either side of it.

The outer layer, the *n*-type oxide, Fe_2O_3 with both cation interstitials and anion vacancies can transmit both iron as Fe^{3+} ions and oxygen as O^{2-} ions, but the latter species dominates.

These considerations lead to the following view on the locations of the growth interfaces:

1. There is virtually no scale growth at the Fe/FeO interface.
2. FeO grows at the FeO/Fe_3O_4 interface.
3. Fe_3O_4 grows at both the FeO/Fe_3O_4 and Fe_3O_4/Fe_2O_3 interfaces.
4. Fe_2O_3 grows predominantly at the Fe_3O_4/Fe_2O_3 interface, supplemented by some growth at the Fe_2O_3/atmosphere interface.

The scale layers are of different thicknesses, reflecting their different rates of formation that depend on the relative diffusion rates of the reacting species, due *inter alia* to the very different defect populations referred to earlier. The Fe_3O_4 is thicker than the overlying outer layer of Fe_2O_3 and for temperatures >570°C, the innermost layer of FeO is the thickest of all. As oxidation proceeds, the total thickness of the scale increases but the ratios of the thicknesses of the different layers remain approximately the same. Quantitative measurements of the actual scale thickness are inconsistent. Values given for the Fe_2O_3/Fe_3O_4/FeO ratios at temperatures >570°C range from 1/10/100 to 1/10/50 and the value for the Fe_2O_3/Fe_3O_4 ratio at temperatures <570°C is approximately 1/10.

7.3.3.3 Oxidation Rates

At all temperatures, the oxidation kinetics is parabolic, confirming that the oxidation is diffusion controlled. The temperature dependence of the parabolic rate constant for the total scale thickening, K_P, obeys an Arrhenius-type relation that is also consistent with diffusion control, as described in Chapter 3. For oxidation in still dry air, the expression is:

$$K_P = 6.1 \exp(-17000/RT)\ \text{cm}^2\,\text{sec}^{-1} \tag{7.21}$$

Diffusion through the outer layer of Fe_2O_3, which is the least defective oxide, is rate controlling. Consequently, there is no abrupt discontinuity in the temperature dependence of K_P at the eutectoid temperature, below which the FeO layer disappears.

7.3.4 Oxidation of Steels

It is usually the oxidation of steels rather than pure iron that is of most interest. Plain carbon steels contain small concentrations of carbon, manganese, silicon and excess aluminium from deoxidation; alloy steels may contain increased concentrations of these elements and nickel, chromium, molybdenum and vanadium as alloying components. The interplay between these elements can be complex, but the following brief summary of the effects of individual elements can be used as an empirical guide.

7.3.4.1 Effects of Alloying Elements and Impurities

Carbon oxidizes to carbon monoxide and hence may be expected to have a disruptive effect on the oxide. For steels with high carbon contents at temperatures >700°C, this prediction is correct but at lower temperatures, carbon is found to be beneficial rather than deleterious. Silicon promotes the formation of glassy silicates in oxide layers formed at high temperatures, for example, $2FeO \cdot SiO_2$ (fayalite), that tend to stifle oxidation and provide some resistance to sulfur-bearing gases that stimulate oxidation. If silicon is present in high concentrations, for example, 3% or 5% silicon steels for electrical applications, it promotes some internal oxidation, that is, some oxide is nucleated in the outer layers of the metal, under the true scale. Manganese and nickel resemble iron in oxidation behaviour and in moderate concentrations have little effect on the oxidation rate, as would be anticipated from Hauffe's valency rules. Although it does not have a significant influence on the oxidation rate, nickel can have a curious effect on the scale structure formed in air. In the early stages of oxidation, FeO and NiO both form because they have oxygen potentials that are close and much lower than that of air. As the scale thickens and the oxygen supply recedes, the FeO controls the oxygen potential at a value for which NiO is reduced back to metallic nickel that is found in the scale. Aluminium can improve the oxidation resistance because it can form a protective layer of $FeAl_2O_4$ under the main scale, but the effect of concentrations normally present is small. Molybdenum and vanadium yield oxides that form low melting point eutectics with other scale components. Thus they can flux away the scale, destroying its protective function and thereby promoting catastrophic oxidation as explained in Chapter 3. The effects can be confusing because molybdenum is sometimes inexplicably beneficial. sulfur and phosphorus in the small residual concentrations normally present have little effect.

Chromium is of such importance and value in formulating iron-base alloys for oxidation resistance as well as general corrosion resistance, for example, stainless steels and chromium irons, that it requires the detailed discussion given in Chapter 8.

7.3.4.2 Influence on Metal Quality of Scales Formed during Manufacture

Scales formed on plain carbon steels are adherent but they are easily detached when the steel surface is compressed and increased in the early stages, hot-rolling at temperatures >1000°C. Thus scaling of ingots during preheating for rolling can *remove* pre-existing surface blemishes. In contrast, scales formed on some alloy steels are only partially detached and become fragmented and imprinted on the steel surface as hot-rolling continues. This effect *creates* surface blemishes. Hence, experience of the oxidation behaviour of particular steels is essential in producing steels for applications where surface finish is important.

7.3.4.3 Oxidation in Industrial Conditions

Irons and steels are used at temperatures up to about 600°C in industrial conditions, but the idealized conditions for which the value of K_P given in Equation 7.21 applies do not necessarily or even usually prevail. The complicating factors include the composition and flow of the gaseous environment and the thermal history.

7.3.4.3.1 Atmospheres

In applications such as flues, furnace structures, steam pipes, oil stills and fume treatment facilities, steels may be exposed to various gases or mixtures of them from among air, carbon monoxide, carbon dioxide, sulfur dioxide and live steam that can modify the composition of the scale, introducing for example, sulfates, sulfides or hydroxides; there may also be abrasion or chemical attack from ash particulates. Further complications are introduced if the gas is flowing or if its composition is variable. These variables cover such a wide range that every case must be considered on its merits. Experience and comparative experimental tests simulating service environments have produced a large body of comparative information which is apparently reliable but difficult to correlate. For an index of specific information, the reader is referred to specialized texts such as the monographs by Kubaschewski and Hopkins and by Shreir, cited at the end of the chapter.

7.3.4.3.2 Thermal History

So far, it has been implicitly assumed that the scale remains intact but if the temperature varies randomly or cyclically, differential expansion and contraction between the metal and the scale can cause it to buckle or flake, reducing the protection it affords. Thick scales produced by prolonged oxidation can also buckle or flake under the shear forces built up within them. For these reasons, it is unsafe to extrapolate results of short-term tests to predict long-term service.

7.3.5 Oxidation and Growth of Cast Irons

Cast irons exposed to air at high temperatures form surface scales as do steels, but by infiltrating the metal structure, oxidation has additional significance in contributing to a form of degradation known as *growth*.

At temperatures above about 500°C in clean air, cast irons, especially gray flake irons are susceptible to dimensional changes that distort the metal, due to two effects.

1. Internal oxidation
2. Graphitization of cementite, Fe_3C

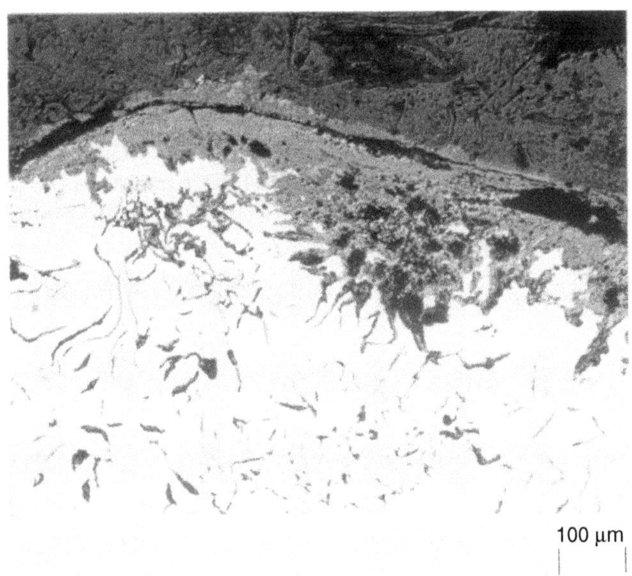

FIGURE 7.6
Oxide penetration along graphite flakes in gray cast iron.

Since these effects are cumulative and operate simultaneously, they are best considered together as different aspects of a single synergistic problem.

7.3.5.1 Internal Oxidation

In a gray iron, scale forms not only on the metal surface but also as intrusions penetrating into the metal interior along the interfaces between graphite flakes and iron, where oxygen can percolate through fissures propagated by the intruding scale. Shallow penetration improves oxidation resistance if the iron surface is stressed in compression, because it keys the external scale to the surface, excluding direct access of air to the metal. Deeper penetration, illustrated in the photomicrograph given in Figure 7.6, is, however, a source of weakness if the surface is under tension because the metal is expanded and weakened by a wedging action of the scale that occupies a greater volume than the volume of metal consumed.

Raising the carbon content increases internal oxidation because there is more graphite in the structure. Refining the graphite flakes has the same effect because it increases the area of the vulnerable graphite/iron interface. Nodular and malleable irons are less susceptible because the modified shapes of the graphite give lower surface to volume ratios and reduce coalescence of the scale intrusions evident in Figure 7.6.

Sulfur dioxide and other atmospheric pollutants that accelerate surface oxidation also accelerate scale penetration into cast irons. This increases the risk of premature failure of components exposed to heavily polluted hot gases such as the products of combustion from coals with high-sulfur contents and off-gases from pyritic smelting operations. Examples are fire grates, furnace doors and supports for furnace roofs.

7.3.5.2 Graphitization

Cementite tends to decompose to iron and graphite because of its instability at temperatures below about 800°C:

$$Fe_3C = 3Fe + C \tag{7.22}$$

In a gray iron as cast, carbon is distributed between graphite in the flakes and cementite as a constituent of the pearlite matrix. The silicon content of the iron that is added to promote the deposition of carbon as graphite in the eutectic transformation during casting also accelerates the conversion of the cementite fraction in the pearlite to graphite when the iron is held at temperatures in the range 500–800°C. The conversion expands the iron because the graphite occupies a greater volume than its cementite precursor. Thermal cycling through a temperature range around 732°C accelerates the growth by mechanisms associated with the austenite/ferrite transformation.

Further Reading

Evans, U. R., Mechanism of rusting under different conditions, *Br. Corr., J.*, 7, 10, 1972.
Honeycombe, R. W. K., *Steels: Microstructures and Properties*, Edward Arnold, London, 1981.
Kubaschewski, O. and Hopkins, B. E., *Oxidation of Metals and Alloys*, Butterworths, London, 1962.
Nicholls, D., *Complexes and First Row Transition Elements*, Macmillan, London, 1981.
Shreir, L. L., *Corrosion and Corrosion Control*, John Wiley, New York, 1985.

8

Stainless Steels

Stainless steels are alloys of iron and chromium usually with other added components. They are the most widely applied and versatile of the corrosion-resistant alloys formulated from the anodic passivating metals described in Chapter 3. Other alloys from within the nickel–chromium–iron–molybdenum system are specialist materials, designed to meet specific difficult environments, and are excluded from general application by their high costs. This constraint does not apply to alloys in which iron and chromium predominate because chromium is available as an inexpensive iron–chromium master alloy, *ferro-chrome,* suitable for alloying with iron. As a consequence, a wide range of corrosion-resistant alloys is available at moderate cost; some of them are mass products, manufactured in integrated steel plants.

For stainless steels, more than for any other group of alloys, structures, properties, corrosion resistance and costs are so closely related that none of these aspects can be properly appreciated without detailed reference to the others. The essential requirement to confer the expected minimum passivating characteristics on a stainless steel is a chromium content of not less than 13 mass %. To modify passivating characteristics or manipulate phase equilibria to produce metallurgical structures matched to various specific applications, other components are required. The need to maintain or enhance the passivating characteristics restricts the most useful additional alloy components to nickel, molybdenum, and carbon. Other elements may also be added, for example, titanium or niobium, to control the effects of carbon, manganese or nitrogen to economize on the use of nickel or copper to induce precipitation hardening. A further constraint is the relative costs of alloy components, summarized in Table 8.1.

Components added to steel can influence the phase equilibria in two respects:

1. They can alter the relative stabilities of austenite and ferrite. Carbon, nickel, manganese, nitrogen and copper stabilize austenite; chromium, molybdenum and silicon stabilize ferrite.
2. They can precipitate any carbon present if their own carbides are more stable than the iron carbide, Fe_3C (*cementite*). Metals that selectively form carbides in steels include chromium, molybdenum, titanium and niobium but not nickel, copper or manganese. Nitrogen, if present, can partly replace carbon, forming carbo-nitrides.

8.1 Phase Equilibria

In the following brief review, equilibria in multi-component stainless steels are approached by considering how successive additions of other components modify equilibria in the iron–chromium binary system. Pickering, cited in Further Reading, gives a more comprehensive treatment.

TABLE 8.1

Raw Material Costs for Stainless Steels

Material	Price $/tonne
Iron (as scrap)...	196
Chromium (as high carbon ferro-chromium)	1350–1
Nickel (as the pure metal)	8835[a]
Molybdenum (as ferro-molybdenum)	10,000[b]
Manganese (as ferro-manganese)	1550
Copper (as the pure metal)	5636[a]
Titanium (as ferro-titanium)	5500
Niobium (as ferro-niobium)	38,000[a]

Note: Costs are for alloy content, for example, the cost given in the second row is per tonne of chromium not per tonne of ferro-chromium.

[a] London Metal Exchange, June 2017.

[b] Trade prices, June 2017.

8.1.1 The Iron–Chromium System

The standard equilibrium phase diagram for the iron–chromium system is reproduced in Figure 8.1. Chromium and α-iron both crystallize in the BCC structure and their atom sizes do not differ much, so that they have extensive mutual solubility and a ferrite, α, phase field extends over most of the diagram. There are two other significant features:

1. A phase field, the *γ-loop*, in which austenite, γ, is the stable phase, spanning the temperature range 840°C to 1400°C, with a maximum chromium content of 10.7 wt-% at 1075°C. The γ-loop is surrounded by a narrow two-phase field in which austenite and ferrite co-exist, with a composition range of 0.8 wt-% chromium at its widest.
2. A phase field with a maximum at 821°C, in which a tetragonal, non-stoichiometric FeCr compound, the sigma phase, σ, is stable. The sigma phase field is flanked

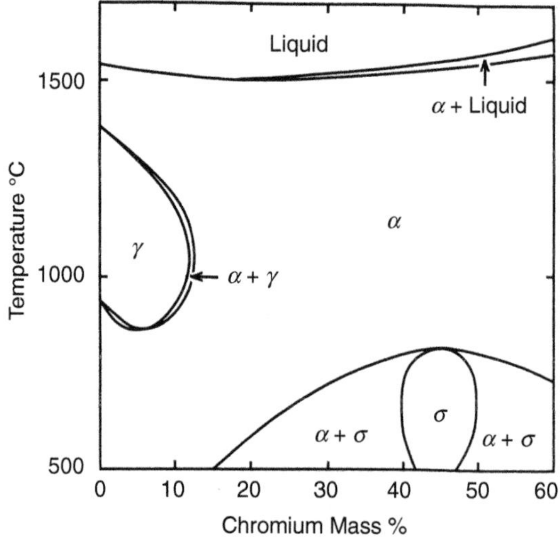

FIGURE 8.1
Equilibrium phase diagram for the iron–chromium system.

by wide two-phase, (α + σ), fields so that the sigma phase can be precipitated in alloys containing 25 to 70 mass % chromium by prolonged heat treatment at temperatures in the range 500–800°C. Permissible chromium contents in stainless steels based on the binary system are limited to the range 12 to 25 mass % to secure adequate passivation without risk of precipitating the sigma phase, which embrittles the steel.

8.1.2 Effects of Other Elements on the Iron–Chromium System

8.1.2.1 Carbon

Carbon is introduced naturally during steelmaking and its content in the finished steel can be adjusted to any required value. It modifies the iron–chromium system in two respects:

1. On a mass for mass basis, it is the most efficient austenite stabilizer, expanding both the γ-loop and the (α + γ) phase field, due to the much higher solubility of carbon in austenite than in ferrite. The effect is limited by the maximum carbon content that austenite can hold in solution, about 0.06% C at 1300°C, which extends the γ-loop and (α + γ) phase field to 18% and 27% Cr, respectively, as illustrated for 0.05% carbon in Figure 8.2.
2. At temperatures below the boundary of the γ-loop, carbon in excess of its very small solubility in ferrite tends to precipitate as the complex carbides, Cr_3C, $Cr_{23}C_6$ and Cr_7C_3. Since the carbides are rich in chromium, the precipitation can severely deplete the chromium content of the adjacent metal.

8.1.2.2 Nickel

Nickel forms a complete series of solid solutions with γ-iron, with which it shares a common FCC structure, but it has only limited solubility in the structurally incompatible BCC

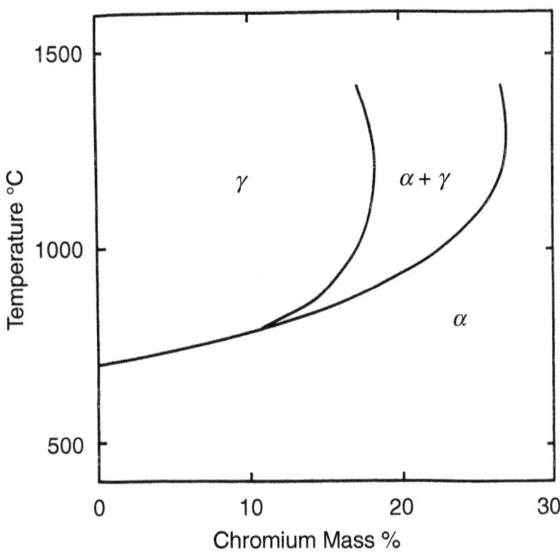

FIGURE 8.2
Section through the phase diagram for the iron–chromium–carbon ternary system at 0.05% carbon showing extended γ and (α + γ) phase fields.

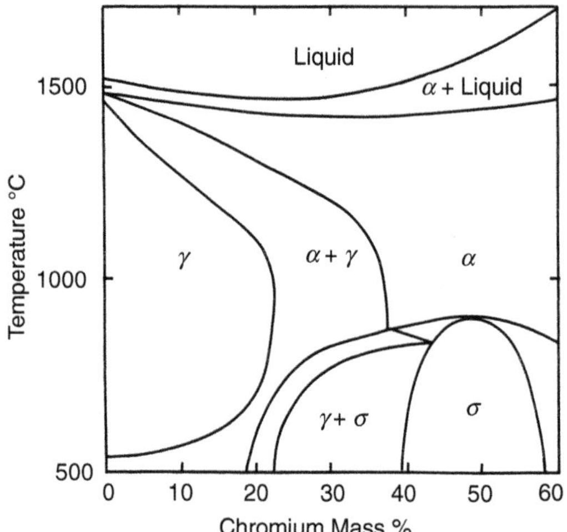

FIGURE 8.3
Section of the phase diagram for the iron–chromium–nickel system at 8% nickel.

α-iron. As a consequence it is a powerful austenizing element and its addition to the iron-chromium system progressively extends the γ-loop, especially towards higher chromium contents and lower temperatures. Comparison of Figure 8.3 with Figure 8.1 shows the extent of the expansion for an addition of 8% nickel. Although the γ-loop does not extend to ambient temperatures, the transformation is sluggish at low temperatures and in practice austenite can be retained as the metal cools. The nickel content to retain a fully austenitic structure depends on the chromium content and has a minimum value of 10–11% at an optimum chromium content of 18%. This balance of chromium and nickel contents is economically important in formulating austenitic steels, as described in Section 8.2.2.2, because the cost of nickel is much higher than the cost of chromium.

A disadvantage of added nickel that must be accepted along with its advantages is that it enlarges the sigma phase field and its associated two- and three-phase fields, as evident in Figure 8.3.

8.1.2.3 Nickel and Carbon Present Together

The austenizing powers of carbon and of nickel are additive, but the assistance given by carbon is limited by its low solubility in iron–chromium–nickel austenites, as indicated by Table 8.2. Even then, to exploit its considerable austenizing power, the carbon content must be retained in solution by quenching the steel from a high temperature, *circa* 1050°C. If the steel is not quenched, carbon contents in excess of 0.03% tend to precipitate as carbides.

Nickel and carbon introduce pseudo-peritectic reactions into the iron–chromium–carbon–nickel system that control the structure of metal solidified from the liquid. The

TABLE 8.2
Solubility of Carbon in Austenite with 18 Mass % Chromium + 8 Mass % Nickel

Temperature °C	700	800	900	1000	1050
Solubility mass %	0.03	0.03	0.04	0.06	0.09

first solid deposited is δ-ferrite that subsequently reacts with residual liquid to form austenite, γ:

$$\text{Liquid} + \delta = \gamma \tag{8.1}$$

The reaction is incomplete in industrial casting and welding practices, leaving up to 10% residual δ-ferrite. This remains as a structural feature of near net-shape castings and in welds, but it is converted to austenite by the mechanical and thermal treatments used in producing wrought materials, so that there is less than 1% of δ-ferrite in plate or sheet products.

8.1.2.4 Molybdenum

Molybdenum stabilizes ferrite and when it is added to improve corrosion resistance of steel formulated to be fully austenitic, its effect must be compensated by reducing the chromium content and increasing the nickel content. Molybdenum also replaces some of the chromium in the complex carbides, $Cr_{23}C_6$ and Cr_7C_3 described in Section 8.1.2.1. Unfortunately, high molybdenum contents promote sigma phase formation in austenitic as well as in ferritic steels, therefore, molybdenum contents in stainless steels are normally restricted to <3.5 mass %.

8.1.2.5 Other Elements

Titanium and niobium can be added as scavengers for carbon in steels designed for welding.
Manganese can be used as a substitute to reduce nickel contents in austenitic steels.
Copper is added to certain steels to participate in precipitation hardening.
Nitrogen can enhance the mechanical properties and assist austenite stabilization.

8.1.3 Schaeffler Diagrams

Schaeffler diagrams, named for their originator, provide means of estimating the proportions of austenite, ferrite and martensite in multicomponent stainless steel from the combined effect of the components expressed as empirical equivalent chromium or nickel contents, defined as:

$$\text{Ni equivalent} = \text{mass \% Ni} + 30\ \text{mass \% C} + 25\ \text{mass \% N} + 0.5\ \text{mass \% Mn} \tag{8.2}$$

$$\text{Cr equivalent} = \text{mass \% Cr} + \text{mass \% Mo} + 1.5\ \text{mass \% Si} + 1.5\ \text{mass \% Ti} \tag{8.3}$$

This reduces the system to a section of the iron-rich corner of a pseudo-ternary system, as in Figure 8.4. The phase fields identify the phases observed in steels quenched to ambient temperature from 1050°C. Fields labelled for martensite represent compositions which pass through a γ-loop on cooling and transform to the metastable phase martensite, α′ instead of the stable phase, ferrite. Schaeffler-type diagrams were devised to assess the structures of welds. Although useful for more general application, they are only approximations and alternative equations for nickel and chromium equivalents are sometimes advocated.

EXAMPLE 1

Using the Schaeffler diagram given in Figure 8.4, suggest the phases probably present in the following steels:

Steel 1 – Analysis (all mass %): 17% Cr, 12% Ni, 2.5% Mo, 0.06% C, 2% Mn, 0.6% Si
Steel 2 – Analysis (all mass %): 12% Cr, 0.5% Ni, 0.15% C, 0.4% Mn, 0.4% Si
Steel 3 – Analysis (all mass %): 25% Cr, 6% Ni, 3% Mo, 0.06% C, 0.25% N

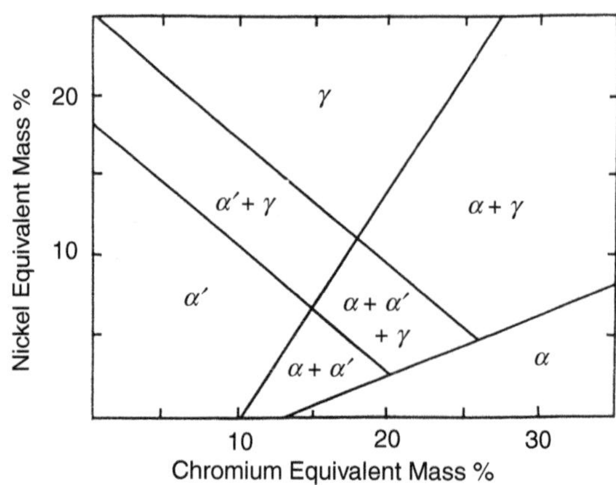

FIGURE 8.4
Schaeffler diagram for structures of stainless steels in terms of Ni and Cr equivalents.
Ni equivalent (mass %) = % Ni + 30% C + 25% N + 0.5% Mn.
Cr equivalent (mass %) = % Cr + % Mo + 1.5% Si + 1.5% Ti.
α = Ferrite; α′ = Martensite; γ = Austenite.

Solution

The probable structures are assessed from the Schaeffler equivalents:

$$\text{Ni equivalent} = \text{mass \% Ni} + 30 \times \text{mass \% C} + 25 \times \text{mass \% N} + 0.5 \times \text{mass \% Mn}$$

$$\text{Cr equivalent} = \text{mass \% Cr} + \text{mass \% Mo} + 1.5 \times \text{mass \% Si}$$

Steel 1

$$\text{Ni equivalent} = 12 + (30 \times 0.06) + (0.5 \times 2) = 14.8 \text{ mass \%}$$

$$\text{Cr equivalent} = 17 + 2.5 + (1.5 \times 0.6) = 20.4 \text{ mass \%}$$

These coordinates are in the γ phase field, so the steel is probably wholly austenitic.

Steel 2

$$\text{Ni equivalent} = 0.5 + (30 \times 0.15) + (0.5 \times 0.4) = 5.2 \text{ mass \%}$$

$$\text{Cr equivalent} = 12 + (1.5 \times 0.4) = 12.6 \text{ mass \%}$$

These coordinates are in the α′ phase field, so the steel is wholly or mainly martensitic.

Steel 3

$$\text{Ni equivalent} = 6.0 + (30 \times 0.06) + (25 \times 0.25) = 14.05 \text{ mass \%}$$

$$\text{Cr equivalent} = 25 + 3 = 28 \text{ mass \%}$$

These coordinates are in the middle of the α + γ phase field, so the steel is duplex, probably with equal quantities of austenite and ferrite.

8.2 Commercial Stainless Steels

8.2.1 Classification

The American Iron and Steel Institute (AISI) specifications use a three-digit code to identify wrought stainless steels by structure. The first digit defines the following classes:

AISI 300 series: Austenitic steels with nickel as the primary austenite stabilizer

AISI 200 series: Austenitic steels with manganese and nitrogen as nickel substitutes

AISI 400 series: Ferritic and martensitic steels with little or no nickel

The other digits identify the steels and letter suffixes indicate other features; in particular L denotes low carbon content. Duplex and precipitation-hardening steels are specified in different formats. Table 8.3 gives specifications for representative standard steels.

8.2.2 Structures

8.2.2.1 Ferritic Steels

Standard ferritic stainless steels with 11 to 30 mass % chromium, <1 mass % carbon and little or no nickel have single phase BCC structures and are less ductile than the austenitic steels that have FCC structures. They are susceptible to grain coarsening on heating to high temperatures and can be embrittled if heated in the temperature range 340°C to 500°C, because the homogeneous ferrite can precipitate a chromium-rich α phase from the main α phase. The effect is known as *'475 embrittlement'*. There is the further risk that some sigma phase can form in ferritic steels with more than 25 mass % chromium on prolonged heating in the temperature range 500°C to 800°C. Ferritic steels with ultra-low carbon and nitrogen contents, >0.01 mass % have improved ductility. Versions with 26 to 29 mass % chromium and 1 to 4 mass % molybdenum have been produced, but they are expensive.

8.2.2.2 Austenitic Steels

Commercial austenitic steels are formulated to realize almost completely austenitic structures at minimum cost, consistent with satisfactory performance. In practice, this implies minimizing the nickel content because nickel is one of the most expensive alloying elements. Carbon is available free of cost because of the nature of the steelmaking process, and if it is retained in solid solution by quenching from a high temperature it can replace a small but economically significant fraction of the nickel content. The maximum carbon content that it is convenient to retain in solution is 0.06 mass %, and the composition for a fully austenitic steel with the minimum nickel content is 18 mass % chromium, 9 mass % nickel and 0.06 mass % carbon.

Versions of austenitic steels are formulated at extra cost for welding, to counter carbide precipitation that can cause local corrosion, as explained later. There are two approaches:

1. Steels such as AISI 304L, AISI 316L and AISI 317L with low carbon contents and compensating higher nickel contents to stabilize the austenite
2. Steels such as AISI 321 and AISI 347 with titanium or niobium as carbon scavengers

TABLE 8.3

Selection of Specifications for Stainless Steels

		Composition, Mass %						
Designation	**Type**	**Cr**	**Ni**	**Mo**	**C**	**Mn**	**Si**	**Others**[a]
AISI 300 Series								
AISI 304	Austenitic	18–20	8–10.5	–	<0.08	2	1	
AISI 304L	Austenitic	18–20	8–12	–	<0.03	2	1	Ti = 5 × %C
AISI 321	Austenitic	18–20	8–10.5	–	<0.08	2	1	Nb = 10 × %C
AISI 347	Austenitic	18–20	8–10.5	–	<0.08	2	1	
AISI 316	Austenitic	16–18	10–14	2–3	<0.08	2	1	
AISI 316L	Austenitic	16–18	10–14	2–3	<0.02	2	1	
AISI 317	Austenitic	18–20	11–15	3–4	<0.08	2	1	
AISI 317L	Austenitic	18–20	11–15	3–4	<0.03	2	1	
AISI 310	Austenitic	24–26	19–22	–	<0.25	2	1.5	
AISI 330	Austenitic	17–20	34–37	–	<0.08	2	1.5	
AISI 200 Series					–			
AISI 201	Austenitic	16–18	3.5–5.5	–	0.15	5.5–7.5	1	0.25 N
AISI 202	Austenitic	17–19	4.–6.	–	0.15	7.5–10	1	0.25 N
AISI 400 Series					–			
AISI 409	Ferritic	10.5–11.7	<1	–	<0.08	1	1	
AISI 430	Ferritic	16–18	<1	–	<0.08	1	1	
AISI 434	Ferritic	16–18	<1	0.8–1.2	<0.08	1	1	
AISI 410	Martensitic	11.5–13.5	<1	–	0.15	1	1	
AISI 431	Martensitic	15–17	2.0–2.5	–	0.20	1	1	
Duplex Steels								
Steel 1		25	6.0	3.0	<0.08	–	–	1.5 Cu + 0.25 N
Steel 2		25	5.5	3.0	<0.08	–	–	
Precipitation hardening								
Steel 1		16	4.2		0.04	0.5	0.5	3.5 Cu
Steel 2		15	4.5		0.04	0.3	0.4	3.5 Cu + Nb

[a] All steels: %P < 0.04%, %S < 0.03%.

Austenitic steels containing molybdenum to enhance corrosion resistance, AISI 316, AISI 316L and AISI 317 have lower chromium contents to avoid the sigma phase and higher nickel contents to offset the effect of molybdenum in stabilizing ferrite.

8.2.2.3 Martensitic Steels

The chromium content of a martensitic stainless steel must be above the minimum needed for passivation but below the maximum of the γ-loop, so that the steel can be converted to austenite for transformation to martensite on cooling. For a nickel-free steel such as AISI 410, a high carbon content is needed both to expand the γ-loop to accommodate enough chromium for passivation and to harden the martensite. The chromium content can be higher if a small nickel content is added to expand the γ-loop further, as for AISI 431. It is usually sufficient for the steel to cool in air to effect the martensitic transformation.

After the martensite transformation, the steels must be *tempered* to alleviate brittleness, that is given controlled thermal treatment in the range 200°C to 450°C that allows the martensite to transform into hard but less brittle decomposition products. Care is needed not to exceed 450°C, because carbides precipitate above this temperature, reducing both corrosion resistance and toughness.

8.2.2.4 Duplex Steels

Duplex steels contain about 28 mass % chromium equivalent and 6 mass % nickel equivalent, as defined for the Schaeffler diagram, to produce structures with about equal proportions of austenite and ferrite. The alloy components are distributed unequally between austenite and ferrite. The austenite has a higher proportion of the nickel, carbon and copper contents and the ferrite has a higher proportion of the chromium and molybdenum contents.

8.2.2.5 Precipitation-Hardening Steels

Precipitation-hardening steels have compositions with a potential for hardening by martensitic or other transformations that can be exploited by thermal and mechanical treatments.

8.3 Resistance to Aqueous Corrosion

8.3.1 Evaluation from Polarization Characteristics

8.3.1.1 Relevance to Corrosion Resistance

Stainless steels resist or succumb to corrosion according to whether or not they succeed in maintaining a passive surface. Information on conditions favouring passivation of particular steels is given by anodic polarization characteristics, that is the current flowing at the surface of a steel as a function of the potential applied to it. These are not inherent properties of the steels but system characteristics, because they are environment-dependent.

In a natural system, the potential of the steel is determined by mutual polarization of the prevailing anodic and cathodic reactions. This potential, the *mixed potential* and the corresponding current exchanged between anodic and cathodic reactions, can be assessed by displaying the net polarization characteristics of the steel and of cathodic reactions in the same format as for corrosion velocity diagrams described in Section 3.2.3 of Chapter 3.

The essential feature of the polarization characteristics for an anodically passivating metal, such as a stainless steel, is a *passive range* over a potential interval, as in the examples illustrated in Figures 8.5 to 8.11, discussed later. The objective in matching a steel to an environment is to ensure that the balance struck between anodic currents from the steel and the currents due to the prevailing cathodic reactions yields a mixed potential that is stable and lies in the passive range.

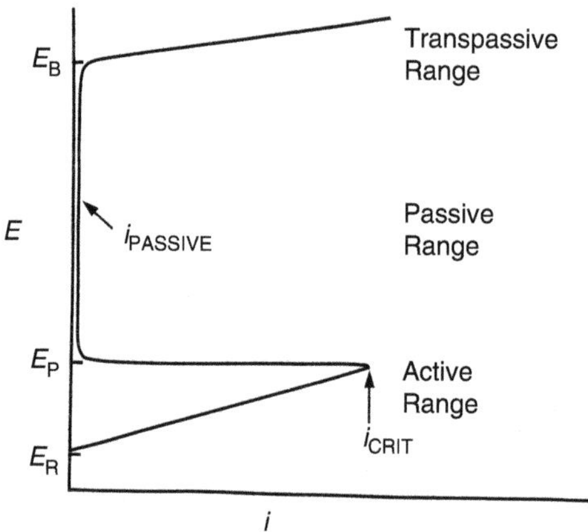

FIGURE 8.5
Schematic anodic polarization characteristics for a stainless steel in acidic media. Current density is plotted on the semi-logarithmic scale generated by a potentiostat, that is linear for $i < 10^{-4}$ A cm^{-2} and logarithmic for $i > 10^{-4}$ A cm^{-2}. E_R = Rest potential; E_P = Passivating potential; E_B = Breakdown potential. i_{CRIT} = Critical current density; $i_{PASSIVE}$ = Passive current density.

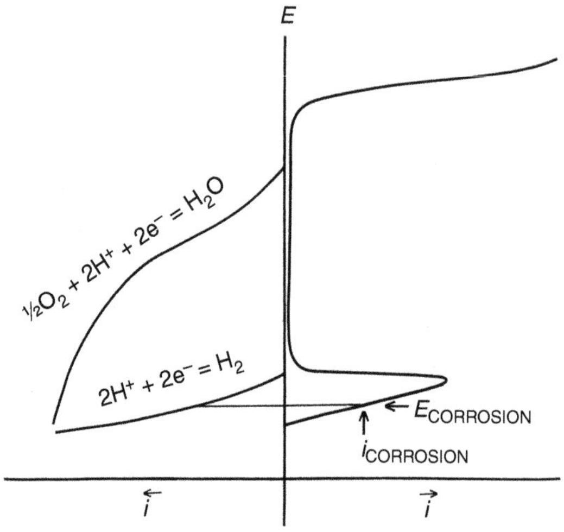

FIGURE 8.6
Corrosion velocity diagram for a stainless steel in dilute acidic media. In oxygen-free acid, the steel corrodes at an active potential, $E_{CORROSION}$, established by hydrogen evolution. In air-saturated acid, it is passivated at a potential in the passive range imposed by reduction of dissolved oxygen.

8.3.1.2 Determination

Polarization characteristics can be determined empirically using a potentiostat, which is a customized instrument designed to maintain a constant potential on a small sample of the steel in any selected aqueous medium irrespective of the current/potential relationships of the circuit. It can simulate any mixed potential that the metal may experience in a natural system. The instrument monitors deviations from an adjustable preset potential and corrects them by applying an external potential to a counter electrode remote from the

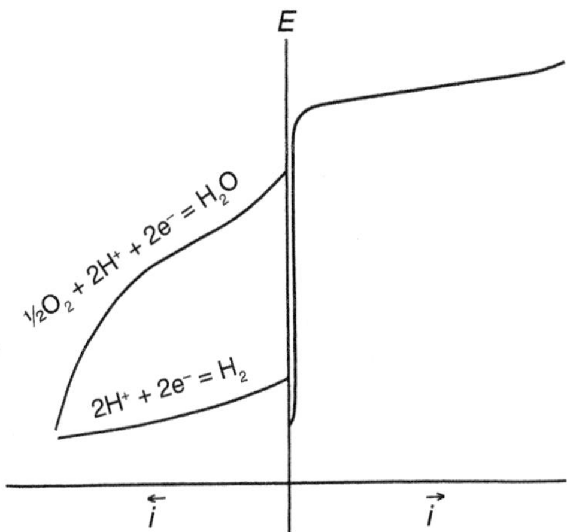

FIGURE 8.7
Corrosion velocity diagram for a stainless steel in neutral aqueous media. The steel is either immune or passive at any potential imposed by hydrogen evolution or the reduction of oxygen dissolved in air-saturated media.

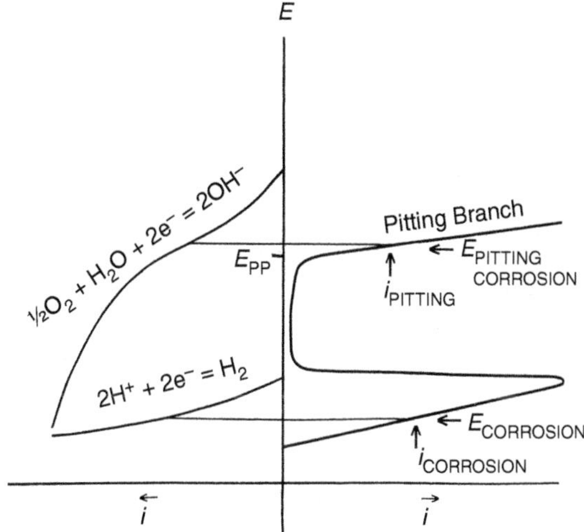

FIGURE 8.8
Corrosion velocity diagram for a stainless steel in dilute acid chloride media. In oxygen-free acid, the steel corrodes at an active potential, $E_{CORROSION}$, established by hydrogen evolution. In air-saturated acid, it corrodes by pitting at a potential $>E_{PP}$ imposed by reduction of dissolved oxygen.

sample. The information required is the current density at the sample surface as a function of potential. The preset potential is scanned through the potential range of interest and the potential/current density relationship is recorded. The procedure is standardized in the ASTM designation G5, which is accepted internationally.*

* Designation G5, *Annual Book of ASTM Standards*, American Society for Testing and Materials, Philadelphia, PA, 2006.

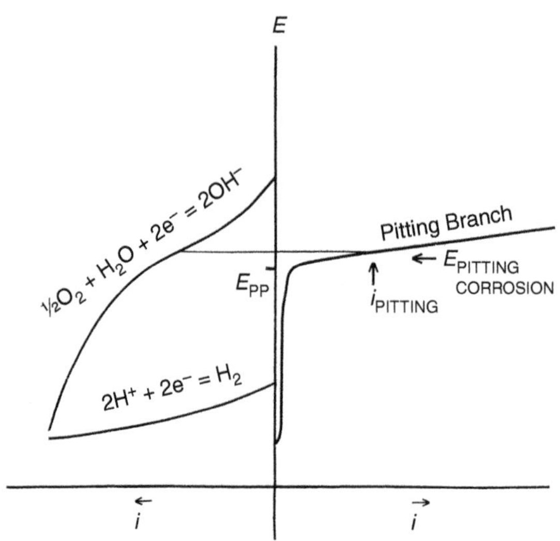

FIGURE 8.9
Corrosion velocity diagram for a stainless steel in neutral chloride media. In oxygen-free media, the steel is passive. In air-saturated media, it corrodes by pitting at a potential $>E_{PP}$ imposed by reduction of dissolved oxygen.

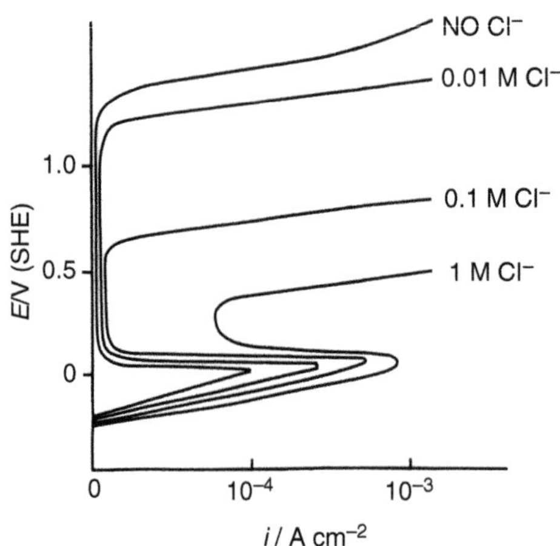

FIGURE 8.10
Typical anodic polarization characteristics for AISI 304 steel in 0.05 M sulfuric acid solutions containing sodium chloride, showing progressive lowering of the pitting potential and expansion of the active peak as the chloride ion concentration is raised. *Note* the current density is plotted on the semi-logarithmic scale generated by a standard potentiostat. This scale is linear for $i < 10^{-4}$ A cm^{-2} and logarithmic for $i > 10^{-4}$ A cm^{-2}.

Fortunately, the active ranges of most stainless steels lie almost wholly above the equilibrium potential for hydrogen evolution, even for acids with low pH, so that the measurement does not suffer significant interference from it. Interference by reduction of dissolved oxygen is circumvented by continuously purging oxygen from the test solution with a stream of very pure nitrogen. Under these conditions, an experimentally determined

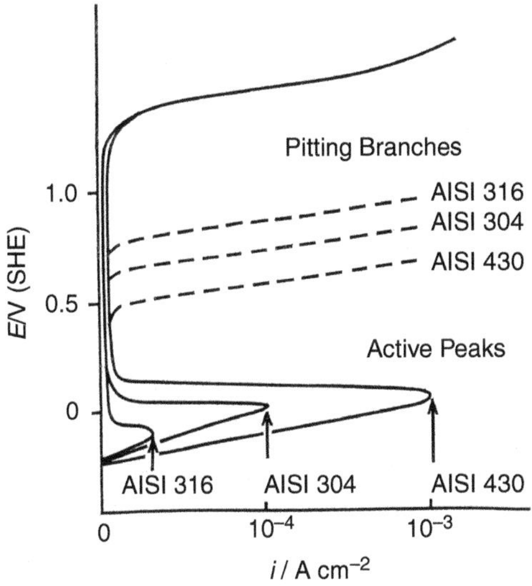

FIGURE 8.11
Typical anodic polarization characteristics in acid and acid chloride media at 20°C for some stainless steels quenched from 1050°C. Expansion of active peaks in acid chloride media omitted for clarity.
Full lines: Polarization characteristics in 0.05 M sulfuric acid.
Dotted lines: Pitting branches for 0.05 M sulfuric acid +0.01 M sodium chloride.
AISI 430 – 17% Cr, 0.05% C (ferritic steel).
AISI 304 – 18% Cr, 9.5% Ni, 0.05% C (austenitic steel).
AISI 316 – 17% Cr, 11% Ni, 2.5% Mo, 0.05% C (austenitic steel).

polarization plot for the steel merges smoothly with plots for hydrogen evolution at potentials below the range of interest and for oxygen evolution at potentials above it.

8.3.1.3 Presentation

The current is usually plotted as a semi-logarithmic coordinate, not for any fundamental reason but to circumvent the difficulty in recording on the same plot both the passive current and the maximum current at the peak of an active range that can differ by three orders of magnitude. The following terms and symbols describing anodic polarization characteristics are marked on the schematic example for a stainless steel in an acid, given in Figure 8.5:

Active range: A potential range where the metal dissolves as ions in a low oxidation state, for example, Fe^{2+} and Cr^{2+}.

Passive range: A potential range where passivity prevails.

Transpassive range: A potential range where the metal dissolves as ions in a higher oxidation state, for example, CrO_4^{2-}.

Rest potential, E_R: The potential for zero current.

Passivating potential, E_P: The lower limit of the passive range.

Breakdown potential, E_B: The upper limit of the passive range.

Critical current density, i_{CRIT}: The active current density at the passivating potential.

Passive current density, $i_{PASSIVE}$: The current density in the passive range.

8.3.1.4 *Influence of the Environment*

Two factors in particular influence the polarization characteristics for a stainless steel, pH and the presence or absence of Cl^- or Br^- ions. Figures 8.6 and 8.9 show schematic polarization characteristics in neutral, acidic, neutral chloride and acid chloride aqueous media, presented on schematic corrosion velocity diagrams together with cathodic polarization characteristics for oxygen absorption and hydrogen evolution for later reference.

8.3.1.4.1 *Acidic Media*

For acidic media, there are active, passive and transpassive ranges in sequence with rising potential, as in Figure 8.6. Typical critical values for an AISI 304 austenitic steel in 0.05 M sulfuric acid are given in Table 8.4.

8.3.1.4.2 *Benign Neutral Media*

There is no active range for neutral media and the steel is either immune or passive at all potentials below the breakdown potential, as in Figure 8.7. The high potentials required for corrosion in the transpassive range are rarely encountered.

8.3.1.4.3 *Acid Halide Media*

Chloride, Cl^-, or bromide, Br^-, ions introduced into an acid modify the anodic polarization characteristics, illustrated for Cl^- ions in Figure 8.8, as follows:

1. The active peak is extended to a higher critical current density, i_{CRIT}.
2. The passive range is prematurely terminated by the intervention of a breakaway current at a potential below the normal breakdown potential, E_B. Steels polarized to potentials above this potential are found to be pitted by intense local attack and on this account, it is called the *pitting potential,* E_{PP} and the line tracing the current densities after the breakaway is called the *pitting branch.*

8.3.1.4.4 *Neutral Halide Media*

Cl^- or Br^- ions introduced into neutral media also produce pitting potentials as illustrated in Figure 8.9, but there is, of course, no active peak for them to influence.

A pitting potential with a value significantly less positive than the breakdown potential is observed if the Cl^- ion concentration exceeds about 0.01 M and it moves to progressively lower values as the concentration is raised. In an acid, the pitting branch intercepts the active range as it approaches 1 M, eliminating the passive range, as illustrated in Figure 8.10.

TABLE 8.4

Polarization Characteristics of AISI 304 Steel in 0.05 M Sulfuric Acid (pH 1.2): Critical Potentials and Current Densities

Rest potential, E_R	−0.2 V (SHE)
Passivating potential, E_P	0.0 V (SHE)
Transpassive breakdown potential, E_B	1.3 V (SHE)
Critical current density, i_{CRIT}	10^{-3} A cm^{-2}
Passive current density, $i_{PASSIVE}$	10^{-5} A cm^{-2}

8.3.1.5 Influence of Steel Composition and Condition

Raising the chromium content confers the following benefits:

1. Lower passivating potentials
2. Reduced passive current densities
3. Higher breakdown and pitting potentials
4. Lower critical current densities

The addition of nickel decreases the critical current density without affecting the passivation potential, but the most beneficial additional alloying element is molybdenum because of its strong effects in lowering critical current densities and raising pitting potentials, as illustrated by the polarization characteristics for representative alloys given in Figure 8.11. Polarization characteristics are inseparable from phase equilibria and microstructures. Austenitic steels have higher nickel and chromium contents than martensitic and ferritic steels and yield lower passive currents and critical current densities. Precipitated carbides deplete adjacent metal of chromium and molybdenum, curtailing the passive range and reducing pitting potentials, as described in Section 8.3.3.4, in the context of an effect called *sensitization*.

8.3.2 A Chemical View of Passivity in Iron Chromium Alloys

Conditions for the establishment and breakdown of passivity on iron/chromium alloys, including stainless steels are related to the standard chemistry of the passivating agent, chromium. To explore this relationship, some reasonable working assumptions are needed as follows:

1. The passive condition is due to a film which may be considered as either a tangible thin multilayer or a chromium-rich monolayer.
2. The film is treated as a three component surface phase comprising iron, chromium and oxygen in which the proportions of iron and chromium are similar to those in the alloy.
3. Chromium species dominate the chemical properties of the film.
4. Cations in the film coordinate with oxygen ions in configurations related to those of corresponding bulk phases, for example, Cr^{III} aspires to octahedral coordination.
5. Cations at the free surface coordinate with water and/or hydroxyl ions as ligands.

8.3.2.1 Oxidation States of Chromium and Related Equilibria

The common oxidation states for chromium are Cr(II), Cr(III) and Cr(VI). Both Cr(II) and Cr(III) form soluble cations and solid hydrous/hydroxide phases. Cr(VI) species exist in highly oxidizing media as the soluble chromate anions CrO_4^{2-}, $Cr_2O_7^{2-}$ and $HCrO_4^-$ in decreasing order of pH values. Equilibria between the ions, Cr^{2+}, Cr^{3+}, CrO_4^{2-} and $Cr_2O_7^{2-}$ in acidic aqueous media are described by the following equations:

$$Cr^{3+} + e^- = Cr^{2+} \tag{8.4}$$

for which:

$$E^{\ominus} = 0.41 \text{ V SHE} \tag{8.5}$$

$$Cr^{3+} + 3e^{-} = Cr \tag{8.6}$$

for which:

$$E^{\ominus} = -0.74 \text{ V SHE} \tag{8.7}$$

$$Cr_2O_7^{\ 2-} + 14H^{+} + 6e^{-} = 2Cr^{3+} + 7H_2O \tag{8.8}$$

for which:

$$E^{\ominus} = +1.33 \text{ V SHE} \tag{8.9}$$

It is convenient to represent these equilibria in a Pourbaix type diagram. The information given in Equations 8.4 to 8.9 is reliable and well established, but information on equilibria in the pH range 5.5 to 11.5 is incomplete and uncertain. However, it is sufficient for the present purpose to know that the most stable phases at the lower values of pH within this range are solid hydrous oxides, usually described as $Cr(OH)_2$ and $Cr_2O_3 \cdot nH_2O$. In these circumstances, the diagram for the chromium–water system, given in Figure 8.12, is inevitably incomplete but it is adequate to indicate the domains in which some relevant species are stable. Although it applies strictly to pure chromium, it provides a working

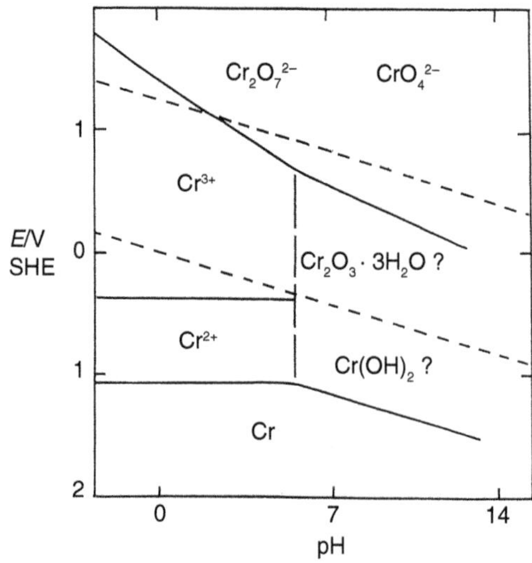

FIGURE 8.12
Schematic partial Pourbaix diagram for the chromium–water system at 25°C.
$a_{Cr^{2+}} = a_{Cr^{3+}} = a_{Cr_2O_7^{2-}} = a_{Cr_2O_4^{2-}} = 10^{-6}$.
Domain for the stability of water shown by dotted lines.

guide with which to explore the behaviour of chromium species in the passive film if, as assumed, they dominate its chemical properties.

8.3.2.2 Relation of Polarization Characteristics to Chromium Chemistry

8.3.2.2.1 Origins of Passivity in Neutral Media

Information given in Figure 8.12 provides the classic explanation of passivation in *neutral media,* that is that at potentials within the domain of stability for water the thermodynamically stable species are solid hydrous oxide products that can form a self-repairing protective film.

8.3.2.2.2 Origins of Passivity in Acidic Media

Different criteria apply to passivation in *acidic media,* for which the stable species are all soluble ions within the domain of stability for water. Thus although the passivating surface film is virtually permanent at potentials between the passivation potential, E_P, and the breakdown potential, E_B illustrated schematically in Figure 8.5, *it is not thermodynamically stable.*

The reason for the permanence of the unstable passivating film is that its chromium content is in the Cr(III) oxidation state which is subject to the exceptionally large kinetic inertia that generally inhibits ligand exchange reactions of Cr(III) species. Hypothetical dissolution of the film would entail an exchange of water and OH^- ligands at the surface of the film by the reaction:

$$-[Cr^{3+} - OH^-]_{FILM} + 6H_2O = -[OH^-]_{FILM} + Cr(H_2O)_6^{3+} \tag{8.10}$$

yielding the soluble solvated cation, $Cr(H_2O)_6^{3+}$, but it is arrested by the inertia.

Table 8.5 illustrates the magnitude of the inertia, showing that the rate constant for ligand exchange in the chromium complex $Cr(H_2O)_6^{3+}$, is a factor of at least 10^{-9} less than corresponding constants for most other $M(H_2O)_6^{n+}$ complexes. The easy passivation of stainless steels implies that Cr(III) can impart its kinetic inertia to films on alloys in which chromium is a significant component. Maximum kinetic stability corresponds

TABLE 8.5

Rate Constants for Ligand Exchange in Some Octahedral Inner Sphere Aquo Ions, $M(H_2O)_6^{n+}$

Central Ion	Half-Life[a], t s^{-1}	Rate Constant[b], K s^{-1}
Cr^{3+}	10^5	2.5×10^{-1}
Cr^{2+}	10^{-4}	2.5×10^8
Fe^{3+}	10^{-4}	1.5×10^8
Fe^{2+}	10^{-6}	1.3×10^6
Ni^{2+}	10^{-5}	1.5×10^4

[a] Burgess J. *Ions in Solution, Basic Principles of Chemical Interactions,* Ellis Horwood, Chichester, Chapter 9, p. 112 (1995).
[b] Eigen, M., *Pure Appl. Chem.* 6, 97–115 (1963).

with the chromium content of the film needed to form a continuous network of Cr^{III}–OH^- octahedra, which is estimated as ~13 mass % from geometric considerations, corresponding to the minimum chromium content required to passivate binary iron/chromium alloys.

8.3.2.2.3 The Active Range in Acids

The active range is confined to acids and corresponds with potentials below the passivation potential, E_P, in Figure 8.5. It can be attributed to dissolution of an intermediate species in the Cr(II) oxidation state, which is free from kinetic inertia for ligand exchange that applies to the Cr(III) state.

Applying the Nernst equation to Equations 8.4 and 8.5 which describe the equilibrium between Cr^{2+} and Cr^{3+}, yields the activity ratio as a function of potential:

$$E = -0.41 - 0.0591 \log(a_{Cr^{2+}} / a_{Cr^{3+}}) \text{ V SHE} \quad (8.11)$$

Rearranging:

$$\log(a_{Cr^{2+}} / a_{Cr^{3+}}) = (-0.41 - E)/0.0591 \quad (8.12)$$

Equation 8.12 yields results showing that the activity of Cr^{2+} in the system is insignificant at potentials above typical passivating potentials, but it rises exponentially with cathodic polarization until at the onset of the active range there are sufficient Cr^{2+} ions to override the kinetic inertia of ligand exchange on Cr^{3+} ions by an alternative fast ligand exchange route, that is:

reduction of Cr(III) to Cr(II): $-[Cr^{3+}–OH^-] + e^- = -[Cr^{2+}–OH^-]$

followed by dissolution: $-[Cr^{2+} - OH^-] + 6H_2O = -[OH^-] + Cr(H_2O)_6{}^{2+}$ (8.13)

Inserting the value $E_P = 0.00$ V (SHE) for the passivation potential in Equation 8.11 yields:

$$a_{Cr^{2+}} / a_{Cr^{3+}} = 10^{-7}$$

Table 8.5 shows that ligand exchange reactions of Cr(II) are a factor of 10^9 faster than corresponding reactions of Cr(III). Therefore, introducing sufficient Cr^{2+} ions to raise the activity ratio to 10^{-7} raises the dissolution current density by a factor of $10^{-7} \times 10^9 = 10^2$. This value is consistent with the ratio of typical active to passive current densities at the passivating potential given in Table 8.4, that is $10^{-3}/10^{-5} = 10^2$. Comparisons between theoretical and empirical values for dissolution current densities cannot be extended to more cathodic potentials because of interference by cathodic currents due to discharge of hydrogen.

The depletion of Cr^{2+} ions in the film is continuously replenished by reduction of further Cr^{3+} ions, maintaining the equilibrium described in Equation 8.4 until the film is consumed. The $Cr(H_2O)_6{}^{2+}$ cations are transient species that are oxidized by water to $Cr(H_2O)_6{}^{3+}$:

$$Cr(H_2O)_6{}^{2+} + H_2O = Cr(H_2O)_6{}^{3+} + OH^-_{aq} + \tfrac{1}{2}H_2 \quad (8.14)$$

8.3.2.2.4 *The Transpassive Range*

The transpassive range extends over all values of pH and corresponds with potentials more noble than the breakdown potential, E_B, in Figure 8.5. It is due to oxidation of Cr(III) to Cr(VI) *in situ* at the surface of the passive film, changing the environment of chromium from the octahedral coordination of Cr(III) to the tetrahedral coordination of Cr(VI). Transpassive dissolution reactions are not ligand replacement reactions so that they are not inhibited by the kinetic inertia associated with dissolution of Cr(III) and depassivation occurs by general dissolution of the film as chromate or dichromate, for example, for acidic media:

$$2Cr^{3+} + 7H_2O = Cr_2O_7^{2-} + 14H^+ + 6e^- \quad (8.15)$$

8.3.3 Corrosion Characteristics of Stainless Steels

The corrosion resistance of stainless steels improves in the ranking order: martensitic, precipitation-hardening, ferritic, duplex and austenitic steels. Martensitic and precipitation-hardening steels are exploited for their strength or hardness, but their limited corrosion resistance restricts their use. Ferritic steels are better but austenitic and duplex steels are required for more aggressive environments. All stainless steels are susceptible to stress-corrosion cracking, but ferritic and duplex steels are less susceptible than austenitic steels. These matters were considered earlier in Chapter 5.

8.3.3.1 *Corrosion Resistance in Acids*

8.3.3.1.1 *Non-oxidizing Acids in General*

To resist corrosion in a non-oxidizing acid, the potential of a stainless steel must be raised from the active into the passive range, either by oxygen or some alternative oxidizing agent, such as Fe^{3+} or NO_3^- ions. Oxygen dissolved from the air is generally sufficient but if it is the only oxidizing agent, access to air must be continuous and unrestricted because even passivated steel slowly reacts with dissolved oxygen and, unless replenished, the oxygen concentration can be depleted to such an extent that the passivity breaks down and the steel corrodes. The role of oxygen can be explained using the critical values for AISI 304 steel given in Table 8.4. In an acid of pH 2 that is *completely free from dissolved oxygen*, the only possible cathodic reaction is hydrogen evolution:

$$2H^+ + 2e^- = H_2 \quad (8.16)$$

for which the equilibrium potential is:

$$E' = -0.0591\,pH = -0.0591 \times 2 = -0.118\ V\ (SHE) \quad (8.17)$$

This potential lies between the passivating potential, $E_P = 0.0$ V (SHE), and the rest potential, $E_R = -0.2$ V (SHE), so that the balance between the anodic and cathodic currents yields a mixed potential in the active range and the steel corrodes. If the acid is equilibrated with ambient air, hydrogen evolution is replaced by oxygen reduction as the dominant cathodic reaction:

$$\tfrac{1}{2}O_2 + 2H^+ + 2e^- = H_2O \quad (8.18)$$

for which the equilibrium potential is:

$$E = E^{\ominus} - \frac{0.0591}{2} \cdot \log \frac{1}{(a_{H^+})^2 \times (a_{O_2})^{1/2}} \tag{8.19}$$

Inserting values, $E^{\ominus} = 1.228$ V (SHE), pH = 2 and $(a_{O_2}) = 0.21$ for equilibrium with air yields the value E = 1.1 V (SHE).

This potential is so much higher than the passivation potential that although the absorption of oxygen is heavily polarized, the current density available at the passivation potential exceeds the critical current density so that a balance cannot be struck between anodic and cathodic currents in the active range. Consequently, the mixed potential rises into the passive range and the steel passivates. The essence of these calculations is illustrated graphically in Figure 8.6.

Provided that these conditions are satisfied, austenitic stainless steels without molybdenum, such as AISI 304, can resist most aerated dilute mineral and organic acids in concentrations up to about 0.5 M at ambient temperatures. Corresponding steels with molybdenum contents, such as AISI 316, can resist acids in concentrations of up to 1 or 2 M, at ambient temperatures and dilute acids at moderately elevated temperatures.

8.3.3.1.2 Oxidizing Acids

Oxidizing acids raise the potential on a stainless steel into the passive range without assistance, therefore, austenitic stainless steels have excellent resistance to nitric acid and mixtures of nitric and sulfuric acids in almost all concentrations and at temperatures up to about 80°C. Attack can occur at carbide sites if titanium is used as a carbon scavenger as in AISI 321 steel formulated for welding.

8.3.3.1.3 Phosphoric Acid

Austenitic stainless steels resist phosphoric acid in all concentrations even at temperatures up to about 80°C. The sparingly soluble iron secondary and tertiary phosphates, referred to in Section 6.4.1, probably assist passivation.

8.3.3.1.4 Hydrochloric Acid

As explained in Section 8.3.1.4, the progressive introduction of chloride ions into an acidic medium curtails and ultimately eliminates the passive range. As a result, the corrosion resistance of stainless steels in hydrochloric acid at ambient temperature is unreliable and limited to concentrations of less than about 0.05 M.

8.3.3.2 Pitting Corrosion

8.3.3.2.1 The Phenomena of Pitting

Pitting corrosion is intense local attack at isolated sites where the passivity has been breached. These sites are very small and they are anodic to the surrounding areas because of the electric field across the passive layer. The resultant active/passive cells stimulate intense attack on the small anodes, driving the growth of the pits and inhibiting the nucleation of new pits in the immediate vicinities of existing ones. This establishes a pattern of attack at discrete sites widely distributed over the metal surface. Pits nucleate after finite induction times and grow progressively deeper into the metal. They can perforate thin gauge metal, rendering stainless-steel containers unserviceable even if general attack is insignificant.

The pitting is induced by halide ions, most commonly the Cl^- ion or its precursor, the hypochlorite ion, ClO^-, but also sometimes the bromide, Br^-, ion. The steels are susceptible to pitting in most aqueous environments containing these ions, including near-neutral media in which corrosion is otherwise not usually serious. Hence seawater, marine atmospheres, acid chloride mixtures in chemical processes and residual bleach-based disinfectants can all attack stainless steels by pitting. The most effective pitting ion is chloride. The bromide ion is less effective but fluoride, F^- and iodide, I^- are not normally pitting ions.

For pits to form, the pitting ion must be present at sufficient concentration in the environment to yield a pitting potential, $E_{PP,}$ and the potential on the metal must exceed it. The potential is imposed by prevailing cathodic reactions and in natural situations, the most likely effective reaction is the reduction of oxygen dissolved in aqueous media from the air.

For AISI 304 steel with polarization characteristics given in Table 8.4, pitting potentials are:

0.1 M solution of sodium chloride (Cl^- ions): $E_{PP} = +0.4$ V (SHE)

0.1 M solution of sodium bromide (Br^- ions): $E_{PP} = +0.7$ V (SHE)

The equilibrium potential for reduction of oxygen dissolved from the air, by Equation 8.19, is well above the pitting potential for the 0.1 M chloride solution and the consequent mixed potential places the current on the pitting branch, as illustrated in Figures 8.8 and 8.9. It is also above the pitting potential for the 0.1 M bromide solution. Hence pitting is expected in both solutions, but inflicts more damage in the chloride solution than in the bromide solution.

Once they have nucleated, some pits grow but others can re-passivate. This is sometimes attributed to changes in the microenvironment within the pits that influence a balance between dissolution and passivation. To arrest pit growth once it has started, the potential on the metal must be reduced to a value significantly below the pitting potential, the so-called *protection potential.* There are sundry other influences on pit growth; for example, a gravitational effect favours pit growth more on horizontal than on vertical surfaces. This is also associated with the microenvironment in pits.

Pits often grow at the sites of microscopic inhomogeneities outcropping at the steel surface, such as carbide and sulfide particles and grain boundaries. Although these sites are preferred, they are not essential, because pits can nucleate elsewhere. These features probably cause weaknesses in the passive surface, due to associated local chromium depletion. Mechanical factors such as heavy cold-work and rough surface finish also encourage pit nucleation.

The tendency of the different halogen ions to induce pitting does not correspond with their increasing electronegativity and their sequence in the Periodic Table, that is iodine, bromine, chlorine, fluorine. From the values of pitting potentials for chlorine and bromine ions given above, a pitting potential for the fluoride ion might be expected at some value less positive than that for the chloride ion, +0.4 V (SHE). However, whereas chloride and bromide ions are *structure breaking,* the fluoride ion is *structure forming* and its ability to interact with a passivated surface is hindered by a solvation shell. Iodide ions are also ineffective as pitting ions for stainless steels but for a different reason. They are easily oxidized to iodine and cannot exist in significant concentrations at high positive potentials, as illustrated by Example 2.

EXAMPLE 2

A stainless steel considered for use in plant processing 0.1 M solutions of iodide salts exhibits the following characteristics at the prevailing pH:

Rest potential, $E_R = -0.2$ V (SHE)
Passivating potential, $E_P = 0.0$ V (SHE)
Breakdown potential, $E_B = +1.3$ V (SHE)

Information is available that 0.1 M solutions of Cl^- or Br^- ions introduce pitting potentials $E_{PP}(Cl^-) = +0.4$ V SHE and $E_{PP}(Br^-) = +0.7$ V SHE, respectively, for this steel. Does this suggest that the steel is likely to suffer pitting corrosion in the application contemplated?

Solution

Pitting potentials in the ascending order Cl^-, Br^-, I^- might be expected, corresponding to decreasing electronegativity, so that a hypothetical pitting potential for a 0.1 M solution of iodide ions would be >+0.7 V SHE. However, the standard electrode potential for the reaction:

$$I_2 + 2e^- = 2I^-$$

given in Table 3.3:

$$E^{\ominus} = +0.536 \text{ V (SHE)}$$

suggests that iodine ions may not survive at such a high potential. Applying the Nernst equation to the reaction for a 0.1 M solution of iodide, assuming unit activity for iodine:

$$\begin{aligned} E &= E^{\ominus} - \frac{0.0591}{2} \cdot \log \frac{(a_{I^-})^2}{a_{I_2}} \\ &= +0.536 - \frac{0.0591}{2} \cdot \log(0.1)^2 \\ &= +0.595 \text{ V (SHE)} \end{aligned}$$

Hence, iodide ions are progressively removed as the potential is raised >+0.595 V SHE, which is well below any hypothetical pitting potential, so that pitting corrosion is not expected.

8.3.3.2.2 A Theoretical Approach to Pitting

In this approach, chloride or bromide ions absorbed by the passive film are envisioned as precursors of alternative soluble species in which the Cr(VI) oxidation state can be realized at anodic potentials below the normal breakdown potential. Cotton and Wilkinson cited in further reading at the end of the chapter summarize relevant aspects of chromium chemistry.

The kinetic inertia of the Cr(III) oxidation state that maintains the integrity of the passive film in acids normally persists to the breakdown potential, at which Cr(VI) becomes the stable state and the film dissolves as dichromate ions, $Cr_2O_7^{2-}$. The absorption of chloride or bromide ions introduces an extra component into the system that modifies the passive film, providing access to the fast kinetics of the Cr(VI) oxidation state via chlorochromate or bromochromate ions, CrO_3Cl^- and CrO_3Br^-. The lower free energies of formation of these ions permit dissolution of the passive film at anodic potentials below the normal breakdown potential. The distribution of isolated breaches in the passive film initiating pits can be explained by selective adsorption of the halide ions at favourable surface sites.

A schematic reaction route is illustrated for chloride ions as follows:

1. Absorption of chloride ions from the environment at selected sites on the surface of the film, replacing hydroxyl ions coordinated with Cr(III):

$$-Cr\text{–}(OH)^- + Cl^- \rightarrow -Cr\text{–}Cl^- + OH^- \tag{8.20}$$

2. Oxidation of Cr(III) to Cr(VI) *in situ* and its realignment in tetrahedral coordination:

$$-Cr\text{–}Cl^- + 3H_2O \rightarrow -CrO_3Cl^- + 6H^+ + 6e^- \tag{8.21}$$

3. Dissolution of the oxidation product from the substrate as the ion, CrO_3Cl^-

The chlorochromate ion is subsequently hydrolyzed by water:

$$2CrO_3Cl^- + H_2O = Cr_2O_7^{2-} + 2H^+ + 2Cl^- \tag{8.22}$$

A similar reaction scheme applies to the analogous species, bromochromate.

There remains for consideration the relation of pitting potentials, to halide ion activities, and steel compositions and selective absorption at favourable sites. Quantitative approaches to these matters, based on adsorption theories and charge densities, require complex aspects of electrochemistry. They are discussed in detail in Section 30.2.3.

8.3.3.3 **Crevice Corrosion**

Crevice corrosion is due to enhanced anodic activity in oxygen-starved crevices in a metal surface as it is for active metals. For stainless steels, the root cause is easier to envision, because local oxygen starvation leads to passivity breakdown and the establishment of active/passive cells in which the large open passive surface stimulates corrosion on the oxygen-starved active area in a crevice. Chloride contamination can assist depassivation where the degree of oxygen starvation would otherwise be marginal.

8.3.3.4 **Sensitization and Intergranular Corrosion**

Standard austenitic steels, such as AISI 304 and AISI 316, are supplied by the manufacturer with the carbon retained in supersaturated solution by rapid cooling from a final heat treatment at 1100°C, and they are intended to be used in this condition. If they are subsequently reheated and cooled slowly through a critical temperature range, they become *sensitized*, a condition in which they are susceptible to *intercrystalline corrosion*, that is corrosion along the grain boundaries. Intercrystalline corrosion due to sensitization in the heat-affected zones close to welds is known as *weld decay*. The effect and suitable countermeasures are well known and it poses no problem to those who are aware of them.

The condition is due to precipitation of chromium or chromium/molybdenum carbides at grain boundaries that produces a thin continuous network of metal so severely depleted in chromium that it cannot passivate. Active/passive cells formed where this

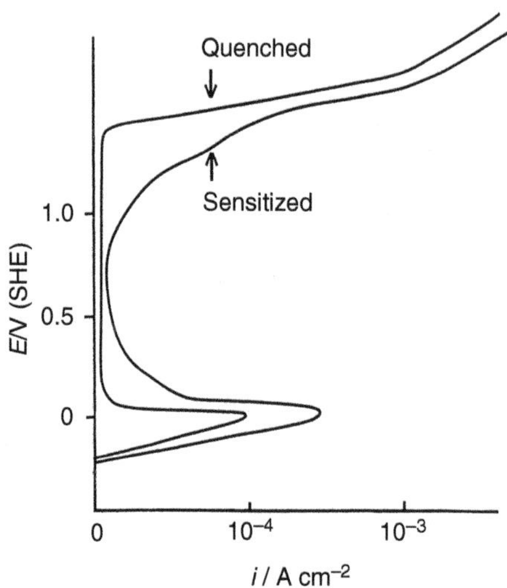

FIGURE 8.13
Effect of sensitizing wrought AISI 304 steel containing 0.06% carbon on its anodic polarization characteristics in 0.05 M sulfuric acid.
(1) Steel as supplied, quenched in water from 1050°C to retain carbon in solution.
(2) Steel sensitized by reheating for 20 minutes at 650°C.

network intercepts the passive surface stimulate intense attack penetrating down the grain boundaries. Sensitized steel can be identified from anodic polarization characteristics, as illustrated in Figure 8.13.

Three factors explain why the chromium depletion is so severe and why it occurs selectively at the grain boundaries:

1. Carbon can trap chromium equivalent to 16.6 times its own mass as $Cr_{23}C_6$ or 10.1 times its own mass as Cr_7C_3.
2. Carbon diffuses in austenite orders of magnitude faster than chromium does and is therefore available from a large catchment volume.
3. Carbides precipitate preferentially at grain boundaries where nucleation is favoured.

Steels with carbon contents above 0.03 mass % are at risk and the temperature range to avoid is 500–800°C. At lower temperatures, carbon diffuses too slowly and at higher temperatures, chromium diffuses fast enough to replenish some of the local depletion.

Molybdenum can partially replace chromium in the carbides as $(Cr, Mo)_{23}C_6$ and $(Cr, Mo)_7C_3$, so that molybdenum-bearing grades of stainless steel are less susceptible to sensitization than the corresponding molybdenum-free steels but they are not immune to it.

Sensitization can be avoided by using the steels customized for welding listed in Table 8.3. There is a choice between steels with low carbon contents, that is AISI 304L, AISI 316L and AISI 317 L, accepting the cost penalty of higher nickel contents, and *stabilized* steels containing titanium or niobium as carbon scavengers, such as AISI 321 and AISI 347. In stabilized steels, precipitation of chromium carbides is pre-empted by a preliminary heat treatment at 900–920°C to trap carbon as TiC or NbC. Niobium is better than titanium as a scavenger for steels exposed to highly oxidizing acids because TiC can be attacked, inducing *knife-line* corrosion.

In theory, sensitization can be eliminated by prolonged post-welding heat treatment to replenish the depleted zones by chromium diffusion from the bulk of the metal, but it is expensive, impracticable for large welded structures and it eliminates strength developed by work hardening that is one of the benefits of using austenitic steels.

8.3.3.5 Corrosion in Aggressive Chemical Environments

In some processes in the chemical industry, particularly aggressive liquors must be handled, including mixtures of undiluted mineral acids and concentrated solutions of aggressive ions, often at high temperatures even approaching boiling. Such media cannot be considered in the same way as aqueous solutions and are difficult to treat theoretically. Alloys based on components selected from among iron, nickel, chromium, molybdenum and cobalt have been customized empirically to meet various application-specific requirements. They are mostly austenitic alloys that include some stainless steels, but more often iron is a minor component or is absent. Information on some of these alloys and their applications is given in Chapter 11, in the context of nickel base alloys.

8.4 Resistance to Dry Oxidation

Stainless steels are among the best standard commercial alloys for resistance to oxidation, an attribute conferred by protective films formed by the preferential oxidation of chromium.

There are five oxides in the iron–chromium–oxygen system, FeO (*wüstite*), Fe_3O_4 (*magnetite*), Fe_2O_3 (*haematite*), Cr_2O_3 and $FeCr_2O_4$ (*chromite*). Chromium can be accommodated in Fe_2O_3 as the mixed oxide, $(Fe, Cr)_2O_3$, and in Fe_3O_4 as the mixed spinel, $Fe^{II}(Fe,Cr)_2^{III}O_4$, but it is only sparingly soluble in FeO.

The relative stabilities of these oxides are functions of alloy composition, as described in Section 3.3.4, but evolution of the scale depends on the intervention of other factors, that is:

1. Transport of Fe^{2+}, Fe^{3+}, Cr^{3+} and O^{2-}, ions within the various oxides
2. The structural incompatibility of FeO with $Fe^{II}Cr^{III}{}_2O_4$ and Cr_2O_3
3. The structural compatibility of Fe_3O_4 with $Fe^{II}Cr^{III}{}_2O_4$ and of Fe_2O_3 with Cr_2O_3
4. Progressive reduction in chromium activity at the metal surface by selective oxidation
5. Stresses developed within and between the layers as the scale thickens

Consider the isothermal oxidation of iron–chromium alloys in clean air, at say 800°C, starting from pure iron and hypothetically raising the chromium content progressively. The structure of the scale characteristic of iron, $FeO/Fe_3O_4/Fe_2O_3$, persists until the chromium content of the steel reaches 2 or 3 mass % and there is little effect on the scaling rate. As the chromium content is raised further, the spinel, $Fe^{II}Cr^{III}{}_2O_4$, appears as discrete islands within the FeO layer, as illustrated in Figure 8.14. At the same time, the total thickness of

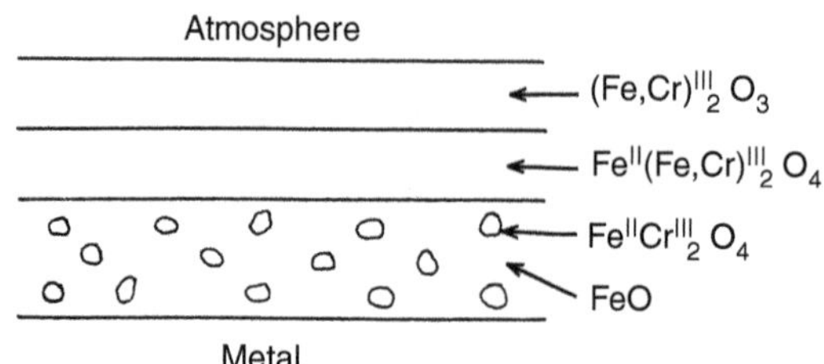

FIGURE 8.14
Morphology of oxide layers formed on iron–chromium alloys in air.

the scale diminishes and so does the relative thickness of the FeO layer. This indicates that the transport of Fe^{2+} ions in FeO is reduced, an effect consistent with a reduction in the composition range over which it is stable. The islands of spinel dissolve in the Fe_3O_4 layer yielding the mixed spinel, $Fe^{II}(Fe,Cr)^{III}{}_2O_4$ when the FeO/Fe_3O_4 interface sweeps over them as it advances into the FeO layer. The chromium is passed on to the outermost Fe_2O_3 layer yielding the mixed oxide, $(Fe,Cr)_2O_3$ as the Fe_3O_4/Fe_2O_3 interface advances in its turn.

As the chromium content of the metal rises, the protection afforded by the chromium-enriched scale improves progressively until at about 13% chromium it is adequate for isothermal exposure to air at temperatures up to about 700°C. For a higher degree of protection, the chromium content must be raised still further. Opinions differ on whether the oxide effective in affording protection is chromium-enriched oxide, $(Fe,Cr)_2O_3$, or the iron-chromium spinel $Fe^{II}Cr^{III}{}_2O_4$. In the long term, stainless steels oxidize by a succession of breakaways when the scale periodically cracks and reseals. The results of short-term tests can therefore be misleading.

Austenitic stainless steels are more oxidation-resistant than binary iron–chromium alloys; for example, AISI 304 with 18 mass % chromium +9 mass % nickel resists oxidation as well as a binary iron-25 mass % chromium alloy. The beneficial influence of nickel has been variously attributed to entry of nickel into the spinel as $(Fe, Ni)^{II}(Cr, Fe)^{III}O_4$ and to nickel enrichment in the metal surface by preferential oxidation of chromium and iron. In moderate amounts, molybdenum in standard stainless steels is reputed to improve the oxidation resistance by formation of an underlay of MoO_2 and manganese is considered harmful if it is high enough to contribute the spinel $MnCr_2O_4$ to the scale.

To serve at high temperatures, an alloy needs not only oxidation resistance but also other essential characteristics, including resistance to creep-rupture and immunity from embrittlement by sigma phase or carbide precipitation during service. Taking all of this into account, special steels with high chromium and nickel contents, such as AISI 310 and AISI 330 listed in Table 8.3, are manufactured for high temperature service. They are, of course, expensive and are used only for the highest temperatures or most onerous conditions, where their special attributes are needed. For temperatures <900°C, steels such as AISI 304 are usually satisfactory.

8.5 Applications

The selection of stainless steels for various applications is an interesting exercise in balancing the requirements of corrosion resistance, mechanical properties, ease of fabrication and economy. The correct steel is always the one that serves its purpose *at minimum cost.*

8.5.1 Ferritic Steels

Ferritic steels have moderate to good corrosion resistance that improves with rising chromium content. Their main attribute is that they are the lowest cost stainless steels. Another merit is that they are less susceptible than austenitic steels to stress-corrosion cracking. They are ductile enough to be cold-formed into simple shapes, but they cannot withstand the severe deformation in forming techniques such as deep-drawing. They are used for moderately demanding service and where retention of brightness or hygiene in mild environments is important. AISI 409 is applied for automobile mufflers and emission control equipment.

AISI 430 is a general purpose ferritic steel with various uses, including uncomplicated plant for handling benign liquors, hygienic surfaces, domestic equipment and automobile bright trim.

The manufacture of domestic sink units illustrates how economies can be made using ferritic steels. The bowl is deep-drawn from a ductile austenitic steel, such as AISI 304, and welded to the draining surface that can be made from a less costly ferritic steel, such as AISI 430.

8.5.2 Austenitic Steels

8.5.2.1 Austenitic Steels without Molybdenum

The austenitic steel, AISI 304, and its derivatives, such as AISI 304L, AISI 321, and so on, are so versatile and popular that the annual US production of them is over a million tonnes. They are some of the most cost-effective mass-produced metallic materials for overall corrosion resistance. A limitation on their use is susceptibility to stress-corrosion cracking when stressed in hot aqueous chloride media, as described in Section 5.1.3. The steels are ductile and can be easily shaped cold. In the form of softened thin, rolled sheets, they can be deep-drawn or pressed into hollowware for which they are especially suitable because they develop high strength from work hardening. Representative applications include:

1. Vessels and other equipment-handling acids for conditions outlined in Section 8.3.3.1
2. Equipment for food processing, such as dairy products, brewing and wine making
3. Domestic equipment including sink basins, cutlery and cookware
4. Cryogenic equipment

The balance of chromium, nickel and carbon contents in AISI 304 and its derivatives yield austenite that is stable at temperatures as low as –196°C, even after the cold-work entailed in shaping them. This accounts for the cryogenic applications. Some variants of these steels, not listed in Table 8.3, are formulated to produce metastable austenite that can be hardened by a martensitic transformation induced by cold-work.

8.5.2.2 Austenitic Steels with Molybdenum

These steels have similar uses to the molybdenum-free steels but their considerable extra costs are justified only when their better corrosion resistance to acids and chlorides is actually needed. It is a mistake to select them for environments where less expensive steels without molybdenum are equally satisfactory. They have many applications in chemical industrial plant handling reducing acids or liquors contaminated with chlorides where the integrity of the plant is of concern. Although they are not immune to pitting induced by chlorides, they are more suitable than molybdenum-free austenitic steels for marine use. A less obvious but important application is decorative external architectural and related use, where loss of visual appeal is a perfectly proper criterion for corrosion failure. Molybdenum-free steels are acceptable in benign atmospheres if regular cleaning is possible, but it is advisable to use a molybdenum-bearing steel, such as AISI 316, if maintenance is difficult or if the atmosphere is industrial, coastal or tropical, because accumulated dirt screens the metal from oxygen and also traps aggressive ions such as chlorides and sulfur compounds.

8.5.2.3 Austenitic Steels with Manganese Substitution

Manganese can be used to reduce nickel contents, but the economic benefit is doubtful because manganese substitution detracts from the good working characteristics and corrosion resistance that are the main recommendations of the nickel–chromium austenitic steels which they replace. They do, however, provide contingency for interruption in the supply of nickel, which is vulnerable to labour or political problems because it is dominated by one principal source, at Sudbury in Ontario, Canada.

8.5.3 Hardenable Steels

Martensitic steels are hard enough for applications to resist wear and to retain cutting edges. They are the least corrosion-resistant stainless steels because of the restricted chromium and nickel contents and the high carbon contents needed for the martensite transformation. They are used for valve seats, turbine parts, knives and scalpels, but their corrosion resistance does not extend to aggressive environments.

The various precipitation-hardening steels were formulated to secure strength and hardness without sacrificing too much corrosion resistance. They contain 14 to 17 mass % chromium, some carbon, nickel, manganese, silicon and selected additions from among cobalt, titanium, copper, molybdenum and aluminium. By appropriate thermal treatment they can yield metastable austenitic or martensitic structures at ambient temperature amenable to precipitation hardening by subsequent thermal and mechanical treatments.

8.5.4 Duplex Steels

Duplex steels are replacements for austenitic steels in applications in which the two-phase structure has advantages. They have better resistance to stress-corrosion cracking because cracks advancing through austenite encounter the more resistant ferrite. They are less susceptible to sensitization because ferrite adjacent to grain boundaries is a good source of chromium to replenish depletion by carbide precipitation. A further advantage is good resistance to pitting in chloride media for duplex steels containing molybdenum. Lower raw material costs for duplex steels than for austenitic steels are offset by higher hot-working costs incurred to avoid sigma phase due to high chromium/molybdenum contents.

8.5.5 Oxidation-Resistant Steels

The discussion on the influence of alloy composition on isothermal oxidation in clean air, given in Section 8.4, is useful in indicating which alloys are possible and which need not be considered for use at a prescribed temperature, but it does not necessarily represent real service conditions that can impose additional, sometimes overriding demands.

High temperature industrial environments include not only air, but also products of combustion, superheated steam, process fumes and abrasive or reactive dust. Products of combustion can be various mixtures of carbon monoxide, carbon dioxide and water vapour, with or without excess oxygen. They can be contaminated with sulfurous gases or loaded with ash. Process fumes from metal extraction and refining, glass making, oil refining and other industrial processes can be encountered. Temperatures may be neither uniform nor steady and in batch operations they may cycle between ambient and operational temperatures. Any of these chemical, physical or thermal effects can undermine an otherwise protective scale.

Examples of the factors that may have to be faced are:

1. The effects of sulfur incorporated in scales formed on steels exposed to products of combustion containing sulfur dioxide, especially if combustion of the fuel is incomplete
2. Catastrophic oxidation, as defined in Section 3.3.3.7, stimulated by vanadium pentoxide, V_2O_5, and other fluxing agents in fuel ash that can flux away protective scales
3. Cracking and spalling of protective scales by differential expansion and temperature fluctuations
4. Attrition of protective scales by abrasive dust

Steels for high temperature industrial uses are therefore selected by experience on an application-specific basis which can include such diverse applications as furnace roof suspension hangers, flues, stator blades in gas turbine compressors, parts for steam turbines and nuclear installations.

8.6 Applications of Cast Stainless Steels

Near net-shape casting often provides the most economical and convenient production route for components used in chemical and related industries. Examples include manifolds, pump bodies, food processing equipment and large centrifugally cast tubes.

The structures of castings are determined as much by the kinetics of crystal growth as by phase equilibria described in Section 8.1. Dendritic crystals growing from the liquid impose their morphology on the structure, manifest in alloys by concentration gradients because the effects of selective freezing are not dispersed by diffusion. In austenitic stainless steels, retention of some δ-ferrite as a metastable minority phase persisting through the incomplete pseudo-peritectic reaction indicates the extent and distribution of the segregation. Since δ-ferrite is enriched in chromium and molybdenum, conjugate interdendritic liquid at the interface is correspondingly depleted and if the retained δ-ferrite is interconnected, a solidified casting is permeated by a network of austenite containing less than the nominal content of passivating elements. Such cast structures are expected to exhibit

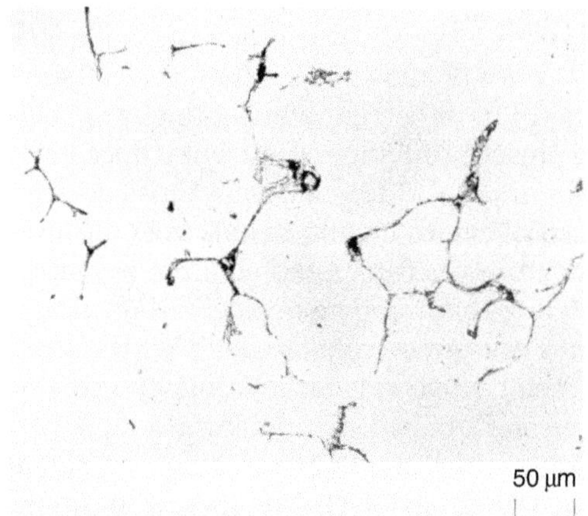

FIGURE 8.15
Photomicrograph of section through AISI 316 casting showing interconnected colonies of δ-ferrite.

different corrosion behaviour than wrought counterparts and to respond to homogenising heat treatments. These effects are reflected in polarization characteristics.

An example is the form of the polarization characteristics determined for samples of a commercially produced AISI 316 sand casting, 20 × 20 × 30 cm, with the analysis, 17 mass % Cr, 10.8 mass % Ni, 2.2 mass % Mo, 1 mass % Mn, 0.04 mass % C. The segregated structure is manifest by the colonies of interconnected δ-ferrite dispersed throughout the casting, illustrated in Figure 8.15. One sample was retained in the cast condition and others were heat treated at 1100°C for successively longer periods and water-quenched to follow changes in the polarization characteristics as the structure progressively approached equilibrium. As expected, the segregation dispersed progressively with time at the heat-treatment temperature.

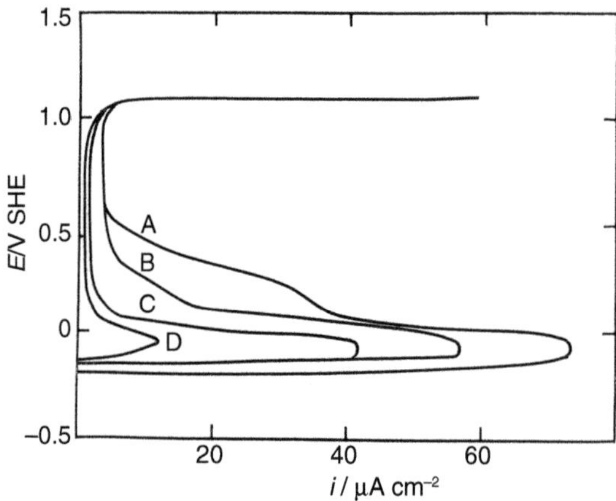

FIGURE 8.16
Polarization characteristics in 0.1 M sulfuric acid for a AISI 316 casting before and after heat treatments and for an equivalent wrought material. *Note* the current densities, *i*, are all $<10^{-4}$ A cm^{-2} and lie on the linear part of the scale generated by a standard potentiostat.
A = As cast material; B = After 1 h at 1100°C and water quenched.
C = After 24 h at 1100°C and water quenched; D = Corresponding wrought material.

Figure 8.16 compares polarization characteristics in 0.1 M sulfuric acid for the cast metal before and after heat treatments with those for an equivalent wrought material. The characteristics for the cast metal exhibit a distorted active loop, severely restricting the passive range. This was partially rectified in as little as 1 hour at 1100°C, suggesting that it was due to solution of carbide precipitates. Full rectification to disperse segregates took much longer, but after ≥24 hours at 1100°C, the passive range and passive current in sulfuric acid approached those of wrought material close enough to suggest acceptable service in acidic media.

The above considerations show that it is unwise to assume that principles established for wrought steels apply also to cast steels, yet it is often tempting to do so for lack of specific information. It is almost certainly necessary to specify post-casting heat treatments for austenitic steels and possibly to change specifications to allow for segregation. For example, there may be occasions for which a ferritic steel is preferred to an austenitic steel for casting because of its shorter freezing range and the absence of a peritectic reaction.

PROBLEMS

Problems for Chapter 8

1. A steel contains 17.0 mass % Cr, 3.5 mass % Mo, 0.05 mass % C and 1.5 mass % Mn, together with nickel. Using the concept of Schaeffler equivalents, estimate the minimum nickel content needed to develop a fully austenitic structure. If a version of the steel with 0.02 mass % C were required for welding, how much more nickel would be required to retain the austenitic structure and how much would this add to the cost of raw materials?
2. Taking into account the functions, costs and probable methods of manufacture, suggest suitable steels from the list given in Table 8.3 for the following applications:

   (a) washing machine drums, (b) exterior architectural fittings got a hotel facing the Atlantic Ocean at Miami Beach, Florida, (c) razor blades, (d) beer barrels made by welding together hollow half-barrels deep-drawn from flat circular blanks, (e) roof supports over a swimming pool disinfected by chlorinating the water, (f) containers for automobile catalytic converters.
3. A particular stainless steel is being considered for use as a container for a dilute neutral aqueous solution of chloride open to the atmosphere. The pitting potential for the steel in the solution is found to be 0.95 V (SHE). Is the steel likely to suffer pitting corrosion? Recall that for the reaction, $\frac{1}{2}O_2 + 2H_2O + 2e^- = 2OH^-$, $E^\ominus = +0.401$ V (SHE).

Solutions to Problems for Chapter 8

Solution to Problem 1

Applying the Schaeffler principle:

$$\begin{aligned}\text{Chromium equivalent} &= \text{mass \% chromium} + \text{mass \% molybdenum}\\ &= 17\% + 3.5\% = 20.5 \text{ mass \%}\end{aligned}$$

The Schaeffler diagram, given in Figure 8.4, indicates that a minimum nickel equivalent of 15.0 mass % is needed for the coordinates representing the nickel and chromium equivalents to lie within the austenite phase field. Excluding the nickel content itself:

$$\begin{aligned}\text{Nickel equivalent} &= 30 \times \text{mass \% carbon} + 0.5 \times \text{mass \% manganese} \\ &= (30 \times 0.05)\% + (0.5 \times 1.4)\% = 2.2 \text{ mass \%}\end{aligned}$$

Hence, the minimum nickel content required is 15.0%–2.2% = 12.8 mass %.

If the carbon content is reduced to 0.02 mass %, its contribution to the nickel equivalent is reduced from 1.5% to 0.6%, so that 0.9% more nickel is needed to retain the austenite. This requires an extra 0.009 tonnes of nickel per tonne of steel, adding $80 to the cost of raw materials per tonne of steel at the price of nickel given in Table 8.1, that is $8835 per tonne.

Solutions to Problem 2

a. *Washing machine drums* The drums are lightly formed thin sheet parts assembled by welding for a competitive domestic appliance market. Exposure conditions are not severe and the main factor is customer perception that the drum retains a hygienic appearance. A ferritic steel, AISI 430, is suitable.

b. *Architectural fittings for a hotel on Miami Beach* The steel experiences severe conditions. It is subject to wind-driven sea spray with intermittent drying, exposing it to depassivation effects induced by chloride, including pitting. Maintenance and cleaning of fittings above ground level is difficult and yet an attractive appearance is an essential element in generating hotel revenue, justifying the capital cost of selecting a molybdenum-bearing austenitic steel such as AISI 316.

c. *Razor blades* The essential feature required of razor blades is the ability to accept and retain a sharp edge. They are disposable items and conditions of exposure are mild so that an inexpensive hardenable martensitic steel, such as AA 410 is suitable.

d. *Beer barrels* The sheet material used for beer barrels must accept severe deformation in the deep-drawing operation by which the two halves are fabricated. It must develop strength by work hardening in the same operation so that the thinnest gauge can be used to minimize weight, facilitating handling in use. These requirements can be met with austenitic steels. Typical pH values for beers are about 4, so that the conditions are slightly acidic but essentially free from chloride ions. The barrels are assembled by welding, but experience shows that sensitization is not a serious problem for austenitic steels with carbon contents <0.06 mass % in the relatively light gauges used that cool quickly after welding. Taking all of these factors into account, AISI 304 is the most suitable choice of steel.

e. *Roof supports for swimming pool* The overriding concern for steels used as load-bearing members of roof supports is susceptibility to stress-corrosion cracking. Chlorine used as disinfectant is the source of the specific agent, chloride ions. Chlorides and water condensate are transported to the roof structure as described in detail in Section 24.7. Steels with the highest molybdenum contents are least vulnerable to stress-corrosion cracking. They include the austenitic steel AISI 317 and duplex steels with 3 mass % molybdenum such as Steels 1 and 2, cited in Table 8.3 that have a further advantage in that the ferrite they contain is more resistant to stress-corrosion cracking.

f. *Containers for automobile catalytic converters* Containers for catalytic converters are part of semi-permanent emission systems and must retain their integrity and remain clean to protect the expensive catalyst assembly. They run hot enough to avoid condensates that would stimulate aqueous corrosion, but must resist dry oxidation at moderate temperatures. Costs bear down on automobile components and an inexpensive steel is required. Ferritic steels provide adequate oxidation resistance and can accept the light forming operations applied during manufacture. Restricting the chromium content reduces susceptibility to *475 embrittlement* at the moderate elevated service temperatures, as described in Section 8.2.2.1. These considerations suggest AISI 409.

Solution to Problem 3

The steel is unlikely to suffer pitting corrosion if the potential imposed by the available cathodic reaction:

$$\tfrac{1}{2}O_2 + 2H_2O + 2e^- = 2OH^-$$

is significantly below the pitting potential, $E_{PP} = +0.95$ V (SHE).

In neutral water in equilibrium with the atmosphere, $a_{OH^-} = 10^{-7}$ and $a_{O_2} = 0.21$, assuming that oxygen is an ideal gas.

Applying the Nernst equation and substituting for $E^{\ominus}$, a_{OH^-} and a_{O_2}:

$$\begin{aligned} E &= E^{\ominus} - \frac{0.0591}{z}\log\frac{(a_{OH^-})^2}{(a_{O_2})^{1/2}} \\ &= +0.401 - \frac{0.0591}{2}\log\frac{(10^{-7})^2}{(0.21)^{1/2}} \\ &= +0.814\,\text{V}\,(\text{SHE}) \end{aligned}$$

This is the maximum potential that the reaction can apply and since it is significantly lower than the pitting potential, the steel is unlikely to suffer pitting corrosion.

Further Reading

Columbier, L. and Hochmann, J., *Stainless and Heat Resistant Steels*, Edward Arnold, London, 1967.

Cotton, F. A. and Wilkinson, G., *Basic Inorganic Chemistry*, Third Edition, John Wiley, New York, 1995, p. 558.

Fontana, M. G. and Green, N. D., *Corrosion Engineering*, McGraw-Hill, New York, 1967.

Hansen, M., *Constitution of Binary Alloys*, McGraw-Hill, New York, 1958, p. 527.

Localized Corrosion, National Association of Corrosion Engineers, Houston, Texas, 1974.

Monypenny, J. H. G., *Stainless Iron and Steel*, Chapman & Hall, New York, 1951, p. 61.

Peckner, D. and Bernstein, J. M., *Handbook of Stainless Steels*, McGraw-Hill, New York, 1977.

Pickering, F. B., The physical metallurgy of stainless steels, *Int. Met. Rev.*, 21, 227, 1976.

Schaeffler, A. L., *Met. Prog.*, 100B, 77, 1960.

Sedriks, A. J., *Corrosion of Stainless Steels*, John Wiley, New York, 1979.

9

Corrosion Resistance of Aluminium and Aluminium Alloys

Aluminium is a relatively expensive metal because its extraction from the mineral, bauxite, is energy intensive. Pure alumina is extracted from bauxite in the *Bayer* process, by dissolution in sodium hydroxide from which it is precipitated, calcined and used as feedstock in the *Hall–Herault* high temperature electrolytic reduction process to recover aluminium metal.

Aluminium has low density, high ductility, high thermal and electrical conductivity, good corrosion resistance, attractive appearance and it is nontoxic. This remarkable combination of qualities makes it a preferred choice for many critical applications in aerospace, automobiles, food handling, building, heat exchange and electrical transmission.

The pure metal is deficient in two respects, mechanical strength and elastic modulus, and aluminium alloy development was driven by the need to improve them without sacrificing other qualities, for example, to improve strength for aerospace, marine and civil engineering applications without losing corrosion resistance. The success of these endeavors has secured the status of aluminium alloys as second only to steels in economic value.

Solid aluminium has an invariable face-centered cubic lattice and there is no counterpart to the structural manipulations exploited for iron alloys, so that different philosophies apply in alloy formulation.

9.1 Physical Metallurgy of Some Standard Alloys

Copper, lithium, magnesium and zinc have fairly extensive solubilities in solid aluminium, but the solubility of many other elements is very limited. Most binary systems with aluminium are characterized by limited solubility of the second metal and a eutectic in which the components are the saturated solution and an intermetallic compound, Al_xM_y. Aluminium is prolific in the number and variety of intermetallic compounds it can form because of its strong electronegativity and high valency. Critical features of some common binary systems are summarized in Table 9.1.

Commercial alloys are based on multicomponent systems that reflect characteristics of the component binary systems but also exhibit features due to interaction between the alloying elements, especially in introducing additional intermetallic compounds and in modifying solubilities.

In alloy formulation, strength can be imparted by:

1. Reinforcement with intermetallic compounds
2. Solid solution strengthening
3. Work hardening
4. Precipitation hardening (aging)

TABLE 9.1
Phase Relationships for Some Binary Aluminium Alloys

System	Maximum Solid Solubility wt % Alloy Element	Eutectic Composition wt % Alloy Element	Eutectic Temperature °C	Eutectic Components
Aluminium–copper	5.7	33.2	548	$Al^a + CuAl_2$
Aluminium–iron	<0.1	1.7	655	$Al^a + FeAl_3$
Aluminium–lithium	5.2	9.9	600	$Al^a + LiAl$
Aluminium–magnesium	14	35	450	$Al^a + Mg_2Al_3$
Aluminium–manganese	1.8	1.9	660	$Al^a + MnAl_6$
Aluminium–silicon	1.65	12.5	580	$Al^a + Si^b$
Aluminium–zinc	82.8	94.9	382	$Al^a + Zn^c$

[a] Containing maximum solubility of second element in solid solution.
[b] Containing 0.5 wt % of aluminium in solid solution.
[c] Containing 1.1 wt % of aluminium in solid solution.

There are more than a hundred current wrought and cast alloy compositions, but they belong to a relatively few series with particular characteristics. A representative selection is given in Table 9.2, using the internationally recognized AA (Aluminium Association) designations.

9.1.1 Alloys Used without Heat Treatment

9.1.1.1 Commercial Pure Aluminium Grades (AA 1XXX Alloy Series)

Commercial grades of pure aluminium are actually dilute alloys. The iron and silicon contents that were the normal impurities in the aluminium product from the Hall–Herault process when it was first introduced raised the properties of the metal to values suitable for a wide variety of general applications, including domestic and catering utensils, packaging foil, some chemical equipment and architectural applications. Although modern practice produces higher purity metal, grades of aluminium with similar iron and silicon contents are still offered as general purpose materials and the most common grade, AA 1100, has typical iron and silicon contents of 0.45 mass % and 0.25 mass %, respectively. The metal is strengthened by dispersed intermetallic compounds, notably Fe_3SiAl_{12}, $FeAl_3$ and the metastable $FeAl_6$, raising the 0.2% offset yield and tensile strengths to the values given in Table 9.3. Additional strength can be and usually is imparted by work hardening as illustrated by the values for 75% cold reduction in thickness, given in the table. The last two digits of the AA 1XXX designation for pure metal grades refer to the aluminium content over 99%, for example, AA1100, AA 1050 and AA 1199 have minimum aluminium contents of 99.00, 99.50 and 99.99 mass %, respectively.

9.1.1.2 Aluminium–Manganese Alloy AA 3003

The alloy, AA 3003, has applications that overlap those of AA 1100, but it has greater strength as illustrated in Table 9.3. It is produced by adding 1.2 mass % of manganese to the AA 1100 composition. The intermetallic compounds are modified to $(Fe,Mn)_3SiAl_{12}$ and $(Mn,Fe)Al_6$. Compounds containing manganese are also present as a fine dispersoid distributed throughout the aluminium matrix.

TABLE 9.2

Nominal Chemical Compositions of Representative Aluminium Alloys

	Alloy Elements Weight %							
Alloy	**Silicon**	**Copper**	**Manganese**	**Magnesium**	**Zinc**	**Others**	**Corrosion Resistance**	**Examples of Applications[a]**
AA 1050	Minimum of 99.50% aluminium – principal impurities iron and silicon						Good	Architectural applications
AA 1100	Minimum of 99.00% aluminium – principal impurities iron and silicon						Good	Cooking utensils, foil
AA 2024	–	4.4	0.6	1.5	–	–	Poor	Aircraft panels
AA 3003	–	0.12	1.2	–	–	–	Good	General purpose alloy, foil
AA 3004	–	–	1.2	1.0	–	–	Good	Beverage cans
AA 5005	–	–	–	0.8	–	–	Good	Architectural applications
AA 5050	–	–	–	1.4	–	–	Good	General purpose alloy
AA 5182	–	–	0.35	4.5	–	–	Good	Beverage can ends
AA 5456	–	–	0.8	5.1	–	0.12 Cr	Good	Transportation, structures
AA 6061	0.6	0.28	–	1.0	–	0.20 Cr	Good	Extrusions, beer containers
AA 7075	–	1.6	–	2.5	5.6	0.23 Cr	Poor	Aircraft stringers and panels
AA 7072	–	–	–	–	1.0	–	Good	Cladding for 7075 alloy
AA 319	6.0	3.5	–	–	–	–	Moderate	General purpose castings
AA 380	8.5	3.5	–	–	–	–	Moderate	Die castings
AA 356	7.0	–	–	0.35	–	–	Good	Age-hardenable castings
AA 390	17	4.5	–	0.55	–	–	Not applicable	Automobile cylinder blocks

[a] Selected applications. Most alloys are multipurpose.

9.1.1.3 Aluminium–Magnesium Alloys (AA 5000 Series)

Magnesium both imparts solid solution strengthening and enhances the ability of the metal to work harden, illustrated for alloys AA 5005, 5050, 5182 and 5456 in Table 9.3. They have practical advantages in their good formability, high rate of work hardening, weldability and corrosion resistance. A reservation on the magnesium content is set by the instability of the solid solution with respect to precipitation of Mg_2Al_3 that can be significant for magnesium contents greater than about 3% during long storage at ambient temperatures. The precipitation is at grain boundaries and can render the metal susceptible to stress-corrosion cracking and to intergranular corrosion. Unfortunately, the instability is increased by work hardening which is one of the most useful attributes of the alloy, but it can be ameliorated with some loss of strength by special tempering procedures.

TABLE 9.3
Typical Strengths of Some Aluminium Alloys at 25°C

Alloy	Condition	0.2% Yield Strength MPa	Tensile Strength MPa
99.99%	Annealed (softened by heating)	10	45
AA 1100	Annealed	35	90
	75% Cold reduction (by rolling)	150	165
AA 3003	Annealed	40	110
	75% Cold reduction	185	200
AA 3004	Annealed	70	180
	75% Cold reduction	250	285
AA 5005	Annealed	40	125
	75% Cold reduction	185	200
AA 5050	Annealed	55	145
	75% Cold reduction	200	220
AA 5182	Annealed	130	275
	75% Cold reduction	395	420
AA 5456	Annealed	160	310
	Fully strain hardened[a]	255	350
AA 6061	Annealed	55	125
	Solutionized at 532°C; Aged 18 h at 160°C	275	310
AA 2024	Annealed	75	185
	Solutionized at 493°C; Aged 12 h at 191°C	450	485
AA 7075	Annealed	105	230
	Solutionized at 482°C; Aged 24 h at 121°C	505	570

[a] Using special temper procedure to minimize structural instability.

9.1.1.4 Aluminium–Magnesium–Manganese Alloy AA 3004

The alloy AA 3004 is strengthened by both magnesium and manganese and has an optimum combination of formability and work hardening for application as beverage cans that are deep-drawn and must have sufficient strength in very thin gages to withstand forces imposed on them by filling and by internal gas pressure.

9.1.2 Heat-Treatable (Aging) Alloys

The strongest aluminium alloys are strengthened by precipitation hardening. The principle exploits the diminishing solubility of certain solutes with falling temperature. The strength is developed by controlled decomposition of supersaturated solid solutions. An alloy is held at a high temperature to allow the solutes to dissolve, an operation called *solutionizing*, and then rapidly cooled to retain them in supersaturated solution. Rapid cooling is usually accomplished by immersing the hot metal in cold water, that is *quenching*. The metal is subsequently reheated to a constant moderate temperature and the unstable supersaturated solution rejects excess solute in the form of finely dispersed metastable precursors of the final stable precipitate. These precursors are of suitable forms, sizes and distributions to impede the movement of dislocations in the lattice that produces plastic flow. The process occupies a considerable time and is called *aging*. The improvement in mechanical properties is illustrated by the values given in Table 9.3 for annealed and age-hardened versions of alloys AA 6061, AA 2024 and AA 7075.

Not all aluminium alloy systems are amenable to strengthening in this way. Standard commercial alloys are based on the aluminium–copper–magnesium, aluminium–zinc–magnesium and aluminium–magnesium–silicon systems. In these multicomponent systems, the aging process depends on the decomposition of supersaturated solutions of more than one solute from which complex intermetallic compounds are precipitated. A typical sequence of events is:

1. Assembly of solute atoms in groups, called *Guinier Preston* (GP) zones, dispersed throughout the matrix at spacings of the order of 100 nm
2. Loss of continuity between the zones and the metal lattice in some but not all crystallographic directions, yielding transition precipitates usually designated by primed letters, sometimes with their formulae in parentheses
3. Complete loss of continuity with the matrix, yielding particles of stable precipitates

The sequence is conveniently described by an equation of the general form:

$$\text{solid solution} \rightarrow \text{GP} \rightarrow \theta' \rightarrow \text{precipitate} \tag{9.1}$$

The detail varies widely from system to system and is more complicated than Equation 9.1 indicates, depending on metal composition, precipitate morphology and aging temperature.

The metal becomes stronger as dislocation movement is inhibited by matrix lattice strains building up around the GP zones and transition precipitates, but as these transition species are progressively replaced by equilibrium precipitate, the lattice strains are relaxed and the metal softens. There is thus an optimum aging period for maximum strength, that is *peak hardness*. Metal that has been aged for less than this period is said to be *underaged* and metal that has passed the peak is *overaged*. These conditions have important implications in corrosion and SCC as explained in Chapter 5.

The formation of transition species and their replacement by equilibrium precipitates are both accelerated at higher temperatures. The consequence is that as the aging temperature is raised, the peak hardness is reached in progressive shorter periods of time but its value diminishes. There is thus an optimum thermal cycle to secure good results economically.

9.1.2.1 Aluminium–Copper–Magnesium Alloys (e.g., AA 2024)

The aging sequence is usually represented by:

$$\text{solid solution} \rightarrow \text{GP} \rightarrow S'(Al_2CuMg) \rightarrow S(Al_2CuMg) \tag{9.2}$$

where $S'(Al_2CuMg)$ and $S(Al_2CuMg)$ are transition and stable precipitates, respectively.

The classic examples of alloys in this system are AA 2024 and its variants that are staple materials for airframe construction. The solutionizing temperature is restricted to 493°C to avoid exceeding the solidus temperature and aging is at 191°C for a period of between 8 and 16 hours to suit the form of the material.

9.1.2.2 Aluminium–Zinc–Magnesium Alloys (e.g., AA 7075)

The alloys based on this system are some of the strongest produced commercially and are also staple materials for airframe construction; the characteristic alloy is AA 7075. Particular compositions, solutionizing treatments and aging programmes have been developed to secure practical benefits of strength, consistency of properties and minimum susceptibility to stress-corrosion cracking. Detailed assessment of the aging sequences is difficult because the transition and final species encountered depend on composition, initial microstructure, speed of quenching and aging temperature. Schemes suggested include:

$$\text{solid solution} \rightarrow \text{GP} \rightarrow \eta' \rightarrow \eta(\text{MgZn}_2) \text{ or } T(\text{Mg}_3\text{Zn}_3\text{Al}_2) \quad (9.3)$$

The heat treatment is critical. For example, Alloy AA 7075 is solutionized at 482°C, just below the solidus temperature, and aged for 24 hours at 121°C.

9.1.2.3 Aluminium–Magnesium–Silicon Alloys (e.g., AA 6061)

The precipitating phase in this system is Mg_2Si and the aging scheme is:

$$\text{solid solution} \rightarrow \text{GP} \rightarrow \beta'(\text{Mg}_2\text{Si}) \rightarrow \beta(\text{Mg}_2\text{Si}) \quad (9.4)$$

The alloys are not as strong as those based on the aluminium–copper–magnesium and aluminium–magnesium–zinc systems, but they have the following compensating advantages:

1. They are easy to hot extrude into sections.
2. In the soft condition they are ductile and accept deep drawing into hollow shapes.
3. They are the most corrosion resistant of the aging alloys.

These attributes recommend them for many applications including architectural fittings such as window frames, food-handling equipment, beer barrels and transport applications. The most widely applied alloy is AA 6061, but there are other related formulations. AA 6061 is solutionized at 532°C and aged for 18 hours at 160°C.

9.1.2.4 Aluminium Alloys Containing Lithium

A range of alloys has been developed for aerospace applications based on alloys containing lithium. One of the objectives is to match the properties of AA 2024 and AA 7075 alloys but with lower density and higher modulus, both of which are promoted by the use of lithium. The basic aging sequence

$$\text{solid solution} \rightarrow \text{GP} \rightarrow \delta'(\text{Al}_3\text{Li}) \rightarrow \delta(\text{AlLi}) \quad (9.5)$$

is complicated by the presence of other alloy components needed to secure the required properties, notably copper and magnesium, that lead to schemes that include additional hardening phases, $T'(Al_2CuLi)$ and $S'(Al_2CuMg)$.

These alloys are expensive and justified only where mass saving is a critical economic factor, as in aerospace applications. Interest in them has temporarily diminished pending reassessment of their long-term integrity, but they remain as options for future exploitation.

9.1.3 Casting Alloys

The most common casting alloys are based on the eutectic aluminium–silicon system and are characterized by fluidity of the liquid metal and low contraction on solidification.

There are many composition variants and Table 9.2 lists alloys AA 319, AA 380 and AA 356 as examples in widespread use. The function of the copper content in AA 319 and AA 380 is to improve strength and machinability at some sacrifice of corrosion resistance. The magnesium content in alloy AA 356 offers the facility of precipitation hardening by Mg_2Si. Alloy AA 390 is a customized special alloy for automobile cylinders.

9.2 Corrosion Resistance

Aluminium exists only in the oxidation state Al(III) in its solid compounds and aqueous solution and in this state it is one of the most stable chemical entities known, as illustrated by the high value of the Gibbs free energy of formation for the oxide:

$$2Al + 1\tfrac{1}{2}O_2 = Al_2O_3$$

for which:

$$\Delta G^{\ominus} = -1117993 - 10.96T\log T + 244.5T\,\text{J} \tag{9.6}$$

and of the standard electrode potentials for formation of ions in acid aqueous solution:

$$Al = Al^{3+} + 3e^-$$

for which:

$$E^{\ominus} = -1.64\text{ V (SHE)} \tag{9.7}$$

and in alkaline aqueous solution:

$$Al + 4OH^- = AlO_{2^-} + 2H_2O + 3e^-$$

for which:

$$E^{\ominus} = -2.35\text{ V (SHE)} \tag{9.8}$$

In the context of Table 3.3, these values show that aluminium is intrinsically more reactive than any other common engineering metal except magnesium and its relative permanence is attributable to protection afforded by the oxide films that form upon it.

9.2.1 The Aluminium–Oxygen–Water System

The Pourbaix diagram for the aluminium–water system, given in Figure 3.4, is constructed on the simplifying assumption that the interactions can be represented symbolically by four simple species, Al, Al^{3+}, AlO_{2^-} and Al_2O_3. In reality, these species are more complex and are now considered in greater detail.

9.2.1.1 Solid Species

9.2.1.1.1 Anhydrous Oxides

Corundum Al_2O_3 is the only anhydrous solid oxide. At temperatures below about 800°C, it is predominantly stoichiometric in the terms set out in Section 2.3.6.1. The structure is hexagonal–rhombohedral, with aluminium ions occupying two-thirds of the octahedral interstitial sites in a close-packed hexagonal structure of oxygen ions. It occurs naturally in some igneous and metamorphic rocks and it can be produced by heating hydrated oxides to high temperatures but once formed, it is kinetically inert to rehydration although it is unstable in water or air at temperatures below about 450°C with respect to hydrated oxides.

9.2.1.1.2 Trihydroxides

The stable phase in contact with water or atmospheric air at ambient temperatures is one of the trihydroxides, $Al(OH)_3$. There are three known variants of $Al(OH)_3$, gibbsite, bayerite and norstrandite. Gibbsite is a common naturally occurring mineral form of $Al(OH)_3$ and is also the form produced during commercial bauxite purification. The structures of all of the variants are stacks of the same basic structural unit, $[Al_2(OH)_6]_n$, comprising aluminium ions in octahedral geometry sandwiched between two layers of hydroxyl ions, similar in some respects to the brucite structure described in Section 2.3.3.5 except that aluminium ions occupy only two-thirds of the available sites. Gibbsite, bayerite and norstrandite differ only in the stacking patterns, for example in gibbsite, the hydroxyl groups in adjacent layers are directly opposed and alternate layers are inverted. These variants have similar energies and which of them is the true thermodynamically stable form has not been unequivocally established.

9.2.1.1.3 Oxyhydroxides

There are two oxyhydroxides with the empirical formula AlO(OH), boehmite and diaspore. Both occur naturally as minerals. Boehmite has a layered structure with a basic structural unit, $[OH–AlO–AlO–OH]_n$ in cubic packing; and the hydroxyl ions of every successive layer nest in the pockets between hydroxyl ions in the underlying layer. It usually contains more than the 15% by mass of water that the formula AlO(OH) would suggest and it is not clear whether the excess is due to intercrystalline water or from the presence of some trihydroxide. Diaspore has a structure of hexagonal close-packed layers of hydrogen bonded oxygen ions, $[–O\cdots H–O\cdots]_n$ with aluminium ions in the octahedral interstices.

9.2.1.1.4 Gels

Trihydroxides are initially precipitated from aqueous solution as gelatinous or colloidal precursors with empirical formulae $Al(OH)_3 \cdot nH_2O$, where n is at least 3. The gels are largely 'amorphous' materials but with vestiges of crystallinity. They are unstable and on aging, they slowly transform to crystalline gelatinous boehmite and finally recrystallize to bayerite.

9.2.1.1.5 Dehydration Sequences

On heating them to progressively higher temperatures, trihydroxides begin to lose water at a little over 100°C. The materials pass through a series of structural changes, ultimately becoming the anhydrous oxide corundum. The intermediate structures depend on the initial particle size, rate of heating and ambient humidity. A representative sequence is:

$$\text{Gibbsite} \xrightarrow{100°C} \text{boehmite} \xrightarrow{400°C} (\gamma\text{-alumina} \rightarrow \delta\text{-alumina} \rightarrow \theta\text{-alumina}) \xrightarrow{1150°C} \text{corundum}$$

The γ, δ and θ forms are the so-called transition aluminas with distorted spinel structures stabilized by a small hydroxyl ion content.

9.2.1.2 Soluble Species

Information on the true nature of the Al^{3+} and AlO_2^- ions and their relationships to insoluble species can be deduced from the viscosity, osmotic properties and electrical conductance of their solutions as functions of pH.

9.2.1.2.1 Aluminium Cations

In acid media with pH < ~4, hydroxides and oxyhydroxides dissolve yielding the cationic species usually written as $Al(H_2O)_6^{3+}$. The discrete ions exist only at very low pH values and in less acidic solutions they associate as polymeric complexes. Their solubility is also influenced by agents that bind aluminium cations, for example some organic acids.

9.2.1.2.2 Aluminium Anions

In alkaline media with pH > ~8, hydroxides and oxyhydroxides dissolve to form aluminate ions, usually described as AlO_2^-, but they also associate as various polymeric species. Discrete aluminate ions do not exist for pH < 13 and even in such strong alkalis, the main species is probably $Al(OH)_4(H_2O)_2^-$ in which a central Al^{3+} ion is in octahedral coordination with four hydroxyl and two water molecules.

9.2.1.2.3 Relation of Soluble to Insoluble Species

Dissolution–precipitation reactions between the soluble ions and solid phases in the system are ill-defined. The tendency of the soluble ions to associate in complexes is progressively more pronounced as pH values tend towards the range within which the solid species are conventionally considered stable, for example pH 3.9 to 8.6 for $a_{Al^{3+}} = a_{AlO_2^-} = 10^{-6}$, and there is no sharp demarcation between the polymeric ions, colloidal material and alumina gels.

9.2.2 Corrosion Resistance of Pure Aluminium in Aqueous Media

9.2.2.1 Passivation

It follows from the invariant valency of aluminium in its ions, that is III, and its very negative standard electrode potential that the passivity protecting it in aqueous media depends only on pH and not on potential. Potential-dependent concepts such as anodic passivity, pitting potentials, and so on, do not apply as they do to passivating alloys such as stainless steels composed of metals that can exercise variable valencies.

In water, the thin air-formed oxide film on aluminium transforms to gelatinous trihydroxide that is then in principle subject both to further growth and to dissolution at the film/water interface. The establishment of passivity, therefore, depends on whether the solubility of the film is low enough for the filmed surface to be a virtually permanent condition. Typical expressions given for the solubility products for the hydrated oxides at 20°C are:

$$K_{Al^{3+}} = (a_{Al^{3+}})/(a_{H^+})^3 = 5.0 \times 10^5 \quad (9.9)$$

$$K_{AlO_2^-} = (a_{AlO_2^-}) \times (a_{H^+}) = 2.5 \times 10^{-15} \quad (9.10)$$

EXAMPLE

Determine the minimum solubility of the trihydroxide film.

By Equations 9.9 and 9.10, $a_{Al^{3+}}$ and $a_{AlO_2^-}$ are opposite functions of a_{H^+}. Their combined minimum activities correspond to the value of a_{H^+} for which they are equal.

From Equation 9.9:

$$a_{Al^{3+}} = 5 \times 10^5 \times (a_{H^+})^3$$

From Equation 9.10:

$$a_{AlO_2^-} = (2.5 \times 10^{-15})/(a_{H^+})$$

For $a_{Al^{3+}} = a_{AlO_2^-}$

$$5 \times 10^5 \times (a_{H^+})^3 = 2.5 \times 10^{-15}/(a_{H^+})$$

Hence:

$$a_{H^+} = 8.4 \times 10^{-6}$$

and

$$\text{pH} = -\log(8.4 \times 10^{-6}) = 5.1 \tag{9.11}$$

Inserting $a_{H^+} = 8.4 \times 10^{-6}$ into Equations 9.9 and 9.10 yields:

$$a_{Al^{3+}} = a_{AlO_2^-} = 3.0 \times 10^{-10} \tag{9.12}$$

Equations 9.9 and 9.10 have been used to construct Figure 9.1 expressing the solubility of trihydroxides as a function of pH, showing that they dissolve readily in strongly acidic and in strongly alkaline solutions, but their solubilities are very low at near-neutral pH values. Adopting the convention introduced for metals in Section 3.1.5.4, that is that a substance shall be deemed to dissolve in the case that the activity of any of its soluble ions exceed 10^{-6}, Figure 9.1 identifies a pH range of about 4 to 8.5, corresponding to the domain for stability of Al_2O_3 in the Pourbaix diagram, given in Figure 3.4.

9.2.2.2 Corrosion Resistance in Natural Waters

9.2.2.2.1 Fresh Waters

The values obtained in Equations 9.11 and 9.12 are instructive because the film is virtually insoluble in an aqueous medium with a pH value of 5.1 that is close to the pH value, 5.6, calculated in Section 2.2.9 for water in equilibrium with atmospheric carbon dioxide. The solubility increases as the pH rises but Figure 9.1 shows that even at pH 8.3, characteristic of hard waters, also calculated in Section 2.2.9, the activity of the soluble ion, AlO_2^-, is still only 5×10^{-7}. These simple calculations suggest that the film should resist dissolution in most fresh natural waters and supplies derived from them. The pure metal has excellent corrosion resistance under these conditions, confirming that the film is a good barrier to reacting species.

9.2.2.2.2 Sea Waters

Pure aluminium is also found to have good corrosion resistance to seawater. This is perhaps unexpected in view of the strong depassivating effect of chlorides on stainless steels described in Section 8.3.2.2. The passive film resists attack by chlorides probably because of the much greater chemical affinity of aluminium for oxygen than for chlorine.

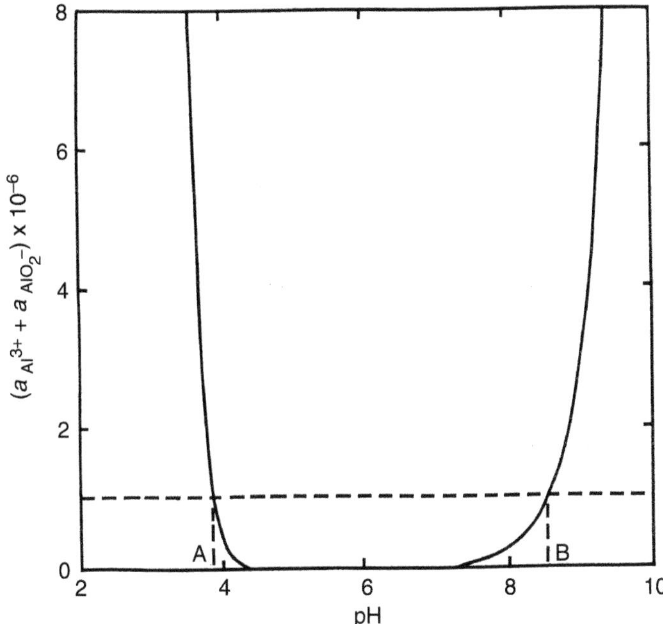

FIGURE 9.1
pH dependence of aluminium trihydroxide, $Al(OH)_3$, solubility expressed as the sum of Al^{3+} and AlO_2^- activities in solution. Intercepts at A and B define the conventional pH range for stability, that is where the activities of species in solution are less than 10^{-6}.

9.2.2.3 Corrosion Resistance in Acidic and Alkaline Media

The susceptibility of the pure metal to corrosion in acids or alkalis depends *inter alia* on how far the acid or alkali shifts the pH outside of the range 4.0 to 8.5. Thus aluminium is expected to dissolve much more readily in 0.1 M solutions of sulfuric acid (pH 0.9), hydrochloric acid (pH 1.1) and sodium hydroxide (pH 13.0) than in 0.1 M solutions of acetic acid (pH 2.9) or ammonium hydroxide (pH 11.1).

The counter ion of an acid or alkali can also influence susceptibility to dissolution of the metal. Thus oxidizing ions such as chromates can inhibit it and ions producing soluble complexes, such as the AlF_4^- ion formed in hydrofluoric acid, can enhance it.

9.2.3 Corrosion Resistance of Aluminium Alloys in Aqueous Media

In use, bare aluminium and aluminium alloy surfaces can develop an unattractive dull gray patina, but it is trivial and can be disregarded unless appearance is important. Incidentally, it is normal and non-toxic on cooking vessels. Damaging corrosion in general purpose applications is not usually a problem and the remarks below apply to the more critical applications and to special circumstances.

9.2.3.1 Chemical Composition of the Metal

Table 3.3 illustrates that except for magnesium and the unreactive metalloid, silicon, all of the common alloying elements in aluminium alloys have much less negative standard electrode potentials than aluminium and are thus capable of stimulating local galvanic attack on the adjacent matrix if distributed heterogeneously. Thus, as a general rule, alloys containing these elements can be expected to be less corrosion resistant than the pure metal.

9.2.3.1.1 *Iron and Manganese*

The corrosion resistance of the various commercial grades of pure metal in the AA 1XXX series improves progressively as the iron content diminishes from about 0.45% in the widely applied AA 1100 alloy through AA 1050 alloy to about 0.002% in AA 1199 (99.99% pure metal), but for most purposes the benefit is seldom sufficient to justify the loss of mechanical properties. The presence of manganese in the form of $(Mn,Fe)Al_6$ is also relatively innocuous. Common experience is that all of these alloys give very good service in everyday situations. Examples include the integrity of kitchen foil as thin as 20 μm and the very long lives of domestic utensils and rainwater gutters and conduits fitted to residences.

9.2.3.1.2 *Magnesium*

The standard electrode potential of magnesium, $E^{\ominus} = -2.370$ V (SHE), is so much more negative than that for aluminium, $E^{\ominus} = -1.663$ V (SHE), that even if segregated as heterogeneities or compounds in an alloy, there is no prospect of local galvanic corrosion on the aluminium matrix. The stable corrosion product of magnesium is the hydroxide, $Mg(OH)_2$, and although this substance is fairly soluble in near-neutral aqueous media, the magnesium contents in wrought AA 5000 series alloys do not detract from corrosion resistance but improve it. The same is true of the medium strength age-hardening alloys based on magnesium and silicon such as AA 6061. Aluminium alloys containing magnesium resist seawater and are applied in the superstructures of large ships and the hulls of smaller ones, taking advantage of their inherent medium strengths and suitability to welded construction.

9.2.3.1.3 *Silicon*

Silicon is a major component of cast alloy formulations to confer good casting characteristics. In binary alloys, such as AA 356, it also enhances corrosion resistance because the elemental silicon in the eutectic is unreactive. The improved corrosion resistance is dissipated in aluminium–silicon alloys that also contain high copper contents, such as AA 319 and AA 380.

9.2.3.1.4 *Zinc*

In dilute solution, zinc contributes $Zn(OH)_2$ *pro rata* to the passive film and does not detract from the corrosion resistance of aluminium, but in multicomponent alloys it participates in intermetallic compounds contributing to some of the structural effects described below.

9.2.3.1.5 *Copper*

It is unfortunate that copper seriously diminishes the corrosion resistance of aluminium alloys because it is an essential component in formulating the high strength aging alloys, characterized by AA 2024 and AA 7075. These alloys must be protected for applications in aircraft construction with which they are particularly associated.

The most effective protection is a thin coating of pure aluminium or an appropriate alloy applied by roll-bond cladding during hot rolling. The cast ingots of the high strength alloy, the *core,* are machined flat and plates of the protective material, the *cover plates,* are placed on top and underneath and secured by steel straps. The composite is heated to the normal rolling temperature for the alloy, the straps are removed, and it is presented to the hot-rolling mill. During hot reduction, relative movement at the interfaces between the core and cover plates due to differential plasticity of the two materials breaches the oxide film and allows a direct bond to form between them under pressure from the rolls. The cladding thickness in the material at final gage must be selected to provide the protection required without sacrificing too much of the strong alloy section. It ranges from typically 5% of the total thickness per side for sheets less than 1.5 mm thick to 1.5% for plates over 4.5 mm thick. The cladding is not only itself corrosion resistant, but is selected to provide cathodic protection at edges or abrasions. Pure aluminium serves this function for aluminium–copper–magnesium alloys such as AA 2024, but not for aluminium–zinc–magnesium alloys in the AA 7000 series, such as AA 7075, which are clad with the 1% zinc alloy, AA 7072 specially formulated for this purpose, listed in Table 9.2. It is easy to produce clad flat products, but impracticable for sections and forgings which must be protected by paint systems applied over chromate or anodic films by the principles explained in Section 6.4. Applications to aircraft are described in Section 21.1.

9.2.3.2 Structure Sensitivity

Premature corrosion of aluminium alloys is unlikely if they are properly selected, applied and, if necessary, protected but when it does occur, it is more structure-sensitive than for most other alloy systems because the highly reactive metal depends on passivity for protection and is vulnerable to reactive pathways provided by the proliferation of intermetallic compounds more noble than the aluminium matrix. For this reason, the form the attack takes is influenced not only by alloy composition but by the influence of heat treatments and of working operations on the microstructure.

9.2.3.2.1 Pitting

When commercial grades of aluminium and aluminium alloys suffer corrosion, they do so by multiple pitting rather than by general dissolution. Various explanations offered for the detailed mechanisms producing pitting are somewhat speculative, but there is general agreement that pits are initiated in the vicinity of small particles of intermetallic compounds outcropping at the metal surface. There is, of course, no concept of local passivity breakdown at characteristic pitting potentials as there is for stainless steels, because the passivity depends only on the value of pH. Initiation may be associated with either weak passivity over the intermetallic compounds or to local galvanic cells between the particles of compound and the adjacent matrix. Once pits are nucleated, their growth can be autocatalytic, that is self-accelerating because of well-known changes that occur in the microenvironment within pits as follows. Metal cations produced at an isolated local anode in the confined space of a pit:

$$Al = Al^{3+} + 3e^- \tag{9.13}$$

coordinate with the existing OH^- population, disturbing the local $a_{H^+} \times a_{OH^-}$ equilibrium, leaving a preponderance of hydrogen ions that shifts the pH in the acidic direction. If it diminishes to a value less than about 4, the base of the pit cannot passivate. The pits either grow into self-arresting shallow depressions or penetrate more deeply.

9.2.3.2.2 Intercrystalline and Exfoliation Corrosion

If intermetallic compounds are present as connected chains around grain boundaries, they can establish paths for corrosion advancing along lines of local action electrochemical cells created by potential differences between the compounds, the matrix and grain boundary zones depleted of the solute lost from solid solution in forming the precipitates. Alloys susceptible to this form of damage include most of the age-hardening high strength alloys, for example AA 2024 and AA 7075. The effect can be controlled by heat treatments designed to minimize solute depletion adjacent to grain boundaries and to induce widespread rather than localized precipitation. Medium strength alloys depending on Mg_2Si for age hardening are less susceptible. Alloys with minimal precipitation or whose precipitates have similar electrochemical properties to the matrix, including commercial grades of pure aluminium in the AA 1XXX series, the aluminium–manganese alloy AA 3003, the can stock alloy AA 3004 and alloys with low magnesium contents in the AA 5000 series are the least susceptible. Alloys in the AA 5000 series with higher magnesium contents are susceptible both to intercrystalline corrosion and its variant, exfoliation corrosion, if Mg_2Al_3 is allowed to precipitate. For this reason, a special stabilized temper is available for alloy AA 5456 used in boat hulls

Susceptible metals can experience an extreme form of intercrystalline attack known as *exfoliation* or *layer* corrosion, in which the metal is opened up along grain or sub-grain boundaries running parallel to the surface. In heavily worked metal such as thin plate and sheet, the grains are elongated in the direction of working and very thin in the short transverse direction, so that they assume a layered morphology with planar grain boundaries parallel to the surface. The voluminous products from intergranular corrosion along these paths separate multiple layers of grains, causing the metal to swell. The layers open away from a neutral axis, that is the metal exfoliates. This supports an idea that there may be a stress-corrosion component due to residual internal stresses in crack opening mode, inherited from mechanical working. The attack can be initiated at cut edges or from the base of pits penetrating from the surface into the system of parallel grain boundaries.

9.2.3.3 Stress-Corrosion Cracking

Principles underlying stress-corrosion cracking of high strength aluminium alloys were introduced in Section 5.1.2, where it was convenient to deal with them in the context of SCC theories of environmentally sensitive cracking in alloy systems in general. Alloy manufacturers and aircraft constructors have a more practical perspective because they have the responsibility of eliminating risk of failure in flight. It must be assumed that the specific agent, chlorides, will inevitably be encountered and since the first failures were identified in the 1930's, immense effort has been expended to avoid the effect.

Advances usually follow experience and a large body of service history is available. Alloy-specific practices have evolved that enable the high strength alloys to be used with confidence by attention to alloy compositions, heat-treatment procedures and limits on design stresses. One guiding principle in the design of heat treatments is to minimize electrochemical differences between grain boundary precipitates by rapid quenching after solutionizing and by aging to an overaged condition. The objective is to minimize susceptibility to stress-corrosion cracking rather than to exploit the maximum mechanical properties that the alloys can deliver.

Cracking normally proceeds in crack-opening mode along grain boundaries and so boundaries normal to the resolved tensile component of the prevailing stress are the most vulnerable. For this reason, stress-corrosion cracking is least likely in thin sheet stressed in longitudinal and long transverse directions and the greatest danger is to thick sections and forgings stressed in the short transverse direction. These matters are taken into account in design stress analyses.

9.2.3.4 Galvanically Stimulated Attacks

The standard electrode potential of aluminium is so negative that when it or one of its alloys is coupled to another engineering metal, either directly or indirectly, it is almost invariably the member that is susceptible to galvanically stimulated attack. Whether it actually suffers attack depends on initial breakdown of its passivity. The causes were explained in Chapter 4 and some examples are illustrated in Figures 4.2 and 4.4. Accelerated corrosion of aluminium alloys from injudicious combination with other metals must be rated as a design fault because, if the hazard is appreciated, it need never be realized. Unfortunately, when such problems occur, they are sometimes mistakenly attributed to faulty manufacture of the aluminium product.

9.2.4 Corrosion Resistance of Aluminium and Its Alloys in Air

9.2.4.1 Nature of Air-Formed Film

The very thin film that protects aluminium in normal air at ambient temperatures can be detected and its thickness measured from its electrical resistance and by electron optical techniques. It grows on a freshly exposed aluminium surface by a version of the Cabrera–Mott kinetics, mentioned in Section 3.3.2, and attains a temperature-dependent limiting thickness after an hour or so. The limiting thickness at 20°C is in the range 2 to 5 nm which is less than 25 atom layers and it does not exceed 100 nm even at temperatures as high as 500°C.

Insofar as it is permissible to relate a bulk structure to such a thin film, the expected oxide species is the aluminium trioxide, bayerite, which is stable in equilibrium with aluminium and normal moist air at ambient temperatures. Electron diffraction fails to disclose the structure and it is usually given the unsatisfactory description, 'amorphous Al_2O_3'. It is probably the gel precursor of bayerite because water can be detected in the film by the following experiment. Hydrogen is removed from a freshly machined sample of aluminium by heating it in vacuum; the degassed sample is exposed to normal air for an hour or so and reheated in vacuum. The water is detected by collecting and identifying hydrogen produced in its reaction with the metal substrate:

$$2Al + Al_2O_3 \cdot 3H_2O = 2Al_2O_3 + 3H_2 \tag{9.14}$$

9.2.4.2 Weathering

The protection the air-formed film affords the pure metal and alloys that have no significant copper content such as commercial grades of pure metal, AA 3003, AA 3004, AA 5050 and other aluminium–magnesium alloys in the AA 5000 series is remarkably good. Metal loss in dry environments is almost undetectable. In weathering in external locations around the world, the metal surface is roughened by shallow pitting and discoloured to dull gray, but the metal does not thin appreciably after as much as ten years. As expected, degradation is worst in marine and polluted industrial atmospheres as assessed by loss of tensile strength of 1 mm sheet samples which is caused by the stress-raising propensity of pitting, but even so the loss of strength is seldom more than 10% in ten years. Surface treatments such as chromating or anodizing are applied for external exposure when preservation of appearance is important, for long-term service as in architectural applications and where thin gage material is used.

The high strength aluminium–copper–magnesium and aluminium–zinc–magnesium alloys typified by AA 2024 and AA 7075 and casting alloys such as AA 319 and AA 380 containing substantial copper contents have considerably worse weathering characteristics and usually need protection by paint systems applied over chromate coatings. Aluminium–zinc–magnesium casting alloys that contain more than 6 mass % of zinc can suffer stress-corrosion cracking during atmospheric exposure, but the influence of metal composition is not fully established.

9.2.5 Geometric Effects

Geometric considerations apply to aluminium and aluminium alloys, as they do to other metals, for example, configurations that trap water from precipitation or condensation extend the fraction of time during which the metal is wet. There is, however, a difference of emphasis between some effects for aluminium and its alloys and many other metal systems.

9.2.5.1 Crevices

9.2.5.1.1 Quiescent Crevices

Crevices in submerged metal are not so deleterious as they are for iron or stainless steels because the passivity does not depend entirely on maintaining a sufficient oxygen concentration to film the surface. Corrosion is accelerated in very fine crevices, perhaps due to depassivation by pH drift caused by local disturbance of the $(a_{H^+})\times(a_{OH^-})$ equilibrium.

9.2.5.1.2 Crevices with Relative Movement

The acceleration of corrosion in crevices between aluminium surfaces in relative motion is an entirely different proposition. It is usually inadvertent and a well-known example is fretting or faying by repeated slight movement between the mating surfaces of riveted joints subject to condensation, It is matter of serious concern in aircraft, as described in Chapter 21. The cause of the damage is continuous mechanical disturbance of the protective passive film so that metal surfaces in the crevice can become active local anodes.

9.2.5.2 Impingement, Cavitation and Erosion–Corrosion

The passive film is vulnerable to damage by energy delivered by water moving across the metal surface. The terminology that describes consequent corrosion damage at the activated surface is related to the mechanical action of the water at the metal surface:

1. Impingement attack is due to deflection of a stream of water or steam at sharp changes in section or direction of channels in pipes and ancillary fittings.
2. Cavitation is due to energy released in the collapse of bubbles generated in turbulent flow. A characteristic example is the output of ill-designed water pumps.
3. Erosion–corrosion is due to energy imparted by water moving over the metal surface at velocities exceeding about 3 m s^{-1}, wearing grooves in the metal by mechanical/chemical joint action. The effect has been observed on aluminium alloy hulls of high-speed vessels.

Damage of the kind described can be serious in its own right, but it can also initiate consequential damage such as fatigue failure due to the creation of stress-raising artefacts.

9.2.6 Oxidation of Aluminium–Magnesium Alloys during Manufacture

Aluminium alloys have little hot strength and can scarcely be regarded as engineering metals at temperatures much over 200°C, and since the air-formed film is very thin, thermal oxidation during service is not an important issue. However, selective oxidation of aluminium–magnesium alloys can produce surface films during manufacture that influence the quality of the product.

9.2.6.1 Enhanced Oxidation by Sulfur Pollution

There is a well-known industrial problem that is sometimes encountered when medium strength aluminium–magnesium–silicon alloys in the AA 6000 series are solution treated in air furnaces at temperatures in the range 450°C to 550°C in preparation for age hardening. It is derived from a thermal oxidation process in furnace atmospheres of typical humidity polluted with sulfur. It is manifest as thick dark gray films on the metal and it is associated with hydrogen blistering and an odor of hydrogen sulfide emitted by any air-cooled metal that escapes quenching. There are sources of sulfur pollution both in furnaces fired indirectly by fossil fuels, for example leaked combustion products and in electric furnaces, for example decomposition of fabrication lubricants carried in on the metal. In the oxidizing conditions inside an air furnace the sulfur species are converted to sulfur dioxide.

The implication of these observations is that the effect is due to a change in the character of the oxidation product in the presence of sulfur which enhances the diffusion both of the species that sustains oxidation, that is Mg^{2+} and of the species that carries hydrogen from water vapour in the furnace atmosphere, that is OH^-. Consider how the oxidation product might be influenced by introducing sulfur dioxide. The first step is to assess the notional replacement reaction:

$$MgO(s) + SO_2(g) = MgS(s) + 1\tfrac{1}{2}O_2(g) \tag{9.15}$$

The Gibbs free energy change for the reaction is derived from the Gibbs free energies of formation for MgS, MgO and SO_2, yielding:

$$\Delta G^{\ominus} = 549442 + 12.3T\log T - 119T\,\text{J} \tag{9.16}$$

Substituting for T in Equation 9.16 yields large positive values for $\Delta G^{\ominus}$ at all reasonable temperatures, showing that MgS is so unstable with respect to MgO in the presence of small concentrations of sulfur dioxide that its formation in preference to the oxide can be discounted. Aluminium sulfide, Al_2S_3 is even less stable than MgS, so that there is no question of replacing the oxide film with a sulfide.

There is, however, an alternative interaction. It is well-known that traces of sulfur dioxide contamination accelerate the oxidation of metals in air, sometimes by orders of magnitude by replacing a minority of the O^{2-} ions in oxide structures with S^{2-} ions. The replacement of O^{2-} ions in MgO with S^{2-} ions is expected to diminish its ionic nature, enhancing its extrinsic character as explained in Section 3.3.3.4, thereby facilitating both the self-diffusion of magnesium and the diffusion of foreign species, including OH^-. This can form the basis of explanations for both an accelerated rate of oxidation and accelerated transport of OH^- ions that is a prerequisite for hydrogen absorption.

Such a kinetic view is consistent with the characteristics of the oxidation product stimulated by sulfur dioxide, for example the presence of S^{2-} ions in the oxide is manifest as the characteristic evolution of hydrogen sulfide in humid air by the hydrolysis:

$$S^{2-} + H_2O = O^{2-} + H_2S \tag{9.17}$$

The enhanced oxidation and its consequences can be inhibited by inducing a surface condition on the oxide that is impervious to the entry of S^{2-} and OH^- ions. The effect is produced by providing small containers of potassium borofluoride, KBF_4 or sodium silicofluoride, Na_2SiF_6 in the hot zone of the furnace. These complex salts decompose at temperatures >350°C, yielding low partial pressures of the active agents, boron trifluoride, BF_3, or silicon tetrafluoride, SiF_4:

$$KBF_4(s) = KF(s) + BF_3(g) \tag{9.18}$$

$$Na_2SiF_6(s) = 2NaF(s) + SiF_4(g) \tag{9.19}$$

Boron trifluoride and silicon tetrafluoride are strong Lewis acids (electron acceptors) and can bind to electron donors, such as unsaturated O^{2-}, OH^- and S^{2-} ions exposed at the surface of the contaminated oxide. Thus the oxide is probably sealed by a borofluoride or silicofluoride based surface condition which is manifest as a light, not unattractive, patina, but it is too thin to identify positively. The particular salts used as sources of BF_3 or SiF_4 are selected because the vapour pressures are of the right order of magnitude at solution-treatment temperatures for alloys in the AA 6000 series.

The nature of the damage that can be afflicted on the metal by hydrogen absorption associated with the oxidation stimulated by sulfur and the industrial practices to control it are not within the scope of the present text. They are considered in detail in another publication, *The Effects of Hydrogen in Aluminium and its Alloys*, cited in Further Reading.

9.2.6.2 Oxide Precursor of Magnesium Carbonate Films

Aluminium–magnesium alloys can experience final temperatures >300°C in the course of manufacture. This applies both to products finished by hot-working and to cold-rolled sheet annealed to soften it typically at 340°C. For alloys with magnesium contents >~1 mass %, the magnesium oxidizes selectively in clean air, as explained in Section 3.3.4.2, producing surface films of magnesium oxide. If it is not removed, the magnesium oxide can be converted to a white deposit carbonate by atmospheric carbon dioxide when the metal is subsequently stored, by the reaction:

$$MgO(s) + CO_2(g) = MgCO_3(s) \tag{9.20}$$

for which:

$$\Delta G^{\ominus} = -117570 + 170T \text{ J} \tag{9.21}$$

Assuming $a_{CO_2} = 0.03$ and $T = 298$ K, the Gibbs free energy of the reaction for exposure of the metal to the ambient atmosphere is negative at temperatures below 834 K (560°C).

The carbonate appears as a white surface deposit that is not deleterious in itself, but it spoils the aesthetic appearance of the metal and gives a misleading image of its corrosion resistance.

Further Reading

Godard, H. P. (ed.), *The Corrosion of Light Metals*, John Wiley, New York, 1967.

Godard, H. P., Examining causes of aluminum corrosion, *Materials Performance*, 8, 25, 1969.

Hatch, J. E. (ed.), *Aluminum Properties and Physical Metallurgy*, American Society for Metals, Metals Park, Ohio, 1984.

Mondolfo, L. F., Structure of the aluminum magnesium zinc alloys, *Metallurgical Reviews*, 16, 95, 1971.

Mondolfo, L. F., *Aluminum Alloys: Structure and Properties*, Butterworths, London, 1976.

Phillips, H. W. L., *Annotated Equilibrium Diagrams of Some Aluminum Alloy Systems*, The Institute of Metals, London, 1959.

Polmear, I. G., *Light Alloys*, Edward Arnold, London, 1981.

Talbot, D. E. J., *The Effects of Hydrogen in Aluminium and its Alloys*, Maney, for the Institute of Metals, Leeds, 2004.

10

Corrosion Resistance of Copper and Copper Alloys

Copper occurs naturally in lean sulfide or oxide ore bodies in association with iron and sometimes also with nickel and the extraction processes by which it is recovered are expensive. Most new metal is produced from mixed copper/iron sulfide minerals concentrated by froth flotation from ore bodies containing <1% copper. It is roasted to oxidize the iron sulfide selectively and smelted with silica to remove the iron as a slag. The residual liquid copper sulfide matte is blown with air to yield an intermediate product, blister copper, containing residual sulfur, lead and zinc. For most purposes, it is refined by injecting oxygen, under a slag to oxidize the impurities and the excess oxygen is reduced either with a hydrocarbon gas, yielding tough pitch copper containing ~0.05 mass % dissolved oxygen, or with phosphorus, yielding phosphorus-deoxidized copper. As an alternative for electrical applications, blister copper is refined by electrolysis, yielding oxygen-free high-conductivity copper. Oxide ores are usually leached with sulfuric acid and the metal is recovered by electrolysis. Recycled scrap is essential for economic production. Accounts of extractive processes are given in texts, for example by Bodsworth and by Pehlke cited in Further Reading. Table 10.1 gives compositions for representative grades of copper identified by the international ISO designations.

Solid copper has an invariable face-centered cubic lattice. It is not particularly strong, but it is tough and ductile and can be drawn to wire and tube or formed into complex shapes. Pure copper has the lowest value for electrical resistivity, 1.7 μΩ cm at 20°C, and the highest value of thermal conductivity 397 W $m^{-1}K^{-1}$ for 0°C –100°C, of all common engineering metals which are exceeded only by corresponding values for silver. About half of world production of copper is used for conductors to supply and utilize electrical energy and much of the rest is used in applications requiring the excellent corrosion resistance of the pure metal and its alloys, including the supply and utilization of water, building applications and marine service.

10.1 Chemical Properties and Corrosion Behaviour of Pure Copper

The chemistry of copper and hence its corrosion behaviour is determined by its position as the penultimate element in the first transition series, given in Table 2.3. From the preceding sequence, scandium to nickel, the configuration of the outer electrons in the free copper atom might be expected to be $(3d)_9(4s)_2$, but because of the stability of the completed *d* shell, the actual configuration is $(3d)_{10}(4s)_1$. Hence, copper commonly exists in the oxidation states Cu(I) and Cu(II). A higher oxidation state, Cu(III), is known but it is of little significance.

TABLE 10.1

Nominal Compositions of Some Representative Grades of Copper

ISO Designation	Copper Grade	Composition, Mass % Copper	Others	Typical Applications
Cu-OF	Oxygen-free high conductivity copper	>99.95%	–	Electrical conductors and components
Cu-ETP 1	High grade electrolytic tough pitch copper	–	<0.0075 Oxygen	High speed strand annealing and enamelling
Cu-FRTP	Fire refined tough pitch copper	>99.90%	<0.05% Oxygen	General engineering and building
CuAs	Tough pitch arsenical copper	>99.20%	0.3–0.5% Arsenic	
Cu-DHP	Phosphorus deoxidized copper	>99.85%	0.013–0.050% Phosphorus	Tube and sheet subject to welding or brazing

The equilibrium between the two common oxidation states:

$$2\text{Cu(I)} = \text{Cu(II)} + \text{Cu} \quad (10.1)$$

can be displaced in either direction according to conditions and the nature of the species present, as illustrated in the calculations given in Section 3.1.5.5.1, showing that

1. In aqueous solution, the equilibrium constant for the reaction:

   $$2\text{Cu}^{+}(\text{aq}) = \text{Cu}^{2+}(\text{aq}) + \text{Cu(s)} \quad (10.2)$$

   is given by:

   $$K = 1.55 \times 10^{6}$$

   so that equilibrium is displaced so far to the right that the Cu^{+}(aq) ion can exist only in very low concentrations. Its instability is due partly to the higher solvation energy of Cu^{2+}(aq).
2. The corresponding reaction for solid oxides:

   $$2\text{Cu}_2\text{O(s)} = \text{CuO(s)} + \text{Cu(s)} \quad (10.3)$$

   is not spontaneous, so that Cu_2O is stabilized by the environment of oxygen ions and the relative stability of the oxides is determined by prevailing temperatures and oxygen potential:

   $$2\text{CuO(s)} = \text{Cu}_2\text{O(s)} + \tfrac{1}{2}\text{O}_2\text{(g)} \quad (10.4)$$

   for which

   $$\Delta G^{\ominus} = 146230 + 25.5T\log T - 185.3T \text{ J } (298\text{ K} < T < 1300\text{ K}) \quad (10.5)$$

10.1.1 The Copper–Oxygen–Water System

The Pourbaix diagram of the copper–water system for $a_{Cu^{2+}} = 10^{-6}$ at 25°C and atmospheric pressure is given in Figure 3.6. The essential features are:

1. The domain of stability for Cu is overlain by a domain for Cu^{2+} for pH < 6.5 and by a domain for Cu_2O for pH > 6.5.
2. There is no domain for Cu^+, consistent with the value of K for Equation 10.2.
3. The upper boundary of the domain of stability for copper lies everywhere at more noble potentials than the potentials for evolution of hydrogen at 25°C, that is copper and water share a common domain of stability throughout the pH range.
4. If pH > 7, copper tends to passivate with the formation of CuO and experience shows that the passivation protects the metal very effectively.

Since copper and water share a common domain of stability, copper cannot corrode with evolution of hydrogen and it is unaffected by quiescent water and dilute non-oxidizing and non-complexing acids in the absence of air. An example of highly oxidizing conditions in which copper dissolves is the familiar reaction with concentrated nitric acid:

$$3Cu + 8HNO_3 = 3Cu(NO_3)_2 + N_2O_2 + 4H_2O \qquad (10.6)$$

Aqueous solutions of Cu(I) salts are virtually nonexistent because they are decomposed by water. Most Cu(II) salts dissolve in water as the solvated cations $[Cu(H_2O)_6]^{2+}$ which can form complexes by successive replacement of up to four of the water molecules by other ligands, notably ammonia or amines, yielding intensely blue ions, for example $[Cu(NH_3)_4(H_2O)_2]^{2+}$. Reactions forming these complexes are significant as a potential cause of stress-corrosion cracking of some alloys, notably brasses. They are also utilized in formulating fluxes to scavenge surface oxide films in soldering copper.

10.1.2 Corrosion Resistance in Natural Environments

Artefacts of copper and simple alloys surviving from antiquity, for example as ornaments, tools and coinage, testify to the excellent corrosion resistance of the metal in natural environments.

10.1.2.1 The Atmosphere

On initial exposure to clean dry air at ambient temperatures, copper forms a protective thin film according to the Cabrera–Mott theory described in Section 3.3.2. In ordinary indoor air in temperate climates it tarnishes and the tarnish is thicker and more unsightly if the air is polluted by sulfur dioxide, probably because sulfur ions incorporated in the film raise the concentration of cation vacancies thereby reducing the activation energy for formation of the oxide and increasing the critical thickness at which continuing oxidation is not detectable.

The oxidation of copper at elevated temperatures is complex and incompletely resolved. Application of Equation 10.5 shows that CuO is stable with respect to Cu_2O and oxygen at atmospheric pressure at temperatures below 1120°C, so that CuO is expected at the oxide/atmosphere interface of scales formed by isothermal oxidation in air. Scales formed at temperatures below 370°C consist *entirely* of CuO. Scales formed at temperatures in the interval 300 to 800°C, are duplex with an underlay of Cu_2O at the metal/oxide interface by the reaction:

$$CuO + Cu = Cu_2O \tag{10.7}$$

The proportion of the total scale thickness occupied by Cu_2O increases as the temperature of oxidation is raised until at temperatures above about 800°C the film is almost entirely Cu_2O. The replacement of CuO by Cu_2O facilitates cation diffusion and hence accelerates oxidation because CuO is an intrinsic semiconductor with low ion transport numbers, whereas Cu_2O is a *p*-type oxide in which the diffusing species is Cu. At intermediate temperatures in the interval 370°C to 800°C, the contributions of the CuO and Cu_2O layers to diffusion control depends on their relative thicknesses. Hence, the temperature dependence of oxidation deviates from the Arrhenius equation, in which the logarithm of the rate constant is a linear function of the reciprocal temperature. These considerations are of mainly academic interest because at elevated temperatures *pure* copper oxidizes too rapidly and does not retain adequate strength for most applications. The required characteristics for service at elevated temperatures are imparted by appropriate alloying, as explained Section 10.2.2.

External exposure of copper in urban environments, for example as sheet for roofing and associated architectural features are intermittently washed by weak solutions of atmospheric carbon dioxide and sulfur dioxide in rain, converting existing oxide surface films into an aesthetically pleasing green patina in which basic copper(II) carbonate and sulfates predominate, without unduly compromising the protection afforded. Nevertheless, the degree of protection depends on conditions prevailing when the metal is first exposed. The metal is better protected if it is first exposed when pollution is low than when it is high, indicating that the composition of the corrosion product immediately adjacent to the metal surface is a critical factor determining subsequent development of the protective film.

10.1.2.2 Natural Waters

Copper is resistant to most quiescent natural waters, both fresh and saline. An advantage in seawater is that despite an insignificant loss of metal, a very low concentration of copper ions maintained in the immediate vicinity of the metal surface is sufficient to keep it free from biological fouling organisms. The resistance depends on the integrity of the protective surface film and is diminished if it is disturbed. The film is vulnerable to mechanical disruption by unusually high velocity water flows and by dissolution in waters containing exceptionally high concentrations of carbon dioxide, organic acids or nitrogenous species.

10.1.2.3 Soils

Information from extensive empirical tests at numerous locations show that the low risk of corrosion extends to copper buried in most soils and clays but there are exceptions including acid peaty soil, tidal marsh and cinder backfill. Unfortunately, experience of the effects of particular physical, chemical or biological characteristics of soils can be unreliable and every case must be considered on the basis of local experience.

10.1.3 Water Supply and Circulation Systems

Typical applications for delivery and circulation of fresh waters include municipal and domestic water supplies, hot water tanks, pumped small bore closed circuit central heating systems and heat exchangers using drawn tubes carrying hot or cold water as required. The corrosion resistance of such installations is generally very satisfactory but there are potential problems associated with faulty manufacture, installation or maintenance. These might include residues from soldering fluxes, carbon films acquired during manufacture or dezincification of brass fittings as explained in Sections 10.1.3.1 and 10.2.1.3, respectively.

10.1.3.1 Surface Films as Cathodic Collectors

Copper tube is manufactured by hot extruding an intermediate product, the *tube shell*, from ingots and then cold drawing it through successive dies with annealing in a protective reducing atmosphere to resoften the work-hardened metal between passes. If the tubes are inadequately cleaned, carbonaceous die lubricant residues leave a thin film of carbon that is abraded against succeeding dies when drawing is resumed. In service, this film can act as a cathodic collector stimulating galvanic attack on bare metal exposed at abrasions in the film. A similar effect is possible if abnormal annealing conditions produce films of semiconducting glassy Cu_2O, which can also act as a cathodic collector.

10.1.3.2 Impingement Attack

Impingement attack occurs when the integrity of the protective surface film is disrupted mechanically by high velocity water flows. The film is particularly vulnerable at geometric features that create turbulence or impose changes in the direction of flow. Perforations in the film become local anodes at which the metal dissolves if depolarized by reduction of dissolved oxygen on the undisturbed surrounding filmed surface. As expected from the direction of the applied forces, the pits are undercut in the direction of the water flow. The severity of attack is a direct function of the dissolved oxygen concentration and is insignificant if it is not available. Prevention is a matter of design to restrict water velocities and avoid adverse geometric features.

10.1.3.3 Pitting

Pitting is localized attack caused by small isolated differential aeration cells under extraneous deposits or at defects in the protective film initiated by incipient impingement. Susceptibility to pitting can be minimized by design and maintenance to prevent accumulation of detritus, restrict water velocities and eliminate features that favour impingement attack.

10.1.4 Stress-Related Corrosion Failure

It is difficult to find firm evidence for failure by stress-corrosion cracking or corrosion fatigue for pure copper, because the metal is ductile and applications for which it is exploited do not often impose high stresses. Some authorities suggest that arsenical or phosphorus deoxidized copper might be susceptible to stress-corrosion cracking but there is little evidence of significant practical problems.

10.1.5 Corrosion of Associated Metals by Galvanic Stimulation

Soft water containing dissolved oxygen and carbon dioxide passing over copper can dissolve minute quantities of the metal as Cu^{2+} ions. If it subsequently passes over a second, less noble metal, traces of metallic copper are deposited at small isolated sites by a replacement reaction, creating galvanic cells that stimulate rapid pitting corrosion of the second metal. As expected from the differences between the standard electrode potentials for Cu^{2+}/Cu and Zn^{2+}/Zn, Fe^{2+}/Fe, or Al^{3+}/Al given in Table 3.3, the effect can be significant for iron, bare steel and galvanized steel that are often encountered in mixed metal water supply and circulation systems and severe for aluminium. The severity of the effect for aluminium is unfortunate because domestic water supplies are usually delivered through copper pipes, so that aluminium cooking vessels suffer rapid perforation at weak spots in the original protective passive film if the water is inherently soft or artificially softened. Closed mixed metal circuits such as domestic central heating systems using steel radiators, cast iron boilers and brass accessories connected by copper pipes are usually trouble free because the water is less aggressive after the initial oxygen and carbon dioxide are depleted. Nevertheless, it is good practice to add appropriate inhibitors to the water when the system is filled.

10.2 Constitutions and Corrosion Behaviour of Copper Alloys

Pure copper lacks sufficient strength, castability, machinability and bearing qualities for general engineering use and applications requiring highly stressed corrosion-resistant components. These deficiencies are addressed by alloys formulated especially for marine, steam raising and chemical applications. They include binary alloys of copper with zinc, aluminium, nickel or tin, but some of the most useful are multicomponent alloys based on combinations of these and other components. There are more than a hundred registered compositions and Table 10.2 gives only a representative selection identified by ISO designations. Some of them are application-specific, developed empirically by successive adjustments of composition to meet problems identified from experience of service and manufacture.

The constitutions of copper alloys are complex and detailed consideration of phase equilibria, system by system, is required to explain the effects of compositions on corrosion behaviour and corrosion resistance. Phase equilibria for binary systems of copper with most metals and metalloids are well established. Corresponding equilibria for some of the multicomponent systems of interest are not fully resolved, but sufficient information for practical purposes can often be deduced from equilibria for binary systems, especially for zinc, aluminium, nickel and tin.

10.2.1 Alloys Based on the Copper–Zinc System–Brasses

Alloys of copper with zinc, the brasses, have been known and used for almost as long as pure copper. They are stronger than the pure metal, easier to machine and can be produced as castings and in all standard wrought forms. They are the least expensive copper alloys and are widely applied where corrosion resistance in natural and moderate industrial environments is required.

TABLE 10.2

Nominal Chemical Compositions of Representative Copper Alloys

		Alloy Elements Mass %						
ISO Designation	**Alloy Type**	**Zinc**	**Tin**	**Aluminium**	**Nickel**	**Iron**	**Manganese**	**Others**
CuZn30	α Brass	28.5–31.5	–	–	–	–	–	–
CuZn40	Low lead 60/40 αβ brass	38–41	–	–	–	–	–	–
CuZn36Pb3	Free cutting 60/40 brass	34–38	–	–	–	–	–	2.5–3.7 Pb
CuZn36Pb2As	Dezincification-resistant brass	36	–	–	–	–	–	2 Pb + 0.1 As
CuZn38Sn1	Naval brass	35–38	1.0–1.4	–	–	–	–	–
CZ110	Aluminium brass	22	–	2	–	–	–	–
CuZn39FeMn	High tensile brass	40	0.2–0.8	–	–	0.3–1.0	0.5–2.0	–
CuAl7	α aluminium bronzes	–	–	6.0–7.5	–	–	–	–
CuAl10Fe3	Duplex aluminium bronzes	–	–	8.5–10.5	1.0 max	1.5–3.5	1.0	–
CuAl10Fe5Ni5	Nickel aluminium bronze	–	–	9.0–10.0	4.0–5.5	4.0–5.5	1.5 max	–
CuMn13Al8Fe3Ni3	Manganese aluminium bronze	–	–	7.5–8.5	1.5–4.5	2.0–4.0	11–15	–
CuAl6Si2Fe	Aluminium silicon bronze	–	–	6	–	0.6	–	2 Si
CuNi10Fe1Mn	90/10 Cupro-nickel	–	–	–	10–11	1.0–2.0	0.5–1.0	–
CuNi30Mn1Fe	70/30 Cupro-nickel	–	–	–	31.0	1.0	1.0	–
CuSn5	5% Tin phosphor bronze	–	4.5–5.5	–	–	–	–	0.02–0.40 P
CuSn7	7% Tin phosphor bronze	–	5.5–7.5	–	–	–	–	0.02–0.40 P
CuSn10Zn2	Gunmetal	2	10					
CuSn5Pb5Zn5	Leaded gunmetal	5	5					2 Pb

10.2.1.1 Phase Equilibria for the Copper–Zinc Binary System

Figure 10.1 gives a partial phase equilibrium diagram for compositions of interest in commercial alloys at the copper-rich end of the system. There are two solid solutions, the α-phase, a solution of zinc in the FCC copper lattice and the body-centered β-phase which appears at higher zinc contents. There is an invariant point, a peritectic reaction at 902°C:

$$\alpha(32.5\%\ \text{Zn}) + \text{liquid}\ (37.6\%\ \text{Zn}) = \beta\ (36.8\%\ \text{Zn})$$

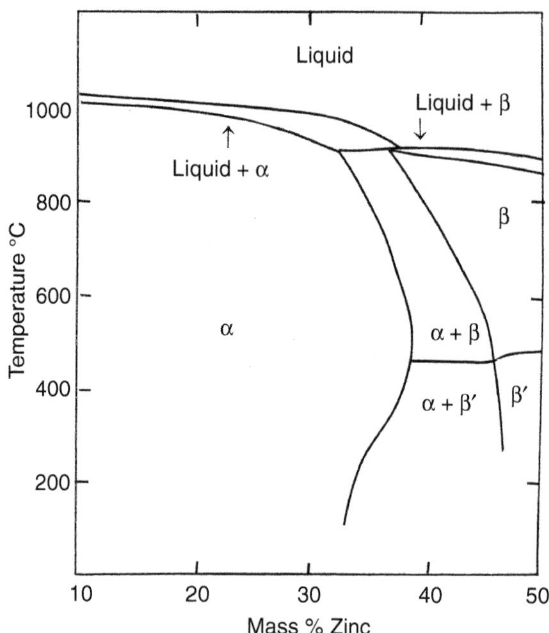

FIGURE 10.1
Partial phase equilibrium diagram for the copper–zinc binary system. β and β′ are disordered and ordered forms of the β phase above and below 454°C.

The α phase field is extensive with a maximum solubility of 39 mass % zinc at 454°C. The range of useful zinc contents is limited by separation of a brittle low-symmetry cubic γ-phase not shown in Figure 10.1. The distribution of β in duplex α/β brass microstructures depends on composition and cooling rate during solidification through the peritectic reaction. The high temperature form of β is disordered and transforms to an ordered form, β → β′, at 450°C.

10.2.1.2 Formulation and Characteristics of Alloys

There are two basic types, the single phase α-brasses and the duplex α/β-brasses. They are general purpose materials as well as having corrosion-resistant applications. The α-brasses, represented by cartridge brass, a 70Cu/30Zn binary alloy, are used for their cold ductility and deep drawing qualities. Typical applications are for military ordnance as the name implies and as tubing for condensers and heat exchangers using fresh water for cooling.

The α/β-brasses, typified by common brass, a 60Cu/40Zn alloy, are less expensive than α-brasses because zinc is less expensive than copper. They are stronger, exhibit good hot formability, are well suited to die casting and easy to machine. Typical products are rolled sheet, extruded sections and forgings for miscellaneous applications and die-cast fittings for copper pipe installations, for example faucets, valve bodies and unions.

Special brasses introduced in Chapter 25 contain additions of other components for application-specific marine service. Examples of multicomponent compositions are given in Table 10.2, for example naval brass, aluminium brasses, high tensile brasses and free cutting brasses.

10.2.1.3 Corrosion Resistance

With the reservations explained later, brasses are generally resistant to corrosion in natural environments. In dry air the metal tarnishes, forming a thin protective film. In extended exposure to external air in temperate climates, a green patina of protective basic carbonates and sulfates is formed, similar to that which forms on copper. It can also contain chlorides in marine atmospheres.

Both α and α/β brasses normally have excellent resistance to corrosion in both cold and hot fresh waters, but in particular specific circumstances they may fail by pitting, impingement, stress-corrosion cracking or *dezincification*, a specific effect in brasses described below. Duplex α/β brasses are more vulnerable to adverse conditions than single phase α brasses.

Pitting can be initiated by the same mechanisms as for pure copper, that is local differential aeration and impingement attack, but brasses are stronger and hence more resistant to impingement. Unlike other causes of failure, susceptibility to pitting is insensitive to zinc content and microstructure. To avoid it, the same precautions are required as for pure copper as described earlier in Section 10.1.3.3.

Dezincification is a distinctive form of corrosion induced by water supplies from particular geographic areas. In this form of attack, zinc is selectively removed from the alloy, leaving the residual copper *in situ* as a cohesive porous mass retaining the original shape and dimensions of the artefact but with no strength. It can afflict brasses with more than 15 mass % zinc, which includes both α and α/β-brasses. In systems delivering or circulating water it can cause failures by blockages with voluminous corrosion products as well as by loss of metal. In general, hard waters do not induce dezincification and it is associated with soft waters, but the critical solute concentrations and pHs that promote or discourage it are somewhat ill-defined. Water authorities can often advise on the tendency of local waters to induce it. Conditions that stimulate it include high temperatures, high chloride contents, stagnant water and differential aeration under detritus or in crevices. There are two kinds of dezincification, *layer type* characterized by an advancing front of dezincified metal and the more common *plug type*, comprising isolated areas a few millimeters in diameter that can perforate tube walls. In duplex α/β-brasses the β phase is the more vulnerable and can suffer selective attack because the zinc content is higher. Opinions differ on whether the mechanism is selective dissolution of the less noble zinc component or complete dissolution and redeposition of the more noble copper component. In recent times, dezincification-resistant brasses containing arsenic and lead have been developed. A typical composition, given in Table 10.2, includes lead and a small addition of arsenic. A proprietary heat treatment is required to develop the resistance. They are strongly recommended for use not only in areas where the character of the water supply is known to promote dezincification, but also for mass-produced fittings supplied wholesale for plumbing in any locations. It is customary to mark them as dezincification resistant.

Stress-corrosion cracking in brasses was the first recorded incidence of the effect which is now recognized in many other metallic systems. It was originally known as season cracking in view of its association with climatic cycles. The specific agents are nitrogenous species, especially ammonia and its amino derivatives in the presence of water or its vapour and oxygen. The crack path is usually intergranular but it can be transgranular in severely deformed metal. Cracking can be initiated under external loading, but it is often the result of unrelieved internal stress in severely cold-worked metal, for example deep-drawn α-brasses. Examples of circumstances in which cracking is produced by internal stress include exposure to traces of agricultural fertilizers and residues of trimethylamine hydrochloride used as a flux for soldering. The cracking mechanism is imperfectly resolved but it is assumed to be associated with the formation of stable copper complexes of the general type, $[Cu(NH_3)_4(H_2O)_2]^{2+}$. Precautions to avoid failure include careful heat treatment to relieve internal stress, for example 1 hour at 350°C is appropriate for 70/30 α-brass, accepting some loss of strength developed by work hardening, restricting external loading and reducing contact with the specific agents by avoiding inappropriate environments.

10.2.2 Alloys Based on the Copper–Aluminium System

Alloys based on the copper–aluminium system are collectively known as aluminium bronzes. They include both simple binary and multicomponent formulations with additions of iron, nickel, manganese or silicon. Their mechanical properties are comparable to those of some medium-carbon steels and they have outstanding resistance to corrosion, cavitation corrosion, erosion–corrosion, and corrosion fatigue in severe environments, including seawater. These characteristics give them a prominent role for critical marine applications. They are also resistant to combustion products, sulfurous atmospheres and sulfuric acid and are used where these conditions prevail. Their range is extended by other characteristics including resistance to oxidation at elevated temperatures, retention of strength and toughness at elevated and cryogenic temperatures, hardness, wear resistance, good heavy duty bearing properties and low magnetic permeability.

Aluminium bronzes are most widely applied as near net-shape castings produced by all of the standard processes except pressure diecasting. Some alloys are also available in wrought forms including forgings, rolled products and extrusions, usually supplied in the hot-worked condition because they cold-work harden rapidly. They are suitable for arc welding.

10.2.2.1 Phase Equilibria for the Copper–Aluminium Binary System

Figure 10.2 gives a partial phase equilibrium diagram for the copper-rich end of the system. The FCC α-phase and the BCC β-phase form a eutectic at 1037°C with conjugate compositions of 7.5 and 9.5 mass % aluminium, respectively. The eutectic composition is 8.5 mass % aluminium. There is a eutectoid reaction at 565°C in which β transforms to α and an aluminium rich phase, γ_2, with conjugate compositions 9.4 and 15.6 mass % aluminium, respectively.

10.2.2.2 Formulation and Characteristics of Alloys

The following brief review describes some standard compositions and introduces constitution features that influence corrosion resistance. Some representative alloys are listed in Table 10.2.

10.2.2.2.1 Aluminium Bronzes

Commercial alloys based on the binary system are formulated with aluminium contents in the range 5 to 10.5 mass %. Alloys with <7.5 mass % aluminium solidify as single phase α without significant segregation because the freezing range is short. Alloys with >9.5% aluminium solidify as β and deposit α in a Widmanstätten structure as the solidified metal cools to the eutectoid temperature. In simple binary alloys, the eutectoid reaction is incomplete because it is sluggish and consequently much of the β is retained as a metastable phase and undergoes a diffusionless transformation to acicular martensite at temperatures below the eutectoid temperature. However, the eutectoid reaction must be *completely*

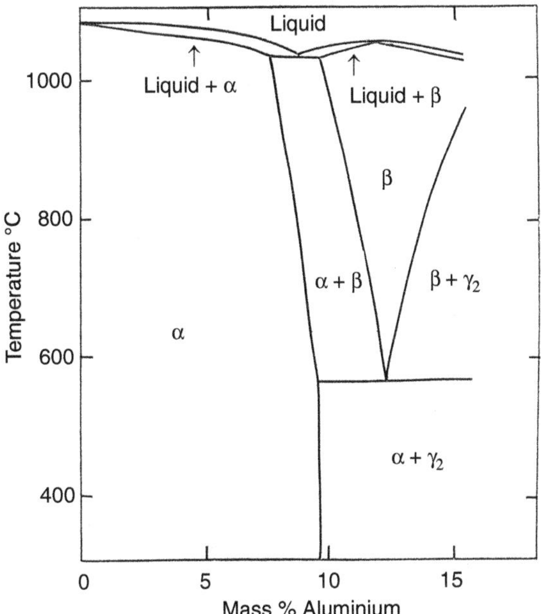

FIGURE 10.2
Partial phase equilibrium diagram for the copper–aluminium binary system.

suppressed in commercial alloys because the γ_2 phase severely impairs the corrosion resistance as explained later. It can be suppressed by restricting the aluminium content and by further alloying with iron, nickel or manganese. Iron is often preferred because nickel is expensive and manganese contents above 1 mass % can reduce the corrosion resistance of the retained β phase. An iron content of 2 mass % is usually sufficient to suppress the formation of γ_2 in castings with sections below 75 mm, but more is required for larger sections. The addition is fail-safe if it is not quite adequate because iron also refines the microstructure so that any residual γ_2 is discontinuous.

10.2.2.2.2 Nickel–Aluminium Bronzes (Copper–Aluminium–Nickel–Iron Alloys)

Nickel–aluminium bronzes are complex multicomponent copper alloys formulated for high strength, containing aluminium, nickel, iron and manganese with a typical composition as for the example given in Table 10.2. They solidify as the β phase from which a new phase, κ, containing aluminium, iron, nickel and manganese, separates as the metal cools below about 900°C. As it cools further, the remaining beta disproportionates completely yielding (α + κ). The alloy is free from the γ_2 phase because it does not normally separate in the presence of the κ phase. Cast and wrought versions are available.

10.2.2.2.3 Aluminium Silicon Bronze (Copper–Aluminium–Silicon Alloy)

An aluminium–silicon bronze containing 2% silicon with 6% aluminium and 0.5% iron is available as forgings, extrusions and castings. This alloy is also normally free from the γ_2 because the aluminium content is too low.

10.2.2.2.4 Manganese-Bronzes (Copper–Manganese–Aluminium Alloys)

Multicomponent copper–manganese–aluminium alloys within the compositions given in Table 10.2 are related to aluminium bronzes even though manganese is dominant. They have duplex α/β structures in which the β phase differs qualitatively from the β phase in aluminium bronzes, with the implications for corrosion resistance described in Section 10.2.2.3. They were originally formulated as casting alloys, but they can also be hot worked.

10.2.2.3 Corrosion Resistance

Copper–aluminium alloys benefit both from the relative nobility of copper and from their ability to form and sustain self-healing protective films based on alumina derived from their aluminium contents. Stable films prevail in many difficult environments, including especially near-neutral aqueous chloride solutions and also some other difficult aqueous and gaseous media. These alloys are particularly well suited to service in contact with sea-water and they are exploited extensively in the marine applications featured in Chapter 25. Premature corrosion failures are usually attributable to inappropriate alloy selection, local destruction of the passive film by unforeseen turbulent environments or poor metal quality.

10.2.2.3.1 Some Effects of Microstructure

Provided that the eutectoid reaction yielding γ_2 is suppressed, duplex α/β aluminium bronzes are equal to single phase α alloys in resisting general corrosion and surpass them in resisting erosion–corrosion and cavitation corrosion.

A major factor in the alloy development described earlier was to eliminate the γ_2 phase because it severely impairs the general corrosion resistance of alloys in which it appears. The mechanism is probably based on differences in electrode potential between the aluminium-rich γ_2 phase and the adjacent β matrix, stimulating local action cells. The most damaging structure is a continuous network of γ_2 phase, providing pathways through which selective corrosion can infiltrate. It is best to control it by alloy selection, but there are also practical steps that can be employed during manufacture. They include rapid cooling, for example water-quenching from temperatures above the eutectoid temperature, after casting or hot working.

The separation of the multicomponent κ phase from nickel–aluminium bronzes during normal manufacture raises the question whether or not it has any adverse effect on corrosion resistance. A definitive answer is unrealistic because the κ phase can exist over a wide composition range, but it is found empirically that it is innocuous in alloys in which the nickel content is no higher than the iron content and the manganese content is below 1.3 mass %.

Apart from the particular problem due to the γ_2 phase, aluminium bronzes rarely experience selective phase attack that can afflict some other copper alloys except for occasional insignificant attack at phase boundaries in crevices or under extraneous surface deposits. The related manganese bronzes are relatively immune to selective phase attack in the turbulent waters that prevail around ship propellers for which they are applied, but they can suffer severe selective corrosion of the β phase in stagnant sea water, especially in shielded regions or if coupled to more noble alloys. The damage can be extensive because the β phase is continuous.

Satisfactory initial structures can be jeopardized by adverse thermal histories following welding or hot working. One example is precipitation of γ_2 phase in the heat-affected zones in duplex aluminium bronzes. Another is retained β phase quenched *in situ* in weld beads on large nickel–aluminium–bronze castings. The original satisfactory structures can be restored by solution heat treatments to dissolve unwanted phases followed by cooling at an appropriate rate.

10.2.2.3.2 General Corrosion Resistance

Aluminium bronzes are valued for their unrivalled resistance to seawater which provides solutions to some of the difficult problems in marine service described in Chapter 25, but their applications extend to service in a wide range of other environments, including chemical process liquors. They are selected where appropriate for tubes, boilers valves, pumps and associated equipment used in chemical engineering and in industry in general. Single-phase α alloys, duplex α/β alloys and nickel aluminium bronzes are all successfully applied in contact with non-oxidizing acids, including sulfuric, hydrofluoric, phosphoric and most organic acids but not hydrochloric or nitric acids. A significant consideration is that oxidizing species such as iron(III) salts, chromium(VI) or dissolved oxygen introduced into acidic media can markedly raise the aggressive nature of acidic media. Another reservation is that strong alkalis such as sodium hydroxide can dissolve the protective alumina film. The effects of aeration, composition, concentration and temperature of the media are too complex to assess *ab initio* and the suitability of aluminium bronzes for use in any particular chemical environment can be determined only from experience. Suppliers' brochures and publications issued by associated industrial development associations are sources of accumulated experience for practical application.

Crevice corrosion is not a significant problem for aluminium bronzes and even when it occurs it is usually so superficial that does not significantly reduce strength or impair surface finish. Aluminium bronzes are, therefore, widely applied in submerged moving parts such as shafts and valve spindles for which other metals, for example stainless steels can fail by crevice corrosion.

In practice, galvanic effects do not significantly accelerate corrosion of either metal in mixed metal systems in which aluminium bronzes are coupled to other copper alloys, the nickel–copper alloy monel, stainless steels or titanium, provided that the exposed area of the other metal is no larger than that of the aluminium bronze. This circumstance can be exploited for example by applying aluminium bronze as tubeplates in condensers fitted with titanium tubes.

10.2.2.3.3 Resistance to Stress-Related Corrosion Damage

Aluminium bronzes and related alloys have low susceptibility to all forms of stress-related corrosion damage, by virtue of their mechanical strength and the chemical stability and physical tenacity of their self-repairing alumina surface films. They seldom suffer stress-corrosion cracking, corrosion fatigue or erosion–corrosion in normal service. Manganese-aluminium bronze is rather more susceptible to stress-corrosion cracking than other members of the group and where appropriate it is advisable to relieve internal stresses in it by suitable heat treatment.

Nickel–aluminium bronzes and manganese–aluminium bronzes are some of the best materials to resist both corrosion fatigue in seawater and cavitation corrosion, that is damage inflicted by collapsing vapour cavities generated in fast moving water. For these reasons, these alloys are exploited in applications for items such as large marine propellers and other rapidly moving parts exposed to seawater, as explained in Chapter 25.

Good foundry practices are essential to exploit the inherent resistance of aluminium bronze castings to stress-related corrosion because outcrops of unsoundness and oxide inclusions can act as stress raisers that nucleate damage.

10.2.2.3.4 Resistance to Atmospheric Corrosion and Oxidation

The alumina-based films that protect aluminium bronzes in aqueous media also protect them from corrosion and oxidation in polluted gaseous atmospheres. At ambient temperatures the alloys serve safety-critical functions in industrial atmospheres, such as suspension arms for overhead electric conductors. At elevated temperatures they can resist oxidation in gaseous media polluted with combustion products, especially sulfurous gases and associated condensates. This attribute, together with the retention of mechanical properties at elevated temperatures and low susceptibility to stress-corrosion cracking extends applications for aluminium bronzes to service in industrial conditions that few other metals can withstand. Examples are hanger bars supporting furnace roofs, ducting for combustion products and chimney stack fittings.

10.2.3 Alloys Based on the Copper–Nickel System

Copper is alloyed with nickel to improve its strength and enhance its resistance to corrosion, in fresh waters, seawaters and brackish or polluted waters. The alloys have excellent resistance to stress-corrosion cracking, corrosion fatigue, erosion–corrosion and cavitation corrosion. They are complementary to aluminium bronzes in the options available for marine applications and the choice between the two alloy groups is often based on specific service experiences. A related valuable characteristic of copper–nickel alloys is their superior resistance to biological fouling. They are applied with discrimination because of their high initial costs due to the expense of the copper and nickel contents.

10.2.3.1 Phase Equilibria for the Copper–Nickel Binary System

Figure 10.3 gives the phase equilibrium diagram for the copper–nickel binary system. The two metals are completely miscible in the solid state, forming a single FCC α-phase. The wide freezing range promotes pronounced dendritic segregation during solidification which is difficult to disperse by thermal and mechanical treatments because interdiffusion of the component metals is slow. It persists as characteristic banding in wrought forms.

10.2.3.2 Formulation and Characteristics of Alloys

10.2.3.2.1 Basic Compositions

Two fundamental alloys with the compositions given in Table 10.2 are based on the copper-nickel system. They contain 10 and 31 mass % nickel together with iron and manganese. The iron is added because it is found to improve resistance to impingement attack by flowing seawater. Manganese deoxidizes the metal during manufacture, improves its working

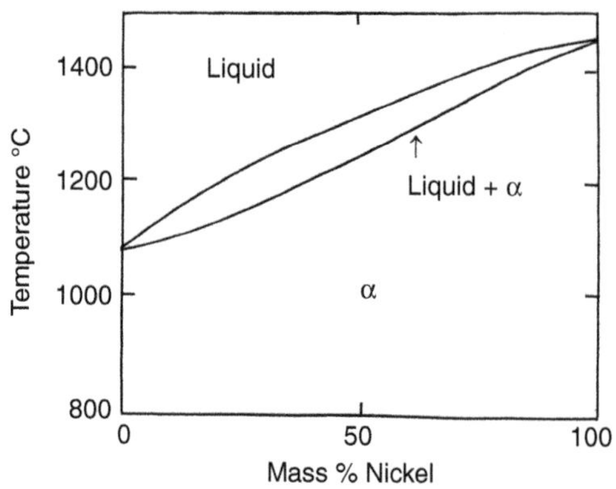

FIGURE 10.3
Phase equilibrium diagram for the copper–nickel binary system.

characteristics by scavenging residual sulfur and also contributes to corrosion resistance in seawater. Some derivatives contain chromium or niobium in place of iron and silicon to enhance strength and casting characteristics.

10.2.3.2.2 Copper–Nickel Clad Steel

Copper–nickel alloys are also available as cladding, 1.5 to 3 mm thick, on large steel plates produced by hot roll bonding, explosive bonding or weld overlaying. These composite materials combine the corrosion-resistant surface properties of copper–nickel alloys with the economy and strength of the steel base. Welding procedures for clad plate are well established. Initial runs of of a 65% nickel–35% copper alloy are applied as a parting layer to discourage embrittlement of the steel by copper penetration.

10.2.3.3 Corrosion Resistance in Water

The corrosion resistance of cupro-nickel alloys is due to a passive surface condition, associated with surface films that develop during a period of several weeks in contact with oxygenated water. Unlike the films formed on aluminium bronzes, they are not determinate chemical substances but a complex surface system that changes continuously as they develop, absorbing chemical species from the metal and the environment.

The initial films formed on the commercial alloys in natural seawater in equilibrium with atmospheric oxygen are based on copper(I) oxide, but as they develop they become indeterminate in both composition and structure. The outer layers containing copper, nickel and iron as oxides and chlorides are adherent in quiescent water but they are removed by turbulence. There remains a thin inner layer containing copper(I) chloride that seems to be the essential protective agent.

10.2.4 Alloys Based on the Copper–Tin System

Binary copper–tin alloys with small phosphorus contents form a class collectively known as phosphor bronzes. It is customary to include the multicomponent gunmetals described later, in the same class. These alloys are also applied with discrimination because of their high initial costs due to the expense of the copper and tin contents.

10.2.4.1 Phase Equilibria for the Copper–Tin Binary System

Figure 10.4 gives a partial phase equilibrium diagram for the copper-rich end of the system. The primary α-phase is a solution of tin in the FCC copper lattice. The phase boundary terminates at the composition of the conjugate solid phase, 13.5 mass % tin at 798°C for the peritectic reaction:

$$\alpha + \text{liquid} = \beta \tag{10.8}$$

The diagram predicts that equilibrium solidification of alloys containing <13.5 mass % tin would yield a single α-phase, but the very wide freezing range induces such heavy segregation that in practice the final liquid and adjacent solid yield some β by the peritectic reaction.

As the solidified metal cools, the β transforms in successive eutectoid reactions:

$$\beta \rightarrow \alpha + \gamma \rightarrow \alpha + \delta \tag{10.9}$$

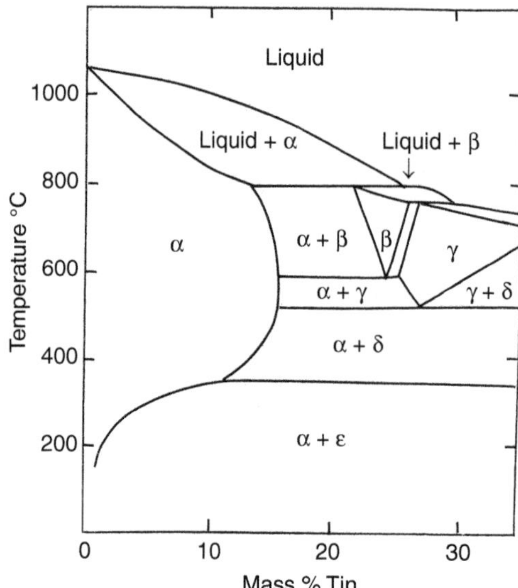

FIGURE 10.4
Partial phase equilibrium diagram for the copper–tin binary system.

In practice, the final transformation to ε indicated in the phase diagram is kinetically inhibited, so that the final microstructure is $\alpha + \delta$. The small phosphorus content in phosphor bronzes introduces a eutectic reaction by which it separates from the final fraction of interdendritic liquid as a skeletal structure of Cu_3P.

10.2.4.2 Formulation and Characteristics of Alloys

Phosphor bronzes are generally used in wrought forms. There are numerous registered compositions within the ranges 3 to 12 mass % tin and 0.02 to 0.40 mass % phosphorus. The phosphorus content raises the elastic modulus and enhances the wear resistance. Two of the more common alloys have the compositions given in Table 10.2.

The gunmetals are ternary and quaternary alloys formulated primarily for castings, with compositions in the copper–tin–zinc–lead system, some with and some without lead. The zinc and lead improve the casting characteristics. Lead also enhances the bearing properties. Examples of compositions for non-leaded and leaded gunmetals are given in Table 10.2.

10.2.4.3 Corrosion Resistance

Phosphor bronzes and gunmetals resist corrosion in a wider range of aqueous media than pure copper because the conditions for immunity and passivity for the pure metal represented in the Pourbaix diagram given in Figure 3.6 are extended by passivity superimposed by SnO_2 derived from the tin component, especially for oxidizing acidic media.

These alloys have roughly the same susceptibility to stress-related corrosion damage as most other copper alloys, but they are noteworthy for good resistance to impingement attack in flowing acidic cooling waters together with resistance to abrasion by suspended particles. It is probably due to the presence of the SnO_2 film together with adequate surface hardness.

Phosphor bronzes and gunmetals are complementary to aluminium bronzes and copper–nickel alloys for corrosion-resistant service in severe environments, including acidic media. Some of the duties for which they are most suitable are where their particular physical characteristics, that is high elastic modulus, wear resistance and good bearing properties are also required.

Further Reading

Bodsworth, C., *The Extraction and Refining of Metals,* CRC Press, Boca Raton, Florida, 1994.
Pehlke, R. D., *Unit Processes of Extractive Metallurgy,* Elsevier, New York, 1973.

11

Corrosion Resistance of Nickel and Its Alloys

Nickel and its alloys provide strong corrosion-resistant materials that are indispensable to sustain modern technologies including chemical engineering, marine applications, aircraft gas turbine engines, electric resistance heating and other high temperature functions. It is a strategic commodity because few ores are sufficiently rich for economic recovery. The principal sources are sulfide deposits at Sudbury, Ontario and the Kola Peninsula in NW Russia that between them produce over half of the world's supply. There are smaller significant lateritic (silicate and oxide) deposits in New Caledonia, Cuba and Oregon. The ores are lean, for example the nickel content of the Sudbury deposits is ~1.5% and recovery entails mining and processing large quantities of ore to extract minerals and recover other metal values, notably copper.

Nickel is expensive not only because workable deposits are scarce and lean, but also because elaborate sequences of operations are needed to recover it. Sulfide ores are crushed and a nickel concentrate is extracted by froth flotation. The concentrate also contains iron and copper which are removed by pyrometallurgy. It is roasted to oxidize the iron sulfide selectively and smelted with silica to remove the iron oxide as a slag, yielding a matte that solidifies as a eutectic containing nickel sulfide, copper sulfide and a small metallic fraction. The nickel sulfide is extracted by froth flotation, roasted to oxide and reduced with hydrogen to metal which is refined electrolytically. Lateritic ores are smelted in a similar way, adding calcium sulfate as a source of sulfur. Full descriptions are given in Boldt's classic text*.

Nickel has typical metallic properties. The solid metal has an invariable face-centered cubic lattice. Its thermal and electrical conductivities are quite high, about 20% of the corresponding values for copper. It is ferromagnetic but less strongly so than iron. It has a high melting point, 1453°C, and retains significant strength at elevated temperatures. It is tough and ductile and is available in semi-finished forms produced by standard metal working techniques, including hot- and cold rolling, forging, casting, machining and welding. For economical applications for heavy equipment it is also available as roll-bonded cladding on steel. Massive pure nickel is less widely applied than nickel base alloys but its applications are important. Its most familiar use is as electrodeposited protective coating for steel.

11.1 Chemical Properties and Corrosion Behaviour of Pure Nickel

11.1.1 The Nickel–Oxygen–Water System

The aqueous chemistry of nickel and hence its corrosion behaviour is determined by its position as the eighth element in the first transition series of the Periodic Table. The configuration for the outer electrons in the free atom is $(3d)_8(4s)_2$. The decreasing stabilities of higher oxidation states in the preceding sequence, scandium to cobalt, continues with

* Boldt, J. R., *The Winning of Nickel*, Longmans, Toronto, 1967.

nickel so that it commonly exists in its compounds and ions in only the oxidation state Ni(II). The higher oxidation states, Ni(III) and Ni(IV), that exist in some hydrous oxides and complexes are unusual and of little metallurgical or industrial significance.

A simplified Pourbaix diagram for the nickel–water system for $a_{Ni^{2+}} = 10^{-6}$ at 25°C and atmospheric pressure is given in Figure 3.8. Domains for oxide species in higher oxidation states than $Ni(OH)_2$ which are sometimes introduced are omitted because their stabilities are uncertain, and their replacements for a small part of the domain for $Ni(OH)_2$ would not significantly influence the extent of the passive zone, which is the important consideration in the present context.

The essential features of the diagram are:

1. For values of pH > 7, the domains of stability for nickel and water overlap.
2. For values of pH > 9, the stable nickel phase over most of the domain of stability of water is a solid species, that is $Ni(OH)_2$.

11.1.2 Corrosion Resistance in Natural Environments

11.1.2.1 The Natural Atmosphere

The corrosion of nickel is very slight in unpolluted air. It is too expensive for use in the natural atmosphere by itself, but it is valuable as a coating to protect steel, electrodeposited as described in Section 6.2.2. For regular corrosion resistance, a layer 30 μm thick is usually recommended. Recalling the discussion in Section 6.2.2, steel is prepared to receive nickel electrodeposits by depositing a copper undercoat from alkaline solution. Associated fittings, for example diecast zinc accessories, may be plated to present an integrated appearance.

There are two potential problems associated with electrodeposited nickel coatings on steel. One is that the nickel tarnishes during long exposure to industrial and urban atmospheres, losing the initial brightness of the deposit and acquiring a green/gray patina of basic sulfate and sulfide. This does not usually amount to corrosion failure in the usual material sense, but it is undesirable in contexts where appearance is important, as in office equipment, domestic kitchen and bathroom fittings and automobiles. It is avoided by electrodepositing a very thin, that is <1 μm thick overlay of chromium, producing a multilayer protective coating usually informally described as 'chromium plating'. The other problem is that the copper underlay, nickel main coat and chromium overlay constitute a mixed metal system with widely different electrode potentials. The coating is satisfactory as long as its integrity is not breached, but small defects in the coating either as faults present in the electrodeposits or abrasions in service can expose latent local action galvanic cells. The design of coating systems to control the effects in automobile bright trim are described in Section 22.4.1.

11.1.2.2 Natural Waters

Nickel is passive in all open fresh waters and has excellent corrosion resistance up to the boiling point. It is also resistant to seawaters, but it is vulnerable to pitting in shielded regions and beneath marine growth. For this reason the cupro-nickel alloys with high copper contents that resist biofouling described in Section 10.2.3 are preferred.

11.1.3 Corrosion Resistance in Neutral and Alkaline Media

The features of the Pourbaix diagram represent the behaviour of pure nickel in neutral and alkaline aqueous media quite well. Corrosion is not expected in deaerated solutions because it cannot be supported by evolution of hydrogen and it is expected to be suppressed in aerated solutions by passivation of the metal surface by $Ni(OH)_2$ or a derivative. The slope of the passivation potential as a function of pH is –0.0591, consistent with a passivating reaction of the form:

$$Ni + 2H_2O \rightarrow Ni(OH)_2 + 2H^+ + 2e^- \quad (11.1)$$

Nickel has excellent corrosion resistance in neutral and alkaline solutions of chlorides, carbonates, sulfates nitrates, acetates and concentrated alkalis up to their boiling points, extending to boiling solutions of 25 mass % of sodium or potassium hydroxides. Corrosion is stimulated by highly oxidizing salts such as strong hypochlorite solutions, but it can be inhibited by sodium silicate. Examples of industrial applications are storage vessels and evaporators for caustic soda, boiling kettles for soap manufacture and equipment for bleaching if the solutions are dilute and contact is brief.

11.1.4 Corrosion Resistance in Acidic Media

Nickel can resist *non-oxidizing* acids and is among the few pure metals with useful resistance to hydrochloric acid and acid chlorides. This effect is inconsistent with the formal Pourbaix diagram, which suggests that the metal should dissolve as Ni^{2+} ions with evolution of hydrogen for values of pH below 7. Investigations to discover an explanation for this apparent anomaly disclose the information that nickel can passivate in acids. The mechanism of passivation is unresolved, but empirical determinations show that it is different from that for neutral and alkaline media because the passivating potentials are independent of pH. A further factor contributing to the corrosion resistance in acids is the moderately high hydrogen overpotential on nickel.

Pure nickel is non-toxic and lacks the catalytic activity characteristic of some metals, such as copper that can degrade organic substances. These qualities combined with its resistance to non-oxidizing acids and acid salts are exploited in the use of nickel in commercial equipment for cooking and processing foodstuffs, including soups, fruit and vegetable products.

Mineral acids containing oxidizing species including chromates, nitrates and peroxides or salts with high redox potentials, such as ferric chloride, mercuric chloride or cupric chloride are highly corrosive to nickel. Pure nickel has limited applications for prolonged contact with concentrated mineral acids because its alloys, for example the nickel/copper alloy Monel, or the nickel/chromium/iron alloy Inconel, are better and less vulnerable to minor contamination by oxidizing impurities that commercially supplied acids may contain. A notable exception is the superior resistance of pure nickel to dilute hydrochloric acid solutions in concentrations up to 30 mass % at ambient temperatures. This is attributable to the very low solubility of nickel chloride in the concentrated acid that can tend to film the metal, compared with the much greater solubilities of copper (I) or copper (II) chlorides derived from the copper component of Monel.

11.1.5 Non-Aqueous Media

The protection afforded by nickel chloride extends to some gaseous and non-aqueous media. For this reason, nickel has applications for handling hydrogen chloride and chlorine gases at temperatures above the prevailing dew points in chemical processing, and for containing chlorinated solvents such as carbon tetrachloride and trichloroethylene as used for industrial degreasing and dry cleaning.

TABLE 11.1

Nickel and Representative Nickel Base Alloys for Corrosion Resistance in Aqueous Media

		Nominal Composition, Mass %										
Alloy Type	**Designation**	**Ni**	**Cu**	**Cr**	**Fe**	**Mo**	**Mn**	**Al**	**Si**	**W**	**C**	**Others**
Nickel grades	AT quality	99.4	0.1	–	0.15	–	0.2	–	0.05	–	0.15	
	Hardenable Ni	94	0.1		0.15			4.5			0.15	0.5 Ta
Nickel/Copper alloys	Monel 400[a]	67	30	–	1.4	–	1.0	–	0.1	–	0.15	
	K Monel 500[a]	66	29	–	0.9	–	0.85	2.75	0.5	–	0.15	
	S Monel[a]	63	30	–	2.0	–	0.75	–	4	–	0.15	
Nickel/Chromium and Nickel/Chromium/ Iron alloys	Nichrome[a]	80	–	20	–	–	–	–	–	–	–	–
	Inconel 600[a]	76	–	15.5	7.5		–	–	–	–	0.08	
Nickel alloys with Molybdenum	Corronel B[a]	66	–	–	6	28	–	–		–	–	–
	Ni-on-el[a]	40	2	20	35	3	–	–	–			
	Hastelloy C-276[a]	56	–	15	5	17	–	–	1	4	0.02	2.5 Co

[a] Monel, Nichrome, Inconel, Ni-o-nel, Corronel B and Hastelloy C are registered trademarks, quoted for identification.

Pure nickel is also used in equipment for producing storing and transporting phenols and phenolic resins especially where contamination or discolouration must be avoided. Condensers, pipes and valves are made from nickel and large storage vessels and tank cars are constructed from nickel-clad steel plate for economy.

A grade of nickel containing aluminium that can be age hardened by an intermetallic precipitate, γ', based on Ni_3Al is listed in Table 11.1 as *hardenable nickel.* It can be applied for components subject to wear such as valve seats.

11.2 Constitutions and Corrosion Behaviour of Nickel Alloys

Nickel forms extensive solid solutions in binary systems with other passivating metals including copper, chromium, iron, molybdenum and cobalt, which form the basis of most alloys. Compositions of interest lie within the solid solution phase fields of the binary systems, but most alloys are multicomponent formulations and contain additional components selected from aluminium, titanium, tantalum, niobium and carbon that may be added to enhance resistance to corrosion or oxidation or to improve mechanical properties. Such constitutions are often too complex to represent in simple phase equilibrium diagrams.

Nickel alloys are of two different kinds:

1. Alloys for general corrosion resistance in aggressive chemical engineering, industrial and marine environments, which are essentially solid solutions with simple structures designed to passivate in prescribed media.
2. Customized alloys for blades and associated components operating at high temperatures in gas turbines for aircraft and marine propulsion, known as *superalloys.* They are complex highly specialized alloys, some consisting of as many as ten components, with chemical compositions and controlled microstructures

designed to maintain mechanical stability and chemical integrity in the unique environments of gas turbine flames derived from aviation or marine fuels and air ingested from prevailing ambient atmospheres.

11.2.1 Nickel Alloys for General Corrosion Resistance

Alloys for chemical engineering, marine and allied applications have evolved by progressive improvements based on information accumulated from service experience and plant and laboratory corrosion tests. Tables 11.1 and 11.2 list compositions for some representative alloys and corresponding mechanical properties typical of the many registered formulations available. The properties are comparable with those of carbon steels.

The list includes alloys that can be matched to applications in acidic, alkaline or neutral aqueous media, including hydrochloric acid, for wide ranges of compositions and temperature, and to reactive gases including chlorine, hydrogen chloride and fluorine. The selection of a reliable and economic alloy for any particular application is usually guided by experience and confirmed by extensive experimental and plant tests because the conditions in which they must serve are often too complex and variable to predict in advance. Environmental factors that must be considered include pH, temperature, speciation, prevailing redox potentials and degree of aeration, all of which may vary in the course of process sequences in chemical or related industries. Other factors may also be important such as acceptable metallic contamination of material in process.

The particular conditions in which a nickel alloy can resist corrosion depend on which of the components, copper, chromium, or molybdenum, is dominant. In general, copper enhances resistance in reducing but not oxidizing acids, chromium promotes resistance in oxidizing media and molybdenum confers protection in unusually severe conditions. The following sections review characteristics of some standard alloys, especially the archetypal nickel–copper alloy Monel and the archetypal nickel–chromium–iron alloy, Inconel, which are the most versatile and widely applied engineering nickel alloys.

11.2.1.1 Nickel–Copper Alloys (Monels)

Nickel–copper alloys in which nickel predominates are collectively known as Monels. The basic composition, Monel 400, usually called simply 'Monel', is nominally 67 mass % nickel with 30 mass % copper 1.4 mass % iron and 1 mass % manganese, with small silicon and carbon contents but there are derivatives, some of which are listed in Table 11.1. The variant, K Monel, is essentially the basic composition with an addition of 2.75 mass % aluminium at the expense of the copper to promote age hardening by the intermetallic γ' phase, Ni_3Al. The improvement in mechanical properties is evident from Table 11.2. Other derivatives are formulated to meet particular requirements, such as S Monel listed in Table 11.1, which contains 4 mass % silicon and a little more iron to improve casting characteristics and erosion–corrosion resistance, but which is brittle.

Monels inherit corrosion-resistant attributes from both main components by combining the relative nobilities and passivation tendencies of copper and nickel. They have a useful degree of resistance to a wide range of aggressive media and to erosion–corrosion, cavitation and wear, which commends them for uses as materials for moving parts. There is no selective dissolution of components from the solid solution as there is, for example, in the dezincification of brass. Some incidents of stress-corrosion cracking attributed to internal stress reported in the field can be reproduced in laboratory tests, so that it is advisable to forestall any problems by heat-treating fabricated items to relieve stress if possible.

TABLE 11.2

Typical Mechanical Tensile Properties at 25°C of Some Representative Nickel Alloys[a]

Alloy	Condition	0.2% Yield Strength MPa	Tensile Strength MPa	% Elongation on Gage Length
Pure nickel, AT	Annealed (softened by heating)	150	400	40
	Cold worked	480	660	25
Monel 400	Annealed	230	530	40
	Cold worked	570	750	25
K Monel 500	Annealed	300	675	40
	Cold worked and aged	750	1110	25
S Monel	As cast	640	640	nil
Nichrome	Annealed	310	700	–
Inconel 600	Annealed	260	620	45
	Cold worked	700	880	20
Ni-on-el	Annealed	250	630	46
	Cold worked	–	1040	
Corronel B	Cold-rolled sheet	420	900	40
Hastelloy-C-276	Cold-rolled sheet	416	768	

[a] Measured on 12.5 mm diameter test pieces with 50 mm gage length to American Standard ASTM E8-78.

11.2.1.1.1 Resistance in Natural Waters

The Monels exhibit negligible corrosion and erosion–corrosion in both hard and soft fresh waters at all temperatures and any rate of flow irrespective of the degree of aeration. A spectacular demonstration of the performance of Monel in natural waters is its application for valve seats controlling the flow of water from the Hoover dam to the turbines it serves.

The resistance to erosion–corrosion and cavitation also applies to flowing or turbulent seawater. Hardened K Monel in particular is a standard material for propeller shafts, pump impellers and condenser tubes considered further in Chapter 25 within the context of corrosion in marine environments. It is less suitable for prolonged exposure to stagnant seawater, because biological fouling can initiate local differential aeration cells, causing shallow pitting.

11.2.1.1.2 Resistance in Neutral and Alkaline Aqueous Media

The corrosion of Monel is negligible in solutions of alkalis and neutral and alkaline salts at all concentrations and temperatures except for strong solutions of hypochlorites. It is one of the most satisfactory metallic materials for handling these solutions and applications include containers, valves and pumps for hot, saturated brines and for refrigerating brines. Despite its vulnerability to *strong* solutions of hypochlorites, Monel can be exposed for short periods to the *weak* solutions encountered in industrial bleaching and sterilization, provided that the concentration of available chlorine does not exceed 0.1 mol kg^{-1} and that the equipment is intermittently rinsed with acid solutions, as is standard practice in many commercial bleaching operations.

11.2.1.1.3 Resistance to Acidic Aqueous Media

The Pourbaix, diagram given in Figure 3.8, shows that the domain of stability for nickel lies <0.4 V below potentials for hydrogen evolution for pH < 7 and the corresponding diagram for copper, given in Figure 3.6, shows that copper and water share a common domain of stability throughout the pH range. Although Pourbaix diagrams apply strictly to pure metals, these factors, combined with hydrogen overpotentials of 0.43 V for copper and 0.30 V for nickel at a current density of 1 mA cm^{-2} suggest that evolution of hydrogen is unlikely to support the dissolution of solid solutions of the two metals in acids. However, potentials for oxygen absorption are well above the lower limit of the zones of stability for both Cu^{2+} and Ni^{2+}, tentatively indicating that the metal is vulnerable to corrosion in acids supported by absorption of dissolved oxygen.

These indications are consistent with the actual behaviour of Monel in acids. It is resistant to corrosion in *non-oxidizing* acids and acidic salts, but is subject to severe corrosion in *oxidizing* acids and salts, including nitric acid and acidic media containing species with high redox potentials for example chromates, nitrates, perchlorates, iron(III), copper(II) and mercury(II).

Monel has good qualitative resistance to most common non-oxidizing mineral and organic acids, including sulfuric, hydrochloric, hydrofluoric, fluosilicic, phosphoric, and acetic acids, but quantitative information on acceptable limiting values of aeration, concentration, and temperature is difficult to present in a consistent form because much of it is derived from specific applications that are not necessarily related. Technical service departments of specialist manufacturers maintain databases for the service experience and results of corrosion tests of their products. Table 11.3 summarizes empirical information extracted from various sources illustrating approximate limiting conditions for resistance in some common acidic media.

A notable exception to the resistance of Monel in reducing acidic media is its failure to resist the so-called 'sulfurous acid', a misnomer that appears in some trade literature for a solution of sulfur dioxide. The solution probably exerts its adverse effect by infiltrating the passive surface condition. Hence, Monel is usually unsuitable for service in systems handling flue gases where concentrated solutions of sulfur dioxide can develop in condensates.

11.2.1.1.4 Applications

Such a versatile engineering metal is widely applied and successful applications are based primarily on long experience. The present context permits only a few well-known examples, given below to illustrate the scope for exploiting its corrosion resistance.

TABLE 11.3

Representative Acidic Media in Which Monel Exhibits Good Corrosion Resistance

Acid	Condition	Temperature °C	Strength
Sulfuric	Air-free solution	25	<~85 mass %
		<95	<~60 mass %
	Air-saturated solution	25	<~85 mass %
		<95	<~60 mass %
Hydrochloric	Air-free solution	25	<~20 mass %
		75	<~5 mass %
	Air-saturated solution	25	<~10 mass %
		50	<~2 mass %
		80	<~1 mass %
Hydrofluoric	Air-free solution	120	All concentrations
	Anhydrous gas	550	Not applicable
Phosphoric	Air-free solution	93	All concentrations
Acetic	Air-free solution	100	All concentrations
	Air-saturated solution	100	All concentrations

11.2.1.1.4.1 Industrial Operations with Sulfuric Acid Monel is often the material of first choice for handling moderately concentrated sulfuric acid. An example is its application for cradles and accessories used to support steel products during *pickling*, that is immersing them in typically ~1 mol kg^{-1} sulfuric acid solutions at 60 to 80°C or an equivalent mixture of sulfuric and hydrochloric acids to remove mill scale from hot-rolled steel strip for onward cold-rolling or to prepare finished steel surfaces for application of protective coatings, as described in Section 6.1.2.1. Evolution of hydrogen by dissolution of the steel maintains the reducing conditions needed to protect Monel. An advantage in using Monel over alternative copper base alloys is that the copper content is insufficient to yield unwanted copper deposition on the steel.

11.2.1.1.4.2 Chemical Processing Using Chlorinated Substances The corrosion resistance of Monel in dilute hydrochloric acid confers corresponding resistance to weakly acidic hydrolysis products of chlorinated solvents and chlorides at temperatures up to 150°C and determines its use to construct equipment for dry cleaning and for solvent distillation and reclamation.

11.2.1.1.4.3 Chemical Processing Using Hydrofluoric Acid Monel is exploited extensively for equipment in which chemical processes are conducted with hydrofluoric acid as a reactant or catalyst and for equipment producing the acid itself, especially for critical components such as pump and valve parts. These processes include organic syntheses such as the HF alkalation process for improving the octane value of gasoline using hydrofluoric acid as a catalyst. The anhydrous gas is admitted to the system and recovered after use with water from feed materials and water cooling for regeneration and recycling. The benefit of using Monel is its resistance to the acid both as anhydrous gas and in solution.

11.2.1.1.4.4 Manufacture of Fluorine Fluorine has an important role in modern chemical industries. It is required in the production of fluorocarbon plastics and refrigerants and for the preparation of uranium hexafluoride required to extract the isotope, ^{235}U, from natural uranium for energy from nuclear fission. Fluorine is produced by electrolysis of liquid fluorides. Suitable electrolytes are the double fluorides, KF.2HF and KF.HF, which melt at 70°C to 100°C and 250°C to 270°C, respectively, imposing a requirement to contain the electrolyte and collect the gas at elevated temperatures.

Fluorine reacts with most other substances including water, often violently, therefore, it is the most difficult substance to contain and handle. Nickel and Monel are some of the best metallic materials to use for parts of the electrolytic cell exposed to hot liquid fluorides, anhydrous hydrofluoric acid and fluorine at temperatures up to 300°C, including the water-cooled cathode box and cell cathode. The corrosion resistance depends on the formation of a passive fluoride film on the metal surface. To ensure the integrity of the passive film, new surfaces of Monel that are to be in contact with fluorine or fluorine compounds must be thoroughly cleaned of all foreign matter, degreased with non-aqueous solvents, dried and passivated with a nitrogen/fluorine mixture before commissioning.

Fluorine gas is often liquified by compression for storage. Although the liquid can be stored under pressure in copper or steel cylinders, Monel is preferred for safety.

11.2.1.1.4.5 Dispensing Chlorine Monel resists corrosion in aqueous solutions of hydrochloric acid and in *dry* chlorine gas, but it is rapidly attacked by *wet* chlorine. At first this may seem to be inconsistent but on reflection the reason becomes clear. In both hydrochloric acid and dry chlorine, the metal is passivated by a film of sparingly soluble nickel chloride as explained earlier for pure nickel. A very different condition develops if the chlorine is wet because it reacts with water, forming hydrochloric and hypochlorous acids:

$$Cl_2\ (g) + H_2O\ (g) = HCl\ (solution) + HClO\ (solution) \tag{11.2}$$

yielding an acidic oxidizing dew that stimulates corrosion of Monel, as described earlier. This explains the apparent anomaly that Monel is the standard material for valves and fittings on chlorine cylinders and tanks and is used in delivery systems for chlorination of water supplies and swimming pools yet the temperature of a system must not fall below its internal dew-point, nor must any water be allowed to enter. For the same reason, moisture in ambient air replacing bromine drawn from Monel drums in which it is stored stimulates corrosion of the metal at the air/bromine interface by forming an oxidizing acidic solution of hydrobromic and hypobromous acids:

$$Br_2\ (l) + H_2O\ (g) = HBr\ (solution) + HBrO\ (solution) \tag{11.3}$$

11.2.1.1.4.6 Food Processing Monel is virtually unaffected by water, dilute organic acids, brine and fats at temperatures prevailing in food preparation, so that processing equipment made from it does not taint the flavor or colour of foods with which it is in contact. The resistance to brine is an advantage over stainless steel which is vulnerable to pitting. Moreover contact surfaces remain smooth and can be kept clean and hygienic. Typical items are dairy equipment, steam heaters, cooking coils, boiling pans, reduced pressure evaporators, stills and condensers.

11.2.1.2 Nickel–Chromium Alloys (Nichrome and Inconel)

Chromium is alloyed with nickel to promote resistance to oxidizing aqueous media by raising the passive range to higher potentials and to protect the metal from oxidation in air at high temperatures by selective oxidation yielding a surface film of chromium oxide, Cr_2O_3, by the principles explained in Section 3.3.4.2. Alloys in general use include the binary alloy Nichrome and the ternary nickel/chromium/iron alloy, Inconel 600, listed in Table 11.1.

11.2.1.2.1 Nichrome

Nichrome is a binary alloy with the composition 80 mass % nickel 20 mass % chromium. Chromium forms an extensive solid solution in nickel but, as with the iron–chromium system, the chromium content must be restricted to 20 mass % to avoid embrittlement by precipitation of the sigma phase, σ. Its principal use is as a high temperature material and its most familiar application is drawn wire for electrical resistance heating elements. It has historical significance as a precursor of alloys for gas turbine blades considered in Section 11.2. Nichrome is resistant to some aqueous media but the related alloy, Inconel, is usually preferred.

11.2.1.2.2 Inconel

Inconel 600 is an alloy with the nominal composition of 76 mass % nickel, 15.5 mass % chromium and 7.5 mass % iron, which is an optimum composition combining resistance to corrosion in reducing conditions characteristic of nickel with passivation in oxidizing conditions characteristic of chromium and iron. Its structure and mechanical properties are suitable for standard metal shaping operations.

Inconel is immune to corrosion in all natural waters and it is unaffected by superheated water. In common with nickel and other standard nickel alloys, it is virtually not attacked by most neutral and alkaline solutions but is vulnerable to corrosion in alkaline hypochlorites.

Inconel has useful resistance to aqueous solutions of mineral and organic acids including hydrochloric acid, but it is not selected in preference to Monel for that reason alone. Its advantage is in its resistance at ambient temperature to many but not all oxidizing species in acidic media. Acceptable oxidizing species include nitrates, chromates and permanganates but not nitric acid or acid chlorides of some cationic species with high redox potentials including, iron(III), copper(II) and mercury(II).

Selective oxidation yields a surface layer of Cr_2O_3, as explained in Section 3.3.4.2, which protects Inconel in clean air or steam at temperatures up to 1100°C, in CO/CO_2 mixtures at carburizing temperatures, 850–950°C and in ammonia at nitriding temperatures, 500–600°C.

Applications of Inconel complement those of Monel, illustrated in Section 11.2.1.1, and a decision on which is the most effective and economic for any particular purpose is not always apparent *ab initio* and must often be settled by experience or test results. Inconel is particularly suitable for applications related to its tolerance of oxidizing conditions, its exceptional resistance to superheated water and steam and its resistance to dry gases at high temperatures.

11.2.1.2.2.1 Applications in Chemical and Process Industries Inconel is a general purpose alloy for constructing equipment for a wide variety of chemical operations. Applications include reaction vessels for plastics and industrial chemicals of various kinds, acid dyeing of wool and sulfate pulp digesters for the manufacture of paper.

11.2.1.2.2.2 Applications for Steam Generation Critical equipment for generating and utilizing steam are made from Inconel tubing, including heating coils, valve trim and accessories for steam turbines. An important application is for tubes in steam generators serving pressurized water nuclear reactors, described in Section 27.3.1.2.

11.2.1.2.2.3 Heat-Resisting Duties Furnace components and accessories for various industrial uses are fabricated from Inconel to exploit its resistance to oxidation. Examples are parts and accessories for furnaces used to surface-harden steel by carburizing in hydrocarbon gas mixtures in the temperature range 850–950°C and by nitriding in ammonia/hydrogen/nitrogen gas mixtures in the range 500–600°C. A related application is for tubes in which anhydrous ammonia is cracked over a nickel catalyst at 850°C to yield a nitrogen/hydrogen mixture for use as an atmosphere for bright annealing stainless steels and copper. Many continuous conveyor furnaces for pottery or enameling employ Inconel rollers to convey the products through the hot zone, supported on Inconel racks. A familiar heat resisting application is the use of Inconel sheaths to protect heating elements in domestic appliances, for example electric cookers and washing machines.

11.2.1.3 Nickel Alloys Containing Molybdenum

Nickel alloys containing molybdenum, such as the examples listed in Table 11.1, are formulated for equipment exposed to severe conditions prevailing in chemical, petrochemical and other industrial operations that are too aggressive for other engineering metals. They include hot concentrated sulfuric, hydrochloric, and phosphoric acids, solutions of sulfur dioxide and acidic solutions of ions with high redox potentials such as iron(III), copper(II), nitrates, chromates and wet chlorine. The introduction of molybdenum improves resistance to these substances and also reduces vulnerability to both SCC and pitting induced by chlorides.

The principal components of these alloys are nickel, chromium, iron and molybdenum. The relative proportions must be adjusted to deliver both the required corrosion-resistant characteristics and favourable microstructures that are not necessarily compatible. For example, the maximum resistance to reducing acids is promoted by a high molybdenum content, but the introduction of chromium to secure resistance to oxidizing ions indirectly diminishes it because the molybdenum content must be restricted to avoid precipitation of the deleterious sigma phase, σ. Moreover, the stability of the γ phase depends on the balance of nickel, iron and chromium contents. Thus the alloys are inevitably environment specific and selection of the most effective and economical alloy for a particular application depends as much on experience and comparative plant tests as on the expected characteristics imparted by the components. The essential features and applications of three representative alloys, Corronel B, Ni-on-el and Hastelloy C are now introduced.

11.2.1.3.1 Corronel B

The nominal composition of Corronel B is 66 mass % nickel, 28 mass % molybdenum and 6 mass % iron Chromium is omitted to permit the high molybdenum content that secures outstanding resistance to *reducing* mineral acids. The following examples illustrate some unusually severe conditions that the alloy can resist:

1. Hydrochloric acid in all concentrations at temperatures up to the boiling points
2. Phosphoric acid solutions <85% by mass at temperatures up to boiling
3. Sulfuric acid in all concentrations at temperatures <100°C
4. Sulfuric acid solutions <60% by mass at temperatures up to boiling

Corronel B has suitable mechanical properties for standard metal working operations but if welded, it is sensitive to intercrystalline corrosion by grain boundary carbide precipitation in the heat-affected zones. Sensitized zones in components for which suitable furnace capacity is available can be rectified by a solution heat treatment in the temperature range 1050–1150°C to dissolve the precipitate, followed by quenching in water.

Corronel does not resist corrosion in oxidizing acidic media for example acidic solutions of ions with high redox potentials such as iron(III), copper(II), nitrates or chromates. Alternative alloys containing chromium such as Ni-o-nel or Hastelloy C are required.

11.2.1.3.2 Ni-o-nel

The nominal composition of Ni-o-nel is 40 mass % nickel, 21 mass % chromium, 31 mass % iron, 3 mass % molybdenum and 1.5 mass % copper. The introduction of substantial chromium and iron contents enables the alloy to passivate in concentrated oxidizing acidic media. The consequent reduction in molybdenum and nickel contents curtails resistance to reducing acids. The following list illustrates environments in which the alloy can be applied in the presence of ions with high redox potentials, for example copper(II), iron(III), nitrates and chromates:

1. Sulfuric acid <40 mass % at the boiling point and <70% at 80°C
2. Hot strong solutions of sulfur dioxide (the so-called 'sulfurous acid')

3. Boiling phosphoric acid in concentrations <85 mass %
4. Boiling acetic acid and many other strong organic acids
5. Nitric acid mixed with sulfuric, phosphoric, hydrofluoric or acetic acids

Ni-o-nel has various applications for arduous duties in the chemical and petrochemical industries, including service in contact with hot strongly oxidizing acidic media for which Monel and Inconel are unsuitable. Other applications exploit its resistance to hot solutions of sulfur dioxide, for example as furnace accessories exposed to products of combustion from fuels with high sulfur contents. It is also an alternative to austenitic stainless steels for applications in aqueous chloride media because it is less vulnerable to pitting and SCC.

11.2.1.3.3 Hastelloy C

The nominal composition of Hastelloy C is 56 mass % nickel, 15 mass % chromium, 5 mass % iron, 17 mass % molybdenum, 1 mass % silicon, 4 mass % tungsten and 2.5% cobalt. It is one of the few metallic materials that can resist wet chlorine and hypochlorites.

11.2.2 Superalloys for Gas Turbine Components

11.2.2.1 Properties of Materials for Gas Turbines

Nickel base *superalloys* were developed for arduous duties in gas turbine engines, notably turbine blades and discs and combustion chambers where the material characteristics required are corrosion resistance in the severe environment and appropriate mechanical properties at the high temperatures. The essential mechanical properties are (1) resistance to *creep* and *creep rupture*, that is progressive deformation of the metal under steady stresses experienced in service ultimately causing mechanical failure by decohesion of the microstructure and (2) resistance to fatigue failure under cyclic stresses. The alloys were developed from the alloy of 80% nickel with 20% chromium that consists of a single FCC phase, γ, which is protected from oxidation at high temperatures by a film of chromium oxide, Cr_2O_3, formed selectively by the principles explained in Section 3.3.4.2.

11.2.2.2 Design of Microstructures

The first modification was to add aluminium to introduce the intermetallic phase γ' based on Ni_3Al, precipitated as a fine dispersion throughout the γ matrix that confers strength by impairing dislocation movement. The characteristics of the precipitate phase and its relationship to the matrix have driven the development of the superalloys. Because both γ and γ' have FCC structures with similar lattice parameters (<1% mismatch), the precipitate is coherent with the matrix and has low interfacial energy and good long-term stability at high temperatures. If the volume fraction is low, dislocations can bypass the precipitate particles by looping around them, but if it is high, the dislocations cannot loop and must pass through them, where they are impeded by the ordered structure of the phase. Titanium, tantalum and niobium can participate in the formation of γ' and by suitable alloying it is possible to obtain volume fractions of γ' in excess of 90%. The matrix can be further strengthened by adding elements such as tungsten or molybdenum that enter solid solution, stiffening the matrix and slowing the diffusion of metallic solutes that could undermine high temperature stability.

TABLE 11.4

Representative Nickel Base Alloys Developed for Gas Turbine Engines

	Composition, Mass %										
Alloy	**Ni**	**Cr**	**Ti**	**Al**	**C**	**Mo**	**Co**	**Fe**	**W**	**Nb**	**Others**
Mar M002	Balance	9	1.5	5.5	0.15	–	10	–	10	–	2.5 Ta + 1.5 Hf
IIN 738	Balance	16	3.5	3.5	0.11	1.8	8.5	–	2.5	0.7	1.6 Ta
Nimonic 105	Balance	15	3.8	5.0	0.15	3.3	13	–	–	–	–
IN 939	Balance	22	3.7	1.9	0.5	–	19	–	2.0	1.0	1.4 Ta
IN 718	Balance	19	0.9	0.6	0.04	3.0	–	20	–	5.2	–
IN 901	42	13	3.0	0.3	0.04	5.7	–	Balance	–	–	–
Waspalloy	Balance	19	3.0	1.3	0.08	4.3	13.5	–	–	–	–
Astralloy	Balance	15	3.5	4.0	0.06	5.2	7.0	–	–	–	–
C 263	Balance	19	2.2	0.5	0.06	6.0	20	–	–	–	–
SRR 99	Balance	8	2.2	5.5	–	–	5	–	10	–	3 Ta
CMSX-4	Balance	6	1.0	5.6	–	0.6	9	–	6	–	7 Ta + 3 Re + 0.1 Hf

Ti = titanium, Mo = molybdenum, Co = cobalt, W = tungsten, Nb = niobium, Ta = tantalum, Hf = hafnium, Re = rhenium.

Strength conferred on the matrix exposes microstructural weaknesses that must be addressed. One of these is relative movement between adjacent grains, allowing creep and creep rupture. This is controlled in polycrystalline materials by incorporating carbon that precipitates metal carbides at the grain boundaries to pin them. The problem is circumvented for blades operating at high temperatures by eliminating grain boundaries, using the directional solidification and single crystal casting techniques described in Section 14.2.5.1.

By exploiting all of these effects, alloys such as Mar M002 can deliver high strength at temperatures approaching 1150°C.

11.2.2.3 Alloy Range

These developments have yielded an extensive range of alloys, illustrated by the examples given in Table 11.4. The introduction of the alloy components needed to deliver alloys with mechanical strength and stability at the highest temperatures diminishes the chromium content to values insufficient to sustain a protective chromia film. Protection is then afforded by alumina derived from the aluminium content, conflicting with its vital structural role and reducing resistance to attack by sulfur contaminated air. The resolution of this dilemma depends on alloy selection which is discussed in detail within the context of corrosion control in aviation considered in Section 14.2 of Chapter 14.

12

Corrosion Resistance of Titanium and Its Alloys

The development of titanium as an engineering metal in the latter half of the twentieth century was driven by the requirements of aerospace constructors for a strong metal with a high strength to weight ratio. The consequent availability of the metal stimulated applications in chemical engineering to exploit its outstanding corrosion resistance in oxidizing media. This uniquely useful combination of properties justifies the commercial operation of the expensive processes required to extract titanium from its ores.

Titanium is one of the more common elements in the earth's crust, with a relative abundance of 0.6 mass %. The ores are ilmenite, $FeO \cdot TiO_2$ and rutile, TiO_2. Titanium cannot be extracted from titanium oxides derived from these ores simply by reduction with carbon because the product is not the metal but the carbide, TiC. Moreover, the extraction process must exclude oxygen, carbon, nitrogen and hydrogen because traces of them in interstitial solution adversely affect the physical properties. The problem is solved by the *Kroll process*, in which titanium is extracted via a chloride intermediary. The oxide is first converted to the carbide by heating a pelletized intimate mixture of the oxide with carbon:

$$TiO_2(s) + 2C(s) = TiC(s) + CO_2(g) \tag{12.1}$$

which is then chlorinated in a stream of chlorine gas, yielding the chloride, $TiCl_4$

$$TiC(s) + 2Cl_2(g) = TiCl_4(g) + C(s) \tag{12.2}$$

Titanium chloride boils at 137°C and is recovered as a gas from which impurities are removed by selective condensation. It is reduced in a batch process by liquid magnesium at 800°C in a reaction vessel previously purged of all reactive gases:

$$TiCl_4 + 2Mg = Ti + 2MgCl_2 \tag{12.3}$$

The product is recovered, the $MgCl_2$ is leached out, excess magnesium distilled away and the residual porous mass, *titanium sponge*, is consolidated by vacuum arc melting to produce ingots suitable for fabrication. The Kroll process is operated on a fairly large scale.

There are two allotropes of titanium, a hexagonal close-packed (HCP) lattice, the α-phase, at temperatures below the transition temperature, 882°C, and a body-centered cubic (BCC) lattice, the β-phase, at temperatures above it. At 25°C, the density of the pure metal is 4.51 g cm^{-3} and the mechanical properties, given later in Table 12.2, are comparable with those of low carbon steels or austenitic stainless steels which have much higher densities, ~7.9 g cm^{-3}. It has a high melting point, 1667°C, and retains useful mechanical properties at temperatures of the order of 400°C. It, therefore, has appropriate physical and mechanical properties required of an engineering metal to exploit its low density and superior corrosion resistance.

The HCP lattice might seem to suggest difficulty in forming titanium, but the *c/a* ratio, 1.587, is not particularly large for an HCP metal so that there are sufficient slip planes to confer adequate ductility at convenient hot and cold metal working temperatures. Titanium and its alloys are available as standard metallurgical products, including castings, forgings, flat rolled products, rod and wire. Scrap arising in manufacture and applications must be minimised because recycling it to recover metal values requires vacuum arc melting for the same reason as for virgin metal.

12.1 Chemical Characteristics and Corrosion Behaviour of Pure Titanium

12.1.1 The Titanium–Water System

Titanium is the second element of the first transition series, with four valency electrons in the configuration, $(3d)_2(4s)_2$. It exists in the oxidation states, Ti(–I), Ti(0), Ti(II), Ti(III) and Ti(IV). The most stable oxidation state is Ti(IV), but the charge density that would result from removing all four electrons is so high that compounds of Ti(IV) are mainly covalent in character and it is doubtful if the simple Ti^{4+} ion exists in aqueous solution.

Figures 3.9 and 3.10 give alternative Pourbaix diagrams for the titanium–water system considering anhydrous and hydrous oxide species, respectively. The essential features of the diagrams are:

1. There is no common domain of stability for titanium and water.
2. There is no domain of stability for a simple Ti^{4+} aquo ion. It probably does not exist.
3. The domain of stability for the Ti(IV) anhydrous oxide, TiO_2, extends across the whole of the pH range in the diagram in Figure 3.9.
4. For values of pH < 2, the hydrous oxide, $TiO_2 \cdot H_2O$ in Figure 3.10 dissociates to yield a soluble species labelled TiO^{2+}, although its independence as a simple titanyl ion is unconfirmed.
5. No domain is given for a 'titanate' ion, for example TiO_3^{2-}, sometimes proposed for high values of pH because most substances described as titanates such as ilmenite, $FeTiO_3$ and perovskite, $CaTiO_3$, do not contain discrete anions but have mixed metal oxide structures.

Titanium is intrinsically a very reactive base metal but a new surface is immediately passivated by acquiring a tenacious film of rutile, TiO_2. The film formed in air at ambient temperatures reaches a limiting thickness of <4 nm after several weeks and the bond to the metal is so strong that the film remains coherent and adherent despite an adverse volume ratio of oxide. It is thermodynamically stable and self-healing in oxidizing media free from complexing anions and is responsible for the outstanding corrosion resistance of the metal.

12.1.2 Corrosion Resistance

12.1.2.1 Passivity

Titanium exhibits anodic passivation behaviour in acids similar in principle to that of stainless steels described in Chapter 8. It is consistent with the form of the Pourbaix diagrams given in Figure 3.9 in which there is an active domain between a domain of immunity for the metal and domains of stability for potentially passivating oxide species. It does not apply to neutral media, because domains of stability for the oxide species lie immediately above the domain of immunity for the metal. Hence, the metal is active or passive according to whether the potential of the corrosion redox system is above or below the appropriate passivation potential, which depends on alloy composition and the nature, strength and concentration of the acid. Passivation potentials for titanium and its alloys determined from polarization characteristics are within the range, −2.5 to +0.05 V SHE.

The principal value of titanium for corrosion-resistant applications is its outstanding resistance to strongly oxidizing media, especially concentrated oxidizing acids, aqueous chloride media including seawater, moist chlorine and hydrogen sulfide gases, which few other metals can resist. Its weakness is failure to resist concentrated non-oxidizing and reducing acids but there are methods of ameliorating this deficiency, described later.

12.1.2.2 Stress-Related Corrosion

Titanium and its alloys are more resistant than most metals to synergistic actions of stress and corrosion. There is little evidence of failures by SCC or by corrosion fatigue in aqueous media, even seawater. A curious phenomenon is that despite its immunity to aqueous media, titanium and most of its alloys are susceptible to stress-corrosion cracking in anhydrous methanol. Moreover, it is inhibited for all but the most severe conditions by 5 mass % of water in the methanol. This suggests that SCC might be related to local depassivation under stress by the formation of a complex such as an alkoxide:

$$TiO_2 + 4CH_3OH = Ti(OCH_3)_4 + 2H_2O \tag{12.4}$$

which is easily hydrolyzed, thereby explaining the inhibition by water.

Titanium and its alloys resist erosion–corrosion due to the resilience of the surface film under stresses applied by impinging or fast flowing water and the ability of the metal to repair breaches in the film instantly. It is negligible for clean water flowing as fast as 18 m s^{-1}.

12.1.2.3 Titanium in Mixed Metal Systems

The behaviour of titanium in mixed metal systems is significant because its engineering applications may require contact with metallic components of different composition either for economy or to meet design requirements. The standard electrode potential for the reaction:

$$Ti^{2+} + 2e^- = Ti \tag{12.5}$$

is

$$E^{\ominus} = -1.63 \text{ V SHE} \tag{12.6}$$

so that when it is activated, titanium is one of the least noble metals. Passivation raises the nobility so that for practical purposes, titanium is placed with other strongly passivated metals in the compatibility group below the group for noble metals given in Section 4.1.3.4. Thus titanium is almost always the noble partner if coupled with active base metals such as steels or less strongly passive metals, such as aluminium and even relatively noble metals, including some copper alloys, stimulating galvanic attack on them. Coupling with some other strongly passivating or relatively noble metals including stainless steels, aluminium bronzes and some nickel alloys including Monel may be acceptable, depending on the relative metal area and the environment, but the suitability of any application of titanium in a mixed metal system must first be confirmed by experience or by extensive *in situ* tests.

12.1.2.4 Resistance to Natural Environments

12.1.2.4.1 The Atmosphere and Fresh Waters

The corrosion of titanium in fresh waters and in the natural atmosphere at temperatures up to at least 400°C is scarcely detectable, showing that the oxygen potential of the atmosphere is sufficient to maintain passivity, but its use in such benign environments is rarely economically justified except for special duties such as devices exposed to superheated water and steam, and for heat exchangers and condenser tubes where the cost is offset by the advantage of titanium over other metals in resisting erosion–corrosion in turbulent or fast flowing water.

12.1.2.4.2 Seawaters

The excellent resistance of titanium to natural waters extends to both static and flowing seawater, and it remains virtually unattacked, irrespective of water velocity from zero to 20 ms^{-1} and metal temperatures as high as 125°C. It is also virtually immune to crevice corrosion in seawater at ambient and moderately elevated temperatures and highly resistant to pitting, attributes that are unequalled by most other metallic materials. This remarkable combination of corrosion-resistant qualities in some difficult applications that few other metallic materials can survive repays the high capital costs of using titanium by low maintenance. An example is its successful use for heat exchangers in desalination plants.

Although it does not deter marine growth, titanium does not suffer enhanced corrosion in the virtual crevices underneath them and it is unaffected by sulfide pollution generated biologically or delivered by sewage outflows. Therefore, it is not vulnerable to stagnant seawater in urban coastal districts or to associated biofouling on submerged surfaces.

12.1.2.5 Resistance to Acids

12.1.2.5.1 Oxidizing Acids

Titanium resists strongly oxidizing inorganic acids and most organic acids in high concentrations and at high temperatures. It resists nitric acid in all concentrations* and at all temperatures up to the boiling points. Extensive information on the resistance of titanium to other aggressive acidic media is available from specific service experience and field tests accumulated by titanium producers and is available on request in their brochures. Mixtures of non-oxidizing acids with sufficient proportions of nitric acid behave as oxidizing acids. Table 12.1 gives examples of severe conditions in which titanium serves particularly well.

---

* Excluding the so-called. 'fuming nitric acid', a solution of nitrogen dioxide in nitric acid with explosive potential.

TABLE 12.1

Examples of Limiting Conditions for Corrosion Resistance of Titanium in Acids

Acid	Conditions	Temperature °C	Mass %
Oxidizing and Organic Acids			
Nitric	Aerated or not aerated	Boiling	100
Chromic	Aerated or not aerated	Boiling	50
Acetic	Aerated or not aerated	Boiling	100
Reducing Acids			
Hydrochloric	Aerated or not aerated	25	5
	Aerated	60	3
Sulfuric	Not aerated	35	5
Phosphoric	Not aerated	30	35
	Not aerated	>35	<35
Hydrofluoric	Not aerated	>25	<1
Acid Mixtures			
20:1 hydrochloric/nitric	Not aerated	35	100
hydrochloric saturated with Cl_2			
5:1 sulfuric/chromic	Not aerated	Boiling	25

Selective attack can occur in nitric acid on welds in pure titanium containing iron as an impurity or as surface contamination. The iron is distributed in the parent metal as isolated particles of iron-rich β-phase, but selective freezing can develop interconnected galvanic cells within the weld bead. For this reason, the iron content of titanium for use in nitric acid is restricted to <0.05% and parent metal surfaces must be properly cleaned before welding.

12.1.2.5.2 Non-Oxidizing Acids

Titanium can resist only dilute solutions of non-oxidizing inorganic acids especially sulfuric, phosphoric, hydrochloric and hydrofluoric acids because the redox potentials of the corroding systems are below the appropriate passivating potentials. Table 12.1 records some limits to conditions that titanium can tolerate.

The introduction of minority foreign species with high redox potentials into a non-oxidizing acid can passivate the metal by raising the potentials of the corroding systems to values above the appropriate passivation potentials. Equilibria for such species encountered in some process liquors include:

$$Fe^{3+} + e^- = Fe^{2+} \quad \text{for which } E^\ominus = +0.77 \text{ V SHE}$$

$$Cu^{2+} + e^- = Cu^+ \quad \text{for which } E^\ominus = +0.15 \text{ V SHE}$$

$$Cr_2O_7^{2-} + 14H^+ + 6e^- = 2Cr^{3+} + 7H_2O \quad \text{for which } E^\ominus = +1.33 \text{ V SHE}$$

$$HClO + H^+ + e^- = \tfrac{1}{2}Cl_2 + H_2O \quad \text{for which } E^\ominus = +1.63 \text{ V SHE}$$

Hence, if present at sufficient activities in otherwise non-oxidizing acids, the species, Fe^{3+}, Cu^{2+}, $Cr_2O_7^{2-}$ or HClO can induce titanium to passivate. It is improbable that such species would be added solely to secure corrosion resistance, but their incidental presence in some process liquors can protect titanium from significant attack.

Where appropriate positive protection in non-oxidizing acids can be applied by impressing an anodic current to raise the prevailing potential to a value above the passivating potential. An applied potential of 2.5 V is usually sufficient. The procedure produces a visible blue anodic film on protected surfaces. A disadvantage is that protection by impressed anodic currents is not fail-safe because raising the potential but failing to reach the passivation potential at any part of the system accelerates corrosion.

12.2 Constitutions and Corrosion Behaviour of Titanium Alloys

12.2.1 Alloy Formulation

Titanium alloys are formulated both for general purposes and to meet the requirements of specific applications, especially aerospace, marine and medical applications. Table 12.2 gives the compositions and typical mechanical properties of some representative alloys. Several different designation codes are in use and the alloys listed are preferably identified by ASTM designations, but some are known by digits representing compositions or producers' codes.

Alloying elements alter the relative stabilities of the α and β phases. Of the common alloying elements, aluminium stabilizes the hexagonal α-phase and vanadium, tin, molybdenum, iron, chromium, zirconium and niobium all stabilize the β-phase. Pure titanium and alloys in which aluminium is the dominant component are single-phase at ambient temperatures and alloys with substantial contents of the other metals have duplex αβ or β structures. The α alloys are generally more ductile and can be more easily welded than the duplex alloys. Vanadium, zirconium and chromium impart strength at the cost of reduced ductility and molybdenum imparts both strength and creep resistance. Molybdenum and zirconium enhance the limited resistance to corrosion in sulfuric and hydrochloric acids. Thus alloy formulation to secure particular combinations of chemical and physical characteristics yields the multicomponent alloys listed in Table 12.2.

12.2.2 Noble Metal Additions for Improved Resistance to Non-Oxidizing Acids

The resistance of titanium and its alloys to corrosion in non-oxidizing acids can be improved by alloying with very small additions of the noble metals palladium or ruthenium. The principle is to create an internal source of a cathodic collector that stimulates the hydrogen evolution reaction to such a degree that the metal is self-passivating. Table 12.2 lists ASTM grades 7 and 26 which are derivatives of Grade 2 pure titanium with noble metal additions and ASTM grades 25 or 29, 28 and 20 which are, respectively, the derivatives of the alloys, Grades 5, 9 and 19.

The noble metal content is distributed throughout the alloy, for example as the eutectoid fraction Ti_2Pd in the titanium–palladium binary system. On immersion in a non-oxidizing aqueous medium, there is a brief induction period before the improvement in corrosion resistance takes effect. This is interpreted as a period during which a subcutaneous layer of the alloy is dissolved in the static liquid layer at the interface and the noble metal

TABLE 12.2
Representative Titanium Alloys

	Nominal Composition, Mass %									Typical Mechanical Properties[a]				
Designation	**Al**	**Sn**	**V**	**Zr**	**Mo**	**Cr**	**Ni**	**Nb**	**Others**	**0.2% Yield MPa**	**UTS MPa**	**Elongation %**	**Applications[b]**	
ASTM grade 2	–	–	–	–	–	–	–	–	–	300	400	35	Commercially pure Ti	G, S, M
ASTM grade 7	–	–	–	–	–	–	–	–	0.2 Pd	300	400	35	Grade 2 Ti + Pd	S, M, R
ASTM grade 26	–	–	–	–	–	–	–		0.1 Ru	300	400	35	Grade 2 Ti + Ru	S, M, R
ASTM grade 5	6	–	4	–	–	–	–	–	–	1000	1100	15	General purpose αβ alloy	G, A, S, M
ASTM grade 23	6	–	4	–	–	–	–	–	–	1000	1100	15	Grade 5 + low interstitials	S, M
ASTM grade 25	6	–	4	–	–	–	0.5		0.05 Pd	1000	1100	15	Grade 5 with Pd	R
ASTM grade 29	6	–	4	–	–	–	–	–	0.1 Ru	1000	1100	15	Grade 5 with Ru	R
ASTM grade 9	3	–	2.5	–	–	–	–	–	–				Medium strength alloy	A
ASTM grade 28	3	–	2.5	–	–	–	–	–	0.1 Ru	–			Grade 9 with Ru	R
ASTM grade 19	3	–	8	4	4	6	–	–	–	–			**β Alloy**	
ASTM grade 20	3	–	8	4	4	6	–		0.05 Pd	–			Grade 19 with Pd	R
10-2-3	3	–	10	–	–	–	–		2Fe	–			Forging alloy	A
550	4	2	–	–	4	–	–		–	1070	1200	14	Heat treatable	A
6-2-4-6	6	2	–	4	6	–	–	–	–	–			Creep-resistant alloy	A
6-2-4-2	6	2	–	4	2	–	–		–	–			Creep-resistant alloy	A
829	5.5	3.5	–	3	–	–	–	1	–	850	960		Creep-resistant alloy	A
834	5.8	4	–	3.5	–	–	–	0.7	–	850	960	12	Creep-resistant alloy	A

[a] Test pieces 12.5 mm diameter 50 mm gage length to ASTM E8-78.
[b] A: Aerospace; S: Seawater; M: Medical; R: Non–oxidizing acids.

fraction is deposited back on the surface, creating the cathodic collector. This implies that the effectiveness of palladium or ruthenium is favoured by stagnant and slow moving liquors. The corrosion resistance of titanium and its alloys with palladium or ruthenium additions is as good as that of the corresponding basic grades in oxidizing media and is significantly better in non-oxidizing media.

12.3 Applications

Demand for titanium and its alloys is driven mainly by aerospace, marine, industrial and medical requirements.

12.3.1 Aerospace Applications

Applications of titanium and its alloys in aerospace are considered in Chapter 21 within the context of airframe design and engine function in relation to corrosion and corrosion control in aviation. For future reference, the list of representative alloys in Table 12.2 includes some alloys formulated for particular aerospace functions.

12.3.2 Marine Applications

Titanium and its alloys complement and to some extent supersede some traditional roles of other metals for marine applications by virtue of their outstanding resistance to general, local and stress-assisted corrosion in seawater. Discussion of their roles is deferred to Chapter 25, which considers material selection from among steels and alloys of copper, nickel, and titanium for corrosion control in marine environments and associated applications using seawater.

12.3.3 Representative Industrial Applications

Industrial activities comprise such extensive and diverse chemical and engineering operations that a few examples must suffice to represent the variety of applications that exploit the particular characteristics of titanium and its alloys.

12.3.3.1 Chemical and Related Process Plants

Commercial purity titanium is a standard material for processing equipment exposed to severe oxidizing acid media in great variety, including concentrated nitric acid, chromic acid and solutions of free chlorine in hydrochloric acid, all potentially at high temperatures, sometimes with heat transfer. The equipment may comprise storage facilities, reaction vessels, heat exchangers and transfer systems with associated impellers pumps and valves. Unalloyed titanium is usually suitable for tubing, tanks and vessels. The general purpose high strength alloy, ASTM grade 5, with the nominal composition 6% aluminium, 4% vanadium, balance titanium is widely applied for accessories that require harder material, such as pumps, impellers and valves.

Commercially pure titanium, with additions of palladium, ASTM grade 7, or ruthenium, ASTM grade 26, and the corresponding variant of ASTM grade 5 with an addition of palladium, ASTM grade 25, are available as alternatives if improved resistance to non-oxidizing acids is required.

12.3.3.2 Bleaching and Bleaching Agents

Titanium is used extensively in equipment used for bleaching in the textile industries because it is immune to boiling solutions of sodium chlorite and hypochlorite in all concentrations and to chlorine dioxide generated during bleaching. The equipment may comprise vessels and heat exchangers made of commercially pure metal and associated accessories including pumps and valves made of a strong titanium alloy, typically ASTM grade 5.

12.3.3.3 Oilfield Equipment

There are opportunities to exploit the outstanding resistance of titanium and its alloys in developing and operating oilfields. An example is for equipment handling sour fluids associated with drilling for oil and extracting it. These fluids are solutions of hydrogen sulfide under pressure with pH values typically as low as ~3.5 and often also containing chlorides. Low pressure lines and wellhead seals can be fabricated from commercially pure titanium, preferably ASTM grades 7 or 26, which have palladium or ruthenium additions for additional protection in non-oxidizing acidic conditions. Accessories such as pumps, valves and springs are fabricated from the medium strength alloys ASTM grades 25, 28 and 29 and the high strength β alloys, ASTM grades 19 or 20 and 6-2-4-6, which also have enhanced acid resistance.

12.3.3.4 Anodes for Cathodic Protection

In view of its outstanding resistance to oxidizing acidic or neutral aqueous environments, titanium would seem to be an ideal material for use as inert anodes in electrochemical systems, but there is a problem. Impressing an anodic current thickens the surface film in proportion to the corresponding potential drop across it, similar to the effect on aluminium in non-solvent electrolytes. The resulting anodic film interposes a resistance between the metal and the electrolyte, insulating it and preventing it from functioning as an effective anode. A practical solution is to use titanium as a support for a thin coating of platinum of the order 10^{-11} m thick. The original air-formed film between the titanium and platinum coating is an *n*-type semiconductor and is so thin that it does not offer significant resistance. The use of platinum-coated titanium anodes in cathodic protection systems is described in Chapter 20.

12.3.3.5 Equipment for Metal Surface Treatments

Titanium alloys are used in ancillary equipment for metal surface treatments, illustrated by the following examples.

12.3.3.5.1 Descaling Steel and Copper

Mill scale on hot-rolled steel is removed by pickling in sulfuric acid. Static pieces or coils are typically descaled in 0.1 mol kg^{-1} acid at 60–80°C. A moving strip descaled on-line is flexed and passed through 0.25 mol kg^{-1} acid at 95°C to undermine the scale. The detached scale dissolves, contributing Fe^{3+} ions to the acid, which raises the redox potential as explained in Section 12.1.2.5, so that any titanium in the system passivates. Therefore, titanium alloys can be used for racks or rollers to support the steel in its passage through the acid and associated wash tanks and for heating coils. The degree of protection can be enhanced by impressed anodic current. Similar considerations apply to descaling copper in acids, in which the redox potential is raised by Cu^{2+} ions.

12.3.3.5.2 Anodizing Aluminium

Titanium jigs to support aluminium workpieces for anodizing in sulfuric and chromic acids are in widespread use. The applied anodic potential protects the titanium and the oxide film on its surface insulates it from the acid bath, preventing loss of current. An advantage over aluminium jigs is that titanium jigs do not need to be stripped before reloading between batches because contact pressure establishes effective electrical connection to the workpieces.

12.3.3.5.3 Electrodeposition

Titanium is widely used for steam heating and water cooling coils for temperature control of electroplating solutions, notably chromic acid for chromium plating and solutions based on nickel sulfate or nickel chloride for nickel plating. It remains free from corrosion products that could interfere with heat transfer. It is unsuitable for coils used in solutions using nickel sulfamate or nickel fluoborate as the nickel source.

12.3.4 Medical Applications

The properties that recommend titanium and its alloys for medical applications are strength, lightness, formability and resistance to general corrosion, SCC and corrosion fatigue. They are biocompatible because their corrosion resistance in chloride media prevents release of metallic ions that interfere with life processes and they do not stimulate allergic reactions that promote rejection. The annual requirement for devices of all kinds exceeds 1000 tonnes.

12.3.4.1 Titanium Alloy Joint Replacements

More than a million joint replacements are fitted annually to replace natural joints degraded by age or accident. The most common replacement is for hip joints and the ideal requirement is for a load bearing universal joint that can serve without maintenance for 40 years or more. The usual design comprises a polished metal sphere with a stem for insertion into the femur, bearing on the mating surface of an insert fitted to the pelvis and held in place either by polymethyl methacrylate cement or by stimulated intergrowth of the bone. The materials in general use for the femur insert are high strength titanium alloys such as ASTM grade 5 or austenitic stainless steels. The pelvic insert is typically formed from high molecular weight low friction polyethylene. Similar considerations apply to replacements for other joints with suitably modified designs.

The femur insert must survive indefinite exposure under stress to tissue and body fluids which are typically near-neutral dilute solutions of chlorides with organic solutes and particulates. Titanium and its alloys are practically inert to such environments and have particular advantages in their immunity to pitting and crevice corrosion in neutral chloride media, so that bearing surfaces remain smooth.

Cyclic stresses imposed on joint replacements by the pedestrian forces applied at limb extremities are magnified by the lever effect and the number accumulated in a human life span can induce fatigue failure of femur inserts produced from materials with insufficient corrosion fatigue strengths. The absence of environmental effects on the fatigue strength of titanium alloys is an advantage over austenitic stainless steels, which are sensitive to chloride media. The lower modulus of titanium alloys also reduces the stress by elastic relaxation.

12.3.4.2 Other Titanium Implants and Prostheses

The outstanding corrosion resistance and biocompatibility apply to other implanted titanium devices in great variety. Representative applications are plates, screws and pins for temporary or permanent support of bone fractures, bespoke components to repair cranial or facial defects and cases for cardiovascular devices especially pacemakers, defibrillators and heart valve supports. These applications are facilitated by the availability of materials ranging from the ductile pure metal and α alloys for forming into complex shapes to the strong α/β alloys for stressed components. External titanium alloy prostheses depending on corrosion resistance include orthodontic calipers and dental accessories. Titanium alloys are also used for components of artificial limbs, but are more for favourable strength/weight ratios than for particular corrosion-resistant characteristics.

13

Corrosion Resistance of Zinc

13.1 Occurrence and Extraction

13.1.1 Sources

Zinc occurs in widespread deposits of zinc blende, ZnS, associated with iron and lead in sulfide ores, notably sphalerite. The ore is ground and the zinc blende is recovered by froth flotation. Most applications of zinc depend on its electrochemical characteristics which are sensitive to its purity and relate to the most common extraction processes, blast furnace reduction and electrolysis. Detailed analyses of these processes are given elsewhere, for example, by Bodsworth, cited in Further Reading, and the following brief summary is sufficient for the present purpose.

13.1.2 Blast Furnace Reduction

For blast furnace reduction, the zinc blende is roasted to oxide and sintered to agglomerate it. The sinter is charged to the furnace with a flux and coke to provide both reductant and fuel:

$$\mathrm{C(solid) + ZnO(solid) = Zn(vapour) + CO(gas)} \tag{13.1}$$

$$\mathrm{2C(solid) + O_2(gas) = 2CO(gas)} \tag{13.2}$$

The products, zinc and carbon monoxide, are produced as components of the same gaseous phase and the zinc is recovered by absorbing it in liquid lead from which it separates as liquid on cooling. A modified blast furnace process can be used to smelt mixed lead zinc ores, in which the reduced lead separates as a liquid in the hearth.

13.1.3 Electrolytic Extraction

For electrolytic extraction, the sulfide is roasted at a lower temperature to zinc sulfate:

$$\mathrm{ZnS + 2O_2 = ZnSO_4} \tag{13.3}$$

and leached with sulfuric acid. The solution is neutralized with zinc oxide to precipitate iron and antimony, and zinc dust is added to precipitate cadmium and copper. It is filtered and concentrated and the metal is recovered by electrolysis. Despite its high standard electrode potential, zinc can be deposited from aqueous solutions because hydrogen evolution is suppressed by the high hydrogen overpotential on zinc. The cell potential is carefully controlled to avoid codepositing impurities, especially iron, yielding a product of 99.9% zinc.

13.2 Structure and Properties

13.2.1 Structure and Physical Properties

Solid zinc has a close-packed hexagonal lattice with a c/a ratio of 1.856. Its melting point is 419.5°C and its density is 7.14 g cm^{-3}. Typical applications exploit its electrochemical characteristics or low melting point, including protective coatings on steel, sacrificial anodes for cathodic protection and die castings. Its mechanical properties are limited but can be improved by alloying with copper and aluminium. Some representative compositions are given in Table 13.1.

13.2.2 Chemical Properties

13.2.2.1 Electron Configuration

Zinc is the first element following the first transition series, with two 2s electrons outside of the filled 3d shell, and normally exists in the oxidation state Zn(II) in its compounds and ions. It is a reactive metal, as illustrated by the Gibbs free energy for formation of the oxide:

$$\mathrm{Zn} + \tfrac{1}{2}\mathrm{O_2} = \mathrm{ZnO}\ \Delta G^{\ominus} = -351874 - 29T\log T + 44.T\ \mathrm{J} \tag{13.4}$$

and the standard electrode potentials of ions in aqueous media at 25°C:

$$\mathrm{Zn^{2+}/Zn}\quad E^{\ominus} = -0.763\ \mathrm{V\ (SHE)} \tag{13.5}$$

$$\mathrm{ZnO_2^{-}/Zn}\quad E^{\ominus} = -0.441\ \mathrm{V\ (SHE)} \tag{13.6}$$

It is protected by cathodic passivation described as follows.

13.2.2.2 Passivation

The initial oxidation of zinc to ZnO in air at normal temperatures is represented by logarithmic laws, implying that it is driven by electric fields in the Cabrera–Mott mechanism.

TABLE 13.1

Nominal Compositions of Some Representative Zinc Grades and Alloys

	Alloying Elements			Impurities		
Grade or Alloy	**Al%**	**Cu%**	**Mg%**	**Fe%**	**Pb%**	**Zn%**
Commercially pure zinc	–	–	–	0.075	0.35	Balance
Electrolytic zinc	–	–	–	–	–	99.99
Sacrificial anode alloy	0.3–0.5			<0.00	<0.006	Balance
Diecast alloy BS1004A	4.0	0.10	0.04	0.075	–	Balance
Diecast alloy BS1004B	4.0	0.75–1.0	0.04	0.075	–	Balance

TABLE 13.2

Representative Composition of Films Formed on Zinc in an Industrial Atmosphere

Substance	Nominal ZnO	Water	Carbonates	Sulfates	Balance
Mass %	37	21	19	16	7

Source: O. Kubaschewski and B. E. Hopkins, *Oxidation of Metals and Alloys*, London, Butterworths, 1962.

It is the precursor of $Zn(OH)_2$, which is the stable species in the zinc–water system between pH 8 and pH 10.5 as indicated in the Pourbaix diagram, given in Figure 3.5. When zinc is exposed to water or condensation within that range it forms as an adherent film on the metal surface. passivating the metal because it is one of the more effective cathodic inhibitors as explained in Section 3.2.6. In natural environments, the initial ZnO film absorbs both water and carbon dioxide yielding a range of basic zinc carbonates, represented by:

$$2ZnO + H_2O + CO_2 = Zn(OH)_2 \cdot ZnCO_3 \tag{13.7}$$

Zinc carbonates are also effective cathodic inhibitors and their introduction into the surface film broadens the passive range from pH 6 through pH 12.5, The corrosion resistance in aerated water is temperature dependent in the range 20–100°C with a minimum at about 70°C, an effect associated with changes in the nature of the passive film.

Accelerated corrosion in industrial and marine environments can usually be attributed to degradation of the passive films by the incorporation of aggressive species, notably sulfates and chlorides introduced from atmospheres polluted with sulfur dioxide or sea spray. Table 13.2 gives a representative analysis of such films formed in an industrial atmosphere.

The corrosion resistance of zinc is sensitive to the purity of the metal. Lead and iron, inherited from blast furnace smelting, are sparingly soluble in the solid and exist as dispersed minority second phases, lead as the element and iron as an intermetallic compound, $FeZn_7$, providing potential nuclei for local action cells at the metal surface. A further effect of the iron phase is to reduce the hydrogen overpotential, allowing local depolarization by hydrogen evolution. These principles are usually cited in explaining the pitting attack that is sometimes observed.

13.3 Applications

13.3.1 Surface Coatings on Steel

The predominant application of zinc is for protective coatings applied to steel as described in Chapter 6 for general purposes in natural environments at ambient temperatures. Well-known examples include systems delivering water, corrugated roofing sheet, fencing, barbed wire and agricultural equipment.

13.3.2 Sacrificial Anodes

The characteristics and properties of zinc commend it for use as sacrificial anodes to protect steel structures in cathodic protection schemes. It is sufficiently electronegative to maintain the potential of a steel structure below the corrosion potential that would exist otherwise. It is non-polarizable and its passive state can protect it from self-corrosion while permitting sufficient anodic activity to supply adequate current to the system. These features yield 85% to 95% efficient use of the metal which is significantly more than for the competing alternative metals magnesium and aluminium. Together with the moderate material cost, this commends it for economic use in ships and for underground utilities in soil of low to medium resistance. Zinc is especially preferred for sacrificial anodes installed in ships' cargo/ballast tanks because it does not present a spark hazard as magnesium does. Small quantities of certain impurities, especially iron, seriously impair the function of zinc as sacrificial anodes, usually attributed to reduction of the hydrogen overpotential by iron, stimulating self-corrosion as explained earlier. The problem is averted by using either high purity electrolytic zinc or less pure zinc alloyed with aluminium as a scavenger for iron. The anodes are cast and are produced in various shapes and sizes between 2 and 100 kg to suit particular installations.

13.3.3 Batteries Based on the Leclanché Cell

The generation of electrical energy by controlled corrosion of zinc in the Leclanché cell:

$$Zn|NH_4Cl|, ZnCl_2|\ MnO_2, \text{Carbon} \tag{13.8}$$

is implemented as a zinc anode set in an electrolyte comprising a paste of ammonium and zinc chlorides containing manganese dioxide packed around carbon as a cathode. The normally passive zinc delivers an anodic current when it is electrically connected by an external circuit to the cathode depolarized by the manganese dioxide.

13.3.4 Diecastings

Zinc alloys are well-suited for mass-produced pressure die castings by virtue of their low temperature melting ranges together with reasonable mechanical properties and clean passive surfaces that provide suitable substrates for decorative nickel/chromium plating. A familiar example is for car door handles.

13.4 Corrosion Characteristics of Zinc

13.4.1 Corrosion in Aqueous Solution

Zinc exhibits good corrosion resistance in natural waters, especially hard waters, that is solutions with a positive Langelier index, as defined in Section 30.5.2.2. The domain of stability is given by the zinc–water Pourbaix diagram, Figure 3.5, as mentioned in Section 3.2.2.1. In quiescent distilled water, pure zinc can occasionally exhibit pitting but the exact nature of this is unknown. In soft water, zinc can be vulnerable to corrosion but its resistance to rainwater is borne out by the fact that it has been successfully used as roofing sheet. However, the most common use of zinc as a building material is as galvanized sheet. This is described in more detail in Section 24.6.1.

In seawater, the corrosiveness of chloride ions is inhibited by magnesium salts and the water hardness. It appears that the natural passive film of basic zinc carbonate as described in Section 13.2.2.2 is reinforced by magnesium and calcium salts. In other chloride solutions, for example, sodium chloride solutions, pitting is observed. These are believed to start by depassivation to form basic zinc chlorides at vulnerable points.

In reinforced concrete structures galvanized steel passivates. This is discussed in Section 24.2.1.4 with respect to zinc coated concrete reinforcement. From the zinc–water Pourbaix diagram it would be expected that in highly alkaline media such as in concrete, zinc should depassivate as soluble zincate. In practice, the reinforcement coating is protected by a film of calcium zincate. There is concern that this layer is vulnerable to chloride ingress from external sources from the concrete, especially from deicing salts. It is thought that this might be due to the extreme solubility of zinc chloride which is soluble at concentrations in excess of 25 mol dm^{-3}. In light of this, it is remarkable that zinc is so durable in natural waters such as domestic supply. The effect of chloride on zinc galvanized reinforcement is not yet wholly understood.

13.4.2 Corrosion in Normal Atmospheres

Zinc and zinc coated steels are ordinarily resilient to corrosion in air. Zinc sheeting can last in unpolluted air for 40 or 50 years. The distinctive feature of corrosion of zinc is *white rust*. This appears in the form of white blemishes on zinc surfaces and is characteristic of condensation on newly fabricated surfaces. A mature or aged zinc surface does not easily exhibit white rust and this is attributed to an underlying resilient coat of zinc oxide at the metal surface.

When a fresh surface is exposed to droplets of water a ring of basic zinc carbonate forms around the droplet. The action is due to differential aeration between the center of the drop and the periphery. The initial corrosion product is a porous oxide–hydroxide, but it absorbs carbon dioxide from the environment to produce the final product. White rust is usually a surface phenomenon and is normally of cosmetic appearance.

White rust is often a symptom of poor storage conditions, for example, items packed in plastic packaging in humid conditions. For this reason care should be taken that after manufacture items made of zinc are stored in a dry and airy place.

13.5 Cadmium

This chapter is devoted to the corrosion of zinc. The metal cadmium is in the same group of the Periodic Table as zinc and is used as a thin applied layer in corrosion control to address the compatibility of mixed metal systems, particularly of aluminium with high strength steels in the aviation industry. Because of its toxicity cadmium is being phased out of other nonessential uses. From its standard electrode potential cadmium should be cathodic to steels, but is in fact anodic to the metal. The element is considerably more basic than zinc and according to the cadmium–water Pourbaix diagram the metal should not be protected. This is the subject of Problem 3 in Chapter 3 of this book. In fact the metal is protected by cadmium carbonate, which in industrial or marine environments may also include sulfate or chloride ions.

One final note is important to mention when using cadmium for corrosion control. As was mentioned previously, cadmium plating is used to solve the problem of galvanic corrosion in mixed metal systems. In this use it should not be used with titanium as it can cause *cadmium embrittlement*. If a layer of cadmium is applied to a titanium alloy, the cadmium slowly works its way along the grain boundaries of the titanium substrate. This may take several years but results in weakening of the surface.

Further Reading

Bodsworth, C., *The Extraction and Refining of Metals*, CRC Press, Boca Raton, Florida, 1994.

14

Corrosion Resistance of Magnesium and Its Alloys

14.1 Physical Properties

Magnesium has a close-packed hexagonal lattice elongated along the six-fold axis with a *c*/*a* ratio of 1.856. Its melting point is 649°C and its density is 1.74 g cm^{-3}. It has specialized applications in engineering by virtue of its useful properties and characteristics. It is the lightest of the common engineering metals, and when applied with discrimination it resists corrosion well. Its mechanical properties are useful but are usually improved by alloying. Other attributes that attract special applications include easy machinability, stiffness, high strength-to-weight ratio and low neutron capture cross section. The compositions and properties of some representative alloys are given in Table 14.1.

14.1.1 Occurrence and Extraction

Sources of magnesium suitable for extraction are dolomite, $CaCO_3 \cdot MgCO_3$, and $MgCl_2$ obtained from seawater by ion exchange. Most metal is extracted by the Dow process in which it is recovered by electrolysis of a molten halide with a typical composition of 25% $MgCl_2$, 15% $CaCl_2$ and 60% NaCl contained in a steel vessel and operated at a convenient temperature, typically 1000°C. Other processes include the Pidgeon process in which magnesium is distilled as a vapour by ferrosilicon reduction of calcined dolomite at 1200°C under reduced pressure and condensed outside the reaction vessel:

$$CaCO_3 \cdot MgCO_3 = CaO \cdot MgO + 2CO_2 \tag{14.1}$$

$$2(CaO \cdot MgO) + 2Si = 2Mg(\text{vapour}) + 2CaO \cdot SiO_2 \tag{14.2}$$

Thus alternative sources of magnesium are available, electrolytic and thermally reduced metal. This distinction is important because of the influence of impurities on the corrosion resistance.

14.2 Chemical Properties

Activated magnesium is inherently the most reactive of the common engineering metals, as illustrated by the Gibbs free energy of formation for the oxide:

$$Mg(\text{solid}) + O_2(\text{gas}) = MgO(\text{solid}) \quad \Delta G^{\ominus} = -603960 - 12.34T\log T + 1424T \text{ J} \tag{14.3}$$

TABLE 14.1

Nominal Compositions of Some Representative Magnesium Grades and Alloys

	Composition, Mass %							
Grade or Alloy	**Mg**	**Al**	**Zn**	**Mn**	**Cu**	**Fe**	**Pb**	**Be**
Electrolytic magnesium	99.95 Grad					<0.016	<0.01	0.005
Distilled magnesium	99.5					<0.016		
AZ31	Balance	3.0	1.0	0.3		<0.016		
AZ61	Balance	6.0	1.0	0.3		<0.016		
AZ80	Balance	8.0	0.5	0.3		<0.016		
Standard anode alloy	Balance	6.0	3.0	0.05	>0.15	<0.003		
Proprietary anode alloy	Balance	>0.01	3.0	0.02	0.5–1.3	<0.003		
Magnox MN70	Balance			0.75				
Magnox AL80	Balance	0.75						0.005

where T is the temperature in Kelvin. The standard electrode potential for the ion Mg^{2+} in aqueous media at 25°C:

$$Mg^{2+}/Mg \quad E^{\ominus} = -2.37 \text{ V (SHE)} \tag{14.4}$$

The electron configuration of the free magnesium atom has two *s* electrons outside of the filled penultimate shell so that it normally exists in the oxidation state Mg(II). Mg^{2+} and Zn^{2+} ions have the same charge and similar radii, 0.065 and 0.069 nm, respectively, so that many of their salts are isomorphous, including carbonates with sparingly soluble calcite structures. Their oxides absorb water to give sparingly soluble hexagonal hydroxides. From these characteristics, the chemistry of magnesium is expected to resemble that of zinc in features that influence its corrosion behaviour.

Magnesium oxide, MgO, is one of the most ionic oxides. From Table 2.9 it can be seen that magnesium has 74% ionic character. It has a simple cubic structure (Table 2.10) and as a result of its ionic character defects are restricted to Shottkey disorder, that is, pairs of vacant cation and anion sites. For these reasons as discussed in Section 2.3.4, magnesium oxide is an excellent insulator and possesses a large band gap.

14.3 Corrosion Resistance

14.3.1 Corrosion Resistance in Aqueous Solution

Although it is one of the most electropositive elements, magnesium can resist corrosion in the natural terrestrial atmosphere and fresh waters because it is passivated by adherent surface films of the magnesium hydroxide phase *brucite*, $Mg(OH)_2$, and magnesium carbonate, $MgCO_3$, formed from the initial air-formed film of magnesium oxide, MgO, on the metal surface. The structure of brucite is described in Section 2.3.3.5. In hard water, the passivity can be reinforced by $CaCO_3$, introduced from calcium bicarbonate in supersaturated solution. Magnesium carbonate is a good agent for nucleating this from solution; as mentioned in Section 30.5.2.3, magnesium promotes the formation of a kinetically favoured scale called low magnesium dolomite which contains 4% magnesium. The corrosion resistance is

TABLE 14.2

Phase Relationships for Some Binary Magnesium-Rich Alloys

System	Maximum Solid Solubility Mass % alloy	Eutectic Composition Mass % alloy	Eutectic Temperature °C	Eutectic Components
Magnesium–aluminium	13	32	437	$Mg^a + MgAl$
Magnesium–zinc	8.5	53	342	$Mg^a + MgZn$
Magnesium–manganese	3.4	Peritectic	652	
Magnesium–iron	0.001	0.006	640	$Mg^a + Fe$
Magnesium–copper	None	30.7	485	$Mg^a + Mg_2Cu$
Magnesium–nickel	None	13	507	$Mg^a + MgNi_2$

[a] Containing maximum solubility of second element in solid solution.

diminished if the passive film is degraded by sulfates or chlorides introduced from sulfur dioxide or saline condensates in polluted industrial environments. Chlorides are the most disruptive and unprotected magnesium artifacts and are not recommended in marine environments.

The corrosion of pure magnesium is sensitive to impurities, particularly iron, copper and nickel, which have vanishingly small solubilities in solid magnesium and appear in the structure as cathodic intermetallic phases some of which are listed in Table 14.2. The most troublesome is iron, inherited from extraction processes for which the tolerance in practice is 0.016 mass %.

In most metals, such impurities segregated at the grain boundaries are *anodic* to the matrix and can establish networks of interconnected selective corrosion paths around boundaries between grains referred to as intercrystalline corrosion. In contrast, magnesium alloys are so strongly electropositive that grain boundary impurities are *cathodic* to the matrix, stimulating selective corrosion of the adjacent grains while the boundaries remain intact. For the same reason, it is self-evident that care is required to avoid galvanic couples in designing mixed metal systems that include magnesium alloys. The design implications discussed in Section 4.1.3 imply that it is unwise to consider allowing magnesium to contact any other metal so that complete insulation is the only safe option.

14.3.2 Corrosion Resistance in Dry Oxidation

At elevated temperature, magnesium exhibits a slow but significant linear oxidation rate. This is indicative of an effective constant oxide thickness. As mentioned previously, magnesium oxide has very good insulation properties and is protective of the metal. However, the oxide occupies a smaller molar volume than the equivalent metal. The ratio between them is known as the *Pilling–Bedworth ratio*. This is defined as

$$R_{PB} = \frac{M_O}{D_O} \cdot \frac{D_M}{M_M} \tag{14.5}$$

where M_O and M_M are the molar masses of the oxide and metal, respectively, and D_O and D_M are the molar densities of the oxide and metal. For magnesium the Pilling–Bedworth ratio is 0.81 which means that the oxide is in tension. Thus at elevated temperature beyond a characteristic thickness it is prone to microcracking and this is responsible for the linear oxidation rate.

14.4 Alloy Formulation

Alloy specifications are devised to improve the mechanical properties and casting characteristics. Magnesium is anodic to all potential alloy components and specifications are constrained by the need to minimize quantities of minority second phases in the structure that can act as cathodic collectors for local action cells. This restricts the compositions of most alloys to low contents of components selected from among aluminium, zinc and manganese, occasionally with zirconium as a grain refiner. Manganese and zinc have a further function in moderating the deleterious effects of iron, probably by converting it into some other form. For the same reason, specifications prescribe strict limits on impurities. Table 14.1 gives compositions of some representative alloys.

14.5 Canning for Nuclear Reactor Fuel

A former important application of magnesium was as cladding material for uranium rods in early designs of Magnox gas cooled nuclear reactors, by virtue of its transparency to neutrons and fairly high melting point combined with adequate corrosion resistance to the gas coolant. Two representative alloys are given in Table 14.1. The role of metals in the evolution of reactor designs is considered in Chapter 27.

14.6 Stress-Corrosion Cracking

Magnesium alloys are less widely applied for critical applications than most other common engineering metals so that the incidence of SCC is correspondingly less. Failures are mainly confined to wrought metal and seldom afflict castings. The source of the stress is more often unrelieved internal stresses from fabrication or welding than from external loading in service. The specific agents are aqueous solutions of chloride and chromate ions and, surprisingly, moist air.

14.6.1 Susceptible Alloys

Most recorded failures by SCC are for alloys in the Mg/Al/Zn system, but some are occasionally reported for alloys in the Mg/Mn and Mg/Zn/Zr systems. The susceptibility to cracking increases with increasing aluminium contents and is often attributed to the aluminium-bearing intermetallic constituents present.

14.6.2 Mechanisms

There is a fairly substantial literature describing experiments designed to identify mechanisms of SCC in magnesium alloys. They encompass alloy compositions, environments, crack paths and grain sizes, but they are generally empirical and uncoordinated. The crack path is variable. In one example, the crack path in AZ61 sheet could be varied between intergranular and trangranular by varying the pH of a $NaCl/K_2CrO_4$ solution to which it was exposed. In another, the crack path in the same alloy could be varied by varying the rate of cooling after a heat treatment. It is uncertain how cracks advance. Varying values of potential and acoustic emissions suggesting that it is discontinuous have been reported on some but not all occasions. It is possible to explain some of these anomalies in terms of structural features.

Most magnesium-rich alloy systems are characterized by limited solid solubilities of solute metals terminated by eutectics in which the components are the saturated solution and intermetallic compounds. Critical features of some influential binary systems are summarized in Table 14.2. The basic concept is that susceptibility to SCC and the nature of the crack path are related to the nature and disposition of intermetallic phases. For reasons given in Section 14.3.1, the matrix adjacent to intermetallic phases is vulnerable to selective corrosion. If this is assisted by an applied stress in crack-opening mode, it could stimulate the advance of a crack tip. On this assumption, the distribution of the phases could define the crack path *viz* intergranular or transgranular. It could also explain effects of composition, of grain size and of various heat treatments that have been reported. There is scope for coordinated investigations of relations between SCC and structure.

The usual precautions to minimize risks of stress-corrosion apply, avoiding exposure to specific agents as far as possible and applying appropriate thermal treatments to relieve stress imparted by fabrication and welding.

14.7 Magnesium Sacrificial Anodes

The high value for the standard electrode potential Mg^{2+}/Mg, given in Equation 14.4, indicates the application of magnesium for galvanic corrosion protection of steel. It is impractical to apply it as an electrodeposited coating because of the low value of the hydrogen overpotential on magnesium, but it is the most widely applied of available sacrificial anodes because it has the highest current yield. This attribute is offset by the disadvantage that it is subject to anode wastage by self-corrosion. A standard anode specification is given in Table 14.1. Minor impurities, especially iron, nickel and copper accelerate self-corrosion for the reasons given earlier and must be rigorously excluded. As it is, the average anode efficiency is only about 50%. A further limitation is that its effective use is restricted to soils with resistivities of less than 4000 Ω cm^{-1}. An alternative specification, given in Table 14.1, without aluminium and with increased copper content is available for service in soils with resistivities up to 6000 Ω cm^{-1}, at the cost of reduced efficiency.

14.8 Protection of Magnesium by Coatings

As was discussed previously, magnesium is the most electropositive of all the engineering metals. This means that it is prone to stimulation by stray residues of almost any other metal, all of which can act as a local cathode. For this reason, when preparing a magnesium surface for treatment, great care is needed to make sure that contaminating residues are not left on the surface. Glass papers, being nonconducting, are a good abrasion medium for cleaning, but even this can reduce the natural passivity on the metal prior to coating.

TABLE 14.3

Solution Compositions for Fluoride Anodizing of Magnesium[a]

Parameter	HAE Process		Dow 17 Process	
Fluoride	NaF	35	NH_4HF_2	360
Acid/Alkaline Reagent	KOH	120	H_3PO_4	96
Oxidant	$KMnO_4$	20	$Na_2CrO_2F_7$	100
Additive	Al_2O_3	34		
Additive	Na_3PO_4	35		
Temperature, °C	21		70	
pH	14		5	

[a] All additives given in units of g dm^{-3}.

14.8.1 Paint Systems

Magnesium artifacts can be painted, if the surfaces are chromated as described in Section 6.4.3.3, to provide a key and to remove alkaline air-formed films on the substrate that would degrade paint. The chromated surface acts as a key to the paint. Acceptable paints for magnesium are epoxy resins and stove enamels.

14.8.2 Anodizing

Magnesium can be anodized electrolytically, which is a feature it shares with aluminium and aluminium alloys. Anodizing of aluminium is described in Section 6.4.2. However, for magnesium the process is conducted in fluoride media and the product is a dense coating of magnesium fluoride, MgF_2. Magnesium fluoride has a low solubility of 1.2×10^{-6} mol dm^{-3} and unlike magnesium oxide, shows better resistance to mildly acidic media. The two main proprietary methods for anodizing of magnesium are the HAE process and the Dow 17 process. There are environmental concerns for both processes because they use oxidizing metal salts of chromium and manganese. Typical anodizing conditions are given in Table 14.3. There has been much recent effort in formulating more environmentally friendly fluoride anodizing solutions, but the finish of the HAE and Dow 17 processes is superior. For example, it can function as the substrate for electroless nickel plating of magnesium.

15

Corrosion Resistance of Tin and Tin Alloys

Tin is expensive, as indicated by the representative prices quoted in Table 1.1, but it is in demand because it has important applications by virtue of its excellent resistance to corrosion, nontoxicity, ease of electrodeposition, low melting point and ability to wet other metals. There are three allotropes, with equilibrium transformation temperatures given in Equation 15.1:

$$\alpha\text{-tin} \xleftrightarrow{13.2^\circ C} \beta\text{-tin} \xleftrightarrow{161^\circ C} \gamma\text{-tin} \xleftrightarrow{232^\circ C} \text{liquid} \tag{15.1}$$

The α-phase is nonmetallic and the term 'tin' usually refers only to the β and γ phases which crystallize in tetragonal metallic structures. β-Tin is kinetically stable at temperatures well below its lower transformation temperature. It is soft with an ultimate tensile strength of only 15 MPa at 20°C and does not work harden because it recrystallizes at ambient temperature. The density is 7.3 g cm^{-3} at 20°C.

15.1 Occurrence Extraction and Refining

Most commercial sources of tin are placer deposits of cassiterite, SnO_2, unevenly distributed in the Far East and Africa and deep mined deposits in Bolivia. Physical methods yield concentrates up to 50% from alluvial gravels but deep mined deposits may need chemical or magnetic separation to remove iron, copper zinc or tungsten. Tin concentrates are smelted with coal in reverberatory or electric furnaces by a two-stage operation described in standard texts, for example, by Bodsworth,* designed to recover a maximum yield of metal that can be refined by liquation to a sufficient purity for many commercial purposes. It is further purified by chemical or electrolysis processes as required, yielding the grades and alloys given in Table 15.1.

15.2 Chemical Characteristics and Corrosion Resistance

Tin is in Group IV of the Periodic Table with four electrons in its valence shell and can yield both divalent and tetravalent oxides and ions, with preference for the tetravalent species

$$SnO_2 + O_2 = Sn \quad \Delta G^\ominus = -586513 + 215.6T \text{ J mol}^{-1} \tag{15.2}$$

* Colin Bodsworth, *The Extraction and Refining of Metals*, Boca Raton, FL, CRC Press, 1994.

TABLE 15.1
Composition and Uses of Tin and Tin Alloys

	Composition, Mass %					
	Sn	Cu	Sb	Ag	Pb	Application
Grade A tin	99.85	–	–	–	–	Tin sheet
Soft solder	62	–	–	–	38	Joints
Lead-free solder	96	0.5	–	3.5	–	Electrical
Babbit metal	89	3.5	7.5	–	–	Bearings
Britannia metal	92	1.0–2.0	6.0–7.5	–	–	Base metal for plating
English pewter	91	1.5	7.5	–	–	Decorative wear
Fine pewter	99	1.0	–	–	–	Decorative wear

$$Sn^{4+} + 2e^- = Sn^{2+} \quad E^\ominus = +0.15 \text{ V (SHE)} \tag{15.3}$$

$$Sn^{2+} + 2e^- = Sn \quad E^\ominus = -0.136 \text{ V (SHE)} \tag{15.4}$$

15.2.1 Corrosion Resistance in Natural Environments

Tin is kinetically stable in unpolluted air because the freshly exposed metal surface immediately acquires a tenacious passive film based on a derivative of SnO_2. Tin dioxide, SnO_2, has the rutile structure given in Section 2.3.3.3. This is an ordered close-packed structure and tin is surprisingly electropositive; Table 2.9 shows that the polarity of a single Sn–O bond is 47% ionic, the same as for copper and not much less ionic than the oxides of other metals such as nickel and iron.

In humid air, the film absorbs water and is best described as $SnO_2 \cdot nH_2O$, a hydrated form of SnO_2, since there is no evidence of discrete hydroxyl ions in the tin–water system. The Pourbaix diagram for the tin–water system at 25°C reproduced in Figure 3.7 correctly predicts corrosion resistance in mild aqueous media, because although the domains of stability for tin and water do not overlap in the pH range 0–14, corrosion currents are limited by the passive surface, which restricts the access of dissolved oxygen:

$$\tfrac{1}{2}O_2 + 2H^+ + 2e^- = H_2O \tag{15.5}$$

and the high hydrogen overpotential quoted in Table 3.5, which prevents hydrogen evolution:

$$2H^+ + 2e^- = H_2 \tag{15.6}$$

There is a second oxide phase, SnO. This has more than one form and some of the structures are complicated, for example, by the presence of lone pairs within a crystal. The Sn(II) oxidation state is generally air sensitive so passivation of tin is often regarded as being due to the rutile structure of the dioxide.

15.2.2 Applications

15.2.2.1 Tin Coatings on Steel for Food Cans

Tin is applied as a corrosion-resistant coating to the inner surfaces of steel food cans, taking advantage of the polarity reversal of iron and tin in deaerated aqueous solutions of organic acids such as fruit juices that form complex ions with tin, as explained in Chapter 4. The preservative function of cans has been overtaken by their convenience as food and beverage dispensers and is now responsible for the greatest area of food/metal contact. The cans are fabricated from sheet steel stock to which the tin coating is applied as it emerges from the rolling mill. It is considered in detail in Chapter 23.

15.2.2.2 Soft Solders

Soft solders are based on the tin–lead eutectic, 61.9% tin – 38.1% lead, melting at 183°C. The simple eutectic is suitable for many *ad hoc* general purposes, provided that a suitable flux is used to dissolve oxides formed in the operation, allowing the solder to wet the substrate. Tin-based solders to suit a wide range of purposes and substrates are available. There are potential corrosion hazards to avoid in designing and creating soldered joints. A joint constitutes a bimetallic system in which the joint is of opposite polarity to the much larger area of substrate. As examples, solders are cathodic to steel, cadmium, zinc and aluminium, but anodic to copper. This suggests that soldered joints in copper plumbing may be vulnerable to galvanic attack in aqueous media, but little trouble is experienced in copper cold water supply systems. This apparent immunity to galvanic attack does not apply to hot water, so perhaps it is due to protection by limescale deposited from cold hard water, as explained in Section 2.2.9.2. Nevertheless, it would seem to be good practice to tin copper in the vicinity of soldered joints to spread the risk. Another hazard is carelessness in operation by not eliminating residual fluxes, which activate the metal surface.

15.2.2.3 Lead-Free Solders

The restrictions on the use of lead based materials due to health and environmental concerns have been responsible for the increased use of lead-free solder, especially in the electronics industry. The main alloy is a tin–silver–copper alloy comprising approximately 96% tin, 3.5% silver and 0.5% copper, named SAC from the elements **Sn**–**Ag**–**Cu**. As it is mainly a tin alloy, the corrosion properties are very similar to that of lead bearing solder that it replaces. The main difference is in the temperature at which components can be soldered.

15.2.3 Bearings

Tin is the matrix of white metal inlays in bearings. The degradation of inlays by corrosion is influenced by engineering parameters including lubrication, wear and compliance and is best approached in specialist mechanical engineering texts.

15.2.4 Pure Tin and Pewter

Pure tin is chemically remarkably stable and inert with respect to moisture. Before the development of ion exchange resins the only means of producing the highest purity water was by distillation using a still entirely made of tin. This 'equilibrium water' was used to determine the ionic product of water, K_w. This is a fundamental property of water and determines pH neutrality. Autodissociation of water is discussed in Section 2.2.4 and values for K_w are given in Table 2.5. To measure K_w, water was first distilled in glass and then repeatedly redistilled in a tin still until the conductivity of the water reached a minimum. From the molar conductances of hydrogen and hydroxide ions (Table 2.6) the ion product of water K_w could be determined. It should be self-evident from this that the passive film on tin has the least effect on water of all the passive metals.

Pewter is essentially tin alloyed with antimony or copper to harden it. It is now mainly used for decorative items but in the past an alloy called fine pewter, consisting of 99% tin and 1% copper by mass, was used for plates and cups. It has excellent corrosion resistance but is expensive. The same resistance is given by hot-dipped tin coatings on other metals, but pewter items can be engraved because obviously there is no coating to breach.

An interesting historical corrosion issue concerns tin. It is sometimes stated that the first major case of corrosion was the rusting of iron nails securing copper sheeting to Royal Naval vessels in the eighteenth century. In fact, electrochemical degradation of tin was observed nearly a century before. It is not common knowledge that the English Royal Mint issued tin coinage from the reign of Charles II to William III. Before the reign of Charles II, no coins had been issued by the Royal Mint except gold and silver ones. The standard denomination was the silver penny which had been issued since the time of King Offa in the eighth century. Beginning in 1675, copper halfpennies and farthings were issued. However, in an age where the value of the coin was reflected in the metal content, it was felt that copper did not have enough value. Therefore, halfpennies and farthings were issued for several years made of tin, as this metal had a higher intrinsic value. It became quickly apparent though, that low melting forgeries could be produced and, therefore, tin was not an ideal coinage material.

To counteract this halfpennies and farthings were then issued which consisted of a tin coin with a copper plug at the center. It was found that for these coins the tin degraded. This was due to a bimetallic effect between the copper, which was cathodic and the tin, which was anodic. The electrolyte was from human handling of these coins. Tin coins from this time are extremely rare. After Sir Isaac Newton was appointed Warden of the Mint in 1696 only copper halfpennies and farthings were issued. He conducted research into the durability of coinage materials and this possibly represents the earliest scientific investigation of an economically important corrosion problem.

16

Corrosion Resistance of Lead

Lead has been known and used from antiquity because of the minimal technology needed for small scale extraction of the metal, its low melting point and ease of fabrication and the easy identification of some lead minerals. The survival of ancient lead artifacts in good condition testifies to its resistance to corrosion in natural environments. As the toxicity of the metal became acknowledged and alternative materials and technologies became available, some earlier applications have been scaled down or discontinued including, domestic plumbing, solders for food cans, some architectural uses, typesetting and the lead chamber process for sulfuric acid manufacture. Most current applications are based on its density, 11.34 g cm^{-3}, its low melting point, 327.5°C, its electrochemical properties and its resistance to several mineral acids.

16.1 Occurrence and Extraction

The principal sources of lead are galena (PbS), cerussite ($PbCO_3$) and anglesite ($PbSO_4$), associated with small quantities of zinc, copper, tin, arsenic, bismuth and antimony. After comminution and concentration, the mineral is mixed with fluxes to form a slag and sintered to agglomerate the particles and convert it to the oxide, PbO_2. The oxide is smelted in a blast furnace designed to reduce lead but to receive iron and zinc impurities in the slag which is discarded. The metal is transferred to a reverberatory furnace where impurities derived from the ore are removed by a sequence of operations, typically selective solidification, selective oxidation and recovery of silver by partition with zinc, yielding 99.99% lead. Detailed information is given by Bodsworth*.

16.2 Chemical Characteristics and Corrosion Behaviour

16.2.1 Chemical Characteristics

Metallic lead exists only in a face-centered cubic structure, with an atomic radius of 0.175 nm based on the closest distance of approach. The electron configuration is [Xe]$4f^{14}$ $5d^{10}$ $6s^2$ $6p^2$ so that there are four valency electrons outside the penultimate filled shell as in its congeners in Group IVB of the Periodic Table, but for lead the divalent state is dominant. The Pb(II) ion has a lower charge density than the Pb(IV) ion and because of this its salts are much less susceptible to hydrolysis. Lead forms the oxides, PbO (*p*-type), PbO_2 (*n*-type) and Pb_3O_4 (mixed) which are all insoluble in water. It is doubtful whether true hydroxides

* Colin Bodsworth, *The Extraction and Refining of Metals*, Boca Raton, FL, CRC Press, 1994.

TABLE 16.1

Solubilities of Lead Oxides and Some Salts in Water at 25°C

Substance	Solubility mol kg^{-1}
Lead monoxide, PbO	Insoluble
Lead dioxide PbO_2	Insoluble
Lead phosphate $Pb_3(PO_4)_2$	Insoluble
Lead chromate $PbCrO_4$	5×10^{-7}
Lead carbonate $PbCO_3$	2.7×10^{-6}
Lead oxalate	2.2×10^{-5}
Lead sulfate $PbSO_4$	1.2×10^{-4}
Lead fluoride PbF_2	1.9×10^{-3}
Lead chloride, $PbCl_2$	3.6×10^{-2}
Lead acetate $Pb(CH_3COO)_2$	1.36
Lead nitrate $Pb(NO_3)_2$	1.64

exist and species quoted as '$Pb(OH)_2$' are more likely to be hydrous oxides $PbO \cdot nH_2O$. With the exceptions of the nitrate [$Pb(NO_3)_2$] and acetate [$Pb(CH_3COO)_2$], most lead salts are either sparingly soluble, such as the chloride [$PbCl_2$] and fluoride [PbF_2] or insoluble, like the phosphate [$Pb_3(PO_4)_2$], chromate [$PbCrO_4$] carbonate [$PbCO_3$], oxalate [PbC_2O_4] and sulfate [$PbSO_4$], given in Table 16.1. These solubilities are determined by the lattice energies and hydration energies of the ions, as considered elsewhere, for example, by Cotton and Wilkinson* and Bockris and Reddy†.

16.2.2 Corrosion Resistance

16.2.2.1 Corrosion Resistance in Air and Water

By the usual convention that $a_{Pb^{2+}} < 10^{-6}$, the Pourbaix diagram for the lead water system, given in Figure 16.1 predicts that lead is vulnerable to corrosion for pH < 9. This may be an academic concept for a new lead surface exposed to *pure* water, but it takes no account of other species in the environment or of the insolubility of lead oxides and lead carbonate. In the natural world, lead resists corrosion within a wider pH range because reaction of the surface with oxygen and carbon dioxide creates and maintains an insoluble oxide/carbonate passive film. A further factor is that the solubilities of lead sulfate and lead chloride are so low that the passive film is resistant to sulfur pollution in industrial and urban environments and generally protective in marine environments, including seawater.

Lead is so dense that to take advantage of its corrosion resistance combined with easy formability in plumbing and architecture, it must be supported to prevent it deforming under its own weight.

16.2.2.2 Corrosion Resistance in Acids

A dominant factor in selecting lead for corrosion-resistant service is its outstanding resistance to mineral acids, especially sulfuric, phosphoric, chromic, oxalic and hydrochloric acids due to the virtual insolubility of lead salts of these acids, described in Section 16.2.1.

* F. A. Cotton and G. Wilkinson, *Advanced Inorganic Chemistry*, 2nd ed., Longmans, 1966.

† J. O'M. Bockris and A.K.N. Reddy, *Modern Electrochemistry*, Plenum Press, New York, 1970.

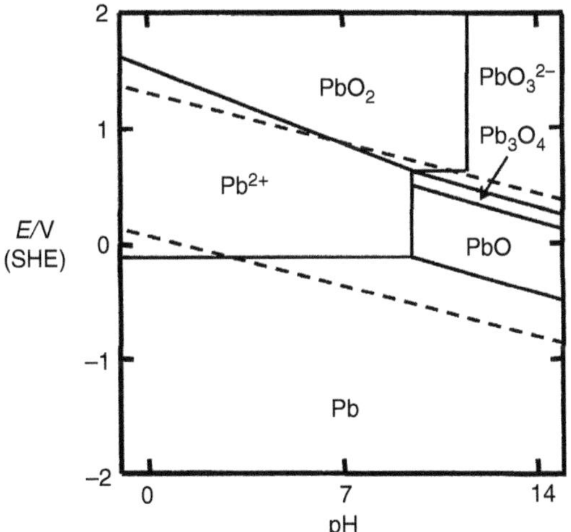

FIGURE 16.1
Pourbaix diagram for the lead–water system at 25°C, $a_{Pb^{2+}} = 10^{-6}$. Domain for the stability of water shown by dotted lines.

As an example, compare the Pourbaix diagram for the lead–water–sulfuric acid system, given in Figure 16.2 with the corresponding diagram for the lead–water system, given earlier in Figure 16.1. The domain assigned to the soluble Pb^{2+} ion in the lead–water system is expanded in the lead–water–sulfuric acid system and reassigned to the insoluble salt, $PbSO_4$. The effect is that almost the whole of the domain in which water is stable is covered by domains in which lead is stable or passivated by the oxides, PbO, Pb_3O_4 or the almost insoluble salt, $PbSO_4$. Practical observations confirm that lead does not corrode in sulfuric acid, except at concentrations greater than 96 mass %.

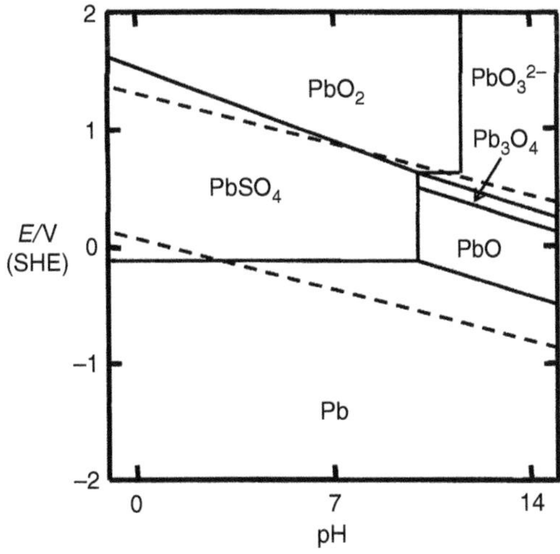

FIGURE 16.2
Pourbaix diagram for the lead–sulfate–water system at 25°C, $a_{Pb^{2+}} = 10^{-6}$. Domain for the stability of water shown by dotted lines.

Similar considerations based on Pourbaix diagrams for the lead–water–chromic acid and the lead–water–phosphoric acid systems are consistent with the observed outstanding corrosion-resistance of lead in chromic and phosphoric acids.

16.2.2.3 Effects of Stress

Lead is not a stress-bearing structural material so that it is seldom subject to stress-corrosion cracking or corrosion-fatigue in the conventional sense, but because it is protected by a passive film, it is vulnerable to erosion–corrosion by high velocity or turbulent liquid flow as described in Section 5.3.1 but with the additional possibility of local deformation of the soft metal underlying the passive film.

16.3 Applications

Lead alloys with small quantities of alloying elements from among tin, antimony, copper and tellurium are available in the usual metal forms to suit particular applications. Some representative compositions are listed in Table 16.2. The usual purpose in alloying is to enable hardness to be improved and to refine the grain. The quantities concerned are too small to have much effect on the corrosion resistance. It is noteworthy that pure lead cannot be hardened by cold-work because its recrystallization and recovery temperatures are below 0°C.

16.3.1 Lead Acid Batteries

The most prolific current use of lead is for lead acid batteries for automobiles manipulating the reversible valency change $Pb^{IV} \leftrightarrow Pb^{II}$ to store and supply electrical energy using the reaction:

$$Pb^{IV}O_2 + 2H_2SO_4 + Pb^0 = 2Pb^{II}SO_4 + 2H_2O \tag{16.1}$$

conducted in an electrochemical cell by partial reactions on separate electrodes in a 30 mass % sulfuric acid electrolyte and coupled by the external circuit supplying energy:

$$Pb^{IV}O_2(\text{solid}) + 4H^+ + SO_4^{2-} + 2e^- = Pb^{II}SO_4(\text{solid}) + 2H_2O \tag{16.2}$$

$$Pb^0(\text{solid}) + SO_4^{2-} = Pb^{II}SO_4(\text{solid}) + 2e^- \tag{16.3}$$

TABLE 16.2

Some Representative Compositions of Lead Grades and Alloys for Chemical Applications

	Nominal Composition Mass %				
Grade or Alloy	**Pb**	**Cu**	**Te**	**Ag**	**Sb**
Pure lead	>99.99				
Copper lead	Balance	0.05–0.07			
Tellurium copper lead	Balance	0.05–0.07	0.02–0.05		
Antimonial lead	Balance				2.5–11.0
Silver copper lead	Balance	0.003–0.005		0.003–0.005	

When current is drawn from the cell, Reaction 16.2 is the cathode and Reaction 16.3 is the anode. On recharging, these reactions are exactly reversed with Reaction 16.2 the anode and Reaction 16.3 the cathode.

In the present context, the significant feature is that the lead does not corrode in contact with the sulfuric acid in idle periods when the external circuit is disconnected.

16.3.2 Equipment for Electrochemical Metal Finishing

Several common electrolytic metal finishing processes require vessels to contain mineral acids in which workpieces and counter electrodes are immersed. They include electrodeposition of metals from acidic electrolytes and anodizing, all of which are key processes in major manufacturing industries. Lead is often the preferred material for fixed equipment in contact with the process liquors. It is robust for industrial use and well-suited by virtue of its outstanding resistance to chromic, phosphoric, sulfuric and oxalic acids.

16.3.2.1 Electrodeposition of Metals

Plating tanks and associated equipment must resist corrosion or other forms of degradation not only to preserve their own integrity but also to avoid contaminating the electrolyte. There is a menu of options from which to select materials to suit particular processes, including wood or steel with or without linings of rubber, bitumen or lead. Lead lining is the most versatile and sometimes the only satisfactory material for acidic electrolytes but it is expensive. It is not suitable for either nitric or acetic acid due to the high solubility of their lead salts as shown in Table 16.2. The following examples illustrate its application.

16.3.2.1.1 Chromium Plating from Chromic Acid Baths

The electrolyte for bright chromium plating is a solution of 400 g dm^{-3} of chromic anhydride with 4 g dm^{-3} of sulfuric acid at 45°C as described in Chapter 6. The concentration of chromic anhydride is reduced to 250 g dm^{-3} for hard industrial deposits. An inert counter electrode is required to serve as the anode because the electrolyte consumed is replenished by adding chromic anhydride for the reasons given in Chapter 6. Lead also fulfills this role when sufficient potential is applied to overcome the resistance polarization due to the passive film.

16.3.2.1.2 Copper Plating from Acid Sulfate Baths

A typical acid sulfate electrolyte for copper plating is a solution of 1.0 molar copper(II) sulfate with 0.75 molar sulfuric acid to maintain the copper in the divalent state. The concentration of the acid justifies the expense of using lead-lined tanks to resist corrosion.

The effective current efficiency of the anode exceeds that of the cathode because the divalent copper deposited at the cathode is resupplied by a consumable anode partly as univalent copper which is oxidized in solution:

$$2Cu^{I}{}_2SO_4 + 2H_2SO_4 + O_2 = 4Cu^{II}SO_4 \tag{16.4}$$

One practice is to use an inert lead auxiliary anode to carry some of the anodic current to correct the effective current efficiency.

16.3.2.1.3 Tin Plating from Acid Sulfate Baths

A typical acid sulfate electrolyte for tin plating is a solution of 0.5 molar tin(II) sulfate $Sn^{II}SO_4$ with 1 molar sulfuric acid, 0.75 molar cresol sulfonic acid, $C_2H_6OH \cdot HSO_3$ and small quantities of other addition agents. The sulfonic acid is a convenient form of cresol to refine the grain of the deposited tin. The tin(II) sulfate can oxidize to tin(IV) sulfate and the strongly acidic solution is needed to prevent the formation of insoluble tin(IV) hydroxide by the hydrolysis:

$$Sn(SO_4)_2 + 2H_2O = Sn(OH)_4 + 2H_2SO_4 \tag{16.5}$$

This imposes the need for lead-lined tanks to resist corrosion.

16.3.2.2 Anodizing Aluminium

General anodizing is carried out in 10–15 mass % of sulfuric acid, but alternative electrolytes including chromic, oxalic or phosphoric acids may be required. Workpieces to be anodized come in a wide variety and can include architectural fittings, cladding, automobile parts and parts for aerospace applications. The use of lead-lined equipment and lead counter electrodes can resist the wide range of process liquors needed to suit customer-specific requirements.

16.3.3 Roofing Materials

Despite efforts to find replacements for lead due to its toxicity, lead remains the premier material for roof flashings. It is malleable for sealing awkward shapes in roof gullies and is reusable. Lead used in this way is protected from exterior corrosion by a passive layer of lead salts. This is discussed in more detail in Section 24.4.2.

17

Corrosion Resistance of Zirconium and Hafnium

Zirconium and hafnium are hard corrosion-resistant metals. Zirconium was developed as a significant engineering metal to satisfy the need for a cladding material for nuclear fuel rods in water moderated nuclear reactors. Its particular attributes for the application are its low thermal neutron capture cross section and corrosion resistance in superheated water. In contrast, hafnium has a very high thermal neutron capture cross section and is applied as reactor control rods in the same superheated water, as described in Chapter 27.

17.1 Occurrence Extraction and Refining

Zirconium occurs in the earth's crust, with a relative abundance of 0.01%–0.1%. The principal ores are zircon, $ZrO_2 \cdot SiO_2$ and baddeleyite, ZrO_2. Zirconium is always associated with 1%–4% hafnium in its minerals because their geochemical properties are nearly identical so that they are not differentiated in geological processes.

The metal is extracted by the *Kroll* process, as described for titanium in Chapter 12, after first converting the silicate or oxide minerals to the carbide by heating pelletized intimate mixtures of them with carbon. The product is *zirconium sponge*, which is consolidated by vacuum arc melting to produce ingots suitable for fabrication.

Zirconium produced by the standard Kroll process contains the hafnium inherited from the minerals with which it is associated. It has little effect on zirconium used for general chemical applications, but it must be almost completely removed from zirconium used for reactor fuel rods because it has a high thermal neutron capture cross section that would otherwise quench the reactor. It is difficult to separate hafnium from zirconium but it is accomplished successfully by ion exchange or solvent extraction fractionation methods applied in the course of the Kroll process. There are thus two grades of zirconium, a commercial grade containing hafnium and the more expensive hafnium-free reactor grade.

The separated hafnium is recovered by a modified Kroll process to use in reactor control rods that are exposed to the same superheated water environment as the zirconium clad fuel rods. In fact this is the principal source of hafnium.

17.2 Some Chemical Characteristics and Corrosion Behaviour

Zirconium and hafnium both have electron configurations with four valence electrons outside the penultimate filled shell and similar atomic and ionic radii, as given in Table 17.1. There are two allotropes of zirconium, a hexagonal close-packed α-phase with a *c/a* ratio of 1.589 at temperatures below the transition temperature, 840°C, and a body-centered

TABLE 17.1

Electron Configurations and Structures of Titanium, Zirconium and Hafnium

Metal	Electron Configuration	Structure	Metallic Radius/nm[a]	M^{4+} Ion Radius/nm
Ti	$[Ar]3d^24s^2$	α < 882°C Hexagonal *c/a* ratio 1.601 β > 882°C Body-centered cubic	0.1467	0.068
Zr	$[Kr]4d^25s^2$	α < 840°C Hexagonal *c/a* ratio 1.589 β > 840°C Body-centered cubic	0.1597	0.074
Hf	$[Xe]4f^{14}5d^26s^2$	Hexagonal, *c/a* ratio 1.587	0.1585	0.075

[a] Linus Pauling, *The Nature of the Chemical Bond*, Ithaca, NY, Cornell University Press, 1960.

β-phase at temperatures above it. Hafnium exists as a hexagonal close-packed phase with a *c/a* ratio of 1.587. The *c/a* ratios for the zirconium and hafnium hexagonal phases are not particularly high and there are sufficient slip planes for hot and cold forming into simple shapes. Thus at temperatures below 840°C, the two metals are both chemically and structurally very similar and similar corrosion resistance is expected. Equivalent information for titanium included in Table 17.1, shows that these features apply to all three congeners in Group 4A of the Periodic Table.

The only important stable oxidation states are Zr(IV) and Hf(IV). The charge density that would result from removing all four valence electrons is so high that compounds of Zr(IV) and Hf(IV) are mainly covalent and simple Zr^{4+} and Hf^{4+} ions do not exist so that in aqueous solutions Zr(IV) ions are polymeric such as the cations $[Zr_3(OH)_6]^{6+}$ and $[Zr_4(OH)_8]^{8+}$ and the anion $[ZrCl_6]^{2-}$.

Zirconium and hafnium each form only one oxide zirconia, ZrO_2, and HfO_2. Only the monoclinic forms, stable at temperatures below 1100°C, are relevant to most corrosion issues.

No true hydroxides, $M(OH)_4$ exist but only the hydrous oxides, $ZrO_2 \cdot nH_2O$ and $HfO_2 \cdot nH_2O$, with variable water contents. On strong heating they lose water, yielding hard refractory monoclinic oxides ZrO_2 or HfO_2, which are insoluble and resistant to attack by acids and alkalis.

Figure 17.1 gives sketches of standard Pourbaix diagrams for the zirconium–water and hafnium–water systems as described in Chapter 3. They show extensive passive domains but they seldom represent the extreme conditions for which the selection of zirconium and hafnium are justified. Values of pH and potentials assumed in standard Pourbaix diagrams do not apply to superheated water in pressurized water reactor (PWR) applications or too high concentrations of aggressive ions that justify the use of zirconium in chemical applications. The corrosion resistance of zirconium in nuclear applications has been well researched and coordinated with extensive well-planned in-plant observations. Corresponding information on alternative applications is less well researched and is often based on uncoordinated bespoke tests.

17.2.1 Corrosion Resistance

17.2.1.1 Oxidation Mechanisms

Zirconium oxidizes in air at elevated temperatures, yielding thin dark gray coherent scales of ZrO_2 tightly adherent to the metal. ZrO_2 is an *n*-type anion deficit semiconductor so that the oxide grows at the oxide/metal interface by inward migration of O^{2-} ions through vacant anion sites, $O^{2-}\square$, in the oxide:

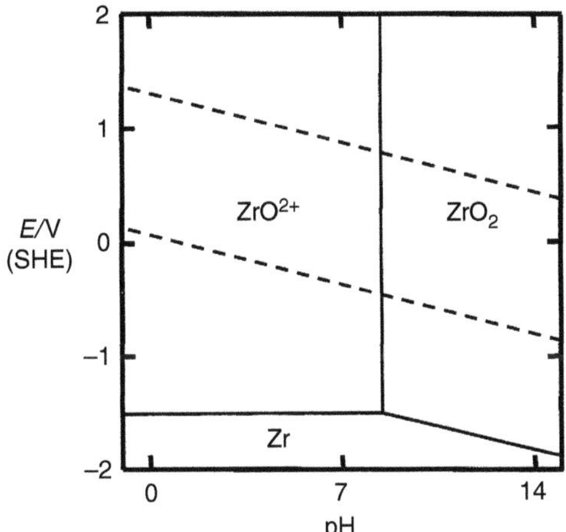

FIGURE 17.1
Zirconium–water diagram. *Note*: The Hafnium–water diagram is virtually identical.

$$Zr = ZrO_2 + 2O^{2-}\square + 4e\bullet \tag{17.1}$$

complemented by the entry of adsorbed oxygen atoms at the atmosphere/oxide interface annihilating anion vacancies and excess electrons

$$O_2 + 2O^{2-}\square + 4e\bullet = 0 \tag{17.2}$$

Oxygen is soluble in solid zirconium, so that it can diffuse into the surface layers of the metal from the oxide growing at the oxide/metal interface if time and temperature permit.

17.2.1.2 Passivity

The outstanding resistance of zirconium and hafnium in aqueous media is due to the formation of passive surface films based on their oxides. The following descriptions of passivating reactions for zirconium apply equally to analogous reactions for hafnium.

Equation 17.1, describing the growth of ZrO_2 at the oxide/metal interface, also applies to the oxidation of zirconium in water. Equation 17.2 for the complimentary process at the atmosphere/oxide interface is then replaced by:

$$2H_2O + 2\,O^{2-}\square + 4e\bullet = 2H_2 \tag{17.3}$$

yielding the overall reaction:

$$Zr + 2H_2O = ZrO_2 + 2H_2 \tag{17.4}$$

The ZrO_2 film formed in water free from aggressive ions at temperatures below 300°C is coherent and adherent to the metal surface and reaches a virtually limiting thickness of a few nanometers forming a self-healing passive film that is responsible for the outstanding corrosion resistance of the metal. This protection extends to benign aqueous media in the pH range 4 to 12.

17.2.1.3 Breakdown of Passivity

In inappropriate environments containing aggressive or complexing anions in which the passive film is disrupted or degraded, the reaction represented by Equation 17.4 continues and the metal corrodes. Some of the hydrogen generated at the atmosphere/oxide interface can be transported through the oxide and precipitate as the interstitial hydride, ZrH_n, the so-called δ-phase, in surface layers of the metal where it can act as stress raisers or decompose, inducing blistering.

The life of the metal is determined by the time taken to reach a critical oxide thickness, which depends on the temperature, the prevailing conditions and the metal composition. A particular cause of concern is the risk of breakaway oxidation in superheated water or steam as explained in Section 3.3.3.8 of Chapter 3. These implications are discussed in Chapter 27 in the context of boiling water nuclear reactors.

17.3 Applications

17.3.1 Nuclear Applications

Zirconium and hafnium are produced primarily for applications in the generation of nuclear power and discussion of their roles is deferred to Chapter 27, which considers some corrosion issues in materials' selection for the prevailing environments in nuclear reactors.

17.3.2 Chemical, Industrial and Medical Applications

Zirconium could not be considered an economic choice for corrosion resistance until it was produced on an industrial scale for nuclear applications. Its current availability assigns it to a menu of options for corrosion-resistant service in severe environments. Its use is restricted to niche applications where its particular qualities are appropriate. In many of its applications it is an alternative to titanium, where the cost is justified by superior performance. In general, zirconium resists corrosion better than titanium in nonoxidizing acids. Applications for which zirconium is especially suitable include plants handling hydrochloric acid and dry but not wet chlorine. Results of numerous empirical tests and observations in the field are available to compare the corrosion resistance of zirconium and titanium in various aqueous environments at ambient and elevated temperatures. This information is usually available from suppliers.

Like titanium, zirconium is biocompatible because it has excellent resistance to corrosion and corrosion–fatigue in body fluids and can provide an alternative material for prostheses. One example is a zirconium–2.5 mass % niobium alloy developed for hip and knee replacements. It has the advantage that it can be given a wear-resistant surface of ZrO_2 by suitable oxidation pretreatment. For the same reasons, the material is also used for surgical instruments.

18

Corrosion Resistance of Beryllium

18.1 Occurrence and Extraction

Although beryllium is one of the lighter elements, it is only the 44th most abundant element in the earth's crust. Therefore there are only a few economic ore bodies for extraction of the element. Because of the health effects of beryllium and its compounds, extreme care must be exercised when processing beryllium from the ore and during manufacturing operations. Beryllium should not be handled without the correct specialist facilities.

18.1.1 Sources

Historically beryllium has been produced from several sources. The principal minerals are beryl and bertrandite. The toxicity of beryllium and its compounds, especially as airborne dust, makes refining and extraction of the metal a difficult process and for the last 30 years there has been only one commercial supplier in the Western world, Materion Brush, Inc. (until 2012 called Brush-Wellman Ltd.) although there are still some limited supplies in the former Soviet Union, principally in Kazakhstan, and there is also some output in China. Materion Brush produces beryllium from open-pit bertrandite mines in the Spor Mountains of Utah, in the United States and provides various grades of metal powder for different uses. The price of beryllium varies according to the form and purity of the metal.

18.1.2 Extraction of Raw Metal by Halide Routes

The ore is refined by crushing the ore, forming a slurry and then treating it with sulfuric acid to produce beryllium sulfate. This feedstock then undergoes a series of chemical treatments to ultimately produce beryllium hydroxide $Be(OH)_2$. This is converted to the chloride, electrolytically reduced and cast into beryllium ingots. The principal impurities in the raw ingots are iron, aluminium and silicon. The most critical issue in metal specification is that the atomic ratio of iron to aluminium in the alloy must be kept close to unity. This is to precipitate aluminium from solid solution in the beryllium matrix as an insoluble intermetallic phase, $FeAlBe_4$, to avoid a phenomenon called *hot shortness* which reduces the ultimate tensile strength of the final hot pressed product.

18.1.3 Purification Swarfing and Consolidation

To improve ductility of the metal the powder metallurgical route is required to produce a metal with appreciable mechanical properties. After a vacuum melting process to remove impurities, the ingots are machined into swarf and this is then pneumatically pulverized into powder by high velocity impaction onto a beryllium target. This produces a powder of uniform rounded grains of around 20 microns in size. The powder is sieved and then vacuum hot pressed as billets in graphite dies at 1025–1125°C. The billets then undergo heat treatment to relieve internal residual stresses and are assayed for full densification of the alloy using a dye penetrant method, ASTM 1417. The hot pressed billets are machined into the required shape and the swarf then recycled for reprocessing into the hotpressed metal.

18.2 Characteristics of Commercially Pure Metal

The powder process route is required because although beryllium has an extremely high specific stiffness it has a low fracture toughness and is difficult to machine. From the Hall–Petch relationship, the yield strength of a material is inversely proportional to the mean grain size of the matrix. Thus to increase the yield strength a small grain size is required and this is the reason for the powder forming method employed.

Due to the powder processing route the metal contains a high proportion of beryllium oxide, BeO, usually about 1% by mass, principally on the surface of the hot pressed metal grains. These metal grains are essentially pure beryllium metal. The other impurities are present as follows:

1. Beryllium carbide is present as individual subsurface grains; beryllium carbide is not stable in the presence of atmospheric water vapour so surface carbide is converted to beryllium oxide and methane. The principal source of carbon in the alloy is from the graphite dies in the hot pressing process.

$$Be_2C + 2H_2O = 2BeO + CH_4 \quad (18.1)$$

2. Iron and aluminium are precipitated as inclusions of $FeBeAl_4$. This is precipitated at beryllium grain boundaries and these inclusions are considerably smaller than the parent beryllium grains. As mentioned previously, these precipitates are designed to remove aluminium from forming a solid solution with beryllium by precipitating it as $FeBeAl_4$ to prevent hot shortness. Since both iron and aluminium are precipitated, the parent beryllium grains are essentially composed of the pure metal.
3. Silicon is present as extremely small inclusions, about the same size as $FeBeAl_4$ inclusions. These are effectively a pure silicon phase due to the complete immiscibility of beryllium with silicon.

The scarcity of the raw material for beryllium production, the necessity for special handling due to the toxic effects of beryllium and the powder processing route of production means that the commercially pure metal has a price of around $7000 (US) per kilogram.

18.2.1 Structure and Physical Properties

Beryllium has a hexagonal close-packed (HCP) crystal structure with an axial ratio of 1.568. At ambient temperatures deformation can occur only with slip on the basal plane.

The metal is also anisotropic because of the HCP phase structure. This results in poor ductility and is the reason for the need to produce metal with a very small and uniform grain size so as to improve ductility. The brittleness of the metal and its stiffness (the ultimate tensile strength is 448 MPa) is a dominating factor in the physical metallurgy properties of the metal. Beryllium is the second lightest engineering metal after magnesium, with a specific gravity of 1.85 kg dm^{-3} and has a relatively high melting point (1278°C) for a light metal. This means that despite the brittleness of beryllium, its high strength to weight ratio marks it out as a niche product with potential future uses in case critical applications where weight is a factor.

18.2.2 Nuclear Properties

Beryllium has exceptional nuclear properties. It is both a moderator and a neutron reflector. It is also nearly transparent to x-rays. For this reason one of the earliest uses of beryllium was as windows for x-ray detectors. For this application very high purity material is needed. The other applications of beryllium take advantage of its neutron moderating and neutron reflecting powers to enhance the effectiveness of chain reaction neutron capture in nuclear reactions, both in civil and military uses.

18.2.3 Chemical Properties and Corrosion Behaviour

The element beryllium occupies the position at the top of the alkaline earth group of the Periodic Table (Group IIA). As expected from this position, all of its chemical compounds contain beryllium in the +II oxidation state similar to that of calcium, magnesium, strontium and barium. However, because the ion is very small in size (the Be^{2+} ion has a higher charge density than Al^{3+}) the chemistry of beryllium is different from that of that of the other alkaline earth elements.

18.2.3.1 Electron Configuration

Beryllium is the fourth element of the Periodic Table and the electron configuration of the element is $1s^2$, $2s^2$ as shown in Table 2.3. The atomic radius of beryllium is 0.11 nm, but the beryllium ion Be^{2+} has an ionic radius of only 0.031 nm, corresponding to an electron configuration of $1s^2$. In the metal the two 2s orbitals are donated to the metallic orbital. One feature of the electron configuration of beryllium and beryllium ions is that the atomic orbitals are entirely *s* orbitals and these are non-directional. Thus beryllium exhibits little tendency for directional bonding.

18.2.3.2 Chemical Characteristics

The chemical properties of beryllium are determined by the electron configuration of the isolated atom in which the two 2*s* valence electrons are outside the filled 1*s* shell so that its normal oxidation state is Be(II). The 2s electrons are not well shielded from the nuclear charge, as the inner shells comprise only the two 1s electrons. This gives rise to a characteristically high ionisation potential and a small ionic radius that is so extreme that the element is disposed to form covalent rather than ionic bonds and the cationic radius is small enough to favour fourfold coordination. It is doubtful whether a simple Be^{2+} ion exists in either solution or crystalline salts.

Beryllium is reactive, as illustrated by the Gibbs free energy for formation of the oxide:

$$\text{Be(solid)} + \tfrac{1}{2}\text{O}_2\text{(gas)} = \text{BeO(solid)} \quad \Delta G^{\ominus} = 1{,}200{,}400 - 13.9T\log T + 235T \text{ J} \quad (18.2)$$

Be^{2+} and Al^{3+} ions have ionic charge densities of a similar order, so that many of their salts are chemically similar, although of different stoichiometry. For example, neither forms a distinct carbonate due to hydrolysis of the cation. Their oxides absorb water to give sparingly soluble hexagonal crystalline hydroxides, and they can also form amorphous gel-like hydroxide phases. These similarities extend also to the more covalent phases of aluminium and beryllium; for example, Be_2C and Al_4C_3 comprise a class of metal carbides by themselves in that they are macromolecular structures that liberate methane from the solid when exposed to water vapour. From these characteristic similarities, the chemistry of beryllium resembles that of aluminium in determining the properties of a passivating layer that describes its corrosion behaviour.

The toxicity of airborne beryllium salts creates special hazards for handling of all beryllium compounds. The generally accepted exposure in air above which it is deemed to be a risk to humans is 2 μg m^{-3}. Manufacturing operations typically operate with exposure limits at around one-tenth or one-twentieth of this. The organs at principal risk are the lungs and the first effect of beryllium exposure is usually *sensitization* which can occur after a single exposure. This means the affected party can develop a future debilitating disease called *Chronic Beryllium Disease* (CBD). Chronic beryllium disease is debilitating and incurable; understanding the corrosion characteristics of beryllium has an added dimension of reducing exposure to the health risks associated with the exposure to airborne degradation products of the metal.

18.2.3.3 Passivation by Surface Oxide Film

The normal passive film on beryllium is beryllium oxide (BeO). This has the hexagonal wurzite structure in which the beryllium and oxygen atoms are both in a tetrahedral conformation. Beryllium oxide is a very good electrical insulator; the electrical resistivity of BeO is about two orders of magnitude greater that of corundum, α-Al_2O_3. Beryllium oxide is attacked by acids below pH 4 with the evolution of hydrogen and does not repassivate appreciably, irrespective of the nature of the acid. Beryllium is amphoteric and is attacked slowly in alkaline aqueous solution above pH 9.5. Between pH 4 and pH 9.5, general corrosion is not observed and the metal is passive although as described below it can be vulnerable to pitting corrosion.

18.2.3.4 Pitting at Local Cathodic Collectors

Pitting corrosion is intense chemical attack at dispersed points on an unshielded passive metal surface, forming pits that can perforate thin gauge metal. They are due to breakdown of passivity at very small isolated sites distributed over the metal surface, creating local active/passive cells. They can be initiated by heterogeneities in the metal surface due to minority phases in the microstructure of the metal as for aluminium alloys or by environmental agents as for stainless steels, notably halide and hypochlorite ions. In particular, the chloride ion often imposes severe limitations on the use of passivating metals for service in seawater and in chemical and food processing where the solutions can contain chloride contents in excess of about 0.01 mol dm^{-3}, equivalent to a 0.06 mass % solution of sodium chloride. Susceptible metals include not only stainless steels and aluminium, but also some copper-based alloys and mild steels in certain environmental conditions.

The highly localized damage can render equipment unserviceable even if attack on the rest of the metal surface is negligible.

Commercial vacuum hot pressed beryllium metal is vulnerable to pitting corrosion with the following depassivating agents; chloride ions, fluoride ions and sulfate ions. In this regard it has some similarity with commercial grade aluminium, for example, AA 1100, although unlike beryllium, aluminium shows little susceptibility to sulfate ions.

Pitting of beryllium is normally found associated with local regions on the metal surface which sustain a cathodic reaction, normally the reduction of oxygen once corrosion is underway.

$$O_2 + 2H_2O + 2e^- = 4OH^- \quad (18.3)$$

These cathodic collectors are principally associated with the intermetallic inclusions, $FeAlBe_4$. The nature of the cathodic properties of the inclusions is not as simple as just considering the hypothetical difference in electrode potential between $FeAlBe_4$ and beryllium metal. As was mentioned, beryllium oxide is an extremely good insulator. Thus a region of the metal surface with a BeO layer cannot support either a cathodic or anodic reaction. On the other hand, the oxide over the inclusion is of indeterminate composition. The exposure of $FeAlBe_4$ to the atmosphere will produce an oxide with a similar stoichiometry to that of the inclusion, that is Fe:Al:Be equals a ratio of 1:1:4. The structure of this phase does not correspond to any known phase but is forced by being formed by quantum tunneling as described by Cabrerra–Mott theory in Section 3.3.2. From their characteristic chemistry, beryllium and aluminium can only exist as Be^{2+} and Al^{3+} ions but iron can exist as both Fe^{2+} and Fe^{3+} ions. The presence of mixed valency in the oxide phase lends a degree of electron conductivity to the hypothetical oxide phase over the inclusion. This is the prerequisite for a phase to act as a cathodic collector; that is, electrons can be conducted through the oxide phase to sustain the oxygen reduction reaction, Equation 18.3.

The cathodic collector properties of $FeAlBe_4$ inclusions can sustain a local action cell, but do not explain the depassivation processes which expose the anodic site, which is the beryllium grains around the cathodic site. The exact mechanism is unclear but it is probably similar to that proposed for the depassivation of aluminium in pitting as described in Section 30.2.4. The difference is that depassivation of beryllium is caused by fluoride, chloride *and* sulfate ions, whereas for aluminium it is initiated only by fluoride and chloride ions.

Artificially induced pitting of beryllium by electrochemical methods produces morphological effects such as parallel striations in the *c*-direction of the metal. This is not normally seen in beryllium samples that have naturally corroded. This has been ascribed to differential bonding energies in the metallic orbital in the metal in the *a*- and *c*-directions of the hexagonal phase but this is probably incorrect. The metallic orbitals of beryllium are formed from spherical (that is nondirectional) 2s orbitals. However, it should be noted that both the oxide, BeO, and the metal are hexagonal phases and have different ratios of *a*- to *c*-dimensions. Assuming the oxide and metal are epitaxially compatible this means that for beryllium interfacial strain between the oxide and metal are different in the *a*- and *c*-directions. This is probably the reason that parallel striations are seen when beryllium is artificially pre-pitted using electrochemical methods.

18.2.3.5 Effects of Carbide Inclusions

The carbide, Be_2C, is present in vacuum hot pressed beryllium, resulting from graphite dies. The carbide is present as inclusions within the matrix of the metal, but any that break out onto a machined surface react exothermically with water vapour, via Equation 18.1, liberating methane. This reaction converts surface carbide inclusions to beryllium oxide, and ultimately to beryllium hydroxide, $Be(OH)_2$. The product is not coherent but often has a somewhat fluffy appearance. It has been proposed that these spots are progenitors of localized corrosion. This has been ascribed in the past to external contaminants of trace metals within the space formerly occupied by the beryllium carbide inclusion which render the area cathodic. Modern fabrication methods often involve ultra-clean facilities to control toxic exposure to beryllium and this reduces the possibility of contamination. Thus the effectiveness of beryllium carbide as cathodic sites to stimulate pitting of the metal is less pronounced than that of intermetallic $FeBeAl_4$ inclusions.

18.3 Applications

Beryllium has some exceptional properties, particularly concerning its strength to weight ratio, its stiffness to weight ratio and its desirable nuclear properties. Due to the toxicity of beryllium it occupies niche applications where these desirable properties outweigh the expense of handling it.

18.3.1 Nuclear Reactors

For fission processes, the original use was in weapons as a neutron reflector surrounding a plutonium core, however, it is also used in certain civil nuclear applications such as brazed beryllium tampers in the CANDU reactor. It has also been used as a neutron reflector in some older experimental designs. In fusion research the neutron reflective power of beryllium is often regarded as of paramount importance in producing a high enough neutron flux to enable a self-sustaining fusion reaction. For example, it is used for this purpose in the Joint European Torus (JET) project.

18.3.1.1 Responses to Reactor Environments

Natural beryllium is composed of only one isotope, ^9Be. This isotope can undergo nuclear reactions with both alpha particles and gamma radiation

$$^{9}\mathrm{Be} + \alpha = {}^{12}\mathrm{C} + \mathrm{n} + \gamma(4.44\ \mathrm{MeV}) \quad (18.4)$$

$$^{9}\mathrm{Be} + \gamma(1.7\ \mathrm{MeV}) = 2\alpha + \mathrm{n} \quad (18.5)$$

Equation 18.4 was the reaction used by Chadwick to discover the neutron in 1932 using polonium as the alpha source. When used as a nuclear reflector or moderator, beryllium can be vulnerable to activation by alpha bombardment. The transmutation of the metal in this way can be a potential source of corrosion initiation, presumably because of particulate damage to the passive film. To eliminate damage to the metal a thin layer of a suitable material such as gold, copper or titanium is inserted between the reflector and the reactor core. The effect of nuclear reactions of beryllium on the corrosion performance of the metal is unclear, but since good design eliminates particle bombardment, corrosion from nuclear sources should not be cause for concern.

18.3.1.2 Integrity of Beryllium as a Structural Material

The brittleness of beryllium can cause limitations to its applicability as a structural material. Beryllium combines high stiffness with poor fracture toughness. This combination of properties means that beryllium has its own standard, ASTM E377, for determining the fracture toughness of the metal. Although beryllium has high thermal conductivity and low thermal expansivity, its poor fracture toughness is a concern regarding its durability in future fusion reactors. There is no evidence that beryllium is subject to stress-induced corrosion, however, for a brittle material this does not preclude that the opposite interaction might occur, that is local corrosion sites can act as stress raisers for a component under thermal cycling stress.

18.3.2 Applications Exploiting High Strength to Weight Ratio

Non-nuclear applications of beryllium are principally concerned with the high strength to weight ratio. As beryllium is not a bulk material, such as aluminium or even titanium, these applications tend to be either for niche components where space is at a premium for strengthening for structural integrity, such as in military aircraft applications or where it is essential to minimize all extraneous mass, for example, in satellites and other space vehicles. For space applications, there is of course no possibility of corrosion due to atmospheric conditions and there is normally no electrolyte to sustain an electrochemical cell. For terrestrial applications the position of beryllium in mixed metal systems needs to be considered. In Section 4.1.3.4 the concept of compatibility groups was introduced. For beryllium, the Standard Electrode Potential of the element is

$$Be^{2+} + 2e^{-} = Be \qquad E^{\ominus} = -1.85\ \text{V (SHE)} \tag{18.6}$$

Comparing this with Table 3.3 places beryllium between magnesium and aluminium in its latent reactivity as a metal. This is reflected in its position in the compatibility group in Section 4.1.3.4. Beryllium is normally placed in Group 2 – the base metals, but is regarded as slightly less noble than aluminium. For this reason when designing engineering applications that use beryllium to take advantage of its high strength to weight ratio, similar precautions must be made to avoid potential galvanic cells as is done with aluminium alloy components.

19

Corrosion Resistance of Uranium

19.1 Occurrence and Extraction of the Natural Metal

Uranium is the last element of the Periodic Table to exist in quantities greater than as transient decay. Because of this it is unsurprising that uranium is a rare element in the earth's crust, although it is no more rare than molybdenum, and is about as abundant as silver. The main ores are the uranium dioxide, *uraninite,* of which pitchblende is a variant, and *carnotite,* which is a uranium vanadate produced hydrothermally in roll-front deposits.

The extraction routes for the metal from the ore are complicated, but all involve the processing of the ore by acid or alkali to form yellowcake, U_3O_8. Purification often involves solvent extraction. The metal is then reduced from the halide salts either electrolytically in a molten salt bath or else with an electropositive metal, for example, magnesium.

For nuclear applications enrichment of the fissile isotope U-235 is required. Natural uranium is composed of 0.7% U-235 and 99.3% U-238. Enrichment increases the proportion of U-235 in the metal, with the byproduct which consists essentially of pure U-238 being termed depleted uranium. Uranium enriched up to 20% U-235 is described as low enriched uranium (LEU) and is primarily used in the civil nuclear industry. Uranium enriched to over 20% U-235 is termed highly enriched uranium (HEU) and is sometimes called weapons grade uranium. Enrichment can be carried out using either of two processes, the older gas diffusion process which uses a semi-permeable membrane, and the newer gas centrifuge method. Both rely on the slight difference in density of the two isotopic forms of the gaseous uranium hexafluoride, $^{235}UF_6$ and $^{238}UF_6$. A cascade system is used to continuously increase the very small enrichment obtained at each stage. Nuclear grade uranium must be very pure and free from impurities such as boron or cadmium which moderate thermal neutrons.

19.2 Uranium Metallurgy and Chemistry

19.2.1 Metallurgy

Uranium has a melting point of 1132°C and exists as three crystal allotropes. The room temperature form is orthorhombic in structure and is, therefore, an anisotropic phase. The finely divided metal is pyrophoric. With many other metals, uranium forms a series of intermetallic phases rather than extensive solid solutions. Thus many of the alloys encountered have compositions in excess of 95% uranium by mass.

For non-nuclear applications, the metal is often alloyed with a low mass percentage of a range of these elements to improve its corrosion resistance. These include titanium, molybdenum, zirconium, niobium and chromium. The physical metallurgy of these alloy phases is very complex, as the processing often involves casting and quenching to retain homogeneity of alloy composition owing to the density difference between uranium and alloying components and the tendency to form intermetallic precipitates. Some alloys can undergo cold- or hot-working to reduce variation in composition or physical properties. Typical features of these alloys, depending on composition, include either single-phase or duplex compositions with cubic, orthorhombic and twinned martensitic phases. Stress–strain relationships are complicated. In evaluating the corrosion performance of these uranium alloys variations in the distribution of atomic composition should be considered.

19.2.2 Chemical Characteristics

Uranium possesses unusual chemical properties which arise from its position within the actinide block of the Periodic Table (thorium to lawrencium). The actinide block is the region of the Periodic Table where, like the lanthanide block (cerium to lutetium), *f* orbitals are added to the electron complement of elements. However, unlike the lanthanide block the available *d* electrons (the 6*d* orbitals) are relatively low lying and are available for occupation as well as the *f* electrons. The relative energy of the 5*f* and 6*d* orbitals actually changes as one moves from the beginning of the actinide group (actinium and thorium) where the *d* orbitals are lower in energy than the *f* orbitals to the later transuranic elements where the reverse is true. This means elements such as thorium have characteristics similar to the transition metals, but the later actinide elements have chemical characteristics more typical of the lanthanide elements.

The elements protactinium–uranium–neptunium–plutonium are the region of the Periodic Table where this change in chemical behaviour occurs. Here the energy of the 6*d* and 5*f* orbitals are similar in energy and this means that in the chemistry of these elements, both types of orbitals come into play and the exact choice of orbital can also be influenced by coordination geometry and by relativistic electron effects. For this region of the Periodic Table much of the chemistry of these elements is unique. Of these elements, only uranium finds use as an engineering metal; the unusual chemistry confers special characteristics of the corrosion behaviour of uranium.

In aqueous solution uranium can exist in four different oxidation states, U(III), U(IV), U(V) and U(VI), but in aerated aqueous solution the system is dominated by one U(VI) species, the uranyl ion. This is a three-atom ion, UO_2^{2+}, a type of species which is only exhibited by elements of the protactinium–uranium–neptunium–americium region of the Periodic Table. The species UO_2^{2+} is linear with very short U–O bonds, such that in older literature it is often represented as having two double bonds, *viz*, O=U=O. The uranyl ion, although triatomic is a discrete ion, for example, the half-life of exchange of the oxygen ions in the O=U=O group (referred to as oxo groups) is of the order of one or two years and the uranyl ion also forms a large range of salts with common anions in the same way as a conventional cation. A peculiar property of some of these salts, for example, uranyl nitrate, $UO_2(NO_3)_2$, is that they are soluble in organic solvents such as ether and such solutions are immiscible with water even though they are inorganic salts. This is a convenient way to separate uranium from other actinide elements. The uranyl ion also has a great affinity with the carbonate ion to form species such as $UO_2(CO_3)_2^{2-}$ and $UO_2(CO_3)_3^{4-}$.

Another unusual feature of uranium is that it forms a discrete and stoichiometric hydride, α-UH_3, at room temperature. This is formed by the reaction of uranium with water or water vapour.

$$7U + 6H_2O = 3UO_2 + 4UH_3 \tag{19.1}$$

The solid uranium hydride at the surface can then further react with more water or water vapour

$$2UH_3 + 4H_2O = 2UO_2 + 7H_2 \tag{19.2}$$

The overall reaction is one which slowly produces hydrogen over time and for this reason uranium metal is traditionally stored in drums with a pressure release valve to vent hydrogen gas. The combined reaction is given in Equation 19.3.

$$U + 2H_2O = UO_2 + 2H_2 \tag{19.3}$$

19.3 Corrosion in Water and Steam

The reaction of water with uranium produces uranium dioxide and hydrogen as given by Equation 19.3. The effect on the metal is a function of the temperature of the water and also the amount of dissolved gases, particularly the amount of oxygen present. When oxygen potential is low, for example, for boiling water, for temperatures in excess of 90°C or when the oxygen has been displaced by inert gases, general corrosion is observed. In aerated solutions at lower temperatures the attack of the metal is manifest as pitting corrosion. When dissolved oxygen is purged, for example, with hydrogen gas, the corrosion rate of uranium increases by over an order of magnitude. Therefore, it seems that the dissolved oxygen in aerated solutions has an inhibitory effect.

Considerable work has been conducted on the corrosion rate of uranium in steam. At temperatures above 450°C the product is uranium dioxide and hydrogen and below this temperature a mixture of uranium dioxide and uranium hydride is formed; the proportion is skewed toward the oxide indicating that some of the hydride is decomposed by more steam to the dioxide as in Equation 19.2. Attack is over the entire surface.

19.3.1 Corrosion Products

The main corrosion product is uranium dioxide, which can then further oxidize to U_3O_8 in air. The oxide structure of the dioxide is described in detail in Section 19.4.2. Uranium dioxide and U_3O_8 are markedly less dense than the metal (10.96 kg dm^{-3} for UO_2 and 8.3 kg dm^{-3} for U_3O_8 versus 19.05 kg dm^{-3} for uranium) so that the corrosion product is under extreme compressive stress. This means that even though the oxides are chemically stable to over 600°C the corrosion products are prone to crack and spall once it has reached an appreciable thickness. The main corrosion products are usually black and powdery, but under damp conditions may have a yellowish colour. The thickness of oxide that can be produced before it spalls from the metal depends on the lateral compressive interfacial strain between the oxide and the metal and this depends to a degree on surface conditioning and topography. The underlying oxide adjacent to the metal is usually uranium dioxide.

The product formed in the reaction with steam is of a friable consistency, producing a fine powdery corrosion product of uranium dioxide. At higher temperatures, above 600°C with steam there is evidence that a more robust scale can be formed. An interesting observation is that if a marker, for example, a chalk mark, is placed on uranium in air and it is then oxidized to give a UO_2 film, the mark remains on the *outside* of the resultant corrosion product until the corrosion product spalls off. This is consistent with transport through the overlying oxide being due to apparent anionic mobility rather than by movement of U^{4+} cations.

19.3.2 Kinetics

The kinetics of uranium oxidation often follows a linear relation in water, steam or air. This is consistent with the observations that the oxide film spalls once the oxide has reached a certain thickness. The fact that transport is by apparent anionic transport means that oxide growth is at the interior metal/oxide interface rather than the external oxide/air interface. The oxidation rate of unalloyed uranium in air or steam is appreciable, especially above 100°C and the metal should not be regarded as protected by a passive layer. The rate of transport through the oxide depends on the exact stoichiometry as described in Section 19.4.3. The oxidation rate is slowed by the presence of elevated levels of carbon dioxide in air or steam and by oxygen in water.

19.3.3 Effects of Hydrogen Absorption

Reactions 19.1 and 19.2 are highly exothermic and the density of UH_3 at 10.95 kg dm^{-3} is considerably less than that of uranium metal (UH_3 at 19.1 kg dm^{-3}). Interestingly, the density of the hydride is almost the same as that of UO_2. Thus, the volume of expansion per kilogram of uranium by Reactions 19.1, 19.2 and 19.3 are virtually identical. The difference is that the exothermicity of hydride decomposition with more water vapour (Equation 19.2) can cause the bulk metal to disintegrate; in extreme cases this can be pyrophoric when uranium is in contact with condensate.

19.4 Oxidation

The kinetics of uranium corrosion was discussed in Section 19.3.2. The dry oxidation rate in ambient air is around 0.4 millimeters per year which is high compared with other metals. It is better to interpret this behaviour as a form of dry thermal oxidation, the interpretation of which is described in detail in Section 3.3, but proceeding at lower temperatures than other metals.

19.4.1 Uranium–Oxygen Phase Equilibria

The uranium–oxygen system is one of the most complicated oxide systems known. There are three principal oxides which are the brown or black dioxide, UO_2, the greenish black U_3O_8, and the orange trioxide, UO_3. The stoichiometry phase diagram encompassing the region of uranium dioxide is particularly complicated. This is because UO_2 can exhibit a continuous range of stoichiometry of U:O from two to one to four to nine without a discrete change of phase. The latter stoichiometry corresponds to a formula of U_4O_9. At this point a distinct phase change occurs to variants of the oxide U_3O_8. A simplified phase diagram is given in Figure 19.1. At elevated temperatures the phase is based on the parent UO_2 with the extra anions in a disordered arrangement. This is given the formula UO_{2+x} and extends in composition from the range U:O of 2.00 to 2.25 and represents a 12.5% non-stoichiometry. The range of tolerated non-stoichiometry by uranium dioxide is very unusual. Of the refractory oxides of electropositive metals it is exceeded only by tantalum oxide, which can have a continuous stoichiometry range of TaO_2 to Ta_2O_5.

At lower temperatures the phase pertaining to UO_{2+x} exhibits superlattice lines where the extra oxygen atoms occupy particular highly aligned locations in the crystal; the oxide thus has characteristics of two superimposed structures. The first is the stoichiometric oxide, UO_2, the second, which accounts for the diffraction lines of the super-stoichiometric oxygen atoms is given the designation U_4O_9. Thus the high temperature phase of UO_{2+x} is replaced by a solid solution of UO_2 plus U_4O_9. The transition temperature between these two domains is given by the line A–B–C in Figure 19.1 and is composition dependent. At this transition the superlattice order disappears as an order–disorder transition similar to other such transformations such as with β-copper zinc binary alloys described in Section 10.2.1.1.

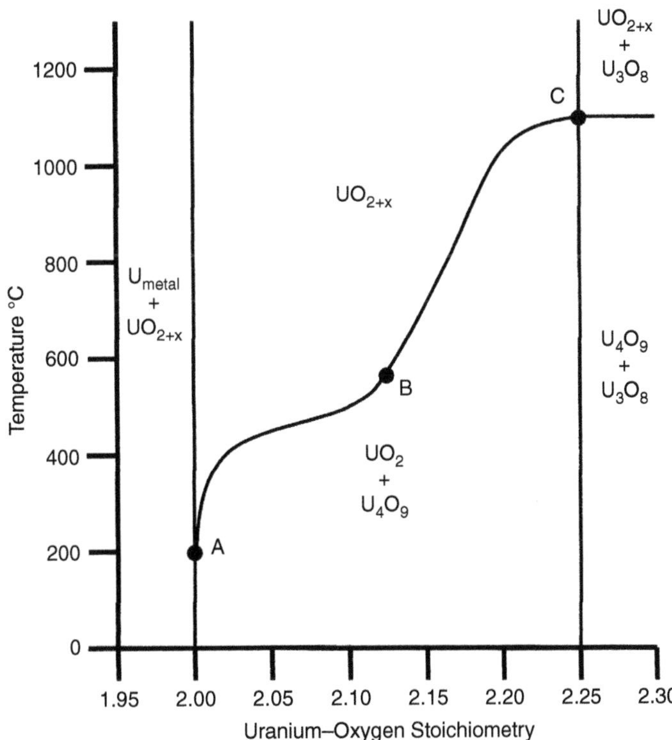

FIGURE 19.1
The simplified phase diagram of the uranium–oxygen system from a stoichiometry range of 2.00 to 2.25.

19.4.2 Critical Review of Uranium Oxide Structures

Uranium trioxide can be thought of as a structure whereby each uranyl center is bridged by oxide anions. Thus it has two short U–O bonds per uranium ion. U_3O_8 has two structures both of which can be derived from the uranium trioxide by removal of oxygen anions from the UO_3, one involving partial replacement at a different lattice coordinate of half of the anions removed.

Stoichiometric uranium dioxide has the fluorite structure. This is the simplest close-packed structure an ionic crystal can possess with a cation to anion ratio of one to two. It is a cubic structure and the small size of the highly charged U^{4+} cation (0.089 nm) means there is anion to anion contact between O^{2-} anions. Therefore, there is minimal room to accommodate extra oxide anions and phases with the fluorite structure are generally very stoichiometric. In the light of this it is surprising that migration through the oxide film during oxidation is via *oxygen transport* and that the UO_2 fluorite structure exhibits hyper-stoichiometry up to $UO_{2.25}$. The location of the excess atoms have been the subject of several investigations, most notably that of Willis. It was found that as well as the basic fluorite lattice, oxygen atoms are located at two other sites as sets of pairs. The first pair is found in the <110> direction and are extra atoms. The second pair is in the <111> direction and can be thought of as displaced atoms from a pair of fluorite lattice anions which are replaced by vacancies. Willis grouped these together in a structure covering several unit cells called a '2,2,2' structure. More recent crystallographic work on the ordered structure, U_4O_9, has extended the crystallographic description of the atom and vacancy locations ascribed in the Willis structure to higher orders to explain the superlattice parameters discussed previously.

There are two difficulties in interpreting the structure of hyperstoichiometric UO_{2+x}. The first one is that the limited space in the fluoride lattice means that the oxygen atoms in the Willis structure have a smaller radius than for the oxide anions in stoichiometric UO_2. The other is that to preserve electroneutrality, the oxidation state of some of the U^{4+} cations must be increased. This, according to Willis, creates sites of U^{5+} cations in the fluorite lattice. This must, however, be regarded as assumption since a discrete U^{5+} is unknown in other structures. In other cases, the metastable uranium (V) oxidation state in an oxo-environment has a similar structure to the uranyl entity with two short equatorial U–O bonds.

19.4.3 Application of Critical Review to Oxidation Kinetics

The discussion of the structures of the uranium–oxygen phase diagram given previously raises difficult issues. From Section 19.3.1, it is known that mass transport is via oxygen transport through the matrix and that this happens at an abnormally low temperature for dry oxidation. The oxidation rate is dependent on the stoichiometry of the UO_{2+x} phase, being fastest at a stoichiometry of $UO_{2.125}$ corresponding to the midpoint, B, of Figure 19.1. In fact the activation energy for oxygen transport within $UO_{2.125}$ is only around 100 kJ mol^{-1}. At the extreme ends of the stoichiometry range, UO_2 and $UO_{2.25}$ transport through the oxide are slowest.

The ease of transport in UO_{2+x} is at odds with that of TaO_{2+x}. In the latter, oxide x can range from 2.0 to 2.5. The tantalum atoms are present as discrete Ta(V) sites and the non-stoichiometry is accommodated by extra tantalum interstitial atoms of Ta(IV) oxidation state. The TaO_{2+x} is an electrical conductor but crucially, there is little anion mobility in the lattice and tantalum is stable to dry oxidation until at least 400°C. The behaviour of uranium to dry oxidation must therefore be regarded as anomalous.

19.4.4 Effects of Alloying Elements on Oxidation Kinetics

Alloying of uranium with small amounts (less than 10 mass %) of many transition metal elements markedly improves the oxidation characteristic of the metal. These especially include elements in groups IV to VI of the transition metals such as titanium, molybdenum, zirconium and niobium. The amount that can be alloyed is typically restricted to a few alloy percent by the propensity of formation of intermetallic phases. These alloys are sometimes known as the 'stainless' uranium alloys, especially uranium–2% molybdenum and uranium–0.75% titanium. The latter alloy is often referred to informally as '3/4Ti'. Such alloys are used in penetrators for ballistic shells and weaponry using depleted uranium.

The rate of oxidation of these alloys in ambient air has been shown to follow an approximate exponential relationship with increasing alloy content; tests were often carried out at slightly elevated temperatures to accelerate results. As an example, at 120°C unalloyed uranium has an oxidation rate of about 0.4 millimeters per year, but adding 0.75% titanium in 3/4Ti reduces this by about 80%. This is still an appreciable rate for dry oxidation that is in the absence of an electrolyte compared with other metals. As mentioned previously, many different alloying components have a beneficial effect on reducing the oxidation rate of uranium and this is not dependent on a particular valency. Thus Hauffe's rules as discussed in Section 3.3.3.7 do not apply; the inference is that the mechanism of dry oxidation of uranium is more complicated than that given by Wagner's theory.

To summarize, the behaviour of uranium and its alloys is not well understood. Considerable effort in the nuclear industry has characterized the structure of the phases of the uranium–water–oxygen system. The thermodynamic stabilities of most of the principal phases are understood. To apply this information to construct a mechanism for the oxidation of uranium is a challenge, especially with respect to how small amounts of alloying elements can retard oxidation. Some thoughts on the anomalous behaviour of uranium are discussed further in Section 30.2.5.

20

Cathodic Protection

Cathodic protection is a method of controlling the corrosion of a metal structure by cathodically polarizing it to a potential more negative than its rest potential in a moist soil or aqueous environment. The objective is to preserve the capital value of expensive buried systems, marine equipment and civil engineering structures and it is widely applied to protect both buried and fully immersed installations. It is based on simple electrochemical principles, but the conditions in the field in which it is implemented are usually complex and sometimes indeterminate. It can be applied by either of two methods:

1. Impressing a cathodic current by applying an external DC potential between a protected structure and an auxiliary anode immersed in the same medium.
2. Providing a sacrificial anode of a less noble metal to polarize the structure galvanically, that is using the corrosion of one metal to protect another as explained in Section 4.2.1.

The appropriate method for a particular situation is the most economical as determined by technical considerations, capital investment, running costs and evaluation of the effects of financial considerations such as discounted cash flow and tax deductible expenditure. Protection by sacrificial anodes incurs the initial costs of the anodes and the deferred costs of labour and materials needed for periodic replacements. Impressed current protection incurs capital costs of power sources and permanent anodes and the running costs to supply power.

Cathodic protection schemes are applied in great variety. The present concise review summarizes the general principles and illustrates how they are exploited in practice by reference to two of the most basic applications, the protection of buried pipelines and distribution systems and the protection of ships and steel structures exposed to open waters.

20.1 Principles

Cathodic protection can be illustrated using the schematic corrosion velocity diagram for iron in aerated neutral water given in Figure 20.1.

The unprotected corroding system is:

Anodic reaction:

$$Fe \rightarrow Fe^{2+} + 2e^- \tag{20.1}$$

Cathodic reactions:

$$\tfrac{1}{2}O_2 + H_2O + 2e^- \rightarrow 2OH^- \tag{20.2}$$

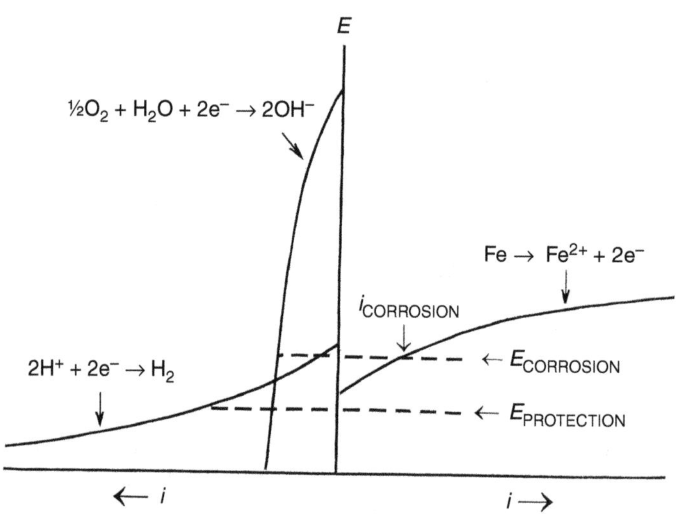

FIGURE 20.1
Schematic corrosion velocity diagram illustrating cathodic protection of iron in aerated neutral water, showing: (1) anodic dissolution current for iron at the corrosion potential in unprotected system sustained mainly by cathodic current due to absorption of oxygen; (2) anodic dissolution current eliminated at the protection potential in protected system with enhanced cathodic current dominated by discharge of hydrogen.

and

$$2H^+ + 2e^- \rightarrow H_2 \tag{20.3}$$

Assuming that the pH of the water is 7, that it is saturated with oxygen from the air and using the conventional criterion for corrosion $a_{Fe^{2+}} = 10^{-6}$, the calculations given in Example 3 of Chapter 3 yield equilibrium potentials for the anodic reaction, $E'_{Fe^{2+}} = -0.62$ V SHE and for the cathodic reactions $E_{O_2} = 0.85$ V SHE and $E'_{H^+} = -0.41$ V SHE. Conservation of electrons determines that the corrosion potential, $E_{CORROSION}$, has a value at which the total anodic and cathodic currents are equal, yielding the corrosion current for dissolution of iron, marked on the diagram as $i_{CORROSION}$, for the unprotected corroding system. The calculations further show that the dominant contribution to the cathodic current at the corrosion potential is the absorption of oxygen and that the contribution from the discharge of hydrogen is small.

Application of cathodic protection either by impressed current or by sacrificial anodes imposes a *protection potential* on the system below the rest potential for iron, at which the anodic current for the dissolution of iron vanishes and the metal is protected. At the same time, the ratio of the contributions to the cathodic current is changed in favour of the discharge of hydrogen. This effect is important in the context of overprotection considered in Section 20.4.

Impressed current protection is generally preferred for large or inaccessible systems such as the hulls of ocean going ships and long buried pipelines. It is also essential where contamination from the corrosion of sacrificial anodes cannot be tolerated in applications in chemical plants and food processing. The current is driven by a potential applied between the structure or system to be protected and inert auxiliary anodes, immersed in the aqueous medium or moist soil as appropriate. The cost of power to protect a large bare steel system by impressed current alone would be prohibitive, so that cathodic protection is usually applied in combination with protective coatings. For a ship's hull it is applied together with a paint system. A buried pipeline is usually protected with tarred wrapping.

20.2 Buried Pipelines and Distribution Systems

Cathodic protection is extensively applied to buried pipelines and distribution systems which must serve for many years with absolute reliability not only for the direct costs of capital investment, but also for consequential costs of disrupted supplies. They include water mains, oil pipelines, natural gas lines, power cables and telephone cables, ranging in scale from local distribution networks to pipelines thousands of kilometers long across the North American and European continents. For convenience, the arrangements required are discussed in terms of the protection of steel pipes conveying fluids, but with appropriate modifications, they apply generally to buried distribution systems.

20.2.1 Protection by Impressed Current

Figure 20.2 is a schematic illustration of an arrangement for a pipe cathodically protected by impressed current. Anodes set in a low resistance *backfill* act as counter electrodes accepting the return ionic current from the structure through the soil. A set of anodes at a particular site and the associated backfill is described as a *groundbed*.

The potential difference between the pipe and soil due to the impressed current is attenuated by the resistance of the pipe material from the point at which the current is impressed. This produces a potential profile along the pipe that can be analyzed by considering an idealized system illustrated schematically in Figure 20.3:

Let x = distance from the point where the impressed current is delivered

I_x = total current in the pipe at x

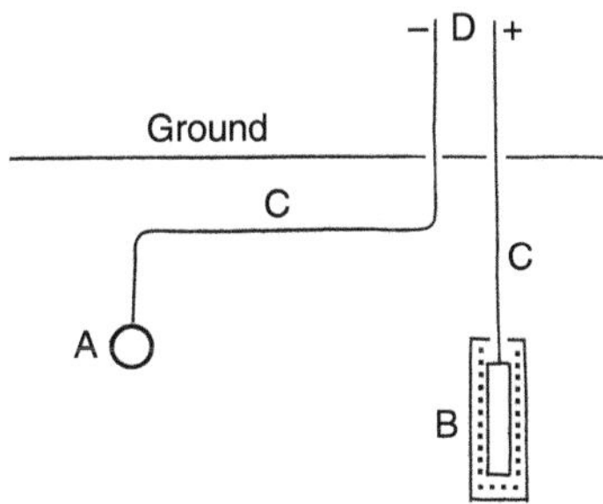

FIGURE 20.2
Buried pipe protected by impressed current. A = Cross section of pipe; B = Anode set in conducting backfill; C = Insulated cables; D = Source of DC potential.

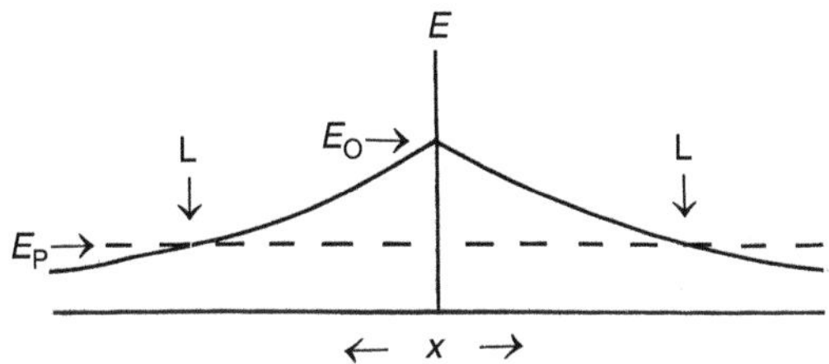

FIGURE 20.3
Idealized potential profile along pipe protected by impressed current. E_o = Applied potential; x = Distance along pipe from applied potential; E_P = Protection potential; L = Limits to protected section of pipe.

$-i_x$ = current decrement at a point x
E_x = potential difference between pipe and soil at the point, x
E_o = potential difference between pipe and soil for $x = 0$
R = resistance per unit length of the pipe

The change in total current at x is $-i_x$, that is:

$$\frac{\mathrm{d}I_x}{\mathrm{d}x} = -i_x$$

Applying Ohm's law:

$$\frac{\mathrm{d}E_x}{\mathrm{d}x} = -Ri_x$$

Assuming i_x is a linear function of E_x:

$$\frac{\mathrm{d}E_x}{\mathrm{d}x} = -R\mathrm{k}E_x$$

Rearranging:

$$\frac{1}{E_x}\frac{\mathrm{d}E_x}{\mathrm{d}x} = -R\mathrm{k}$$

Whence:

$$\frac{d\ln E_x}{\mathrm{d}x} = -R\mathrm{k} \tag{20.4}$$

Integrating between the boundary conditions $E_x = E_o$ at $x = 0$ and $E_x = 0$ at $x = \infty$ and converting to an exponential:

$$E_x = E_o \exp(-R\mathrm{k}x) \tag{20.5}$$

An idealized profile of the potential difference along a pipe according to Equation 20.5 is illustrated in Figure 20.3. Obviously, the potential difference at the point where the impressed current is delivered must be greater than the protection potential to allow for attenuation along its length, that is the pipe is overprotected along much of its length. There is a limit to the acceptable degree of overprotection both because it wastes power and because it promotes the unacceptable side effects described later in Section 20.4. This determines the maximum length of pipe that can be protected from one groundbed. A single groundbed can protect a pipe run of 1 to 80 kilometers, depending on the resistance of junctions between pipe lengths, the quality of protective coatings, soil conductivity and geographic and environmental features. Long pipe runs require repeated applications of impressed current with associated groundbeds at suitable intervals. The current can be supplied by DC generators but it is usually more convenient to supply it from local rectifiers fed from an AC line. In some remote windy locations it is possible to use wind turbines to supply the current, relying on slow dispersion of concentration polarization to cover intermittent calm periods.

20.2.1.1 Schemes for New Buried Pipelines

In principle, the theoretical analysis can be extended to predict the potential profile generated by an impressed current along a proposed new pipeline but it is seldom reliable as a design criterion because the idealized assumptions on which it is based may be unrepresentative of nonuniform soil conditions, the integrity of protective coatings and of pipe connections. The design of the system must take account of information from surveys of the compositions and resistances of soils and the conditions prevailing along the proposed route, including adverse features such as marshy ground, industrial contamination and climatic cycles. Resistances can be measured by the four electrode method, described in Section 20.5.2. The results of such a survey yields information from which the grade of protective coating and the impressed current requirements can be assessed from experience with similar pipelines and local knowledge. Hence, the arrangements for current supplies and the optimum nature and spacing of groundbeds can be determined to ensure that remote parts of the system are fully protected without excessive overprotection elsewhere.

20.2.1.2 Retrofitting to Existing Buried Pipelines

Cathodic protection can be applied retrospectively to existing pipelines. The nature of the soil and the conditions of the metal and any protective coatings may be unknown and inaccessible for comprehensive inspection, so that information on power requirements and groundbed dispositions must be acquired empirically. This can be done by pilot tests in which the system is polarized with respect to temporary groundbeds using a portable DC generator. The cathodic overpotentials developed are determined as explained later in Section 20.5.1 at progressive intervals from the drainage points, that is the points at which current is applied. Permanent arrangements for power supplies and the nature and disposition of groundbeds can then be determined. Series of tests at different locations are often required in established urban or industrial locations because there may be interference from pre-existing cathodic protection schemes or electric fields around power cables and adverse soil conditions derived from pollution and chemical contamination.

20.2.1.3 Anodes and Groundbeds for Impressed Current

In principle, any conducting material can be used for anodes but for economy and reliability the most common materials are steel, cast iron, 14.5 mass % silicon iron and graphite. They are arranged vertically or horizontally as convenient. The backfill, illustrated in Figure 20.2, is usually coke mixed with a slurry of gypsum and salt.

Anodes for impressed current are not sacrificial in the usual galvanic sense, but steel and cast iron anodes are progressively consumed by the reactions converting ionic to electronic current. For ferrous anodes the possible reactions are dissolution of iron as divalent or trivalent ions:

$$Fe \rightarrow Fe^{2+} + 2e^- \quad \text{for which } E^{\ominus} = +0.440\,\text{V SHE} \tag{20.6}$$

$$Fe \rightarrow Fe^{3+} + 3e^- \quad \text{for which } E^{\ominus} = -0.331\,\text{V SHE} \tag{20.7}$$

and discharge of oxygen:

$$H_2O = \tfrac{1}{2}O_2 + 2H^+ + 2e^- \quad \text{for which } E^{\ominus} = -1.228\text{ V SHE} \tag{20.8}$$

The consumption of iron depends on the Faradaic equivalent of the contributions of Reaction 20.6 and 20.7 to the total current. If all of the current were carried by Reaction 20.6, the consumption would be ~19 kg per ampere per year. In practice, the consumption is about half of that value, showing that there are significant contributions from the more anodic Reactions 20.7 and 20.8.

Graphite is consumed at a much lower rate than steel or cast iron, that is ~1 kg per ampere per year but it is more expensive and imposes greater power requirements because it functions as an electron collector and can sustain only Reaction 20.8 which not only has a higher redox potential but also has a high oxygen overpotential. Moreover graphite is brittle.

The 14.5 mass % silicon iron shares some of the advantages and disadvantages of graphite. It is more resistant than cast iron and steel to corrosion in soils so that the current it sustains is also predominantly due to Reaction 20.8 and is consumed at a rate of only ~1 kg per ampere per year, but it is expensive and brittle. A more durable proprietary silicon iron derivative containing 3 mass % of molybdenum is available at higher cost.

20.2.2 Protection by Sacrificial Anodes

Sacrificial anodes fulfill the dual purpose of auxiliary electrodes and galvanic sources of the electrical energy supplying the cathodic polarizing current. They are used where external sources of electrical energy are inappropriate or not readily available. The requirements for a sacrificial anode are:

1. It must establish a long range galvanic cell with a mixed potential below the rest potential of the protected metal in the prevailing environment.
2. It must resist self-corrosion due to local action cells.
3. It must resist passivation by film formation or accumulation of corrosion products.
4. It must be inexpensive and readily available in suitable forms.

Sacrificial anodes used to protect iron and steel are castings of magnesium, zinc and aluminium and alloys based on them. The anodic reactions driving the protected systems are:

$$\mathrm{Mg} \rightarrow \mathrm{Mg}^{2+} + 2\mathrm{e}^- \quad \text{for which } E^{\ominus} = -2.370\,\mathrm{V\ SHE} \tag{20.9}$$

$$\mathrm{Al} \rightarrow \mathrm{Al}^{3+} + 3\mathrm{e}^- \quad \text{for which } E^{\ominus} = -1.663\,\mathrm{V\ SHE} \tag{20.10}$$

$$\mathrm{Zn} \rightarrow \mathrm{Zn}^{2+} + 2\mathrm{e}^- \quad \text{for which } E^{\ominus} = -0.763\,\mathrm{V\ SHE} \tag{20.11}$$

These metals are inherently more expensive than the ferrous metal they protect and the economic benefit is the preservation of the added value of expensive structures and inaccessible systems. By suitable choice of the protecting metal, the ratio of the exposed area of the protecting metal to that of the protected metal and the number and disposition of sacrificial anodes, a structure or system can be completely protected. The choice of the metal is determined by the nature and severity of the environment. Some typical compositions are given in Table 20.1.

Alloys based on magnesium are the most effective because of the high driving force and current output, but the commercial metal is susceptible to self-corrosion by local action

TABLE 20.1

Characteristics of Some Typical Sacrificial Anodes

	Composition, Mass %								
Material	**Mg**	**Zn**	**Mn**	**Al**	**Sn**	**Impurities**	**Efficiency %**	**Resistivity of Soil, Ω cm**	**Potential V SHE**
Magnesium alloy 1	Bal	3.0	–	6.0	–	<0.05	50	<4000	1.55
Magnesium alloy 2	Bal	3.0	1.0	–	–	<0.05	40	<6000	1.75
High purity zinc	–	Bal	–	–	–	<0.01	90	<2000	1.1
Aluminium alloy 1	–	3.0	–	Bal	–	<0.5	50	<2000	1.1
Aluminium alloy 2	–	–	–	Bal	0.2	<0.5	30	<2000	1.4

cells that reduce its efficiency. Traces of copper, nickel and iron intensify self-corrosion by forming intermetallic compounds that are relatively noble with respect to the magnesium matrix and for this reason their concentrations as impurities are strictly limited. Their effects can be mitigated by alloying with a small addition of manganese which acts as a scavenger for them. Aluminium and zinc, introduced as significant alloy components, reduce the susceptibility to self-corrosion. These considerations led to the formulation of the magnesium–6 mass % aluminium–3% zinc alloy for soils of moderate resistance and the more active but less efficient magnesium–1 mass % manganese–3 mass % zinc alloy for soils of higher resistance, both with specified limits for impurity contents.

In principle, zinc appears to offer some advantages over magnesium for sacrificial anodes to protect iron and steel. It is less expensive, very efficient because it does not polarize significantly and less susceptible to self-corrosion. A mandatory requirement is that the metal must be pure, and in particular the iron content must be $<1.5 \times 10^{-3}$ mass % to avoid passivation by adherent films. The lower driving potential is insufficient to sustain protection over extensive systems in highly resistant soils but where soil conditions are suitable, it is possible to derive cost benefit by installing zinc anodes to protect medium sized distribution systems, such as steel gas mains in some large cities.

Aluminium alloy anodes are less widely used for buried structures than magnesium and zinc because they tend to passivate in soils and can cease to function as efficient sacrificial anodes. A depassivating backfill containing chlorides can be provided to counter this tendency. Despite these limitations, aluminium may be preferred if there are objections to magnesium or zinc corrosion products in runoff waters.

20.3 Cathodic Protection in Open Waters

The resistivities of seawaters are typically 20–30 Ω cm, permitting the good current distribution essential for cost effective use of cathodic protection and it is applied extensively to the hulls of seagoing ships, coastal installations and offshore structures. Appropriate schemes are selected from a menu of options including impressed current or sacrificial anodes, depending on the size of an installation, the availability of power supplies, the accessibility and the expected service life. In contrast, the resistivities of natural freshwaters are in the range 1000 to 100,000 Ω cm, which restricts current distribution so much that cathodic protection is usually uneconomic in river systems and freshwater lakes.

20.3.1 Protection of Ships' Hulls

Cathodic protection and paint coatings are complementary for protecting ships' hulls and it is standard practice to use them in conjunction. An impressed cathodic current of $\sim 10^{-1}$ A m^{-2} is required to protect bare steel in seawater but it is reduced to only 10^{-2} to 3×10^{-2} A m^{-2} for steel coated with a typical ship paint system. Large ships are protected by impressed current but sacrificial anodes are often more suitable for smaller ships without suitable power supplies but which can be serviced in dry dock. The protection is effective only below the water line, so that the upper hull above the water line must be protected by special attention to the provision and maintenance of the protective coating.

20.3.1.1 Protection of Hulls by Impressed Current

The impressed current is supplied by onboard generators applying appropriate potentials between the hull and totally immersed auxiliary anodes mounted outboard and insulated from the ship. The anodes are not consumable items, but good ones that are both efficient and resistant to attack are very expensive. The most popular contemporary anodes are of titanium with a very thin electrodeposited coating of platinum, as described in Section 12.3.3.4. The oxide on bare titanium protects it from corrosion but resists the passage of the protecting current from an aqueous medium. It will, however, freely pass electrons into another metal, so that the application of the platinum coating produces an efficient anode with a substrate that is protected in the event of local failure of the platinum film. However, the titanium thereby exposed passivates, reducing the effective area of the anode. The more expensive metals, tantalum and niobium, are sometimes preferred as the substrate for platinum because they are less susceptible to loss of efficiency in the event of damage to the platinum coating. The traditional anode material is lead, which is protected by a film of lead dioxide, which is sufficiently conducting to permit its function as the effective anode. A problem arises for lead electrodes used in chloride media, because the lead dioxide film can be undermined and detached by chloride ions penetrating through it to the lead surface. The problem is resolved by the lead/platinum bi-electrode in which multiple platinum microelectrodes are introduced into the lead surface. The net result is to stabilize the lead dioxide film but the electrochemical mechanism is complex and incompletely explained. Lead/platinum bi-electrodes have been largely superceded by platinised titanium.

In designing a system, the total anode area is matched to the current requirements for protection, within the maximum current densities given in Table 20.2. For example,

TABLE 20.2

Characteristics of Some Typical Anodes for Impressed Current

Material	Maximum Current Density A m^{-2}	Consumption kg A^{-1} per Year
Cast iron and steel	5	9
Iron – mass % Si – 0.85 mass % C	40	1
Graphite	20	1
Plain lead	200	–
Lead/platinum bi-electrode	10,000	–
Platinum-coated titanium	10,000	–
Platinum-coated tantalum	10,000	–

assuming a current density requirement of 2×10^{-2} A m^{-2} for steel coated with a typical ship paint coating, the total impressed current required for a vessel of 15,000 tonnes is of the order of 100 A. The sizes, number and dispositions of individual anodes are customized for particular ships and their assignments based on experience. Anodes are typically disposed along the keel with extra anodes at the stern to counteract the adverse bimetallic effect of propellers, which are often cast from manganese bronze. The requirement for extra current in the vicinity of the propellers varies with the electrical continuity between the hull and the propeller which can depend on lubrication and on whether it is static or running.

The degree of protection required varies with the usual factors that influence corrosion, including the composition, temperature and flow of seawater past the hull and time-dependent factors such as the build up of extraneous deposits and degradation of the paint coating. These factors are determined by locations, seasons, ship assignments and speeds so that automatic control of the impressed current based on signals from potential sensors is desirable to ensure that the hull is always adequately protected without incurring the penalties of overprotection described later in Section 20.4.

20.3.1.2 Protection of Hulls by Sacrificial Anodes

As with impressed current protection, cathodic protection with sacrificial anodes and paint coatings are complementary. Sacrificial anodes are simpler to install than impressed current schemes but they cannot be adjusted or automatically controlled so easily in service. Representative compositions for anodes based on magnesium, zinc or aluminium are given in Table 20.1. The compositions selected for anodes to suit particular conditions and ship assignments are based on experience. They are attached to the hull in electrical contact with it below the waterline at suitable intervals and left unpainted. The total anode area is determined from the current output required and the total mass required is assessed from the expected consumption between planned maintenance intervals determined from related experience, allowing for inefficiency, using information of the kind given in Table 20.1.

20.3.1.3 Protection of Internal Hull Components

Cathodic protection applied externally to a ship's hull does not provide protection to internal hull components and separate provision must be made for them where necessary. This applies especially to oil cargo tanks with seawater ballast and if neglected the damage can be very expensive for fleet operators. Internal impressed current schemes would present fire risk from sparks, so protection is applied by sacrificial anodes. Magnesium is the most effective anode material, but concern over a remote but tangible incendiary risk favours the use of zinc as a safer alternative despite the greater number of anodes required to compensate for its lower driving potential and its greater weight. As much as 20 tonnes of sacrificial zinc anodes is needed to protect the cargo tanks of a 20,000 tonne tanker. The anodes are disposed so that they are decontaminated in the course of normal tank cleaning procedures to reduce the risk of fouling by oil which reduces their efficiency.

20.3.2 Protection of Coastal and Offshore Installations

The measures required to protect coastal and offshore installations are generally similar to those for ships with some essential differences. The structures that may need protection include piles, lock gates, jetties, dry dock facilities, oil platforms and offshore pipelines These structures are usually supported on the seabed and immersed to levels that vary cyclically with the rise and fall of the tide. The part of a structure that is fully wetted intermittently is called the *tidal zone*, and there is a zone above the tidal zone that receives spray driven by waves and wind called the *splash zone*. The permanently immersed parts can be fully cathodically protected but the protection is ineffective in the tidal and splash zones. These regions must rely on other strategies for protection, but in contrast with the deeply immersed parts, they are readily accessible for monitoring and maintenance. Electrical continuity throughout a protected structure is essential to ensure that it is completely covered and to avoid inadvertently setting up differential cells that can accelerate corrosion of unconnected parts. A related precaution is to survey the site of a proposed scheme for possible mutual interference with unrelated pre-existing cathodic protection schemes. Sacrificial anodes are used to protect small scale installations and impressed current schemes are installed to protect large systems such as submerged pipelines and platforms servicing offshore oil fields. Further discussion of these specialized systems is deferred to Chapters 25 and 28, where it is more useful to consider them in the general context of strategies for corrosion control in marine environments.

20.4 Side Reactions and Overprotection

Natural waters contain dissolved oxygen and some also contain significant quantities of calcium, magnesium and bicarbonate ions. Cathodic currents, either impressed or generated by sacrificial anodes, can produce alkaline conditions and calcareous deposits:

$$\tfrac{1}{2}O_2 + H_2O + 2e^- = 2OH^- \tag{20.12}$$

$$Ca^{2+} + 2HCO_3^- + \tfrac{1}{2}O_2 + 2e^- = \downarrow CaCO_3 + CO_2 + 2OH^- \tag{20.13}$$

$$Mg^{2+} + 2HCO_3^- + \tfrac{1}{2}O_2 + 2e^- = \downarrow MgCO_3 + CO_2 + 2OH^- \tag{20.14}$$

The generation of OH^- ions raises the pH of the electrolyte at the metal surface tending to passivate the steel and calcareous deposits are also protective in principle, but the alkalinity can degrade standard paint coatings and imposes the use of special alkali-resistant paints.

If the impressed current is excessive, the cathodic polarization promotes hydrogen discharge to an unacceptable degree, by the reaction:

$$2H^+ + 2e^- = H_2 \tag{20.15}$$

Iron and steel are unique among engineering metals in that they are permeable to hydrogen at ambient temperatures and some of the cathodically produced hydrogen can enter the metal:

$$2H^+ + 2e^- = 2H\,(\text{solution in iron}) \tag{20.16}$$

Hydrogen absorbed from Reaction 20.16 accumulates at points of triaxial stress, where it can initiate cracks by the well-known phenomenon of *delayed failure* if the stress is sufficient. Iron or steel rendered susceptible to embrittlement in this way is said to be *overprotected*. An important aspect of the design of a protection system is to arrange the anodes to give a reasonably uniform potential over the surface, so that the entire structure is fully protected without significantly overprotecting any part.

20.5 Measuring Instruments

20.5.1 Measurement of Potentials

The general criterion for commissioning and monitoring a cathodically protected system is that the potential difference between a protected structure and its environment satisfies the theoretical requirements implicit in the design. The prevailing actual values are determined by measuring potential differences between the structure and a suitable reference electrode.

20.5.1.1 Buried Pipes and Utilities

The reference electrode almost universally used for buried pipes and other utility networks is the copper/(saturated copper sulfate) electrode ($Cu/CuSO_4$) because it is tolerant of rough handling in the field, easily serviced and inexpensive. A robust design is illustrated in Figure 20.4. The equilibrium is:

$$Cu^{2+} + 2e^- = Cu \quad \text{for which } E^\ominus = +0.337\,\text{V SHE} \tag{20.17}$$

Its potential relative to the standard hydrogen scale is given by applying the Nernst equation to Equation 20.17, inserting an appropriate value for the activity of Cu^{2+} in saturated copper sulfate solution, for example, $a_{Cu^{2+}} = 0.20$ at 25°C:

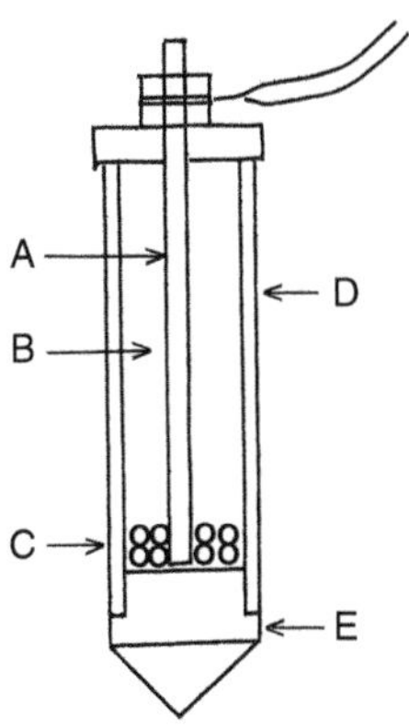

FIGURE 20.4
Schematic section through copper/saturated copper sulfate reference electrode. A = Copper rod; B = Saturated solution of copper sulfate; C = Copper sulfate crystals; D = Plastic container; E = Porous plug.

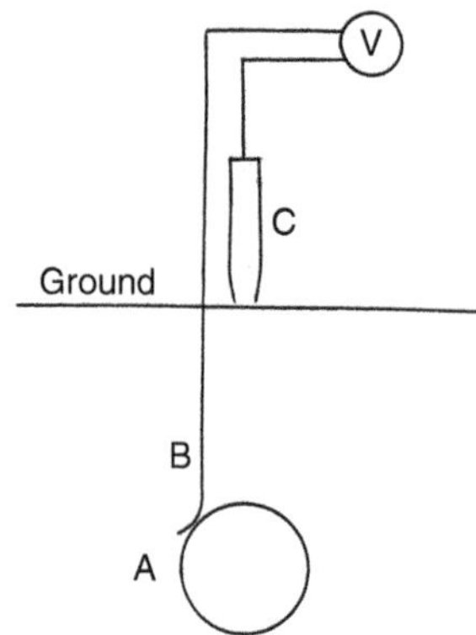

FIGURE 20.5
Measurement of potential of buried pipe. A = Cross section of pipe; B = Probe; C = Copper/copper sulfate reference electrode; V = High resistance voltmeter.

$$\begin{aligned} E &= +0.337 - \frac{0.0591}{2}\log\frac{1}{0.20} \\ &= +0.316 \text{ V SHE} \end{aligned} \tag{20.18}$$

Potentials are measured between the buried structure and the reference electrode, as illustrated in Figure 20.5. Contact with the pipe is made with a probe. Strictly, the reference electrode should be placed close to the structure to avoid error from potential drops in the soil but this is impracticable for buried structures and the reference electrode must be placed in contact with the surface of the soil. This practice does not introduce unacceptable errors, provided that the reference electrode is directly over the pipe and the potential difference is measured by an instrument that draws insignificant current, such as a high resistance voltmeter or preferably a potentiometer.

20.5.1.2 Immersed Structures

The reference electrode usually employed to measure the potential of an immersed structure is the silver/silver chloride electrode, described earlier in Section 3.1.4.3. Its potential relative to the standard hydrogen scale is +0.222 V at 25°C. Contact between the electrode and the aqueous medium is through a porous plug and since its position is not constrained by the liquid medium, it can be located close to the structure.

A length of high purity zinc rod immersed in an aqueous medium is sometimes employed as an alternative to recognized standard reference electrodes. It has practical utility in its simplicity and ability to maintain a constant potential characteristic of its local environment but it is not a universal standard.

20.5.2 Measurement of Soil Resistivities

The resistivity of a soil can be measured using four colinear electrodes inserted in the soil surface with the same distance, a, between consecutive electrodes, as illustrated in Figure 20.6. The outer pair are of metal and the inner pair are copper/saturated copper sulfate reference electrodes. The potential difference between the reference electrodes, E, is measured while a constant DC current, i, is passed between the outer electrodes.

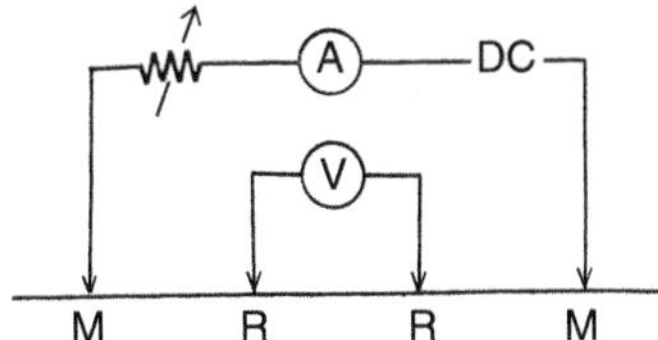

FIGURE 20.6
Measurement of soil resistivity. M = Metal electrodes; R = Copper/copper sulfate reference electrodes; DC = Source of DC potential; A = Ammeter; V = high resistance voltmeter.

The resistivity of the soil, ρ, is given by the following expression derived by application of Ohm's law to current densities centered on the outer electrodes, assuming isotropic resistivity and hemispherical symmetry:

$$\rho = 2\pi a \frac{E}{i} \tag{20.19}$$

The value accepted is the average of measurements for DC currents in both directions to cancel any effect of stray currents in the soil.

Further Reading

Martin, B. A., *Cathodic Protection Theory and Practice*, Ellis Horwood, Chichester, UK, 1986.

21

Corrosion and Corrosion Control in Aviation

A commercial airplane is designed and operated for maximum yield, that is capacity to deliver (e.g., expressed as passenger km) as a function of operating and capital costs. Among attributes contributing to an aircraft's yield are its size, shape, speed, mass and fuel efficiency. Some of these are constrained by aerodynamic considerations and external factors such as runway capacity and traffic patterns, but mass and fuel efficiency are strongly influenced by the materials of construction.

The importance of minimizing mass is illustrated by a specific example. The takeoff mass of a typical fully laden wide-bodied passenger aircraft is 370 tonnes, comprising 160 tonnes of structure including engines and 170 tonnes of fuel but only 40 tonnes of payload. Mass saved in the structure can be either transferred to the payload or used to secure some alternative economic benefit, such as enhanced fuel economy or extended range. Since the mass ratio of the structure to the payload is four, the effect of increments in mass on yield is magnified in proportion, for example, a 1% change in mass theoretically produces a 4% change in yield. Therefore, all contributions to mass are scrutinized. For example, metal sections and thicknesses are the reasonable minima needed to carry the maximum anticipated stresses safely and excessive structural reserves to compensate for *avoidable* corrosion are unacceptable. Even protective coatings, such as paints, contribute some mass and must be applied with discrimination. It is within this context that resistance to corrosion and other degradation agencies must be considered. The control of corrosion has various implications:

1. Safety is paramount and corrosion damage must not initiate structural failure in flight.
2. Aircraft represent heavy capital investment that must be protected. The costs of degradation and rectification must be within allowances made in planned amortization.
3. Airline schedules must not be disrupted through unplanned grounding.

These requirements must be met in the competitive context of airline operations and under the supervision of aviation authorities acting in the public interest, that is the United States Federal Aviation Authority (FAA), the United Kingdom Civil Aviation Authority (CAA) and equivalents in other countries. This imposes strict disciplines on the selection, fabrication, surface treatment and service history records of the structural materials.

21.1 Airframes

21.1.1 Materials of Construction

For economy in mass, cost benefit analysis reveals that the most desirable properties of airframe materials are low density, high modulus and damage tolerance. Following years of experience, the materials that best meet these requirements are the age-hardening aluminium alloys in the AA 2000 and AA 7000 series specified in Table 9.2 in Chapter 9, particularly AA 2024 and AA 7075 alloys. Typically, fuselage skins are fabricated from rolled sheet of the more damage-tolerant alloy, AA 2024, supported on frames usually made from the stronger alloy, AA 7075. Load-bearing members such as landing gear and wing beams may be fabricated from the titanium alloys. A modern practice is to machine wings to shape from AA 2024 alloy plate to produce integral stiffeners and to optimize mass by varying the thickness to keep stress levels consistent. There is some replacement of aluminium alloys by carbon fiber and composites where the imposed stress system is suitable. Aluminium alloys containing lithium offer density and modulus advantages for future use but they are not generally accepted for aircraft at present, although they are applied in space vehicles.

21.1.2 Corrosion and Protection of Airframes

21.1.2.1 Aluminium Alloys

21.1.2.1.1 Protective Coatings

The high copper and zinc contents that confer strength on AA 2024 and AA 7075 alloys reduce the protection afforded by the natural passivation of aluminium and they must be protected by surface coatings. Rolled sheet metal is protected by roll-bonded cladding applied to both sides by the metal manufacturer. AA 2024 and AA 7075 are clad with commercially pure aluminium (AA 1100 in Table 9.2) and an aluminium-1% zinc alloy, respectively, to confer galvanic protection, using a cladding coating of 2½% to 10% of the sheet thickness on each side depending on the gage. Cladding is not possible on machined surfaces and paint protection systems based on chromate-inhibited primer must be used, applied by electrostatically guided spray over anodized coatings produced in sulfuric or phosphoric acid. The same applies to panels that have been formed by shot peening.

21.1.2.1.2 Internal Corrosion

Contrary to casual observations, corrosion of airframes is more severe on the inside than on the outside. Water causing corrosion on the inside of the fuselage structure is accumulated from vapour from human sources condensing against the cold fuselage skin where insulating materials can obstruct drainage. The corrosion is insidious, developing out of sight and eating into structural reserves. The decisive criterion is area loss, reducing the ability of the material to transfer loads correctly within the structure. It is unusual to find substantial damage on open flat panels and corrosion is generally associated with specific features of the structure.

Corrosion around faying surfaces, rivets, joints and crevices is particularly troublesome. The incidence and severity of corrosion in these places is sensitive to details of both design and construction. Errors that encourage corrosion include:

1. Crevices between fuselage skins and frames
2. Dry, that is unsealed, joint assembly
3. Gaps where sealant fails to protect interfaces between multiple skins applied to reduce stress levels around cutouts such as door frames
4. Sections disposed so as to form gravity traps for water

Based on experience, designs and construction techniques have been modified to reduce corrosion damage. These expedients include:

1. Eliminating multiple skins at cutouts by providing the extra thickness needed for reinforcement from thicker sheets, profiled by chemical milling
2. Providing drainage holes closed by spring-loaded valves to prevent accumulation of water at critical places
3. Assembling joints with extra care in applying sealants to prevent ingress of water

Certain zones are corrosion free. Examples are the insides of pylon box sections that are heated by the engines. Supersonic aircraft are exceptional in that the whole airframes are virtually corrosion free inside because they fly warm from frictional heating at cruising speed.

The whole of each wing structure forms a fuel tank and it inevitably collects water through the temperature and humidity cycles experienced. If the water remained, there would be a possible biological corrosion hazard by the growth of a fungus, *Gladisporum resoni*, at the water/fuel interface that generates acids. In modern aircraft, any such problem is eliminated by *sump pumping* the fuel from low remote points in the tank where any water could collect. This extracts the water and delivers it harmlessly to the engines. Provided that this is done, there are no residual corrosion problems on the *insides* of wings.

21.1.2.1.3 External Corrosion

At low altitudes, the airframe is exposed to the natural atmosphere. It can become more aggressive if chlorides are acquired in flight from runway deicing salts or marine atmospheres. Nevertheless, general corrosion is less on the outside than on the inside except in particular areas where it can be severe on some aircraft. Vulnerable areas include the forward faces of the front and rear wing spars and undercarriage bases, associated with erosion–corrosion induced by the airflow. External damage is easier to detect and rectify than internal corrosion.

21.1.2.2 Titanium Alloys

Components fabricated from titanium alloys serve vital functions as load-bearing members and accessories in airframes by virtue of the unique combination of properties characteristic of the metal, including high strength/mass ratios, easy formability and outstanding resistance to corrosion and environmentally sensitive cracking in natural environments. They enhance airframe strength and contribute to the high standards of integrity required for safety in flight.

Some representative alloys for aerospace applications are indicated in Table 12.2 in Chapter 12. They are especially valuable where resistance to corrosion fatigue and stress-corrosion cracking is required and they can account for as much as 10% of the tare weight of a modern commercial airplane. A particular safety-critical application is for landing gear trucks of wide bodied passenger airplanes, which are subject to repetitive cyclic loading on landing, sometimes in the presence of chlorides acquired in flight. The β forging alloy, designated 10-2-3 in Table 12.2 is well-suited to this application by virtue of its high strength combined with outstanding general corrosion resistance and insensitivity to corrosion fatigue and stress-corrosion cracking in the presence of chlorides. The high strength/mass ratios available with β and α/β titanium alloys are also exploited as internal load-bearing members to reinforce airframes, including wing beams, for which resistance to environmental cracking in the presence of condensed moisture is also essential.

There are less obvious but important applications for the medium strength predominantly α alloys which require a high degree of corrosion resistance combined with ductility and weldability. An example is the Ti3Al25V alloy, quoted in Table 12.2, which can be drawn to tube for hydraulic systems.

21.1.2.3 Other Metals

21.1.2.3.1 Steels

Some airframe fittings are made from high strength steels, protected by electrodeposited coatings of cadmium for bimetallic compatibility with aluminium alloys. They include flap-track fittings and pylon pins together with their terminating attachment points that unite the engines with the wings. Corrosion of these components is particularly serious and must be monitored diligently and rectified promptly because of interference with vital functions.

21.1.2.3.2 Aluminium Bronzes

The resistance of aluminium bronzes to atmospheric corrosion combined with their load-bearing capacity at low rubbing speeds without undue wear or distortion makes them particularly suitable for bearing bushes in aircraft frames. The nickel aluminium bronze, designated CuAl10Fe5Ni5 in Table 10.2 of Chapter 10, is widely employed for this purpose.

21.1.3 Environmentally Sensitive Cracking

21.1.3.1 Fatigue

Fatigue cracking is a matter of great concern. Although not necessarily a corrosion problem, the possibilities of enhancement by an aggressive environment, as described in Chapter 5, is recognized. The cyclic stresses are imposed in various ways, by flight maneuver loads, gust loads, landing loads and pressurization and the maximum loads are in landing and banking. Therefore, the fatigue cracks are predominantly propagated by imposition of variable high stress, low-cycle loading, with occasional peak loads. The effects of these loads can be traced in postmortems from markings on fractures.

21.1.3.2 Stress-Corrosion Cracking

If heat treatment is incorrectly applied, age-hardened aluminium aircraft alloys are susceptible to stress-corrosion cracking when exposed to sources of chlorides acquired in flight from marine atmospheres or deicing salts, but the problem is averted by assiduous attention. The high strength AA 7075 alloy is much more susceptible to stress-corrosion cracking than AA 2024 type alloys but a newer modified alloy, AA 7079, is more tolerant.

The characteristics of stress-corrosion cracking are well-known for the standard alloys and the risk can and must be eliminated. In airframe design, stress analyses must confirm that stresses in service will be below the threshold stresses for stress-corrosion cracking with an adequate safety margin. In manufacture, heat-treatment procedures are optimized for minimum susceptibility and must be meticulously controlled to ensure that the threshold stresses assumed in design are actually realized in the products. In assembly, care is required to avoid adding internal stresses. Minimizing susceptibility to stress-corrosion cracking takes priority over exploiting the maximum strength of an alloy. Stress-corrosion cracking in aluminium alloys is structure-sensitive because of the intergranular crack path so that they are most susceptible in the short transverse direction of forged components.

Consequently, the stress thresholds are reduced in this direction and must be taken into account in design.

21.1.4 Systematic Assessment for Corrosion Control

Incidents in flight due to corrosion and related degradation in airframes are rare but in one well publicized event, a large section of a fuselage was lost; fortunately the airplane survived. This alerted the industry and stimulated collaboration that established mandatory standards of prevention, monitoring and control with special reference to aging aircraft.

The US Congress considered legislation to ground aircraft as they reach design-life goals. However, design-life goals based on flight cycles, calendar time or flight hours are guiding concepts in aircraft design and applying them as definitive operational safety limits would have grounded a significant part of the world's aircraft fleet without necessarily contributing to safety because operators find that fail-safe rather than safe life criteria determine structural integrity. A meeting of industry leaders identified six issues relevant to aircraft aging:

1. Local fatigue cracking
2. Corrosion
3. Repairs
4. Maintenance programmes
5. Significant structural inspection details
6. Widespread fatigue damage

Corrosion assessment was assigned to a separate stand-alone programme formulated after a four year study by a structures working group drawn from leading operators, manufacturers and regulators, in which operators assumed a vitally important role due to their accumulated experience derived from observations during maintenance and innumerable service bulletins on aging aircraft. Corrosion control was deemed to be too important to be conducted on an uncoordinated basis and a systematic industry-wide inspection programme based on a zonal system was imposed for all aircraft.

In this zonal system, the whole structure, irrespective of aircraft type, is regarded as an assembly of zones and sub-zones. Every zone in which there is a corrosion concern is accessed at mandatory intervals and inspected specifically for corrosion. The inspection frequency is prescribed zone by zone. The extent of corrosion and the action to be taken are recorded by the following industry standard reporting levels:

Level 1. Corrosion within prescribed tolerable limits
Action — Clean corroded area and resume flying service

Level 2. Corrosion outside of tolerable limits but rectification is permissible
Action — Repair

Level 3. Corrosion too extensive for repair
Action — Refer for assessment of airworthiness

A further part of the programme is the application of proprietary corrosion prevention fluids, conforming to industry standards. One kind is wax-based with joint-penetrating and water-repellant properties. The other kind contains inhibitors for long-term protection.

A minor disadvantage of the practice is that these fluids encourage the accumulation of dust on the surface, obscuring observations in subsequent inspections.

Compliance with the programme in its entirety is mandatory and subject to verification. The structures working group meets annually to review it.

21.2 Gas Turbine Engines

Whittle's pioneering work on gas turbine engines in the 1940s was inevitably based on the relatively uncomplicated nickel/chromium alloys that were available at the time. The most exacting requirement was for turbine blade and disc materials to withstand the stresses and corrosive environments to which they were exposed and the engine efficiencies were limited by the relatively low temperatures at which these materials could be used. This imposed low thrust/mass ratios that restricted aircraft performance and inefficient conversion of thermal to mechanical energy. Since then, continuing commercial and military demands for more efficient engines have stimulated the development of customized nickel alloys with requisite mechanical properties and corrosion resistance at ever higher engine operating temperatures. The characteristics of these alloys are discussed in Section 21.2.2, where corrosion resistance and strength are considered together as different aspects of overall high temperature performance.

21.2.1 Engine Operation

Air collected at the front of the engine is compressed by a ratio of up to 30:1 reaching a pressure of 3 MN. It is delivered into a combustion chamber, mixed with vapourized liquid fuel and burned, yielding a high pressure gas stream that enters a turbine which extracts energy to drive the compressor through shafts. The balance of energy in the gas stream is discharged as a reactive thrust through a nozzle at the rear of the engine.

There are four basic engine arrangements:

1. *Turbojets* in which the turbine drives only the compressor and most of the energy is delivered as jet thrust. These are high performance engines with military applications.
2. *Turbofans* in which the turbine extracts additional energy to drive a large fan at the front of the engine, generating a low pressure airflow that partially bypasses the core of the engine and supplements the hot exhaust gases. This reduces the total jet velocity, increasing the efficiency of propulsion and reducing noise and fuel consumption.
3. *Turboshaft* and *turboprop* engines in which the turbine is designed to extract as much energy as possible and deliver it to a power output shaft. Turboprop engines have been used to drive propellers for slower fixed wing aircraft and turboshaft engines provide power for helicopters.

Figure 21.1 schematically illustrates the layout of a large three-shaft turbofan that is a standard engine type in large commercial aircraft. The turbines and compressors comprise blades of aerofoil sections attached to discs mounted on the shafts. Rows of static guide vanes are interposed between the successive sets of rotating blades. There are three successive sections

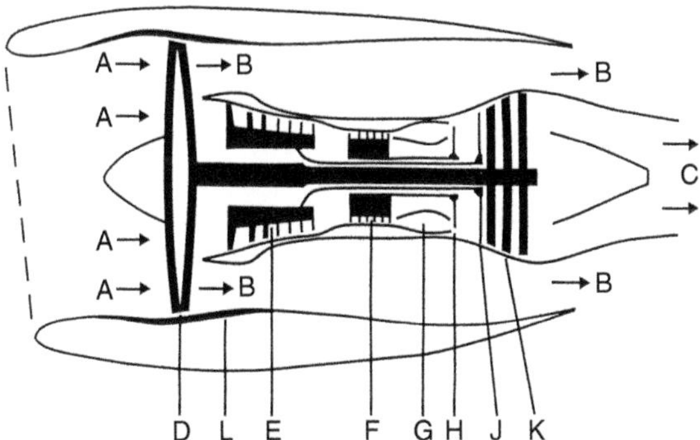

FIGURE 21.1
Basic layout of a three-shaft turbofan. Guide vanes omitted for clarity. A = Air intake; B = Bypass thrust; C = Jet thrust; D = Fan; E = First compressor section; F = Second compressor section; G = Combustion chambers; H = High pressure turbine; J = Intermediate pressure turbine; K = Low pressure turbine; L = Fan casing.

in the turbine, the high, intermediate and low pressure sections. Every section has a dedicated function in the compressor. The low pressure section drives the fan, the intermediate pressure section drives the first compressor section and the high pressure section drives the final compressor section. The rotating assembly of a turbine section, its counterpart in the compressor and the drive shaft carried on bearings is a *spool*. The drive shafts of the three spools are concentric. Turbine and compressor sections are multistages, for example, the low pressure turbine section driving the fan may have as many as five stages.

21.2.2 Application and Performance of Nickel Alloys

The design of gas turbine engines has always been limited by the prevailing development of materials from which various engine components can be fabricated to suit the conditions in which they must operate. Alloys based on nickel with stable microstructures and the requisite mechanical properties at engine operating temperatures have been developed by manipulating classical metallurgical principles as described in detail in Chapter 11. These materials, the *nickel superalloys*, are represented by the more commonly applied alloys specified in Table 11.4. Their performance in service must now be considered in relation to conflicting demands on alloy content to secure the mechanical properties and resistance to corrosion at engine operating temperatures. There are two distinct forms of high temperature corrosion that can be encountered, oxidation in unpolluted combustion products and hot corrosion with sulfidation in the presence of $NaCl/Na_2SO_4$ dew derived from sulfur in the fuel and chlorides in ingested air. These factors are influenced both by engine operation and by aircraft assignments.

21.2.2.1 Oxidation in Unpolluted Combustion Products

In products of combustion free from sulfur and chlorides the high temperature components, turbine blades, turbine discs and combustion chambers resist oxidation by virtue of protective surface films formed by selective oxidation. Depending on the alloy composition, the films are chromia, Cr_2O_3, or alumina, Al_2O_3, and alloys are distinguished by the terms, *chromia formers* and *alumina formers*.

As explained in Section 11.2.2.2 of Chapter 11, the strongest superalloys have high aluminium–titanium ratios and high concentrations of other components for solid solution strengthening. Because of the requirements for γ' stability and the need to retain alloy components in solution, the chromium contents are consequently reduced to below the critical compositions needed to stabilize Cr_2O_3 as the protective film and its role is assumed by a film of Al_2O_3, which is thermodynamically stable with respect to the alloy bulk composition. However, the alloy is unable to maintain the protection unaided because the supply of aluminium by relatively slow diffusion from the interior is insufficient to replenish that which is consumed in forming and repairing the film. The anticipated deficiency is pre-empted by *aluminizing* the surface, that is diffusing aluminium into the surface to produce an aluminium–nickel alloy layer capable of supporting the protective alumina film.

21.2.2.2 Aluminizing for Oxidation Resistance

The simplest and most common process is *pack aluminizing*. Blades are cleaned by blasting with an abrasive and then packed and sealed in a mixture of aluminium powder, alumina, Al_2O_3, and a halide, typically ammonium chloride, NH_4Cl. The pack is heated for a few hours at a temperature in the range 750°C to 1000°C, typically 900°C. The alumina has no active part and is there to prevent coalescence of the aluminium powder particles when they melt. The ammonium chloride generates aluminium monochloride by the reaction:

$$2NH_4Cl(s) + 2Al\ (\text{powder}) = 2AlCl(g) + 2NH_3(g) + H_2(g) \qquad (21.1)$$

initiating a cyclic process that transports aluminium from the powder to the alloy through the vapour phase:

$$3AlCl(g) = 2Al\ (\textit{at alloy surface}) + AlCl_3(g) \qquad (21.2)$$

$$AlCl_3(g) + 2Al\ (\textit{powder}) = 3AlCl(g) \qquad (21.3)$$

An aluminized coating can form without a halide but it is uneven and unsatisfactory.

The coating is an intermetallic compound, β-NiAl, containing up to 40 weight % aluminium. It confers good protection and is sometimes used in that condition, but it is brittle and can crack. Lively blades that experience flexing are given a subsequent diffusion anneal for a short time at a higher temperature, typically 1100°C, to disperse the aluminium further into the metal, conferring some ductility in the coating. The diffusion anneal doubles the coating thickness and reduces the surface aluminium content to between 20 and 30 mass %. The final coatings are 25 to 75 μm thick and increase the blade dimensions by about half as much.

Considerable effort has been expended in a search for improved coatings. Combination of chromium with aluminium yields only marginally better coatings, but additions of silicon significantly improve the sulfidation resistance and enrichment with platinum by preplating the alloy surface with a thin layer of platinum prior to aluminizing enhances the high and low temperature performance of the coatings by introducing aluminium–platinum compounds.

21.2.2.3 Hot Corrosion and Sulfidation

Blade and other engine components exposed to moderate temperatures can be attacked by *hot corrosion* and *sulfidation*, which are specific terms describing attack initiated by a liquid phase containing sodium sulfate and sodium chloride, derived from salts ingested in air contaminated with sea salt. Sea salt already contains a significant amount of sodium

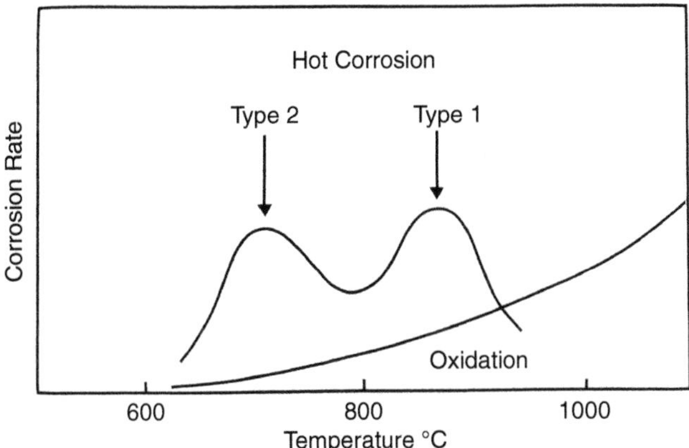

FIGURE 21.2
Temperature dependence of oxidation and hot corrosion rates (schematic).

sulfate, Na_2SO_4, and more can be produced by interaction between sodium chloride and sulfurous oxides in the combustion products from the fuel:

$$2NaCl + SO_3 + H_2O = Na_2SO_4 + 2HCl \tag{21.3}$$

At the highest engine temperatures, the sodium sulfate is vapourized but at temperatures in the order of 650°C to 900°C, the liquid phase condenses as a dew on the metal surfaces, attacking protective oxides leaving the metal vulnerable to accelerated oxidation and *sulfidation*, a form of corrosion in which sulfur penetrates into the metal locally precipitating chromium as chromium sulfide, Cr_2S_3.

There are two regimes of hot corrosion, designated Type 1 and Type 2 that operate in different temperature ranges, as illustrated schematically in Figure 21.2. In Type 1 corrosion at the higher temperatures, the sodium sulfate has an acidic character imparted by the absorption of oxides of alloy components such as molybdenum and tungsten. In Type 2 corrosion at the lower temperatures, it has a basic character due to the release of O^{2-} ions when sulfur is absorbed by the metal. In either condition, it can attack the amphoteric oxides, Cr_2O_3 and Al_2O_3, but Al_2O_3 is the more resistant to acid fluxing and Cr_2O_3 is the more resistant to basic fluxing.

To limit the attack, aircraft fuels are standardized with low sulfur contents but hot corrosion is always a potential problem to a greater or lesser degree.

21.2.3 Engine Environments

The thermal and chemical environments within an engine depend on the functions of the various parts and on aircraft assignments, maneuvers and flight patterns.

21.2.3.1 Factors Related to Engine Operation

The fuel is burnt with excess oxygen so that the gaseous environment is oxidizing. The highest temperature is experienced in the high pressure turbine stage, where at full power the gas temperature approaches 1500°C and the high pressure turbines that first receive the flame may reach a peak metal temperature of 1100°C or even 1150°C. The gas cools in passing through the succeeding intermediate and low pressure turbine stages, where the lowest blade temperatures are typically 700°C. There are temperature gradients within the individual blades, a topic taken up later in relation to fatigue. The rims of the turbine discs on which the blades are mounted can experience temperatures as high as 600°C and the walls of the combustion chambers can reach a temperature of 1000°C.

Compressor blades are not exposed to the flame but they are heated by a combination of the thermal ambience and the adiabatic compression of the air intake. Blades in the second section can reach temperatures between 300°C and 600°C.

21.2.3.2 Flight Pattern Factors

The severity of the environments within engines depend on aircraft assignments and are conveniently considered within the following categories:

1. Long haul civil applications, for example, transatlantic and non-stop transcontinental flights
2. Short haul civil applications especially on island routes, for example, linking Hawaiian Islands
3. Military aircraft flying low over water

Long haul flights impose the least demands on the corrosion resistance of engine components because they comprise long single flight cycles of takeoff, cruising and landing, Full engine speed is required for takeoff and climb and the blade temperature rises rapidly through the ranges for hot corrosion to their peak values. On reaching the cruising altitude, typically 10,000 m, the engine speed is reduced and the temperature falls to a steady lower value where it remains until landing. After landing, the engines accelerate for a few minutes to apply reverse thrust. The temperature cycle is represented in Figure 21.3. Aluminized high strength superalloys resist oxidation in clean air at the constant temperatures below their peak values experienced during cruise. Ingestion of air contaminated with salt is possible only during a short interval immediately after takeoff and during descent.

In short haul applications, the engines experience multiple flight cycles with frequent thermal cycling through the temperature ranges for hot corrosion and if the flights are

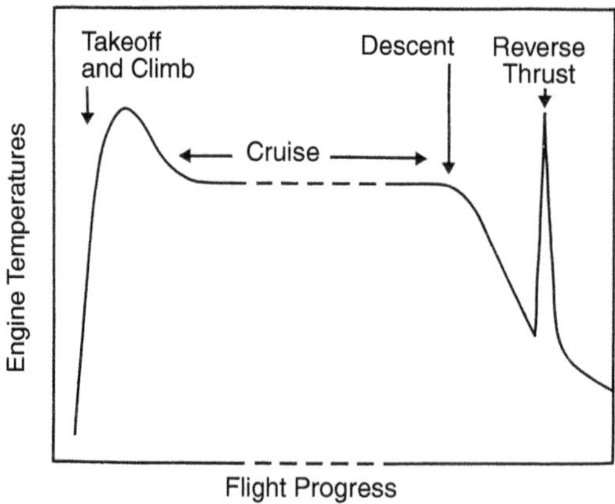

FIGURE 21.3
Engine temperatures during progress of a long haul flight (schematic).

over seawater there are frequent episodes in which air contaminated with salt is ingested at low altitudes during climb and descent, creating conditions for hot corrosion.

Engines on military aircraft flying low over the sea are exposed to continual thermal cycling in salt-laden air. Helicopters create spray over the sea with the down draught from their rotors and their engines are particularly vulnerable.

21.2.3.3 Fatigue

Flight patterns and maneuvers also influence fatigue life. Engine components, especially turbine blades and discs experience a combination of cyclic stresses. Mechanical stresses are imposed by centrifugal force and bending moment in the gas stream. Further stresses are superimposed by cyclic variations in engine temperature due to the application of power for takeoff, reverse thrust and maneuvers. Thus, the expected fatigue lives are longer for long haul than for short haul and military assignments.

21.2.4 Alloy Selection

21.2.4.1 Turbine Blades

Turbine blades that approach temperatures of 1150°C in civil applications are usually produced from one of the high strength multicomponent alloys with high aluminium and low chromium contents, such as MAR M002, which have good oxidation resistance. Where environments or aircraft assignments impose greater exposure to hot corrosion, alloys with higher chromium contents may be preferred. These high strength alloys are difficult to fabricate and they are produced by investment casting, a method that is capable of yielding the complex shapes needed with precision. Wax patterns of the blades are assembled on a wax support that serves as a pattern for the runners to fill the mould with liquid metal. The assembly is coated with successive layers of a ceramic slurry dried in an oven to produce a self-supporting shell. The wax is melted out leaving cavities of the required shapes. The shell is insulated and the alloy is cast into it. Some alloys can be cast in air but those containing reactive elements are cast in vacuum. Blades for low pressure sections can be polycrystalline but blades for the high pressure turbine section are cast by directional solidification to eliminate grain boundaries normal to the principal stress axes. This is done by establishing a thermal gradient through the metal as it solidifies in the mould, using chills and auxiliary heat sources. It can be modified by crystal selection to produce whole blades as single crystals, eliminating all grain boundaries. Some standard high strength alumina-forming alloys, such as MAR M002 have been used extensively for directionally solidified blades and other alloys have been have been customized for these advanced casting techniques. Two examples given in Table 11.4 in Chapter 11 are SRR 99 for directional solidification and CMSX-4 for single crystals.

Critical surfaces of the cast blades are precision machined or ground. The allowable peak temperatures for the blades are significantly lower than the flame temperatures so multiple channels are cut into the blades to serve as ducts for cooling air supplied by the compressor. The channels are cut by electro-discharge machining or lasers because these methods can be controlled to cut channels with uniform overlying metal thicknesses, minimizing the thermal gradients experienced in service to improve resistance to thermal fatigue.

21.2.4.2 Turbine Discs

Materials for turbine discs must resist oxidation at the moderately high temperatures experienced at the rims and they must also be suitable for forging to shape. Hot corrosion is not usually a significant problem with turbine discs because cooling air is used to keep the contaminated gas stream away from the rims. Disc life is determined mainly by fatigue resistance and it is maximized by carefully controlling the microstructure. This requires alloys that are not unduly subject to segregation in the ingots cast for forging and careful control of the deformation sequence and subsequent heat treatment. The high strength superalloys do not meet these requirements. The best-suited materials are the chromium-rich alloys, IN 718, IN 901, Waspalloy and Astralloy, specified in Table 11.4.

21.2.4.3 Combustion Chambers

Combustion chambers are annular structures formed from sheet. The chamber walls experience thermal cycling with peak temperatures above 1000°C. These factors impose requirements for alloys with good fabrication and welding characteristics combined with resistance to oxidation and thermal fatigue, such as C263 specified in Table 11.4.

21.2.4.4 Compressor Assemblies

The first compressor section is a welded assembly comprising a drum carrying the blades. It is usually fabricated from the titanium alloy ASTM grade 5 with 6% aluminium and 4% vanadium, listed in Table 12.2, that combines high strength with low mass. It is relatively cool and resists corrosion in the air it collects. The blades in the final compressor section are produced from a creep- and oxidation-resistant alloy because for the reasons given earlier in Section 21.2.3.1, they attain temperatures in the range 300°C to 600°C. Titanium alloys suitable for moderate temperatures are specified in Table 12.2, but for the highest temperatures a nickel alloy may be required, for example, IN 718, specified in Table 11.4

As a precaution against a remote fire risk associated with the use of titanium, the stationary guide vanes between the stages are made from a martensitic stainless steel to act as fire breaks. Martensitic stainless steels resist corrosion in the atmosphere well and the guide vanes do not normally need protection, but they can corrode in heavy condensates from humid air when aircraft are grounded in tropical locations, especially if the engine contributes salt contamination acquired in flight. If this is anticipated, they are protected with paint films containing aluminium powder as a pigment for sacrificial protection. To serve this purpose, the paint is made conductive by peening and heat treating it to establish electrical contact between the aluminium particles.

21.2.4.5 Cool Components

Titanium alloys are selected for the front fans of turbofan engines to exploit their outstanding resistance to erosion–corrosion in fast flowing turbulent natural atmospheres and their high strength/mass ratios that can be used to minimize centrifugal forces. Fan containment casings are fabricated either from titanium alloy sheet or from aluminium alloy sheet matching that is used in airframes and protected by chromic acid anodizing.

Shafts made from heat-treated high carbon low alloy nitriding steels* are partly protected by lubricating oils and the exposed non-oiled surfaces can be protected by sacrificial or barrier paints filled with aluminium pigments. Corrosion problems for bearings are not so much during service as in the storage of spares. To meet stringent requirements for freedom from pitting they must be packed in wax preservatives with vapour phase inhibitors.

Manufacturers give warranties on component lives but expect to exceed them. They cooperate with customers to prescribe monitoring procedures during service and make arrangements for engines to be inspected for condition on the wing. Suitable facilities are incorporated in engine design, for example, by providing fiber optic paths to examine critical areas.

Successive engine designs are upgraded to meet increasing demands for safety, fuel economy, noise abatement, mass reduction and care of the environment. Manufacturers have strategies to anticipate and prepare for future needs. So much investment is at stake that new materials or coatings can be incorporated in engines only after they have been proven on demonstrator rigs with full instrumentation, designed to simulate service environments.

Further Reading

Crane, F. A. A. and Charles, J. A., *Selection and Use of Engineering Materials*, Butterworths, Boston, Massachusetts, 1984.

Flowers, H. M. (Ed.), *High Performance Materials in Aerospace*, Chapman & Hall, London, 1995.

Forsythe, P. J. E., The fatigue performance of service aircraft and the relevance of laboratory data, *Journal of the Society of Environmental Engineering*, 19, 3, 1980.

Meatham, G. W. (Ed.), *The Development of Gas Turbine Materials*, Applied Science Publishers, London, 1981.

The Jet Engine, 5th Edition, Rolls-Royce, Derby, UK, 1996.

Thomas, M. C., Helmink, R.C., Frasier, D. J., Whetstone, J. R., Harris, K., Erickson, G. L., Sikkenga, S. L. and Eridon, J. M. Alison manufacturing, property and turbine engine performance of CMSX-4 single crystal airfoils. In *Materials for Advanced Power Engineering, Part II*, Kluwer Academic Publishers, Liege, Belgium, 1994.

* Steels containing chromium or aluminium that can be surface hardened by heating in ammonia/hydrogen mixtures.

22

Corrosion Control in Automobile Manufacture

22.1 Overview

Automobiles are built for an internationally competitive market that is demanding not only in vehicle performance and price but also in subjective customer perception. Corrosion-resistant coatings for the body therefore have both protective and cosmetic functions. The cosmetic aspect applies both to initial showroom appearance and to premature development of blemishes. For these reasons, the quality and consistency of paint coatings has a crucial influence not only on vehicle durability but also on sales potential.

Corrosion protection of the body must be durable but not to the point where it incurs excessive costs to maintain it beyond the vehicle life expectancy due to other limiting factors, such as degradation of moving parts, obsolescent technology and changes in fashion. All of these aspects of durability are synchronized in a *design life goal* that for many automobile volume manufacturers is six or seven years. Safety concerns due to loss of structural reserves by corrosion should not be an issue within the design life, assuming normal maintenance.

Premature corrosion in the engine and associated systems can cause unreliability and expense. Corrosion can interact with the mechanical degradation of moving parts by wear, fatigue or erosion to an extent influenced by the compositions of fuels, oils and products of combustion. The protection of recirculating water-cooling systems is complicated by the use of antifreeze additives. The integrity of exhaust systems that have hitherto been replaceable items, is now a serious concern because of widespread adoption of catalytic converters for emission control, reduced noise tolerance and a trend by manufacturers to extend warranties.

22.2 Corrosion Protection for Automobile Bodies

22.2.1 Design Considerations

Basic protection is by a paint system applied over a phosphate coating that also gives the vehicle an aesthetically pleasing appearance. It is reinforced by plastic or bituminous underseal where experience shows it to be needed. This is quite adequate on most panels, but attention is needed for vulnerable features revealed in service experience and destructive whole vehicle corrosion assessments using standardized tests such as exposure to salt sprays and in humidity chambers. It is common observation that the patterns of premature corrosion in automobile bodies are model specific. For example, one model may suffer at the skirts of doors but another is vulnerable at the joint between the front wings and bulkhead. These patterns are related to particular geometries or to methods of assembly. Some of the features that can lead to problems and measures available to counter them are itemized below.

22.2.1.1 Front and Side Panels

These areas receive the most intense exposure to grit and spray. Modern practice is to use panels pressed from steel supplied by the steel mill coated with zinc for galvanic protection. The zinc may be electrodeposited (EZ steel) as outlined in Section 6.2.6 or hot-dip coated (IZ steel) as described in Section 6.3.1 and annealed to allow zinc and iron to interact, producing a surface iron–zinc compound that provides a better base for phosphating. The decision of which kind of zinc coating to use is, however, usually made for the more mundane reason that EZ steel is least expensive for gauges of 2 mm and above and IZ steel for thinner gauges. Some authorities advocate building the whole shell from zinc-coated steel despite the penalty in increased material cost.

22.2.1.2 Wheel Arches

Wheel arches are vulnerable to paint damage by road stones and grit. Despite this, the rear arches have smooth contours and when undersealed give little trouble. The front arches are attached to the frame supporting the engine, front wheels and front fender and may carry inset lamp fittings. The irregular contours offer traps for wheel splash that may accumulate as *mud poultices*. These poultices can remain damp, stimulating corrosion long after rain has fallen. The problem is exacerbated by the high conductivity and aggressive nature of chloride-bearing deicing salts laid down by highway authorities in winter. Such poultices can be more active in a heated garage than in the cold open air. The solution to the problem is quite simply to deflect the splash by fitting smooth plastic internal arches over the wheels.

22.2.1.3 Joints

Joints are of various kinds, for example, lap joints between front wheel arches and bulkhead, clinched joints between the front and back panels of doors, gaps between double metal thicknesses, and so on. These joints can act not only as water traps but also as water conduits, so that the site of corrosion may not coincide with the joint. Part of the difficulty is that the joints often have to be made during the shell assembly before the application of paint. The solution is to apply beads of plastic sealer when the shell has received an undercoat of paint.

22.2.1.4 Rainways

Rainways are built into the shell to deflect rain falling on the roof clear of the doors. It can happen that due to some unforeseen circumstance, the rain can be collected and inadvertently directed into a water trap, such as the gap between the hood and bulkhead. This is a matter for minor modification.

22.2.2 Overview of Paint Shop Operations

A body shell presented for painting is a large intricate workpiece presenting surfaces with many different orientations, cut edges and internal and re-entrant features to which it is difficult to gain access. The paint shop has an input in the design to ensure that all parts of the shell are accessible to the entry and drainage of process liquors. Furthermore, as produced, the surfaces of the body shells are unsuitable to receive paint because they are contaminated with lubricants and detritus from the various fabrication operations. Cleaning and phosphating to provide the key for the paint are, therefore, an integral part of the painting operations. Failure to fulfill these preparatory tasks meticulously is not necessarily apparent in the superficial appearance of the finish, but is manifest by premature corrosion in service.

A paint shop receives the bare metal shells from the vehicle build production line, typically at a rate of 30 to 60 per hour and painting and related operations are organized to match. Pre-cleaning, phosphating, painting, sealing, paint baking and associated operations are arranged in sequence in the same production line, shielded by tunnels to contain liquor splashes. The body shells complete with doors, hoods and trunk lids braced slightly open are suspended from an overhead conveyor that carries them through and between all of the processes through the line.

The paint line represents a capital investment of more than $220,000,000 and must be adaptable to accept different automobile models in any order, to anticipate future models and be capable of treating steel, zinc and aluminium surfaces. The throughput must be variable to synchronize with the output of the vehicle build lines and since this influences the treatment times in the various processes, consistent performance and quality is maintained by adjusting other parameters, using arrangements to set, control and monitor them automatically.

22.2.3 Cleaning and Pretreatment of Body Shells

The body shells are manually wiped with a safety-approved hydrocarbon solvent to remove grease and loose contamination carried over from press-forming and vehicle assembly. This and joint sealing operations after application of the paint undercoat are the only labour intensive part of the whole sequence. The shells are then cleaned in an alkaline solution that is essentially 0.2 to 0.4 M sodium or potassium hydroxide with a surfactant, usually supplied as a proprietary formulation. To reach all parts of the intricate assembly, including undersides and box sections, it is applied in three sequential stages (1) a deluge at ambient temperature to reduce contamination of subsequent stages and to dislodge swarf, (2) a high pressure (100 kN) spray at 50°C from strategically placed multiple nozzles and (3) complete immersion of the body shell in the solution, also at 50°C. There follows two rinses in water to prevent transfer of alkali to the following acidic phosphating process.

22.2.4 Phosphating

The phosphate coatings are produced by applying the principles described in Section 6.4.1 on a large scale. The body shells are completely immersed in the phosphating liquor for a period typically of 3 to 5 minutes, entering and leaving it on the conveyor. The capacity of the phosphating liquor tank, about 100,000 liters, gives an impression of the large scale of the operation. On emerging, the shells have a smooth matt gray appearance. They are thoroughly rinsed twice, once by complete immersion in water to remove all soluble materials that could subsequently cause the paint to blister and again in a 0.005 M chromic acid solution to passivate metal exposed at any microscopic gaps and to impregnate the coating with the active corrosion inhibitor, hexavalent chromium, Cr(VI). Even though the next operation is coating with a waterborne paint, the phosphated shells are dried in hot

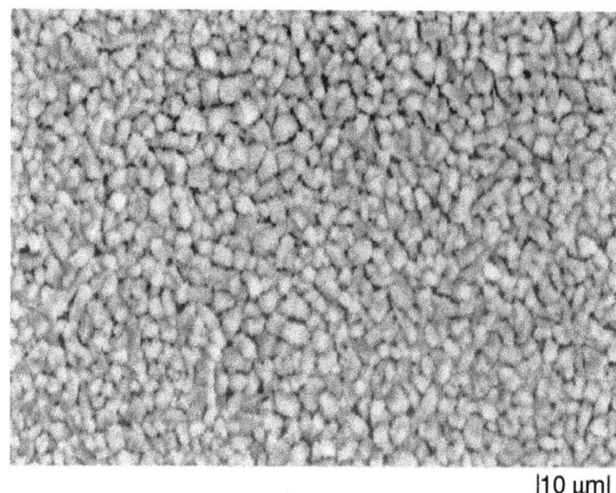

FIGURE 22.1
Structure of phosphate coating on cold-rolled steel, showing equiaxed crystals.

air as a precaution against rusting during temporary hold-ups that would adversely affect paint performance.

The solutions are usually proprietary formulations and the coatings produced are mixed hydrated tertiary phosphates in which, zinc, manganese, nickel and iron cations all participate. The relative proportions of the cations are chosen to meet the requirements that the coatings spread rapidly over the surface and that their chemical and physical nature provide the optimum base for the paint coatings. Figures 22.1 and 22.2 show the appearances of the surfaces of satisfactory coatings at high magnification, illustrating their crystalline structure.

The physical features that characterize good coatings are:

1. The coatings are thin and cover the metal surface completely. If the coatings are too thick, they detract from the smooth finish required for the paint. The coating thickness is expressed conventionally as a *coating weight,* typically within the range, 2.2 to 3.0 g m^{-2} on bare steel and 2.5 to 3.5 g m^{-2} on zinc-coated steel.

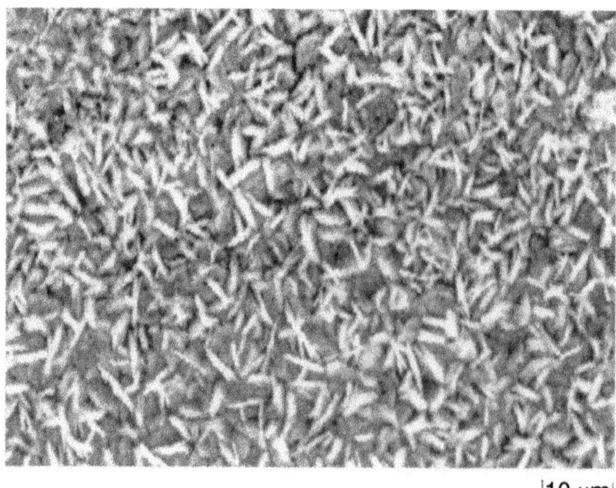

FIGURE 22.2
Structure of phosphate coating on zinc-coated steel, showing acicular crystals.

2. The crystals are equiaxed, regular and of a uniform optimum size. Crystals in satisfactory coatings have maximum dimensions as viewed on the surface of 2 to 8 μm on bare steel and 4 to 10 μm on zinc-coated steel. The higher range for zinc allows for the acicular morphology illustrated in Figure 22.2. If the crystal size is larger, the coating can tend to delaminate, undermining the adhesion of the paint. If it is smaller, the coating does not offer such a firm base for paint.

The phosphate liquor is formulated to suit mixed metal body shells that may contain steel, zinc and aluminium components. The combination of zinc and manganese cations in the formulation produces thin, smooth coatings on steel. The contribution of the nickel cation is to control the growth of phosphate crystals on zinc. The formulation may contain a small quantity of sodium fluoride to limit contamination of the liquor with aluminium, because an aluminium content >0.3 g dm^{-3} inhibits the deposition of zinc tertiary phosphate. It eliminates aluminium by complexing it as AlF_6^{3-} which precipitates from the solution as cryolite, Na_3AlF_6. The small uniform crystal size required is promoted by nuclei provided by a titanium-based surface conditioner added to the second rinse following alkali cleaning.

Strict controls are needed to maintain consistent good coatings. Daily test panels representing the materials in the body shells are processed and used to monitor coating weight and crystal size. The phosphating liquor is continuously circulated through heat exchangers to control temperature and filters to remove sludge. Excessive ionic content in the water supply, especially in the rinse water, can reduce the corrosion resistance of the coating and paint system subsequently applied to it. This requires particular attention if the water is 'hard,' with bicarbonate $>4\times10^{-3}\,mol\,dm^{-3}$ or if the sulfate and chloride concentrations are higher than usual. These characteristics are related to the hardness of the water discussed in Section 2.2.9. It is usually dealt with by matching the composition of the water to the quality of the coating and deionizing only a sufficient fraction of the water to meet the required standards, using ion exchange resins.

22.2.5 Application of Paint

The phosphate coat keys the paint to the metal and prevents undermining by corrosion creep, but it is the paint system and associated operations that provide protection. The following sequence of operations is typical.

1. A priming coat is applied by electropainting
2. Crevices are sealed
3. Undercoat, colour and gloss coats are applied by electrostatic spray systems
4. Supplementary protection is applied to critical areas

22.2.5.1 Primer Coat and Sealing

The primer coat is applied by electrodeposition of a waterborne paint of the kind described in Section 6.5.1.3. An electrical contact is attached to every body shell moving on the conveyor as it descends into a tank containing the paint. The shell is completely immersed and a negative DC potential of the order of 200 V is applied. The current entering a medium sized body shell is about 900 A, and builds a paint film 23 to 35 μm thick in the few minutes allowed in the production line programme. Although the insulating properties of the paint tends to promote a uniform coating by diverting current away from thickly coated areas, the throwing power is finite, extending to about 25 cm, and strategically placed cathodes are needed to obtain an even coating on such a large convoluted workpiece as a body shell. The paint is coherent and free from noxious solvents. The body shell is rinsed to remove loose particles, dried and baked by passing it through a continuous oven. The electrodeposited coat does not penetrate crevices >0.4 mm deep and seams and hemming flanges under wheel arches and in the floor are sealed with a mastic filler after baking.

22.2.5.2 Undercoating Colour and Gloss Coats

Undercoat, colour and gloss coats dispersed in traditional organic solvents are applied sequentially by dedicated electrostatic guided spray systems in clean dust free spray booths. The paint is delivered through *spinning bells,* that is spray heads revolving at 400 to 500 revolutions per second that atomize the paint and charge it to a high DC potential, for example, 90,000 V. The charged paint particles are attracted to the earthed body shell, minimizing overspray. The spinning bells move on computer controlled tracking systems, executing predetermined paths following the body shell contours.

The undercoat is applied, baked in a continuous oven, manually sanded to remove small blemishes, automatically dusted down, and passed to the colour spray booth. The colour coat is automatically selected by a computer programmed with order information. It is applied in its own dedicated spray booth by automatic electrostatic spray equipment, supplemented by manual spraying where needed. The body shell with its colour coat is passed to an adjacent spray booth to receive the gloss coat, also applied by electrostatic spray. Finally, the painted shell is baked.

22.2.5.3 Supplementary Protection

PVC based underseal is applied over the paint inside the wheel arches and under the sills as defenses against abrasive wheel splash. As much as 1.5 kg of warmed fluid wax with a corrosion inhibitor is injected into underbody box sections, where it solidifies to form a thick coating, reinforcing the protection against water accumulated from road wash. The finish is inspected in bright light and minor blemishes are touched in.

22.2.6 Whole Body Testing

No matter how carefully a body shell is designed, there are sometimes areas vulnerable to corrosion that come to light only when it is built and whole body testing is needed. As might be expected, manufacturers test bodies in conditions simulating extremes encountered in service, notably wash from highway deicing salts, high humidity, and exposure to UV radiation. The tests are standardized and can include the following procedures:

1. Spraying the whole automobile with 5% sodium chloride solution with frequent intermittent drying cycles
2. Storing the vehicle continuously in air with 100% relative humidity at 30°C
3. Continuous exposure of a vehicle coated with only a primer coat to artificial UV radiation at a prescribed intensity

Survival to a prescribed acceptable degree of degradation for 960 h is usually required.

22.3 Corrosion Protection for Engines

22.3.1 Exhaust Systems

Exhaust systems were formerly maintenance items, easily replaced when perforated by corrosion, but the advent of catalytic converters and a desire by manufacturers to give extended warranties reassigns them as long service items. Another factor is abatement of noise from perforations. Therefore, corrosion problems are now addressed more urgently.

A catalytic converter is an expensive and delicate device, easily damaged mechanically or by contamination. The active agents are contained in a ceramic foam supported on anti-vibration mountings. Installation of the converter has repercussions in both the front section of the exhaust between the engine and the converter and the rear section leading from the converter to the tailpipe.

The environment in the exhaust front section is more aggressive because it must run hotter to admit gases to the converter at a high enough temperature and at the same time the consequences of corrosion are more serious because oxide flakes can block the catalyst and perforation can allow combustion gases to bypass it. For these reasons, the front end and the catalyst containment casing are formed from AISI 409 stainless steel, listed in Table 8.3.

During short trips, the effect of the converter increases the condensation of acidified water in the exhaust rear section and mufflers associated with it, so that these items must be well protected. They are made from seam welded mild steel protected by hot-dip aluminizing.

22.3.2 Cooling Systems

The cooling system in an automobile is a mixed metal closed circuit. The waterways in the engine block are through the iron or aluminium–silicon alloy from which the block is cast, the heat exchangers may be made from aluminium or copper sheet, thermostats can be of soldered copper bellows and the whole system is connected by rubber hoses. The system differs from most other closed circuit mixed metal systems, for example, as used for central heating described later in Section 24.5.4, in that the coolant in winter is not water but an antifreeze mixture of water and typically 25% ethylene glycol, $(CH_2OH)_2$.

The presence of glycol precludes the use of oxidizing inhibitors such as chromates and nitrites that attack it. The possibility of attack on rubber hoses eliminates some others. A mixture of inhibitors is needed to cope with the mixed metals system. A common system is 1% borax, $Na_2B_4O_7 \cdot 10H_2O$, to act as a mild alkaline buffer, pH 9, to passivate iron and steel together with 0.1% mercaptobenzothiazole to inhibit cuprosolvency that can deposit copper on steel causing indirect galvanic stimulation.

The inhibitors are included in antifreeze sold at gas stations to make up the cooling fluid at the beginning of winter. The system is at its most vulnerable at the end of the winter, when owners can change the coolant for water that is not inhibited. For this reason, there is some merit in using antifreeze in summer as well as winter.

22.3.3 Moving Parts

Moving steel parts are protected by lubricating oil but with some reservations. The oil is not without some effects of its own and it has been shown to influence fatigue under cyclic loading. There is a form of damage sometimes found on mating surfaces of cams and

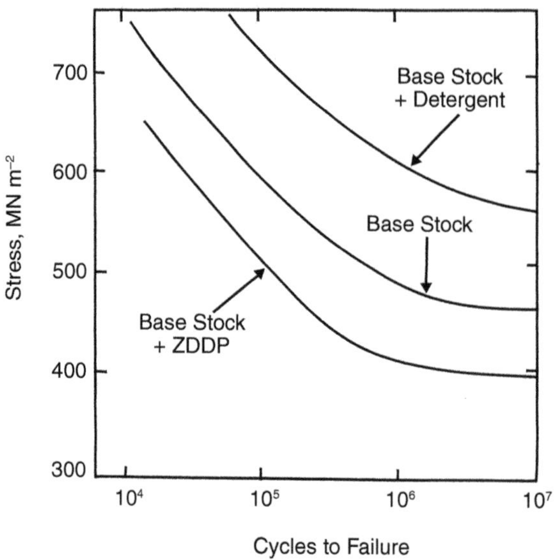

FIGURE 22.3
Effect of additives on the fatigue life of a case-hardened steel in an automotive oil. Anti-wear additive – 1.5 volume % zinc dipentyl dithio phosphate (ZDDP). Detergent – 5 volume % calcium sulfonate overbased with calcium carbonate.

tappets that is called 'pitting', but it is in fact surface fatigue damage. Automotive engine oil is formulated from a hydrocarbon base stock with additives for various purposes. To justify suspicion that surface fatigue was associated with oil formulation, laboratory fatigue tests on rotating steel samples were conducted in the hydrocarbon base stock of an automobile engine oil with separate additions of an alkaline detergent, a soot dispersant and an anti-wear agent*. The tests showed that fatigue is as sensitive to the composition of an oil environment as it is for aqueous media. The dispersant has no significant effect but the detergent improves the fatigue life and the anti-wear agent diminishes it, as shown in Figure 22.3. The detergent probably acts as a scavenger for unidentified acidic agents in the base stock persisting through oil refining, a view supported by observations that fatigue life in the base stock oil is diminished if components are added simulating contamination of the oil by *blow-by* combustion products from the fuel, for example, water, sulfuric acid and chlorides, and restored if detergent is added. The effect of the anti-wear agent is probably associated with its film forming capability that is responsible for its anti-wear function, perhaps introducing local action electrolytic cells at cracks in the film.

22.4 Bright Trim

Bright trim is not as widely used for automobiles as it was some years ago. It was popular for decoration and gave an integrated appearance to disparate accessories like steel fenders and die-cast zinc door handles. Replacement of many of these items with plastics and

* C. Chandler and D. E. J. Talbot, Corrosion fatigue of a case-hardening steel in lubricating oil, *Materials Performance*, p. 34, 1984.

change of fashion have reduced the applications of metal trim but it is still used to a significant extent for radiator grills, lamp bezels, wheel embellishment and side trim. Even so, plated plastics have made inroads on metal trim except for up market automobiles. The standard for acceptable appearance of bright trim is electrodeposited bright nickel/chromium systems or matching alternatives. One is customized brightened and anodized aluminium alloys. Another is polished ferritic stainless steels that needs no explanation.

22.4.1 Electrodeposited Nickel Chromium Systems

In these systems, corrosion protection is afforded by one or two thick coats of nickel. Although protective, nickel tends to dull on outside exposure and is notoriously prone to tarnishing in urban atmospheres polluted by sulfur dioxide. A very thin overlay of chromium electrodeposited on the nickel confers brilliance and tarnish resistance. Recalling discussions in Section 6.2.2, nickel deposited from standard acidic baths, developed from the Watts solution, do not adhere to steel or zinc, and must be preceded by copper strike coats deposited from alkaline solutions.

A problem in the design of the coating system is that the passive chromium is a galvanic stimulant to attack on the nickel. Bright chromium coatings are porous with exposed nickel sites at the bases of the pores that become anodes in a nickel/(passive chromium) cell. Formerly, only one chromium coating was applied and it was not uncommon to see scattered small brown spots of corrosion product on nickel/chromium plated fenders. The problem can be addressed by applying two chromium coatings; a standard bright porous coat followed by a coating applied by proprietary process that causes the plate to crack on a microscale. The pores and cracks do not coincide and so the coatings cover the others defects.

A further feature is that both nickel and copper stimulate galvanic attack on a steel substrate, if exposed. To ameliorate the consequence of breaching the chromium coating, two nickel coats can be applied with small additions of other metals to make the outer nickel coat cathodically protective to the inner coat. This deflects any corrosion attack along the interface between the two nickel coats so that it does not reach the steel. A typical complete system is:

1. A strike copper deposit from dilute alkaline cyanide solution
2. A second thicker copper deposit from a more concentrated cyanide solution
3. A deposit of pure nickel from a standard acidic solution
4. A second deposit of nickel, cathodically protective to the first
5. A deposit of porous chromium
6. A deposit of microcracked chromium

22.4.2 Anodized Aluminium

Acceptance standards for aluminium alloys for bright automobile trim are high and they are critical products based on relatively high purity, >99.8% aluminium to avoid dulling due to elements that yield intermetallic compounds, especially iron. A typical alloy contains 1% Mg with 0.25% Cu to give some ability to strengthen the otherwise soft pure metal by work hardening it. The metal must be homogeneous otherwise the differential brightness obtained over an inhomogeneous structure gives a banding effect called *bright streaking*. To avoid it, the cast ingots from which wrought products are fabricated must be held for a long time, for example, 24 h at a high temperature, >600°C, to disperse segregates by diffusion. Another requirement for anodizing is efficient filtering of the liquid metal before casting to remove oxide inclusions that lead to disfigurement by the so-called *linear defects* which appear as scattered, short, fine lines parallel to the direction of fabrication. The final fabricated products are chemically brightened and anodized in sulfuric acid, using low acid concentration, cool temperature and high current densities to produce anodic films that are bright yet hard enough to resist abrasion, as explained in Section 6.4.2.2.

Further Reading

Crane, F. A. A. and Charles, J. A., *Selection and Use of Engineering Materials*, Butterworths, Boston, Massachusetts, 1984.

Fenton, J., *Handbook of Vehicle Design Analysis*, Institute of Mechanical Engineers, London, 1996.

Ettis, F., *Automotive Paints and Coatings*, VCH, New York, 1995.

Freeman, D. B., *Phosphating and Metal Pretreatment*, Woodhead Faulkner, Cambridge, UK, 1986.

23

Corrosion Control in Food Processing and Distribution

23.1 General Considerations

Food processing and distribution are competitive enterprises that must preserve the capital value of plants and minimize the costs of products and packaging. To that extent, their needs to control materials degradation are similar to those of other industries, but additional factors introduce differences in approach:

1. Supply and distribution of food is subject to scrutiny by public health authorities and sensitive to consumer confidence.
2. Food products have special characteristics as corrosion environments.
3. The life expectancy of plants is many years, but the life required of metallic packing depends on short-term retail turnover.

23.1.1 Public Health

Surfaces specified for contact with food products must not introduce toxic substances nor influence flavor. Materials that fulfill these requirements include a restricted range of metals.

Long experience of tin coated cans, aluminium alloy cooking utensils and foil has established the non-toxic nature of tin and of aluminium in both the short- and long-term. Coated steels are also used for processing plants where appropriate and there is little health hazard from iron exposed locally by breakdown of thin protective coatings. Stainless steels are widely applied for equipment in large-scale food processing, so that nickel, chromium and molybdenum as alloy components are considered safe. The same is true for magnesium as an alloying element in aluminium alloy beverage cans. Copper was formerly used for food vessels, but it is vulnerable to catalytic activity accelerating rancid degeneration of oils and fats. Health and safety regulations prohibit precursors of toxic substances including lead and cadmium from contact with food.

23.1.1.1 Control of Hygiene at Food/Metal Contacts

Corrosion products and degraded protective coatings must not be allowed to accumulate because they offer sites for harmful bacteria to colonize. In this respect, austenitic stainless steels and aluminium alloys are well-suited to the construction of process plant and permanent containers and require no coatings for service in most food environments. They are strong, ductile, have acceptable casting characteristics and can be formed into large complex components with smooth easily cleaned surfaces, joined by welding to eliminate crevices.

A standard disinfectant, sodium hypochlorite, NaClO, kills biological material by oxidation:

$$NaClO \rightarrow NaCl + [O] \tag{23.1}$$

where the symbol [O] indicates nascent oxygen.

Residues must be washed away to avoid pitting corrosion induced by chlorides and hypochlorites concentrated by evaporation, as explained in Section 8.3.1.4. The chloride ion is the primary pitting agent and the hypochlorite ion can impose a redox potential anodic to prevailing chloride pitting potentials for stainless steels:

$$ClO^- + H_2O + e^- = Cl^- + OH^- \quad \text{for which} \quad E^{\ominus} = +0.90\text{ V (SHE)} \tag{23.2}$$

23.1.2 Food Product Environments

Foods are usually associated with aqueous media in the pH range 3–8, as illustrated by examples given in Table 23.1. The generally satisfactory behaviour of aluminium, tin and stainless steels within this pH range is consistent with the Pourbaix diagrams for aluminium and tin given in Figures 3.4 and 3.7 and with the corrosion velocity diagrams for stainless steels given in Figures 8.6 and 8.7.

Features of food and associated materials that distinguish them from many other corrosion environments are the activities of microorganisms, especially bacteria, the presence of emulsions and interplay between organic and inorganic components.

23.1.2.1 Bacteria in the Environment

Bacteria are primitive single cells bounded by rigid walls, typically spheres 0.5–1 μm in diameter or rods 1–5 μm long, that live in aqueous media. They may exist as isolated individuals, clumps or sheets. They obtain their nutrients directly from solution, transforming them by

TABLE 23.1

pH Ranges for Some Common Foodstuffs

Fruits	pH	Vegetables	pH
Apples	3.0–3.5	Potatoes	5.5–6.0
Oranges	3.0–4.0	Cabbage	5.0–5.5
Lemons	2.0–2.5	Carrots	5.0–5.5
Tomatoes	4.0–4.5	Beet	5.0–5.5
Raspberries	3.0–3.5	Pickles	3.0–3.5
Dairy Products	**pH**	**Beverages**	**pH**
Cow's milk	6.0–6.5	Beers	4.0–5.0
Butter	6.0–6.5	Fruit drinks	2.0–4.0
Cheeses	5.0–6.5	Wines	3.0–4.0
Meat and Fish	**pH**	**Bakery**	**pH**
Fish	6.0	Bread	5.0–6.0
Meats	7.0	Wheat flour	5.5–6.5

their metabolism into other substances returned to solution. The nutrients needed include a source of energy such as glucose and small quantities of minerals for the cell structure. The pH of a medium and concentrations of ions, for example, Cl^- within it influence the growth, activity and death of the microorganisms. For example, *Lactococcus lactis*, associated with milk, grows best at pH 3–4 and will not grow in a medium of pH > 5.

Bacteria promote chemical effects corresponding to the vast number of varieties. Some reduce sulfates to sulfides and others oxidize sulfides to sulfates, some thrive in chloride media and others do not, some reduce nitrates, some utilize hydrogen, and so on. These transformations can occur rapidly because bacteria reproduce by successive binary division which stops if nutrients are exhausted, waste products accumulate or the available space is occupied. Colonies of bacteria have high surface to volume ratios because the individuals are minute. According to type, bacteria are associated with decay, putrefaction, soil fertility, and some diseases of plants and animals. Their intervention in corrosion and protection are summarized as follows:

23.1.2.1.1 *Environmental Changes*

Bacteria can change the environment by replacing one substance with another. For example, *Streptococcus lactis* sours milk by converting lactose to lactic acid, reducing the pH.

23.1.2.1.2 *Local Action Cells*

By their physical presence in clumps or films, bacteria can partially screen metal surfaces, creating differential aeration and concentration cells, for example, through local oxygen starvation.

23.1.2.1.3 *Depolarization*

Some sulfate-reducing bacteria produce an enzyme, *hydrogenase*, that enables them to utilize hydrogen in a process capable of depolarizing the hydrogen evolution reaction summarized by the equation:

$$SO_4^{2-} + 4H_2 \rightarrow S^{2-} + 4H_2O \tag{23.3}$$

23.1.2.1.4 *Degradation of Protective Mechanisms*

Foods can contain ascorbic acid (vitamin C) and folic acid, both of which are corrosion inhibitors for iron and other metallic materials, but they are susceptible to depletion by bacterial metabolism. Bacteria can also degrade protective coatings containing cellulose materials.

23.1.2.2 Emulsified Environments

Many foods subject to industrial processing, such as milk, milk products and soft margarine, are two-phase emulsions containing small micelles of fats and oils in water. They can sustain electrochemical activity in the water phase but they also have characteristics conferred by the micelles that carry electric charges. A further factor is that in some food processing the emulsions can separate into bulk phases, as in curds and whey formed from milk.

23.1.2.3 Interaction between Inorganic and Organic Solutes

Food solutions contain organic solutes that can modify the behaviour of inorganic ions. One example, quoted in Section 4.1.3.6, is polarity reversal of iron–tin bimetallic couples.

From their relative standard electrode potentials, given in Table 3.3, tin might be expected to stimulate corrosion on iron in a bimetallic couple and this is found to be true for some purely *inorganic* solutions such as sodium chloride solutions, but in many solutions of *organic* acids such as fruit juices and milk products the polarity of the couple is reversed so that tin coatings galvanically protect steel exposed at defects. Another example, described later, relates to the intensification of pitting corrosion of stainless steels due to the simultaneous presence of chloride ions that stimulate corrosion and organic solutes such as folic and ascorbic acids that inhibit it.

23.1.2.4 Practical Aspects of Corrosion Control

Corrosion presents few problems during short-term contact of foods with suitable metals as in cooking and associated preparation, provided that practices are hygienic. Commercial purity aluminium, AA 1100, and the aluminium–manganese alloy, AA 3003, specified in Table 9.2 in Chapter 9 and austenitic stainless steels, AISI 304 and AA 323 specified in Table 8.3 in Chapter 8 are standard materials for utensils and contact surfaces. Martensitic stainless steels, for example, AISI 431, also specified in Table 8.3, are standard for knives. Care, especially in materials selection, is required in commercial processing include dairying, brewing, baking, production of sugar, margarine, soft drinks and pre-cooked convenience foods. Dairying and brewing are featured later in this chapter because between them they exemplify the contexts within which corrosion control must be exercised.

23.2 Production of Tinplate for Food Cans

The preservative function of tin-coated steel cans has been overtaken by their convenience as food and beverage dispensers, representing the largest area of food/metal contact. The significant feature is that the cans are fabricated from thin steel sheet with the corrosion-resistant coating already in place, the *can stock,* which must survive the extensive deformation.

23.2.1 Historical

The non-toxic nature, inherent corrosion resistance in neutral and mildly acidic aqueous media and its low melting point, 232°C, facilitating hot-dip coating made tin a natural choice as a protective coating for steel. Historically, single steel sheets or continuous steel strip were passed through a flux to remove oxide and thence into a bath of molten tin and the surplus tin was wiped off by pads on exit. The first successful mass-produced can was a three-piece can formed by rolling a tin-coated steel sheet blank into a cylinder with an overlap side seam sealed by a lead–tin soldered joint. The base was clinch-sealed to the cylinder and the top was also clinch-sealed to it after filling the can.

23.2.2 Modern Tinplate Cans

Modern tinplate is produced from cold-rolled steel strip with the typical composition limits:

C 0.13 mass %, Mn 0.6 mass %, P 0.02 mass %, S 0.05 mass %, Si 0.02 mass %.

As produced, the cold-rolled strip is unsuitable for can stock because it is work hardened and must be annealed to soften it and cleaned appropriately to receive the tin.

23.2.2.1 Electrodeposition, Flow Melting and Passivation

The acid sulfate and halide baths are used extensively and the fluoborate bath is used occasionally. Electrodeposition, flow melting and passivation are carried out continuously in sequence in the same strip line with looping storage devices to allow continuous operation and the finished tinplate is re-coiled on exit.

There are three can types (1) the three-piece can evolved from the original concept, (2) the two-piece draw/redraw (DRD) can and (3) the draw/wall-ironed (DRW) can, referring to the methods of fabrication. Tinplate for these cans is customized to economize tin use, consistent with its corrosion resistance, strength and suitability to can fabrication. The mass of tin on each side of the tinplate is 1.4–11.2 g m^{-2} of steel surface, corresponding to tin thicknesses 0.2–1.5 μm. Can manufacturers use the thinnest tin coatings where possible, for which all of the tin is consumed in flow melting, so that the protective layer is wholly $FeSn_2$.

Tinplate is passivated in aqueous Cr(VI) solutions, usually cathodically, that is by application of a negative potential to the strip to improve the corrosion resistance and provide a key for lacquer applied by spray to the inside surfaces of finished cans to improve it further.

23.2.2.2 Three-Piece Cans

Three-piece cans remain in service for non-pressurized food contents using overlap welding to seal the side seams. Lacquer is applied to cover the narrow heat affected zones, 1 to 2 mm wide, where corrosion protection is compromised by welding.

23.2.2.3 Draw/Redraw Cans

Circular blanks sheared from the sheet are pressed by a punch through a die, the *draw-ring*, to produce the hollow shape, while partially restrained by a blank holder to avoid wrinkling. Radii, clearances and friction between the punch and the metal are interacting parameters, which are optimized to ensure that a smooth shape is produced. The cylindrical walls of the hollow shape are strengthened by work hardening as the blank is drawn over the draw ring, enabling it to resist fracture under the load on the punch. Friction is needed at the base of the punch to restrain the unhardened metal below it from slipping into the cylindrical walls where it can fracture under the applied load. The cans are produced by successive drawing operations through dies of progressively smaller diameter; hence the term draw/redraw. The can lid is crimp sealed to the can after filling. The corrosion protection conferred by the passivated tin coating must survive this severe fabrication sequence.

23.2.2.4 Draw/Wall-Iron Cans

Draw/wall-iron steel cans are thin-walled pressurized beverage containers, deriving rigidity from the gas pressure of the contents. For matching support, their bases are of thicker metal formed into an internal dome. A shallow cylindrical hollow shape is first drawn from a flat blank. The basic form of the can is developed by forcing it through a series of dies of progressively smaller diameter while supported internally by a punch.

23.2.3 Tin-Free Steel for Food Packaging

Tinplate is still the principal steel-based material used for processed food and other products, but alternative coatings are becoming available. Since very thin tin coatings are completely converted to the intermetallic compound, $FeSn_2$, on flow melting, requiring chromizing passivation and lacquering for corrosion resistance, it is natural to consider eliminating the electrodeposition of tin and chromizing the steel directly to provide a substrate for lacquer.

This reasoning stimulated developments which led to electrodeposition of chromium-bearing coatings. The dominant operation is a proprietary process delivering a chromium coating consisting of chromium metal and non-toxic chromium trioxide, Cr_2O_3 in the ratio 2:1 or 3:1. The chromium metal is adjacent to the steel with the oxide in a layer above it, forming a key for lacquer. An advantage of the process is that the steel base for the coating and the processing requirements are similar to those for tinplate, so that passivated tinplate and chromium-coated steel can be produced as complementary products in the same plant.

The electroplating solution is derived from industrial chromium metal plating solutions based on chromic acid anhydride, CrO_3. It is more dilute and is operated at a much higher current density. Although well-suited to its purpose for can stock, the deposit is different from chromium metal deposits produced for decoration or industrial hard facing.

A typical formulation with operating parameters is:

Chromic acid anhydride, CrO_3	50–180 g dm^{-3}
Sulfuric acid, H_2SO_4	1 g dm^{-3}
Fluoboric acid, HBF_4	0.8 g dm^{-3}
Temperature	40–55°C
Current density	50–100 A/dm^{-2}

A continuous strip speed of 6 m s^{-1} can be accommodated, allowing a process time of 1 to 8 s. The coatings are 0.01–0.03 μm thick, which is much thinner than the 0.4 μm thick low-tin coatings they replace. The corrosion resistance of the more economic $FeSn_2$ and tin-free coatings is not as good as for metallic tin, but it is sufficient for many food products.

23.3 Dairy Industries

Dairy industries treat and distribute milk, use it as the feedstock for other foods, principally butter, cheese and yogurt and extract residual values as by-products. Corrosion control of the plant depends on the nature of these products and on the processes by which they are produced.

23.3.1 Milk and Its Derivatives

23.3.1.1 Constitution

Cows' milk is a complex aqueous medium containing fats, proteins and carbohydrates, with small but important contents of minerals, vitamins, enzymes and bacteria; the composition varies and Table 23.2 is a simplified example. The *fats* are present as an emulsion

TABLE 23.2

Representative Composition of Natural Milk

Category	Mass %	Principal Constituents
Fats	3.8	Trihydric glycerol esters Phospholipids Other fatty acid derivatives
Proteins	3.3	Casein Lactoalbumin Lactoglobulin
Carbohydrate	4.7	Lactose
Mineral ions	0.75	Ca^{2+}, Mg^{2+}, Na^+, K^+, PO_4^{3-}, Cl^-, $C_6H_5O_7^{3-}$ (citrate)
Vitamins	Traces	A (retinol) B_1 (thiamin) B_2 (riboflavin) B_{12} (cyanocobalamin) C (L-ascorbic acid) D (a steroid derivative) M (folic acid)
Enzymes		
Bacteria		
Water		Balance

of minute globules. *Casein*, the most abundant protein, is combined with calcium and is dispersed in the liquid phase as a colloidal suspension of gelatinous particles. *Lactose* is a sugar in aqueous solution and is the nutrient for bacteria in fermented products. The bacteria *Lactococcus lactis* and *Lactococcus cremoris*, naturally present in milk, progressively convert it to lactic acid during souring. *Calcium and phosphorus* are present as minerals vital for nutrition and bone formation. *Vitamins A and D* are fat soluble and *vitamins* B_1, B_2, B_{12}, *C and M* are water soluble. *Vitamins* B_1 *and C* are unstable when the milk is heated.

23.3.1.2 Processing

Operations using milk as a feedstock are summarized in the typical flow sheet given in Figure 23.1. They start by fractionating the milk by one or the other of two methods:

1. Centrifugal separation based on differences in the densities of milk components. This yields a fat-rich fraction, cream and fat-free *separated milk*.
2. Curdling, induced by adding *chymosin* at 32°C, a synthetic copy of the active component in natural rennet. This breaks down casein, yielding a colloidal mass, *curds* and a watery liquid, *whey*.

23.3.1.2.1 Traditional Dairy Products

The fat-rich fraction from centrifugal separation contains 50% fat and the fat soluble vitamins with 50% of retained liquid. It is marketed as creams or churned into butter releasing the retained liquid as buttermilk. Curds contain most of the fat, casein, colloidal calcium phosphate and bacteria. It is extracted from the whey, pressed and processed to cheese. These traditional dairy activities are not particularly demanding on the corrosion resistance of metallic materials in contact with them.

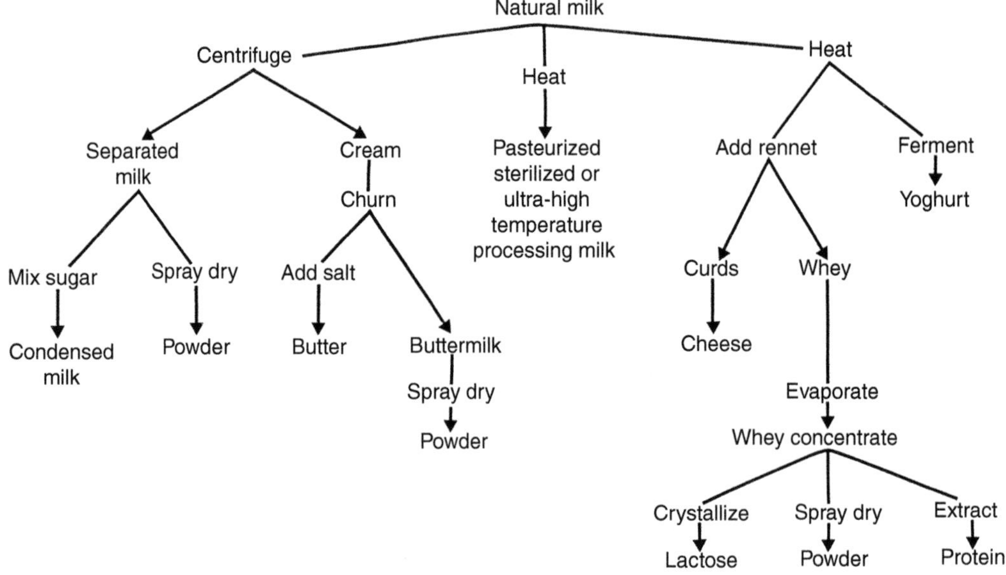

FIGURE 23.1
Simplified flow sheet for milk processing.

23.3.1.2.2 Evaporation of Separated Liquors

Separated milk can be sold as a product, concentrated by evaporation and sweetened for sale as condensed milk or spray-dried to powder. Buttermilk is spray-dried to powder.

23.3.1.2.3 Recovery of Proteins from Whey

Whey retains the proteins lactalbumin and globulin in colloidal solution, the lactose, the water soluble vitamins L-ascorbic acid (vitamin C), folic acid (vitamin M), thiamin (vitamin B_1), riboflavin (vitamin B_2) and vitamin B_{12} and the minerals as listed in Table 23.3. It can be spray-dried and disposed of as low value material for such uses as animal feed. A better alternative is to recover the high quality protein content and the lactose. In one

TABLE 23.3
Representative Composition of Whey

Category	Mass %	Principal Constituents
Fats	0.01	–
Proteins	0.6–0.7	Lactoalbumin Lactoglobulin
Carbohydrate	4–5	Lactose
Mineral ions	0.3	Ca^{2+}, Mg^{2+}, Na^+, K^+, PO_4^{3-}, Cl^-, $C_6H_5O_7^{3-}$ (citrate)
Vitamins	Traces	B_1 (thiamin) B_2 (riboflavin) B_{12} (cyanocobalamin) C (L-ascorbic acid) M (folic acid)
Enzymes		
Bacteria		
Water	Balance	

process the whey is treated with hydrochloric acid, concentrated by reverse osmosis and filtered through a membrane with pores of the order of 1 μm.

23.3.2 Materials Used in the Dairy Industry

To limit contamination from external bacteria, product contact surfaces of processing equipment must be hygienic. Materials, surfaces, construction of plant, cleanability and inspection are considered in publications by the International Dairy Federation*. The recommendations include the use of surfaces that are resistant to corrosion and robust enough to withstand cleaning and sterilization. Organic coatings such as paints are not approved because of degradation by bacterial attack and difficulty in cleaning.

23.3.2.1 Surfaces in Contact with the Product

Metal surfaces in contact with the product must satisfy the following requirements:

1. Low toxicity and physiological indifference
2. Corrosion-resistance to the products and to cleaning and disinfecting solutions
3. Availability in standard forms at acceptable cost

Associated materials contacting metals, including rubbers, plastics, ceramics and carbon for gaskets, rotary seals, bearings and other adjuncts are subject to the same provisos. Care is required to ensure that seals and gaskets fit well and do not offer sites for crevice corrosion.

Numerous materials have been suggested but stainless steels are in general use for the reasons given in Section 23.1.1.2. AISI 304 type austenitic steels are standard materials in general use for milder environments, such as milk storage vessels and separators. The stabilized versions, AISI 321, AISI 347 or AISI 304L, specified in Table 8.3, are clearly preferable for on-site welded structures where weld decay might otherwise be expected.

In filtration, evaporation and spray-drying plants, various combinations of higher temperatures, higher pressures, diminished solubility for oxygen, higher concentrations of chlorides in feed fluids and oxygen starvation under deposits can produce environments that are too aggressive for AISI 304 steels and must be resisted by using the more expensive molybdenum-bearing austenitic steels, AISI 323 or the low carbon version, AISI 323L, for welded structures.

One particular environment that is too aggressive for even AISI 323 is the intermediate liquor obtained after treating whey with hydrochloric acid to recover the proteins. Stainless steels do not exhibit an active range at the prevailing pH, 4 to 5, but there is a chloride concentration of 0.07 M that introduces pitting potentials below the potential for dissolved oxygen in equilibrium with air. The problem is made worse by the presence of folic acid in whey that obstructs repassivation of pits. It does so by increasing the potential difference between active pit sites and the general passivated surface. This is manifest in marked hysteresis in the polarization characteristics, illustrated and explained in Figure 23.2. One approach is to replace AISI 323 with duplex stainless steels that do not exhibit such marked hysteresis. Another is to use an alternative process to recover proteins that does not introduce chloride ions.

* *International Dairy Federation Bulletin*, No. 218, 1987.

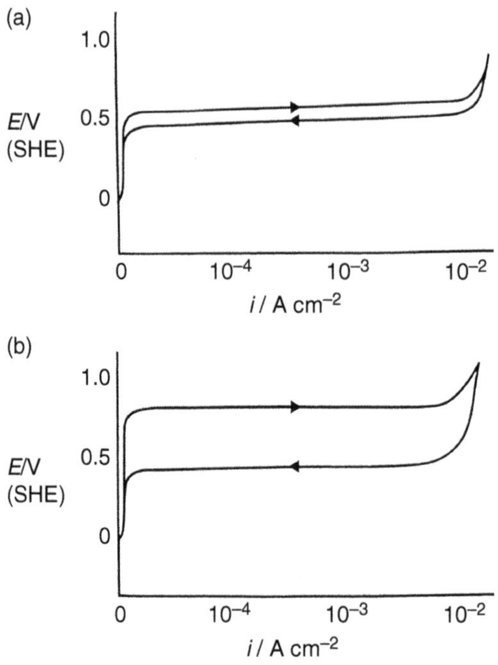

FIGURE 23.2
Polarization characteristics of AISI 323 stainless steel in 0.07 M sodium chloride solutions in (a) water and (b) whey, both of pH4, for rising (→) and falling (←) potentials. The steel exhibits a higher pitting potential with more hysteresis in the whey solution than in the corresponding water solution.

23.3.2.2 Surfaces Not in Contact with the Product

Stainless steels are also preferred over less expensive materials for some surfaces not in contact with the product. For example, the enclosed space between double walls in jacketed and insulated equipment must be sealed against entry of water vapour and bacteria. Mechanical joints do not provide such reliable seals as welding and this implies that the outer wall is of the same metal as the inner wall in contact with the product. If, for economy, other metals are used in proximity to stainless steels, precautions must be taken to avoid galvanic stimulation. Passivated stainless steels are more noble than most engineering metals and stimulate attack on them. For example, uninformed use of other steels, such as plain carbon steels for less critical parts contacting stainless steels is a false economy.

23.4 Brewing

23.4.1 The Brewing Process

The raw materials for beer are barley, hops, yeast and water. Important by-products are surplus yeast processed for health foods and other products, spent barley sold for animal feed and spent hops used for fertilizer. A flow sheet for a typical large brewery is given in Figure 23.3. The barley is *malted*, that is allowed to germinate, during which it secretes an enzyme, *diastase* that is needed for conversion of starch in the grains to the sugar, *maltose*

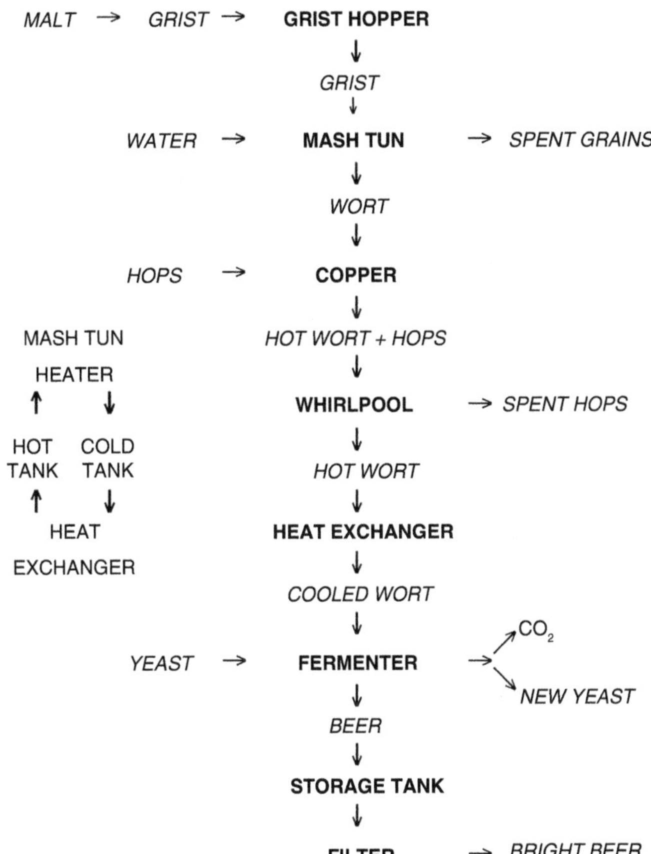

FIGURE 23.3
Schematic brewery flow sheet.

in the next stage. The malt is pulverized to *grist* that is accumulated for use in grist hoppers, from which it is fed as required. The grist is *mashed,* that is steeped in hot water for an hour or two in *mash tuns,* to effect the conversion of starch to maltose, producing a sweet liquid called *wort.* The water for the process is filtered through activated carbon to remove pesticides and its content of ions such as nitrates is reduced by reverse osmosis or ion exchange resins.

The wort is drained off into a large vessel called the *copper* and the spent grains are retrieved and conveyed to large hoppers, awaiting transport to manufacturers of animal feed. Hops are added to the wort in the copper and the mixture is boiled for an hour or two by heat supplied through steam coils. This serves two purposes; the boiling stops the starch to sugar reaction and infusion of hops gives the beer its characteristic flavor. It is discharged into a *whirlpool* separator that retrieves the spent hops by suction to the center of the vortex. The separated liquid wort is cooled through a heat exchanger to ambient temperature, aerated to sustain the growth of yeast and transferred to fermenting vessels, where the yeast is added and the brew is allowed to ferment at about 17°C for 40 to 60 hours. This generates the alcohol and liberates carbon dioxide gas that is recovered, purified and liquified for use in pressuring beer in cans or barrels. The beer is cooled and the yeast is removed and recovered by centrifuging. Part of the yeast is recycled for use in subsequent brews and the surplus is sold as a valuable ingredient rich in protein for human and animal foods. The beer is allowed to mature in tanks then filtered, yielding bright beer that is stored in holding tanks from which it is drawn as required, pasteurized by heating and distributed in cans, bottles, barrels or mobile tanks.

An essential adjunct to the process is a water circulation system to conserve energy by recycling heat. Cooling water, running countercurrent to the hot wort in the heat exchanger is pumped to heating coils in the mash tuns from which it is returned to the heat exchanger in a closed circuit. The system contains a large quantity of water that is accumulated in thermally lagged hot tanks (*hot liquor tanks*) at a temperature of 80°C. After heating, the mash tuns it is again accumulated in cold tanks (*cold liquor tanks*). The hot tanks are potential sites for the SCC problem described in the next section.

23.4.2 Materials Used for Brewing Plants

Originally, breweries were small and their operations were carried out in wooden vessels. The first metal replacements were cast iron for the mash tuns and copper for boiling wort – hence the name, *copper*, for the vessel. When stainless steels became available with their superior hygienic surface qualities, they progressively replaced all other materials for contact with the beer and its precursors and also for many non-contact surfaces.

Plant design is regulated in much the same way as in the dairy industry. Materials for product contact surfaces are required to be non-toxic, resistant to corrosion and able to withstand cleaning and sterilization. Construction is by welding to eliminate crevices and the welds in contact with the product are usually subject to 100% radiographic inspection.

23.4.2.1 General Corrosion Resistance

Beers and their precursors are in the pH range 4 to 5 and AISI 304 stainless steel and its low carbon and stabilized variants, specified in Table 8.3 in Chapter 8, are resistant to them if they are cold and fully aerated. Hence, welded sheet and plate materials of these steels are used for grist hoppers, fermenting vessels, holding tanks and associated pipes.

Hot or boiling liquids are more aggressive, both because of the higher temperatures and because they are depleted in oxygen. Consequently, the more strongly passivating (and more expensive) molybdenum-bearing steel, AISI 323, also specified in Table 8.3, is used for the mash tuns, copper, whirlpool separators, and heat exchanger plates in the wort cooler. These vessels are lagged with mineral wool to conserve heat and AISI 304 is satisfactory for the outer casings holding the lagging in place.

23.4.2.2 Stress-Corrosion Cracking

Several cases of SCC have been observed in brewing vessels handling hot liquids and, as a consequence, precautions against it are taken in design and operation. There are several sources of stress including the weight of large masses of liquid and contraction stresses after welding. Two distinct sources of the specific agent, chloride ions, have been found, one external to the vessel and the other internal.

The external source is chloride leached from the lagging by condensing water vapour. This is now controlled by applying a coat of a proprietary anti-chloride barrier paint on the external surfaces of all vessels to be insulated, specifying low chloride contents in the lagging material and interposing vapour sealing barriers of aluminium foil between successive layers of lagging.

The internal source is the chloride content of the water used for brewing or residues from disinfectants and from acids used in descaling. Though small, these chloride contents can become concentrated locally by thermal cycling or evaporation, an effect well recognized in other cases of SCC. This source is controlled by de-ionizing the water and prohibiting the use of hypochlorites as disinfectants.

Despite these precautions, there remains a risk of stress-corrosion cracking in hot liquor tanks constructed of AISI 323, induced by chloride leached from lagging. The risk is eliminated by constructing them from an expensive duplex steel, for example, steels 1 or 2, specified in Table 8.3, in which the less susceptible ferrite phase acts as a barrier to stress corrosion.

23.4.2.3 Biologically Promoted Corrosion in Spent Grain Hoppers

Spent grain hoppers are not directly associated with the beer production sequence and are simply used to accumulate spent grains awaiting disposal. They are essentially large cylindrical bins made from plain carbon steel, set at a height suitable for discharging the grains into trucks.

The spent grains are moist and contain sugar residue forming a solution that drains down and accumulates at the bottom of the hopper, where bacteria can convert the sugar into an acidic medium that eats into the structural reserves of the steel. In one occurrence, the base of a hopper was so weakened that the bottom dropped away under the heavy load of grains inside but fortunately, it caused no injury.

23.4.2.4 Cleaning and Sterilization

The system is cleaned periodically by flushing with 0.5 to 1 M sodium hydroxide solution to loosen and scavenge organic deposits. It is neutralized with a benign dilute acid and well rinsed.

Sterilization with hypochlorite solutions is discouraged to reduce risks of stress-corrosion cracking from an internal chloride source. Sterilization by live steam is an unacceptable alternative, because it can leach chlorides from lagging, increasing the risk of stress-corrosion cracking from an external chloride source. A better practice is to use chemical sterilization based on non-chloride reagents, such as peracetic acid, that degrade to safe reaction products.

23.4.3 Beer Barrels, Casks and Kegs

Although, for convenience, a large proportion of the beer currently brewed is distributed in cans, many discerning consumers, especially in Europe, prefer it drawn from barrels. *Casks* and *kegs* are large unpressurized and small pressurized barrels, respectively. The introduction of modern barrels provides a useful insight into the relative merits of two corrosion-resistant metals, a stainless steel, AISI 304 and an aluminium alloy, AA 6061, specified in Table 9.2.

23.4.3.1 Historical

Traditionally, beer in bulk was distributed in barrels made from wooden staves bound by steel hoops, produced by a labour-intensive highly skilled craft. Some 70 or 80 years ago, they were replaced by metal casks and kegs whose production could be mechanized and which can better contain the carbon dioxide pressure in modern effervescent beers. Stainless steels, with which the industry was familiar, were a natural initial choice but at the time, techniques had not been developed to exploit their full strength and consequently the barrels were thick-walled and heavy. They were replaced by lighter age-hardening aluminium alloys, but stainless-steel barrels were subsequently re-introduced when techniques for producing and fabricating them were improved.

23.4.3.2 Design

Barrels must be corrosion resistant to cold beers, robust to withstand rough handling and volumetrically stable under internal pressure from dissolved carbon dioxide and sometimes also nitrogen introduced under pressure to generate effervescence. The brewing industry imposes standard drop and pressure tests to ensure these attributes. The barrels are expected to last for a very long time and there are now millions of both aluminium alloy and stainless-steel barrels in service. Since every one costs up to $150 to produce, the capital invested is very considerable and the refurbishment of existing barrels, at intervals of about ten years, is an ongoing project, as essential as the production of new ones.

The basic form of a metal barrel is illustrated schematically in Figure 23.4. It is formed by deep-drawing the top and bottom halves from rolled sheet. They are trimmed to size and joined by a type of automatic arc welding called TIG welding. Before welding, while the internal walls are easily accessible, grease and detritus from deep-drawing are removed to present either a clean stainless-steel product contact surface or a clean aluminium substrate to receive a corrosion-resistant coating to the inside of the finished barrel. The *chimbs* that support the barrel are shaped from sheet by roll-forming and these too are attached by welding. Accessories, such as bosses to reinforce orifices for filling and dispensing, are attached in the same way.

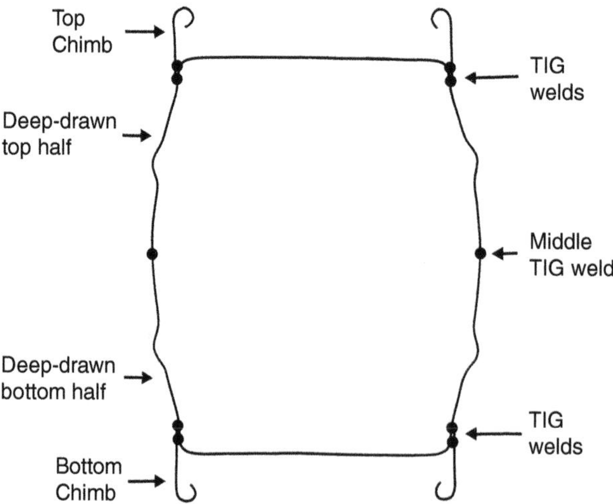

FIGURE 23.4
Typical construction of a metal beer barrel. The top and bottom halves are deep-drawn from sheet and joined by tungsten inert gas (TIG) electric arc welding. The chimbs are each attached by two parallel TIG welds.

23.4.3.3 Materials

23.4.3.3.1 Aluminium Alloys

Medium strength aluminium–magnesium–silicon alloys, as listed in Table 9.2, for example, AA 6061 are used because they are amenable to deep-drawing and the higher strength alloys in the AA 2000 and AA 7000 series have lower inherent corrosion resistance and provide less satisfactory substrates for protective coatings. Barrels are fabricated as described above and then solution-treated, water quenched and age hardened as appropriate for the alloy. Sheet 2.6 mm thick develops sufficient strength for 50 liter barrels. Degreasing before welding is by alkaline cleaning as described in Section 6.1.2.2. The internal surfaces may be anodized or a protective chromate coating is applied by spray, using orifices in the barrels for access. A final coat of lacquer may be applied by spray. Originally, the alloy was supplied clad with a layer of pure aluminium on the side corresponding to the inside wall of the barrel to enhance the corrosion resistance, but for economy, this practice has been discontinued. It does, however, place greater emphasis on maintaining the integrity of the protective coating. This requires close control of the heat treatments because if the full hardness is not developed, the barrels are vulnerable to denting in rough handling which cracks the internal coating, allowing the exposed metal to corrode in contact with the beer, producing nodules of corrosion product.

The refurbishment of existing aluminium alloy barrels is costly, that is $20 to $30 each and this has contributed to the re-introduction of stainless-steel barrels. On receipt, an aluminium alloy barrel is cleaned to the bare metal in nitric acid and rinsed. Dents are removed by shock waves generated by hammering them with the barrel filled with water. Repairs are made by welding where needed and the barrel is re-heat treated. Finally, new protective coatings are applied, using the same procedures as for new barrels.

23.4.3.3.2 Austenitic Stainless Steels

Austenitic stainless-steel barrels are shaped and welded in essentially the same way as for aluminium alloy barrels. Cleaning grease and detritus from the deep-drawn material is more difficult because of the inherent resistance of the material to dissolution. An aqueous solution containing 10% nitric +2% hydrofluoric acids at 60°C is applied by spray and rinsed off.

AISI 304 can resist cold beers almost indefinitely without further protection, but this advantage over aluminium is offset by greater difficulty in meeting other criteria. The sheet thickness is restricted to about 1.5 mm to minimize weight and material cost and the steel develops its strength by work hardening during the deep-drawing operation. This requires careful attention to the drawing operation because the work hardening is not uniform but depends on the degree of deformation the metal receives from place to place. It is particularly important to ensure that the strength is adequate where needed to prevent elastic distortion of the barrel shape under internal pressure that would offend the criterion for a stable volume.

Though unlikely, some corrosion problems are possible. Regular AISI 304 is preferred over the low carbon version, AISI 304L, listed in Table 8.3, because it is slightly stronger and has better deep-drawing characteristics. Sensitization to intergranular corrosion is not usually a problem because such thin sheet cools rapidly after welding, but instances have been observed when welding techniques have been unsatisfactory. The strength imparted by work hardening is accompanied by internal stress which could threaten SCC in the unlikely event of exposure to an aqueous environment containing chlorides.

The periodic refurbishment of stainless-steel barrels is less costly than for aluminium alloy barrels. All that is required is to clean them in a nitric acid/hydrofluoric acid spray and to make repairs by welding. No heat treatment or renewal of a protective coating is needed.

Further Reading

Briggs, D. E., Hough, J. F., Stevens, R. and Young, C. W., *Malting and Brewing Science,* Vol. 1, Malt and Sweet Wort, Chapman & Hall, New York, 1981.

Briggs, D. E., Hough, J. F., Stevens, R. and Young, C. W., *Malting and Brewing Science,* Vol. 2, Hops Wort and Beer, Chapman & Hall, New York, 1982.

Foster et al., *Dairy Microbiology,* Prentice Hall, London, 1957.

Guidelines for Use of Aluminum with Food and Chemicals, Aluminum Association, Washington, DC, 1983.

Hosford, W. F. and Duncan, J.L., The aluminum beverage can, *Scientific American,* 1994, Chap. 48.

Little, B., Wagner, P. and Mansfield, F., Microbiologically influenced corrosion of metals and alloys, *International Materials Reviews* 36, 253, 1991.

Moll, M. (ed.), *Beers and Coolers Definition Manufacture and Composition,* translated from French by Wainwright, T., Intercept Ltd., Andover, UK, 1991.

Morgan, E., *Tinplate and Modern Can Making Technology,* Pergamon Press, London, 1985.

Porter, J. W. G., *Milk and Dairy Foods,* Oxford University Press, London, 1975.

Sekine et al., *The Corrosion Inhibition of Mild Steel by Ascorbic and Folic Acids,* Pergamon Press, London, 1988.

24

Control of Corrosion in Building Construction

24.1 Introduction

Grand buildings in historic cities like Rome, London and Venice were constructed under the private initiatives of ecclesiastical, royal, municipal or rich mercantile patronage. The contemporary buildings for dwellings, farms and trade were built for relatively small populations, using available local resources. All were built without the concept of economic life spans by craftsmen using natural materials, such as stone, well-seasoned woods and baked clay bricks that had acclimatized to the local environments. Examples are found in most European cities and older American cities, including Philadelphia, Boston and New Orleans. The good condition of many of them testifies to the resilient qualities of these materials.

New buildings are entirely different in concept; they are mostly the products of large commercial enterprises developing and redeveloping built environments to satisfy the diverse needs of large populations in industrial societies. They are constructed to meet design life goals specified by clients according to use, characterized by three general time scales:

1. Indefinite life, a concept that is confined to prestige buildings and vital civil engineering works. Examples are buildings of national significance, such as legislative assemblies, centers of culture and strategically important bridges.
2. 60- to 120-year life, applies to most building of semi-permanent character such as city center developments, residential estates financed by loans and freeway structures.
3. 20-year life, characteristic for buildings that are sensitive to redefinition of land use, replacement of declining technologies and changing lifestyles. Examples are supermarkets, light industrial units, warehouses and leisure premises.

Construction is driven by funding within budgets and clients are often inclined to prefer expenditure that enhances outward appearance and internal amenities over that which contributes to maintaining the integrity of the building shell. In addition, competitive tendering induces constructors to economize wherever possible.

Modern construction relies on the application of metals. Some uses are overt, such as external cladding and siding, roofing, plumbing and central heating installations, but the most vital roles are concealed in their application as load-bearing members in steel reinforced concrete and in steel framing for high-rise structures. In selecting metals to retain their integrity against corrosion, different general approaches apply to rural, urban, industrial, marine or tropical locations and other specific environmental factors may need to be addressed, including the chemical nature of the soil and subsoil, mean temperature and temperature variations, humidity, rainfall and water quality.

Taking design life, economic and environmental considerations into account, there is not a universal set of standards for corrosion control but a menu of options from which to select to suit particular projects.

24.2 Structures

Three methods of construction predominate:

1. Erection of a frame with precast reinforced concrete sections connected mechanically to support the structure that is subsequently filled with masonry or clad with reinforced concrete, glass or metal panels
2. Erection of a frame with heavy steel sections, welded or bolted together to support the structure, clad with concrete or metal panels
3. Traditional methods with bricks or cement blocks bonded together with mortar, where the walls both bear the load of the structure and contain the internal environment

Traditional building is suited to low-rise, often residential, development with a long design life to cover financing by savings and loans. Framed buildings are better suited to commercial and industrial use. An important consideration is speed of erection and commercial clients often prefer 'fast track' steel framed buildings because they can be completed weeks ahead of equivalent buildings constructed with reinforced concrete.

24.2.1 Steel Bar for Reinforced Concrete Frames

24.2.1.1 Reinforced Concrete

Concrete is a macroscopic composite material in which an aggregate of carefully sized gravel and sand is bound together by a matrix of hardened cement. Cement is a calcined mixture of limestone and clay containing calcium oxide, CaO that sets to a hard mass when mixed with water. This is due to the formation of a hydrated cement gel in which calcium hydroxide, $Ca(OH)_2$ is present.

The theoretical quantity of water for complete hydration of the material is not enough to confer sufficient lubrication in the concrete to enable it to be shaken or vibrated into place and in practice additional water is required. The surplus eventually dries out, leaving pores distributed through the concrete after setting, some of which are interconnected.

The composite is strong in compression but weak in tension and can fracture if the tensile components of resolved stresses imposed on it are excessive. This weakness is remedied by laying the concrete around 0.4% carbon steel rods to which the tensile component is transferred. Although the steel is not exposed to the environment, it is susceptible to corrosion in water penetrating from the environment through the concrete pore structure.

24.2.1.2 Chemical Environment in Concrete

Calcium hydroxide is strongly alkaline and has a significant solubility in water, that is 2 g dm^{-3} at 20°C and the interior of moist newly hardened concrete has a pH in the range 12.5 to 13.5, depending on the quality of the cement. Reference to the iron–water Pourbaix diagrams, given in Figures 3.2 and 3.3, indicates that this environment is well-suited to protect steel reinforcement bar by passivation. There are, however, two agencies that can degrade this environment during service and depassivate the steel initiating corrosion, the penetration of chloride ions through the concrete and loss of alkalinity due to the absorption of carbon dioxide. It might be supposed that sulfur dioxide in industrial or otherwise polluted atmospheres could also enter the concrete and attack the steel but this does not happen. Instead, the sulfur dioxide is dissolved in water and falls as acid rain that is neutralized by attacking the cement itself at the concrete surface. Old concrete in industrial atmospheres often shows this form of attack as aggregate standing proud of the surface.

Failure of the composite due to corrosion of steel reinforcement is by cracking and *spalling* of the overlying concrete under tensile stresses induced by a large expansion in volume that occurs when the steel transforms into corrosion products. There is rarely a risk of failure of the reinforcement itself because damage to the concrete is so obvious that the problem is identified long before there is significant loss of structural reserves in the steel.

24.2.1.2.1 Chloride Penetration

Until the late 1970s, it was permissible to add calcium chloride to concrete to promote setting in cold weather and, although this practice is no longer acceptable, some buildings with this internal source of chloride ions still exist. External sources of chloride include marine atmospheres and highway deicing salts. The chloride ions can migrate through water penetrating the porosity in the concrete to the surfaces of the steel reinforcement bar. The permeation can be estimated using models based on standard diffusion theory and in practice the rate decreases with a parabolic or a cubic time constant for a given environment.

24.2.1.2.2 Loss of Alkalinity

Carbon dioxide from the atmosphere is carried by water through the pores in the concrete, reducing the pH by the reaction:

$$Ca(OH)_2 + CO_2 = CaCO_3 + H_2O \tag{24.1}$$

A plane of reduced alkalinity advances into the concrete from the surface and initiates corrosion when it reaches the steel. The effect is called *carbonation* and it proceeds at a significant rate for atmospheric relative humidities in the range 50% to 70%. At lower humidities there is insufficient water in the pores to sustain the process and at higher humidities the water content is so high that it blocks the ingress of carbon dioxide. The damage caused by carbonation is entirely due to a reduction in the ability of the concrete to protect steel; it does not harm the concrete itself and in fact it slightly increases its strength.

24.2.1.3 Protective Measures Applied to the Concrete

The onset of damage depends on the time taken for chloride or carbon dioxide to penetrate to the steel and, other factors being equal, this depends on the minimum depth of concrete cover over the outermost reinforcement bars. This ranges from 15 mm for thin concrete sections in benign environments to 75 mm for marine exposure.

Two options are available to modify the pore structure and hence reduce the rate at which carbon dioxide and chloride penetrate concrete:

1. Reducing the quantity of surplus water in the cement mix by adding plasticizers such as melamine or naphthalenes both to reduce the overall volume of porosity in the hydrated cement gel and to increase the proportion that is discrete rather than interconnected.

2. Replacing some of the cement in the concrete mixture with certain grades of ground granulated blast furnace slag or fly ash. The advantage that this has in controlling corrosion of the steel is offset by some loss in strength of the concrete. The practice is environmentally beneficial because it provides a use for a waste product.

Another approach is to apply paint coatings to the concrete surface. The paint is specially formulated to have a structure that serves as a filter, preventing the ingress of carbon dioxide but allowing the escape of water from the concrete that would otherwise accumulate at the interface and break the bond between paint and concrete. A typical paint is based on a polyurethane binding medium.

24.2.1.4 Protective Measures Applied to the Steel

Using good quality concrete to avert the degradation mechanisms described above, bare steel bar reinforcement gives a good life, relying on the protection of the alkaline environment and it is satisfactory for most building projects. Other options are available to meet particular conditions. These are zinc-coated steel, epoxy resin coated steel and AISI 304 or AISI 316 stainless steels. All of them incur additional expense that must be justified.

24.2.1.4.1 Zinc-Coated Steel

Zinc galvanically protects steel but reference to the zinc–water Pourbaix diagram, given in Figure 3.5, might suggest that because zinc is unstable at $pH > 10.5$, it would be rapidly attacked in moist concrete where the pH is 12.5 to 13.5. In practice, the zinc is not attacked and it protects the steel very well from attack due to loss of alkalinity following carbonation. This illustrates the caution needed in making predictions from Pourbaix diagrams that may not include all of the pertinent information. In the present context, the basic diagram does not include other species in concrete, notably calcium ions. The effect of the calcium is to form the insoluble product calcium zincate, $CaZnO_2$ that passivates the zinc surface.

Protection against chloride depassivation is more equivocal. One view is that zinc confers protection against external but not internal chloride sources, such as the calcium chloride setting agent. Experience with zinc-coated bar has not produced evidence of sufficient superiority over bare steel to justify its widespread application.

24.2.1.4.2 Epoxy Resin Coated Steel

Epoxy resin coated steel bar was pioneered in the United States, where it proved to be a solution to the corrosion of reinforcement in bridge decks subject to road wash containing deicing salts. Originally it was manufactured on a small scale, the steel rod was simply cleaned by shot blasting and the epoxy resin powder was applied by spray then cured. As production expanded, a regular phosphate conversion coating was introduced as a base for the epoxy resin. Greater care is needed to store and protect the coated bar on building sites, because the coating is vulnerable to damage by rough handling and degrades by prolonged exposure to extremes of weather.

Stray current corrosion is an issue of some concern for concrete structures in close proximity to electric railroads and streetcars. The effect is due to current leaking from the power lines into the reinforcement bar, anodically polarizing the metal and stimulating enhanced corrosion at the point of entry. If it is expected, it can be averted by using epoxy resin coated bar in which the steel is electrically insulated.

24.2.1.4.3 Stainless Steels

The high cost of stainless steels deters their general use for concrete reinforcement bar. AISI 304 and AISI 316 steel bars are more expensive than plain carbon steel bars by factors of eight and ten, respectively. They are used only where the performance is justified and the cost is commensurate with the value and permanence of the project. Examples are prestige buildings such as the Guildhall in London, England, and the decks of strategic bridges.

As produced, stainless-steel bar, *black bar,* is coated with mill scale from the high temperature exposure it experiences when it is hot-rolled. It can corrode in concrete if used in this condition and it must first be descaled by pickling in a nitric acid/hydrofluoric acid mixture.

24.2.1.4.4 Cathodic Protection

Cathodic protection is rarely a viable option because it is expensive to install and run and requires attention throughout the life of the structure. Nevertheless there are occasional difficult situations where it is prudent to make provision for it. Examples are piers immersed in seawater or parts of a structure below ground with high chloride content. The provision entails ensuring that all of the relevant bars are in electrical contact and are connected to terminals for application of an impressed cathodic current.

24.2.1.5 Stress-Corrosion Cracking of Pre-Stressed Reinforcement Bar

Pre-stressed reinforcement bars apply compression to the concrete that offsets subsequent tensile loading. There are two methods of pre-stressing the bar:

1. The reinforcement is stretched elastically, the concrete is cast in a mould around it and allowed to harden. The stress is imposed through the bond between the steel and concrete. The hardened product is cut into sections. A typical application is for lintels over doors and windows in brick built residential properties.
2. Steel tubes are cast into concrete laid *in situ* and when it has hardened, the reinforcement is threaded through the tubes and stretched. The ends are capped to hold the stress and cement grouting is pumped into the tubes for protection.

Stress-corrosion cracking of the stressed reinforcement bar is sometimes encountered. The cause is rarely the hydroxide content of the cement, as might be expected, but usually an adventitious alternative specific agent that should not be present. Cases are known where the agent was contamination by nitrate ions, NO_{3^-}, from biological sources at agricultural building sites and of thiocyanate ions, SCN^-, introduced into the cement by certain kinds of plasticizer. In other cases water was sealed into cavities around reinforcement through incorrect grouting.

Stronger steel is needed for pre-stressing than that used for normal reinforcement. Pearlitic steel with 0.8% carbon, as used in the United Kingdom, is not so vulnerable to stress-corrosion cracking as quenched and tempered martensitic steels often used elsewhere.

24.2.2 Steel Frames

Steel frames are protected from premature corrosion by painting and successful protection is mainly a matter of good geometric design and good practice. Conditions at a building site are not conducive to refinements such as chemical precleaning, conversion coatings and automatic paint application. A rugged approach is inevitable and the quality of the results depends on the skill, conscientiousness and supervision of those who carry out the work.

24.2.2.1 Design

The corrosion protection starts with good design. Whether painted or not the less time the metal spends in contact with water, the less is the chance of corrosion. Water can accumulate from rain and snow and from internal sources by condensation. To avoid trapping it, angled sections must be orientated to drain freely, and box sections are end capped or fitted with drainage holes. Crevices must be eliminated to avoid oxygen depleted water traps for the reasons given in Section 3.2.3.2. This entails ensuring full penetration of butt welds, double sided welding for lap welds and the application of sealant between the interfaces of mechanical joints. Traps in which dust and debris can accumulate and absorb condensate must also be eliminated.

24.2.2.2 Protection

As purchased, bare rolled steel sections may carry patches of strongly adherent mill scale from hot-rolling. It must be removed and the easiest way is to leave the steel in a stockyard open to the weather before assembly. If any patch of mill scale remains, it can absorb water through the paint, forming an electrolyte that stimulates corrosion of the steel underneath the paint. Pre-treatment for painting consists of shot-blasting the assembled steel and applying priming paint to the fresh surface. Shot-blasting should be delayed to within an hour or two before painting to avoid formation of rust. It is obvious that the paint is likely to be more durable if applied during a spell of fine weather than if preceded by rain or frost.

Painting is expensive, especially because it is labour intensive, and no more is applied than is needed. The treatment varies with the position within the structure. In contact with the external leaf of a building, where conditions are most aggressive, it may be necessary to use galvanized steel sections overlaid with a thick paint coating, but concealed steel sections in the interior of a warmed air-conditioned building can sometimes be left uncoated. The thickness of paint is adjusted to suit intermediate situations. Paints for steelwork are described in an international standard, ISO 12944. There is a current movement towards using *high-build* paints that can give coatings 400 μm thick.

24.2.3 Traditional Structures

Traditional buildings with load-bearing walls of bricks or cement block masonry bonded by mortar are less dependent on metals but there are some critical applications.

24.2.3.1 Wall Ties

Exterior masonry is built with cavity walls, that is two skins of brick or block work with a space between. The skins are held together at intervals with metal wall ties inserted in the mortar between bricks or blocks. Mortars are rich in lime or cement providing an environment similar to that in concrete. The atmosphere in the cavity is frequently moist and the ties are designed to resist corrosion from water condensing on them. A typical tie is made from flat steel bar of 20 mm × 3 mm cross section, protected by a thick coating of zinc, 970 mg m^{-2}. It is splayed at the ends to anchor it in the mortar and has a double twist within the cavity, providing a vertical edge from which the condensate drips away so that it does not collect on the flat surface. An alternative tie design is a thick wire loop with a twist directed downwards to provide the drip facility.

24.2.3.2 Rainwater Goods

Traditionally, roof gutters and down pipes were cast from iron with a high phosphorus content to confer the fluidity needed to flow into thin sections. They are heavy, brittle and require regular repainting. Some remain on older buildings but they have been mainly superseded by the lighter plastic or aluminium alternatives. Aluminium is protected by chromate/organic coatings. A common cause of corrosion failure in metal gutters is neglect to repaint the inside and clear away accumulated debris that retains water and locally screens the metal from oxygen, setting up differential aeration.

24.3 Cladding

Framed buildings can be enclosed in masonry but they are more often clad with panels that are hung on the frame externally. Two kinds of panel depending on metals predominate, reinforced concrete and aluminium alloy sheet. Plain carbon steels supplied with coloured polymer coatings applied during manufacture are less expensive alternatives, which respond to environments as do polymer coated steels generally. Glass is a competitive material.

24.3.1 Reinforced Concrete Panels

The same considerations apply in principle to reinforced concrete panels as to reinforced concrete frame sections. To reduce weight, thinner sections are needed that have less concrete cover but even so bare steel rod is usually adequate, provided that sufficient attention is paid to the quality of the concrete, the care with which it is cast and the application of suitable external coatings. With best practice good lives are obtained but there are examples of careless work where the steel corrodes, promoting premature concrete failure.

24.3.2 Aluminium Alloy Panels

Aluminium alloy panels are formed from rolled sheet and protected from corrosion by anodizing, a surface treatment that is exploited to produce a wide range of attractive finishes. The alloys used are AA 1050 and AA 5005 listed in Table 9.2, both of which develop their strength from the cold-rolling applied during manufacture. The towers of prestige buildings in New York City are examples of the impressive results that can be produced. Besides their primary application in new buildings, aluminium alloy panels are also used to refurbish depreciating exteriors of older buildings.

The sheet is cleaned in alkaline solutions, chemically brightened and anodized, usually in sulfuric acid, applying the principles described in Sections 6.1.2.2 and 6.4.2. The architectural use of aluminium alloys is a highly critical application, requiring material that yields anodized finishes free from surface blemishes with prescribed reflectivity and colour. To meet the standards required, the surface finishing procedures and the structure of the metal must both be carefully controlled.

Deficiencies introduced in the surface finishing operations are usually not difficult to recognize and correct. They can usually be traced to loss of control of solution compositions, temperature or electrical parameters or to lack of care in cleaning and rinsing. Provided that the metal is suitable, a first class anodizer can produce anodic films with consistent thickness, hardness and transparency.

Deficiencies in the metal cannot be rectified at the metal finishing stage because they are caused by faulty manufacture of the metal product and often can be traced back to features of the direct chill (DC) cast ingot from which the sheet was rolled. Porosity due to excessive hydrogen contents dissolved in the metal and aluminium oxide particles or films allowed to remain in the liquid metal from which the ingots are cast persist through rolling and form elongated blemishes on the anodized finished sheet. A more subtle deficiency is associated with the metallurgical structure of cast ingots. It is well-known that the surface zone of a regular semi-continuously cast (DC) aluminium alloy ingot has a non-equilibrium metallurgical structure that is different from the structure of the rest of the ingot. This surface zone is undulating due to the solidification mechanism and when the ingot is prepared for rolling by machining away the irregular cast surface, that is *scalping*, areas of both kinds of structure outcrop at the surface. The two kinds of surface structures persist to the rolled sheet, where they respond differently to brightening and anodizing, yielding objectionable streaks. Casting and other techniques have been developed to ensure that the surface zone is either thick enough to accept the surface machining without exposing the underlying different structure or so thin that it is all removed. Both procedures yield a uniform appearance but which is used depends on the appearance preferred by the architect. All of this means that aluminium alloys for architectural use are special products that must be purchased from reputable aluminium producers that appreciate the problems.

Aluminium panels are often used in their attractive natural silvery metallic appearance but some are coloured to suit the requirements of architects. The colour can be imparted either by dying the anodic film before sealing it as described in Section 6.4.2.3, or by using a self-colouring anodizing process that can produce shades ranging from yellow through bronze to black, without the need for dying. There are several proprietary processes, mostly based on anodizing in organic acids, controlled to produce the colour required. Where appearance is unimportant, aluminium cladding is protected by less expensive chromate/organic coating systems, as for aluminium roofs considered in Section 24.4.1 following.

24.4 Metal Roofs, Siding and Flashing

24.4.1 Self-Supporting Roofs and Siding

Two materials are commonly used for self-supporting roofs, galvanized steel sheet and aluminium alloy sheet, profiled by corrugating them to confer longitudinal stiffness. These roofs are suitable for buildings with design lives of the order of 20 years, such as supermarkets, light industrial premises, and so on. The basic surface protection, galvanizing for steel and application of conversion and baked paint coatings on aluminium, is applied to the sheet by the metal manufacturer when flat and it must withstand the subsequent deformation in profiling.

The galvanized steel sheet is typically coated with 275 g m^{-2} of electrodeposited zinc and then further coated with a 200 μm thick film of polyvinyl difluoride on the outside and a 25 μm thick film of lacquer on the inside.

The alternative material, aluminium alloy sheet is produced from a strain-hardening alloy, such as AA 3004 in medium hard temper. The alloy selected must be free from copper to avoid exfoliation corrosion. It is protected by a chromate or chromate–phosphate conversion coating as described in Section 6.4.3.1, and supplemented by a baked paint coating. Similar material is used for siding, that is cladding on the exterior of low rise domestic property. If the cut ends are left untreated, as they often may be, corrosion working in from the ends gradually undermines the protective coatings and they peel back progressively.

24.4.2 Fully Supported Roofs and Flashings

Pure lead and copper sheet are traditional roofing materials used for buildings with a long life. The sheet is not rigid enough for unsupported spans and is supported on timber or another suitable substrate. A related use of supported lead sheet is for flashings to seal valleys in pitched tiled roofs and for joints between roofs and chimneys or vents. It is well-suited to this function because it is soft and easily shaped to conform to awkward profiles.

Lead roofs exposed to the outside atmosphere develop films composed of lead carbonate, $PbCO_3$, and lead sulfate, $PbSO_4$, that are insoluble and electrically insulating so that protection can be established even in atmospheres polluted with sulfurous gases. In contrast, the film formed on the underside from condensing water vapour is predominantly the unprotective oxide, PbO, so that most failures of lead roofs are from the *inside*. Because lead is so soft, it can also suffer erosion–corrosion from constantly flowing or dripping water laden with grit. Copper roofs, similarly exposed, exhibit the familiar green patina of basic copper carbonate and sulfate, $CuCO_3 \cdot Cu(OH)_2$, $CuSO_4 \cdot Cu(OH)_2$, which is protective and aesthetically pleasing.

The lives of all roofs that depend on the establishment of a natural protective coating on originally bare metals are determined *inter alia* by the initial and early conditions of exposure. Aggressive species such as chloride ions contaminating the carbonate, sulfate or oxide layers during their evolution reduce their protective powers.

24.5 Plumbing and Central Heating Installations

Supply waters vary considerably, depending on sources, contact with substrates, biological activity and artificial treatment. They may be hard or soft as described in Section 2.2.9 with pH values usually in the range 6 to 8. They contain various concentrations of dissolved oxygen and carbon dioxide and other soluble species. The corrosion resistance of metals used in plumbing and central heating systems depends critically on all of these aspects of composition and different metals are selected to suit different localities.

24.5.1 Pipes

Galvanized steel, copper and austenitic stainless steels are all used for pipes. The choice between them is based mainly on experience of what works and what does not in particular localities.

24.5.1.1 Galvanized Steel

Zinc coatings are unreliable in soft acidic waters and galvanized steel is best suited to hard waters which it resists well due to precipitation of a tenacious calcareous scale supplementing the natural passivity of zinc. More failures of galvanized steel pipe in hard waters are due to furring, that is reduction of internal diameter by accumulated scale, than by corrosion. In cold water, zinc sacrificially protects steel exposed at gaps but this does not apply to hot water because there is a polarity reversal at 70°C, of the kind described in Section 4.1.3.6 and at higher temperatures, the zinc coating can stimulate attack on exposed steel.

24.5.1.1.1 Copper

Copper is a current standard material for tube used in plumbing and central heating circuits, usually with 1 mm wall thickness. It is tolerant of most water supplies but there are certain recognized causes of corrosion failure, type 1 pitting, type 2 pitting and dissolution in certain waters that can slowly dissolve copper.

Type 1 pitting occurs in cold water and is associated with a very thin carbon film on the inside wall formed due to lack of care in manufacturing the tube, as described in Section 4.1.3.4. The film acts as a cathodic collector stimulating the dissolution of copper exposed at gaps. The effect is well known and responsibility for it lies squarely with the manufacturer, who accepts liability, typically by guaranteeing the product for 25 years.

Type 2 pitting occurs in hot water and is associated with particular locations, where the water contains traces of manganese. A deposit of manganese dioxide accumulates during several years, forming a cathodic surface that stimulates corrosion of copper exposed at gaps.

Soft acidic waters with low oxygen contents can dissolve copper, that is they are *cuprosolvent*. If the effect is small the copper is not impaired but any less noble metals over which the water subsequently flows can suffer indirectly stimulated galvanic attack by the mechanism described in Section 4.1.3.5. This can cause failure of downstream galvanized steel in the system and of aluminium cooking utensils that are filled from it. This is a good example of where care must be taken not only in laying out a system so that water does not flow from more noble to less noble metals, but also in advising clients who use it.

24.5.1.1.2 Austenitic Stainless Steels

Where waters are so cuprosolvent that they can damage copper pipes, austenitic stainless-steel pipes are used instead. The cost differential is not prohibitive but it is more difficult to make joints in stainless steel.

24.5.2 Tanks

Many older installations used galvanized steel for both cold and hot water tanks. Causes of premature failure of cold water tanks could often be attributed to differential aeration, either at the water line or at the sites of debris that had fallen in. It is now usual to install reinforced plastic cold tanks. Cylinders formed from copper sheet are now standard for hot water tanks. They are, of course, compatible with the copper tubing used in modern systems.

24.5.3 Joints

One of the advantages of copper tubes is the ease with which joints can be made, either by fittings containing rings of solder or by compression fittings. Soldering is the most reliable and least expensive method and is preferred where the heat does no damage; current trends are towards lead-free solders. The copper must be fluxed with a material that enables the solder to wet the metal. There are various fluxes but since their function is to dissolve the copper oxide that covers and protects the metal, they must be rinsed away; corrosion can sometimes be observed in the tracks of flux that trickles from joints in vertical pipes.

24.5.4 Central Heating Circuits

A water circuit in a central heating system is almost inevitably a mixed metal system because of differences in the functions of the components and the most economic means of manufacturing them. The boiler in which water is heated, usually by gas or oil flames, is an iron casting; radiators are constructed by welding pressed steel panels that are painted on the outside but are uncoated inside; brass castings serve for pump and valve bodies. The circuit is connected by copper tubing for ease of installation.

The mixed metal system survives because the circuit is closed. Oxygen in the charge of water is depleted by initial corrosion but is not then replenished. If the water is hard, a thin calcareous scale also affords protection. The system can usually run uninhibited but, if necessary, inhibitors can be added to the water. Since the system has more than one metal, a mixture of inhibitors is required such as sodium nitrite and mercaptobenzothiazole to protect iron and copper, respectively.

Most failures occur through inadvertent and probably unsuspected aeration during service due to poor maintenance. The most common fault is an improperly balanced circulating pump that continuously expels some of the water through the overflow; another fault is neglect in sealing slight leaks that drain the water charge. Either of these faults opens the closed system to a constant supply of fresh aerated water to replenish that which is lost.

24.6 Corrosion of Metals in Timber

Building entails extensive use of metals in contact with and in close proximity to woods. Woods can promote corrosion in two different ways:

1. Providing an aggressive environment for metals in contact with it, especially fasteners, for example, nails, screws and brackets
2. Emitting corrosive vapours

24.6.1 Contact Corrosion

Woods are botanical materials that vary in properties both between and to a lesser extent within species. One of their chief characteristics is the ability to absorb and desorb water with corresponding dimensional changes. They are neutral or acidic media with pH values generally in the range 3.5 to 7.0. Among other solutes they can contain acetic, formic and oxalic acids and carbon dioxide solutions derived from bacterial transformation of starch and sugars. Although woods vary in chemical characteristics even within the same species, there is a recognized hierarchy in their ability to promote corrosion. Generally, harder woods are more acidic and more corrosive than softer woods. Some qualitative examples are given in Table 24.1. Electrochemical processes causing corrosion of contacting metals proceed in the aqueous phase in the wood and the more water that is present the more damage ensues. Woods are at their most corrosive when they are damp, when they are

TABLE 24.1
Qualitative Comparison of Environments in Some Common Woods

Material	Representative pH	Corrosive Influence
Oak	3.6	Strong
Sweet chestnut	3.5	Strong
Red cedar	3.5	Strong
Douglas fir	3.8	Significant
Teak	5.0	Significant
Spruce	4.2	Mild
Walnut	4.7	Mild
Ramin	5.3	Mild
African mahogany	5.6	Mild

new and when the atmosphere is humid. It is advisable to maintain the moisture content of timber below that in equilibrium with 60% to 70% relative humidity. New oak and sweet chestnut are among the more aggressive woods and ramin, walnut and African mahogany are among the least. Iron, steel, lead, cadmium and zinc are the most susceptible metals and stainless steels, copper and its alloys, aluminium and its alloys and tin are less vulnerable.

Treatments given to woods in contact with metals can exacerbate their aggressive nature. Some preservatives with which they are impregnated to protect against biological attack are waterborne and increase the electrolytic conductivity. Alternative formulations based on oxides or organic solvents are less harmful. Fire retardant preparations based on halogens used to impregnate wood can also be aggressive to metal fixings.

When steel nails, screws or bolts, corrode in wood, there are two concurrent damaging processes that weaken the fixture. Not only does the steel lose cross section but the voluminous corrosion products, iron hydroxides and iron salts, disrupt and soften the wood, an effect sometimes called *nail sickness*. For this reason, unprotected steel should not be in contact with wood exposed outside. Nails used to secure battens and clay tiles to wooden roof trusses should at least be galvanized, but it is better to use stainless steel or brass.

24.6.2 Corrosion by Vapours from Wood

Some woods emit acidic vapours that can corrode metals in their vicinity. There are several situations in building where problems can be anticipated and appropriate precautions taken. Red cedar is a popular material for use as shingles, that is wooden tiles, on roofs or walls but its emissions are particularly aggressive to metals in the immediate vicinity. New oak is an attractive wood for interior fittings such as paneling, shelving and window surrounds, but its vapours can damage associated unprotected metal fittings and the metal parts of adjacent equipment and furnishings.

24.7 Application of Stainless Steels in Leisure Pool Buildings

Stainless steels are applied extensively in swimming pool buildings, both for structural members and for accessories like balustrades and ladders. The austenitic stainless steels,

AISI 304 and AISI 316, have a good service record in traditional unheated swimming pools providing facilities for exercise and sport. Public swimming pools are now evolving into more comprehensive leisure centers based around the water. More people use them and spend longer times in the water imposing the following changes in the environment that have increased its hostility towards materials of construction:

1. The water is heated to temperatures in the range 26–30°C.
2. The water is turbulent in features such as water slides and fountains.
3. Higher concentrations of chlorine-based disinfectants are used.

The first two factors stimulate evaporation and hence condensation on cooler surfaces, particularly when the pool is closed.

Greater use of disinfectants increases the aggression of condensates by reaction with organic species in body fluids discharged into the water. Chlorine and some disinfectants containing chlorine interact with urea and other substances to produce chlorinated nitrogenous substances of the generic type, *chloramines*, based on the simplest member chloramine, NH_2Cl. In more complex chloramines, the hydrogen atoms are replaced by organic radicals containing carbon and hydrogen atoms. They are formed by overall reactions represented tentatively by:

$$CO \cdot (NH_2)_2\ (urea) + 2Cl_2 + H_2O = 2NH_2Cl\ (chloramine) + CO_2 + 2HCl \qquad (24.2)$$

The chemistry of these interactions is complicated and the nature of the particular products formed is sensitive to the pH of the water. Chloramines are very volatile and unstable. Their presence is manifest by a pungent odor characteristic of swimming pools. Two aspects of the problems they cause, safety-critical damage and area degradation of the building have stimulated reassessment of the selection and use of the steels.

24.7.1 Corrosion Damage

24.7.1.1 Safety-Critical Damage by Stress-Corrosion Cracking

A particular concern is stress-corrosion cracking. Attention was drawn to the problem by the collapse of a suspended concrete ceiling in Switzerland through failure of the stainless-steel supporting structure. The volatile chloramines can carry chlorine species to condensates in parts of the building remote from the pool, where they decompose into more stable species that can be concentrated by repeated evaporation, for example:

$$NH_2Cl + H_2O = NH_3 + HClO \qquad (24.3)$$

Typical structures at risk are roof supports, wire suspensions and bolt heads. The stress may be applied by external loads or imparted internally by fabrication or pulling up and tightening bolts. The danger is the insidious progress of incubation preceding crack initiation.

24.7.1.2 Area Damage

Area damage is due to depassivation of the steel by the chloride condensate. On open panels, the effect is unsightly rust staining from dissolved iron. Undetected pitting on hidden surfaces can develop into perforation of sheet in ventilation ducts and other services. Corrosion is confined mainly to areas where evaporation can concentrate condensates or fine spray. Metal that is fully immersed or frequently washed is less vulnerable.

24.7.2 Control

As with other structures, corrosion control begins with good geometric design to eliminate not only traps for liquid water, but also traps for condensate remote from the pool with special attention to load-bearing structures and devices. Where possible and appropriate, the materials should be stress-relieved after shaping.

Steels can be selected to suit different situations. The less expensive austenitic steels, AISI 304 and AISI 316 still have a useful role in non critical applications in direct contact with the pool. More specialized steels are needed for critical structures and some other areas sensitive to condensation. Steels with higher molybdenum contents are less vulnerable to stress-corrosion cracking. These include AISI 317, an austenitic steel with 3% to 4% molybdenum, and duplex steels with 3% molybdenum, listed in Table 8.3. Duplex steels have an advantage in the more resistant ferrite that they contain, but AISI 317 may prove to have the best pitting resistance.

Condition monitoring of the structure is now strongly recommended, especially for older buildings that were erected before the full extent of the problems were fully appreciated. The first concern is safety and although stress-corrosion cracking cannot be anticipated during its incubation period, the onset of cracking can be detected before it becomes catastrophic, provided that detailed targeted inspection is carried out at short intervals. Other damage can be reduced by inspection for condensation on open and hidden surfaces and cleaning them regularly to remove aggressive substances.

Further Reading

Berke, N. S., Chaker, V. and Whiting, D. (eds.), *Corrosion of Steel in Concrete*, ASTM, Philadelphia, Pennsylvania, 1990.

Butler, J. N., *Carbon Dioxide Equilibria and Their Applications*, Addison-Wesley, Reading, Massachusetts, 1982.

Franks, F., *Water*, The Royal Society for Chemistry, London, 1984.

Glaser, F. P. (ed.), *The Chemistry and Chemistry Related Properties of Cement*, British Ceramic Society, London, 1984.

Oldfield, J. W. and Todd, B., Room temperature stress corrosion cracking of stainless steels in indoor swimming pool atmospheres, *British Corrosion Journal*, 26, 173, 1991.

Page, C. L., Treadaway, K. W. J. and Barnforth, P. B. (eds.), *Corrosion of Reinforcement in Concrete*, Elsevier Applied Science, London, 1996.

Portland Cement Paste and Concrete, Macmillan, London, 1979.

Short, E. P. and Bryant, A. J., A review of some defects appearing on anodized aluminum, *Transactions of the Institute Metals Finish*, 53, 169, 1975.

Standards for Anodized Architectural Aluminum, Aluminum Association, Washington, DC, 1978.

Wernick, S., Pinner, R. and Sheasby, P. G., *The Surface Treatment of Aluminum and Its Alloys*, ASM International, Metals Park, Ohio, 1990.

25

Corrosion Control in Marine Environments

25.1 The Nature of Marine Environments

25.1.1 The Sea

The integrity of metals responds to the sea as a complex complete system and not simply to an equivalent saline solution in isolation. The natural characteristics of the sea that simultaneously influence the effects of seawater on metals and their protective coatings include not only composition but depth, temperature, dynamics, biology and pollution. The system is complex and successful development of alloys and protective systems for various applications owes as much to empirical observations as to scientific principles. Simplistic interpretation of results from laboratory tests on small coupons of metals submerged in quiescent samples of water from the open sea can give a misleading impression that general corrosion is not markedly worse than in corresponding tests with fresh waters.

25.1.1.1 Composition of Seawaters

Seawaters are solutions of ionic species, predominantly Na^+ and Cl^-, and variable quantities of oxygen and carbon dioxide. The total quantity of ionic species varies only within the narrow range 3.0 to 3.7 mass % in the open oceans worldwide and the quantity of oxygen varies as described later. It is slightly alkaline and the normal value of pH is 8.2 ± 0.1. This information on composition is sufficient for many purposes but for critical applications, a detailed analysis of the ionic species present may be required, as given for a typical sample in Table 25.1.

Obsolete descriptions, 'chlorinity' and 'salinity', given in some publications are incompatible with the SI system and are best discontinued. The values given in Table 25.1 refer to the open oceans and do not necessarily apply to particular geographic features where the solution can be diluted by fresh water from melting ice in the Arctic and Antarctic and from the outflow of rivers as in the Gulf of Bosnia, concentrated by evaporation in enclosed seas such as the Mediterranean and Red Sea or polluted by human activities, for example, by industrial contamination and sewage outflows.

The oxygen content is a particularly important component because it is the principal depolarizer of anodic corrosion reactions in seawater. Volume fractions for equilibrium with atmospheric oxygen pressure as a function of temperature are given in Table 25.2. They are slightly lower than corresponding values for pure water.

The oxygen content at the surface of the open sea approaches equilibrium with the atmosphere because agitation by wind and waves facilitates dissolution to resupply the

TABLE 25.1

Composition of a Typical Sample of Seawater from the Surface of the Open Ocean

Species	Cl^-	SO_4^{2-}	HCO_3^-	Na^+	Mg^{2+}	Ca^{2+}	K^+
Mass %	1.90	0.26	0.01	1.05	0.13	0.04	0.04

Br^-, F^-, Sr^{2+}, H_3BO_3 all <0.01.

TABLE 25.2

Oxygen Content of Seawater in Equilibrium with Air at Atmospheric Pressure

Temperature, °C	−2	0	5	10	15	20	25
Volume fraction	0.0085	0.0081	0.0072	0.0064	0.0059	0.0054	0.0049

normal biological demand, but it can be very different in coastal waters for the reasons explained in Section 25.1.1.4.

25.1.1.2 Temperature

The average surface temperatures of the seas and oceans range between the temperature at which ice begins to form, −2°C in polar regions to about 30°C in the tropics. Seasonal variations around the average at any particular location depend on the elevation of the sun and also on surface currents but they are usually <±5°C due to the large thermal capacity of the immense body of water.

25.1.1.3 Dynamics

The oceans are in constant motion both long range under the influences of tides and currents and short range including swell and spray. These effects produce tidal zones and splash zones that are significant considerations in the corrosion of partially immersed fixed structures used in harbor installations and oil platforms. Tidal action increases the corrosion rate of partially immersed structural steel. The splash zone above high tide and a zone just under low tide levels are particularly vulnerable. Explanations lack detail but are based on differential aeration effects.

The tides are the periodic rise and fall of ocean waters due to the net gravitational pull of the moon and sun. There are two tides per day and the times of high and low water depend on the positions of the moon and sun relative to the earth. The highest (spring) tides occur at full moon and new moon when the earth, moon and sun are aligned and the lowest (neap) tides occur at first and third quarters, when the sun and moon subtend an angle of 90° to the earth. The maximum tidal range is 14.5 m off the coast of Nova Scotia, but in most places it is much less. In the present context, the significance of tides is in subjecting the zone between high and low water on fixed structures to intermittent wetting and drying that are significant considerations in the corrosion of partially immersed fixed structures used in harbor installations and oil platforms. Tidal action increases the corrosion rate of partially immersed structural steel. The splash zone above high tide and a zone just under low tide levels are particularly vulnerable. Explanations lack detail but are based on differential aeration effects.

Surface water in the oceans circulates easterly in the Northern hemisphere and westerly in the Southern hemisphere, driven by prevailing winds and deflected by the earth's rotation. These currents transport heat from tropics toward polar regions so that coastal waters remain liquid at higher latitudes than would otherwise be expected. In deep water, the travelling waves simply undulate the water surface, but on approaching the coast they break on the shore and on structures as surf or spray. The region of any structure above the general water level that is wetted is called the splash zone. Its significance in the present context is that it is efficiently aerated.

25.1.1.4 Biology

Biological organisms can modify corrosion processes by their oxygen demand, by anaerobic activities and by their physical presence on metal and other surfaces, whether active or as residual detritus.

25.1.1.4.1 Biological Oxygen Demand

Replenishment of oxygen by solution from the atmosphere is normally sufficient to meet the moderate biological oxygen demand in the open sea but in tidal estuary and stagnant harbor waters, biological oxygen demand can deplete the water of dissolved oxygen to the extent that anaerobic bacteria can thrive, especially if resupply from the surface is restricted by oil films or other discharges from shipping or onshore service facilities. The significance in the present context is that these bacteria can depolarize anodic reactions by their metabolism, thereby stimulating corrosion.

25.1.1.4.2 Biologically Induced Differential Aeration

Live marine fouling organisms can establish themselves on metal surfaces, especially in tropical waters but also in temperate zones. Their effects depend on the nature of surfaces on which they form and how complete is the coverage. Thus their roles can be protective or destructive according to circumstances. Their physical presence can selectively deny access of dissolved oxygen to metal surfaces, thereby initiating macroscopic differential aeration cells.

25.1.2 Marine Atmospheres

Marine atmospheres of most interest in the context of corrosion are associated with local environments of shipping, coastal activities and offshore installations. They are subject to prevailing winds and close contact with seawater and can be expected to contain saline particles from evaporation of spray as well as the usual contaminants from emissions associated with industrial and human activities including products of combustion from internal combustion engines. Saline particles either alone or in combination with sulfates derived from oxidation of sulfur dioxide yield a potent source of hygroscopic deposits on metal surfaces that can reduce the values of relative humidity required to sustain atmospheric corrosion. A further effect is the promotion of hot corrosion and sulfidation in marine gas turbine engines creating problems like those experienced with aircraft engines, discussed in Section 21.2.2.3.

25.2 Ships

25.2.1 Paints for Steel Hulls

The first priority in selecting a steel for the hull of a welded ship is to avoid premature failure by brittle fracture. All plain carbon steels are brittle at low temperatures so that it is essential to select one with a ductile to brittle transition well below any temperature that the ship is likely to experience in service. Factors that contribute to low transition temperatures are low carbon content, ultra low phosphorus content and small grain size. The selection of suitable steels is guided by experience.

Once the mechanical integrity of the hull is assured, attention can be directed to corrosion protection, which is almost invariably by a combination of paint coatings and cathodic protection. The application of cathodic protection was considered in detail in Chapter 20 with only brief reference to the need for complementary paint coatings. The present discussion examines the functions and nature of the paint coatings.

The surface of a hull comprises four parts which require different approaches to paint systems according to the conditions of service, as follows:

1. The ship's bottom which is always immersed in the sea
2. The top and superstructure that are permanently exposed to air but subject to spray
3. The waterline zone immersed when the ship is loaded and exposed to air when it is not
4. Interior surfaces

Although paint formulation is guided by scientific principles, it is essentially an empirical art depending on the skill and experience of the formulator. This is especially true of paints formulated for marine service and the following discussion is a brief summary of the factors to be considered, bearing in mind that there is often more than one solution to a problem.

25.2.1.1 Ships' Bottoms

Paints used to protect ships' bottoms against corrosion must be formulated to meet conditions quite different from those encountered for structural steelwork. Seawater is slightly alkaline and becomes more so by the application of cathodic protection, promoting the degradation of binding media that are susceptible to saponification. Air-drying paints are required because ships must obviously be painted on site, but standard priming paints formulated for onshore structural steels are unsuitable because they contain long fractions of linseed oil, which is readily saponified and are pigmented with red lead which has insufficient resistance to ionic conduction for continuous immersion in seawater. Suitable binding media are based on phenolic resins, bitumen, chlorinated rubber and non-saponifiable resins. Lead sulfate, zinc chromate or aluminium flake are preferred pigments. Five or six coats are required to build a coating thickness of 0.15 to 0.20 mm. Finishing paints for ships' bottoms contain *anti-fouling* agents, usually copper compounds to prevent the accumulation of marine biological growth which increases friction between the hull and sea, thereby reducing the ship's speed and increasing fuel consumption. The binding medium is modified or selected to maintain slow release of the anti-fouling agent into the boundary layer at the interface with the seawater. A useful component for this purpose is *rosin*, a slightly acidic material that dissolves very slowly in the slightly alkaline environment. Some representative formulations are given in Table 25.3.

25.2.1.2 Ships' Tops and Superstructures

Paint formulation for exterior steel surfaces above the waterline, including superstructures is approached in the same way as for onshore structural steelwork and fulfills the dual

TABLE 25.3
Some Examples of Ships' Paints

Application	Formulation	No. of Coats	Thickness/mm
Ships' bottoms	Bitumen with aluminium flake pigment	6[a]	0.20–0.25
	Phenolic binding medium with lead sulfate and aluminium flake pigments	5[a]	0.20–0.25
Ships' topsides and superstructures	Primer – alkyd binding medium with red lead pigment	3	0.10–0.15
	Black finish – alkyd binding medium with carbon black pigment	2	0.05–0.08
	White finish – alkyd binding medium with titanium dioxide pigment	2	0.06–0.10
Ships' interiors	Primer – alkyd binding medium with zinc chromate pigment	2	0.06–0.10
	Finish – alkyd binding medium with titanium dioxide pigment and stainers	2	0.06–0.08
Cargo/ballast tanks in oil tankers	Cold-cured epoxy resin with titanium dioxide pigment	5	0.15–0.20

[a] Including final coat containing anti-fouling agent.

function of aesthetic appeal as well as protection. A typical system comprises a priming paint based on a quick drying alkyd resin binding medium pigmented with red lead or zinc chromate inhibitors, followed by an undercoat and a gloss top coat, both suitably pigmented with stainers if necessary to provide a specified decorative finish. Some representative formulations are given in Table 25.3.

25.2.1.3 Waterline Zones

Waterline zones present the paint formulator with the combined problems of both ships' bottoms and tops. They experience cathodic protection when submerged when ships are loaded, but are otherwise exposed to view. Thus they must match the aesthetic appeal of surfaces permanently above the waterline and yet offer the same protection as the specially formulated marine paints for ships' bottoms. In addition, they are susceptible to mechanical damage from contact with quays and barges. Unlike ships' bottoms, they are accessible without recourse to dry docking so that one solution is to use similar paint systems as for tops, accepting more frequent repainting.

25.2.1.4 Interior Surfaces

Interior surfaces in passenger ships and crew accommodations are treated in the same way as tops and superstructures, but cargo and ballast tanks of oil tankers present a particular costly corrosion problem if not specially protected. Renewal of corroded tanks can cost hundreds of thousands of dollars.

The cargo and ballast tanks are filled with oil on outward voyages and with fresh or seawater ballast on return. If the tanks were unprotected, the pattern of corrosion would depend on the nature of the oil cargo. Refined oil does not leave residues on the surfaces and subsequent corrosion in seawater ballast is uniform. Crude oil leaves incomplete oil or wax deposits that restrict access of oxygen to parts of the surface when seawater is subsequently admitted, generating differential aeration cells that result in local area corrosion or macroscopic pitting. Some crude oils also contain significant sulfur contents that can yield sulfuric acid.

The principal strategy for corrosion control is the application of cathodic protection by sacrificial zinc anodes described in detail in Chapter 20 together with a suitable protective coating system. Supplementary measures can include conditioning the tank environment by reducing humidity in the air space, purging oxygen with an inert gas and adding inhibitors.

The protective coating must fulfill the following demanding conditions:

1. It must protect the metal when exposed to the ballast water and tank cleaners.
2. It must resist degradation in contact with oil cargos.
3. It must not contaminate oil cargos.
4. It must be suitable for application in shipyards.

Epoxy resin paints and cementiferous coatings, respectively, represent organic and inorganic coatings that meet these conditions. Meticulous surface preparation, presenting clean bare metal, is essential to ensure good adhesion of either kind of coating.

Epoxy resin paints pigmented with rutile, TiO_2, dispersed in organic solvents are chemically cured with amines or polyamides, yielding complexes with outstanding chemical resistance. The resin and curing agent are supplied separately in dual packs and are mixed in prescribed proportions immediately before painting. Three or four coats are applied, yielding a film ~0.15 mm thick. Oil cargos or ballast cannot be admitted until the paints are fully cured which takes several days at 12–16°C.

Cementiferous coatings comprise an alkali silicate binder pigmented with zinc powder. The binder is mixed with an aqueous solution of the silicate just before application. The final film is a complex structure of silicates and zincates.

25.2.1.5 Surface Preparation and Application

Painting costs for ships include not only the costs of materials, labour, supervision and docking facilities, but also the costs of downtime which are very substantial through lost earnings while ships are out of service especially when dry docking is required. This provides powerful incentives to maximize periods between repainting and minimize the time required. The life of a coating depends primarily on the rigor of surface preparation and the skill of application.

25.2.1.5.1 Surface Preparation

The first priority for new ships is to remove millscale remaining from hot-working the steel as supplied by the producer, because it can provide conditions for corrosion underneath paint by absorbing seawater diffusing through it. Moreover, the high conductivity of the seawater environment can sustain long range differential cells between scaled and scale-free areas. It can be removed completely by pickling in sulfuric acid by the steel producer or at the shipyard before erection as described in Section 6.1 of Chapter 6. Alternatively, it can be removed by dry or wet shot or grit blasting either before or after assembly depending on available facilities. Unfortunately, many months may elapse after cleaning the steel before painting it so that good storage facilities or temporary protection is desirable but not always available or effective.

On-site wire brushing followed by chemical pretreatment is used to promote adhesion of priming paints to bare steel surfaces. It may simply be a phosphoric acid wash, emulating on-site phosphating, but more often it is implemented by a proprietary etch primer comprising a vinyl resin solution containing a chromate pigment and an optimum quantity of phosphoric acid. This produces a thin resinous film as a key for the regular paint system.

25.2.1.5.2 Application

Paints can be applied by all of the usual on-site methods, that is brushing, spraying and roller coating. To reach the large external areas of ships' bottoms requires brushes and rollers on handles many meters long, so that the viscosities of paints are reduced by thinning with extra solvent to enable operators to apply them. For operator safety, paints pigmented with zinc chromate are preferred to paints pigmented with lead for internal work and face masks and protective clothing are provided, especially for work with anti-fouling paints.

To avoid excessive costs of dry docking and downtime, repainting of ships' bottoms must occupy as little time as possible. This means that paints may sometimes be applied to damp, rusty or oily surfaces irrespective of weather conditions, risking occasional adhesion failure. Otherwise, if applied in reasonably good conditions, paints have lives comparable with corresponding onshore paint systems, ultimately failing by exhausting the finite supply of anti-fouling agents or by eventual degradation of the binding media.

25.2.2 Aluminium Alloy Superstructures

The attraction in using aluminium alloys for superstructures of large ships is the low density, $\sim 2.7 \times 10^3$ kg m^{-3}, compared with that of steel, $\sim 7 \times 10^3$ kg m^{-3}, yielding a lower center of gravity and reducing overall mass. This gives greater freedom in design and the benefit that accrues can be taken in various ways, for example, improving stability, or permitting the addition of an extra deck on passenger cruise ships.

The appropriate alloys must have adequate strength, good corrosion resistance to marine atmospheres and sea spray, fire resistance to satisfy regulatory authorities, seaworthy criteria and good weldability. Alloys that meet these criteria are the aluminium-magnesium alloy, AA 5456, and the aluminium–magnesium–silicon alloy, AA 6061, with compositions and properties given in Tables 9.2 and 9.3. Most of the structure is composed of alloy AA 5456 plate, welded to avoid crevices that could trap spray and initiate corrosion. Information in Table 9.3 shows that AA 5456 is the strongest alloy when annealed, as it is in the heat-affected zones near welds. This attribute combined with good corrosion resistance in marine environments is why it is well-suited to the application. Metal arc inert gas (MIG) welding which yields sound welds is almost invariably used. The age-hardened alloy AA 6061 is used where greater strength is required, for example, deck pillars, provided that it is joined mechanically because, as Table 9.3 shows, welded joints would be weak if softened by annealing in heat-affected zones.

An aluminium alloy superstructure is a potential partner in mixed metal systems and the design of the vessel must eliminate the risks of galvanic stimulation. The steel structure in contact with the aluminium alloy must be coated with a compatible metal, usually zinc preferably in the form of hot-dip galvanizing or alternatively as a sprayed coating. This also applies to steel fasteners, including bolts and cold-formed rivets. Copper and copper alloys can inflict severe corrosion on aluminium superstructures, not only in direct contact but also by indirect stimulation. Therefore, alternative plastic or aluminium or zinc coated fittings are preferred if possible, and where copper or copper alloy components are used, for example, for electrical or other utilities, they must be isolated or shielded so that copper cannot be transferred to the aluminium structure by cuprosolvent waters.

25.2.3 Propellers

The duty of a ship's propellers is the efficient conversion of the rotational energy generated by the engines to the translational energy of the ship. A propeller is, therefore, a large heavily stressed multi-bladed component of complex shape with sufficient strength to transmit the energy. This imposes the requirements for high strength-to-mass ratio and good casting characteristics on the materials from which it is produced. A further essential requirement is resistance to all forms of stress-related corrosion effects induced by the rapid relative motion and turbulence of the disturbed seawater surrounding the rotating propeller, that is corrosion fatigue, erosion–corrosion and cavitation–erosion. An additional practical consideration is tolerance of welding and local reshaping to repair damage sustained in service.

Few metallic materials meet all of these requirements and the selection of alloys for the manufacture of large propellers is predominantly from the aluminium bronzes, included in the specifications given in Table 10.2 of Chapter 10, that is nickel–aluminium bronze Cu Al10 Fe5 Ni5, and manganese aluminium bronze, Cu Mn13 Al8 Fe3 Ni3, with the reservation that in some circumstances, manganese aluminium bronze can be susceptible to stress-corrosion cracking. The aluminium bronzes have superseded high tensile brass Cu Zn39 Fe Mn as standard materials for propellers because they have exceptionally high resistance to erosion–corrosion and cavitation, they are lighter and they have extended service lives. Nickel–aluminium bronze is also suitable for the blades, hub body, hub cone and bolts of variable pitch propellers.

Welding to repair damage sustained in service can render manganese aluminium bronze propellers susceptible to stress-corrosion cracking, and it is advisable to apply a post-welding stress relief heat treatment. Corresponding welded nickel–aluminium bronze propellers are not susceptible to stress-corrosion cracking in seawater, but the alloy requires careful welding techniques to avoid cracks during welding either in the weld bead or the parent metal.

A significant consequential effect of aluminium bronze propellers is that they are large masses of metal that are noble with respect to steel. The galvanic effect is opposed to cathodic protection applied to the hull and must be compensated by selective disposition of sacrificial anodes or distribution of impressed current as appropriate.

Some small vessels are fitted with propellers cast from titanium alloys which are virtually immune to stress-related corrosion effects and from molybdenum-bearing duplex stainless steels, which have good general corrosion resistance and better resistance to stress-corrosion cracking than austenitic steels by virtue of their ferrite content.

25.2.4 Marine Gas Turbine Engines

Gas turbine engines derived from aircraft engines power many of the world's warships, especially frigates and destroyers. Aspects of corrosion and its control that apply to aircraft engines apply in modified form to engines in marine service. Alloy selection for the turbine blades must take account of greater vulnerability to hot corrosion and sulfidation due to the ingestion of salt-laden air as explained in Section 21.2.2.3 and of the lower temperatures at which marine gas turbine engines operate.

Whereas Type-1 hot corrosion is of most concern for the high temperatures in aircraft engines, the lower temperatures in marine engines promote Type-2 hot corrosion in which the sodium sulfate dew has a basic character. However, the very high temperature strength and stability of alumina-forming superalloys for aircraft engine turbine blades is not required at the lower temperatures so that chromia-forming superalloys, listed in Table 11.4, can be used to exploit their greater resistance to basic fluxing.

25.2.5 Miscellaneous Components

Nickel–aluminium bronze, manganese bronze and high tensile brass are used for external cast underwater components such as propeller shaft brackets and rudder fittings. The biocidal effect of copper ions keeps these alloys free from biological fouling when openly exposed to the sea. The 90/10 cupro-nickel alloy, Cu Ni10 Fe1 Mn, is used for seawater piping on ships.

Fans are used on board oil tankers to maintain an inert gas blanket over the oil cargo to eliminate fire and explosion risks, using exhaust gas from the engines scrubbed with seawater. This environment is variable and so aggressive that only fans made of titanium or aluminium bronze survive to give satisfactory service.

25.2.6 Heat Exchangers

On-board steam condensers, oil coolers and other heat exchangers operate on seawater. A typical heat exchanger is a nest of tubes starting and terminating at a *tube plate* enclosed within a cylindrical shell. A fluid to be cooled or steam to be condensed passes inside the tubes and a cross flow of cooling seawater guided by baffles passes over the outside. The system is contained in a cast iron or steel shell appropriately protected, for example, with chlorinated rubber.

Materials for heat exchangers have evolved from practical experience over a prolonged period. The tubes must resist corrosion not only to retain their integrity for the design life, but also to remain free from surface corrosion products that interfere with heat transfer. Alloys that can give satisfactory service are aluminium brass, and the 70/30 cupro-nickel alloy, Cu Ni30 Mn1 Fe and especially the 90/10 cupro-nickel alloy, Cu Ni10 Fe1 Mn. These alloys are specified in Table 10.2 and their characteristics are reviewed in Sections 10.2.1 and 10.2.2 in Chapter 10. Titanium tubes are used where severe conditions or an extended design life justifies the cost.

The tube plates must be strong enough to support the tube bundle at operating temperatures and galvanically compatible with the tube material. The longest established material for tube plates is rolled naval brass, which is compatible with aluminium brass and 90/10 cupro-nickel alloy but not with the 70/30 copper–nickel alloy, which causes dezincification of the brass. The duplex aluminium bronze, Cu Al8 Fe3 and the nickel–aluminium bronze, Cu Al10 Fe5 Ni5 are more reliable materials which are very widely used with cupro-nickel tubes in condensers and heat exchangers cooled with seawater. Titanium alloys, aluminium bronze or nickel–aluminium bronze tube plates are used to support titanium tubes.

25.3 Offshore Platforms

Many of the considerations that apply to corrosion protection of ships apply also to the structure of offshore oil production platforms, but there are significant differences:

1. The structure is static and cannot relax the full forces of heavy seas and wind.
2. Greater use is made of seawater.
3. There may be sour service environments, that is fluids containing hydrogen sulfide.
4. The installation is remote from service bases so that maintenance is less easily available.

25.3.1 Structures

The first concern is the integrity of the supporting platform legs, which are protected by a combination of alkali-resistant paint as described for ships hulls and cathodic protection by impressed current, using multiple auxiliary anodes of platinum coated on substrates of titanium, tantalum or niobium. Unlike a ship, a fixed platform does not ride with the water surface and suitable arrangements are required to absorb cyclic stresses imposed on the protection system by wave motion and wind. A convenient arrangement is to attach multiple slim cylindrical anodes to insulated wire ropes slung from the structure between tensioning springs above the waterline and heavy concrete anchor weights on the sea bed. The anodes are served individually by separate leads with waterproof insulation.

As with ships' hulls, the cathodic protection is effective only for the parts of the steel platform legs that are permanently immersed. It does not protect the steel in the tidal and splash zones, where the corrosion rate is typically ten times greater than that above or below this level. This is attributed to the high oxygen potential available to corrode the wet steel, aggravated by abrasive action of waves which removes corrosion products continuously, exposing fresh metal surfaces. One method of resolving the problem is to increase the steel thickness in the vulnerable zone, which extends from about 2 meters below the lowest tides to about 5 meters above the highest tides, but this adds considerable weight to the structure. An alternative approach is to clad the steel in the vulnerable zone with a corrosion-resistant alloy, the 90/10 copper–nickel alloy, specified in Table 10.2, either by welding the alloy plate to existing structures or by using steel clad with the alloy by hot-rolling during manufacture for new structures. An additional consideration is the selection of suitable steels for offshore oil production platforms because commercial carbon steels for non-critical applications with sulfur contents of up to 0.05% are susceptible to embrittlement by cathodic hydrogen in the presence of hydrogen sulfide. Steels with ultra-low sulfur contents <0.002% are more resistant than regular steels and are specially manufactured by secondary steelmaking processes for oil rigs.

25.3.2 Systems Using Seawater

Seawater is used very extensively in offshore oil platforms for many purposes, including injecting water back into the well, cooling heat exchangers, washing down decks and equipment, firefighting and sewage disposal. The supply systems comprise pipes with diameters of 50 to 350 mm, centrifugal pumps and valves, all of which require materials that resist corrosion and stress-related corrosion damage in seawater.

25.3.2.1 Pipes for Seawater

The material that most nearly meets the broad requirements of a wide range of pipeline applications for unpolluted seawater is the 90/10 copper–nickel alloy, specified in Table 10.2 of Chapter 10, following its long history of satisfactory use in diverse marine applications. It is ductile with adequate strength and it is not susceptible to stress-corrosion cracking. Soon after commissioning, corrosion of the alloy in static, quiescent or flowing seawater virtually ceases due to the formation of the protective film. It resists marine biofouling by maintaining a minute population of copper ions at the interface with the water and it remains clean indefinitely in contact with untreated seawater. One reservation is local damage on the *outside* of the pipes by differential oxygen cells created under lagging if through negligence it is wetted by seawater.

25.3.2.2 Seawater Pumps

Seawater is delivered and circulated by centrifugal pumps. The essential parts are the pump body, an impeller and its drive shaft. To resist damage by erosion–corrosion, cavitation and wear the impeller is cast from either nickel–aluminium bronze specified in Table 10.2 or the hardened nickel–copper alloy K Monel, specified in Table 11.1. For long life and reliability, the pump body and impeller drive shaft are also made of the nickel–aluminium bronze, especially since it resists pitting in the gland through which the impeller shaft is driven and any crevice corrosion associated with the gland takes the form of selective phase attack with minimal effect on the surface finish of little practical significance. Gunmetal, specified in Table 10.2, is an acceptable alternative material for pump bodies and for low speed impellers with less arduous duties. Cast austenitic stainless steels have sometimes been used, but they are less satisfactory because of their liability to cavitation damage and pitting in aqueous chloride media.

25.3.2.3 Valves for Seawater

The critical parts of valves are the discs seats and stems and for the highest integrity, all of these parts are made from nickel–aluminium bronze or K Monel, exploiting the combination of corrosion resistance and wear resistance. The discs and seats are usually cast and the stems are fabricated from wrought rod. Alternative materials for stems are phosphor bronze and 70/30 cupro-nickel. Incidentally, the spark-resistance of copper alloys contributes to safety in offshore plants by reducing the risk of ignition of flammable atmospheres.

25.3.3 Sour Service Environments

In the reducing conditions in which oil and natural gas are formed, sulfur is converted to hydrogen sulfide which appears as a pollutant of the products recovered from oilfields. Liquids and gases contaminated with hydrogen sulfide derived from this source constitute *sour service environments.* A formal definition according to National Association of Corrosion Engineers (NACE) international specification MR0175 is *fluids containing water and hydrogen sulfide that is at a total pressure of 0.4 MPa or greater and if the partial pressure of hydrogen sulfide is greater than 0.0003 MPa.* These environments are acidic with pH values <3.5.

The copper and nickel alloys that are satisfactory in general marine service suffer rapid corrosion in seawater polluted by hydrogen sulfide because the passive films on which their corrosion resistance depends are infiltrated or replaced by sulfides and fail to protect. The problem is resolved by replacing them for sour service with appropriate titanium and titanium alloys, which are virtually unattacked by aerated aqueous media containing both sulfides and chlorides in addition to their outstanding resistance to local and stress-assisted corrosion, especially stress-corrosion cracking.

The scope to choose titanium alloys matching the copper and nickel alloys that they replace is indicated in the representative specifications and properties given in Table 12.2 of Chapter 12. Low pressure sour water piping systems and wellhead seal rings are made from the ductile but relatively low strength pure metal, ASTM Grade 2. Offshore sour service taper joints for highly stressed dynamic production riser systems are produced from ASTM Grades 25, 28 and 29, because they are of sufficient strength yet flexible and the noble metal additions improve resistance to non-aerated acidic solutions. Components for valves, springs and pumps are made from the high strength β alloys, for example, ASTM Grade 19.

25.3.4 Submerged Oil and Gas Pipelines

Gas and oil extracted from offshore wells are delivered to landfall or loading terminals by steel submerged pipelines. They are protected by a combination of coatings and cathodic protection by impressed current applied using ground beds in the sea floor with silicon, iron or graphite auxiliary anodes and waterproof insulated cables. The design must be robust to withstand wave action and movement of the sea bed and the coating materials must resist marine fouling. Installation costs are high but commensurate with the value of the pipeline and the costs incurred by disruption of supplies.

Flanges on submerged oil or gas pipes are often fastened together by bolts made from standard alloys resistant to corrosion and stress-corrosion cracking in seawater, especially Monel or an aluminium bronze, relying on cathodic protection to offset adverse galvanic effects. An alternative is to use specially designed flanges in which steel bolts are covered by grease. Monel is more commonly used for bolting flanges on large diameter undersea oil or gas pipes.

25.3.5 Submerged Fasteners and Fittings

Bolts used to secure submerged fittings must be made from materials that can maintain their integrity indefinitely when exposed to seawater under tension. Some alloys can be rejected immediately because they are susceptible to stress-corrosion cracking and/or pitting in aqueous chloride media. They include high tensile brass, manganese–aluminium bronze and stainless steels. Suitable materials include the copper alloys, nickel–aluminium bronze, aluminium–silicon bronze and phosphor bronzes specified in Table 10.2 of Chapter 10 and the nickel alloy, K Monel, specified in Table 11.2 of Chapter 11.

Further Reading

Webb, A. W. O. and Capper, H., Propellers. In *Materials for Marine Machinery*, ed. S. H. Frederick and H. Capper, The Institute of Marine Engineers/Marine Media Management Limited, London, 1976.

26

Corrosion Control for Fossil Fuel Boilers for Steam Raising

Large watertube boilers provide the evaporative capacity to raise the quantities of superheated steam needed to drive steam turbines generating electricity for general distribution. The integrity and efficiency of boilers and the systems they serve must be maintained both to preserve their high capital value and to avoid consequential loss through interruption of power supplies to public utilities and industrial concerns.

Methods of corrosion control are based on the principles discussed in Chapters 3 and 4, but are implemented empirically to suit particular conditions of service that prevail and the nature of the heat supply. Fossil fuel boilers can be fired by gas, oil or pulverized coal. They are complex systems designed to maximize thermal efficiency and differ considerably according to the nature of the fuel, power output, steam temperature and local experience. Figure 26.1 summarizes the progression of fuel, flue gases, water and steam through a representative system.

Water is evaporated in banks of parallel tubes suitably disposed in a combustion chamber. The steam is separated from water in a horizontal drum and superheated to maximize the thermodynamic efficiency of the system. The superheater is a separate chamber containing banks of tubes through which the steam passes countercurrent to the hot flue gases issuing from the combustion chamber. The steam, at temperatures in the range 350–650°C according to the design and at corresponding pressures in the range 4–10 MPa, is delivered to high and intermediate pressure turbines in sequence with intermediate reheat and having performed its work, it is condensed and the condensate is returned to the boiler for recycling. The reserve of water in the system corresponds to about half an hour's steam supply. Ancillary equipment includes a heat exchanger to preheat the feed water, a recuperator to preheat air for combustion and an electrostatic precipitator to remove dust.

There are two aspects of corrosion control to consider, waterside corrosion, including the treatment of boiler water, and fireside corrosion.

26.1 Waterside Corrosion Control

It is common practice to manufacture the evaporating tubes, the steam separating drum and most of the superheating tubes, for example, those operating at temperatures $<\sim$450°C from mild steel. This material is inexpensive and strong enough at moderate elevated temperatures to contain the stresses imposed on it. Superheater tubes operating in the temperature range 450–550°C, are manufactured from low alloy steels to supply sufficient hot strength and at still higher temperatures they are manufactured from heat-resistant alloys, as described in Sections 26.1.4 and 26.2.4, to control both waterside and fireside corrosion.

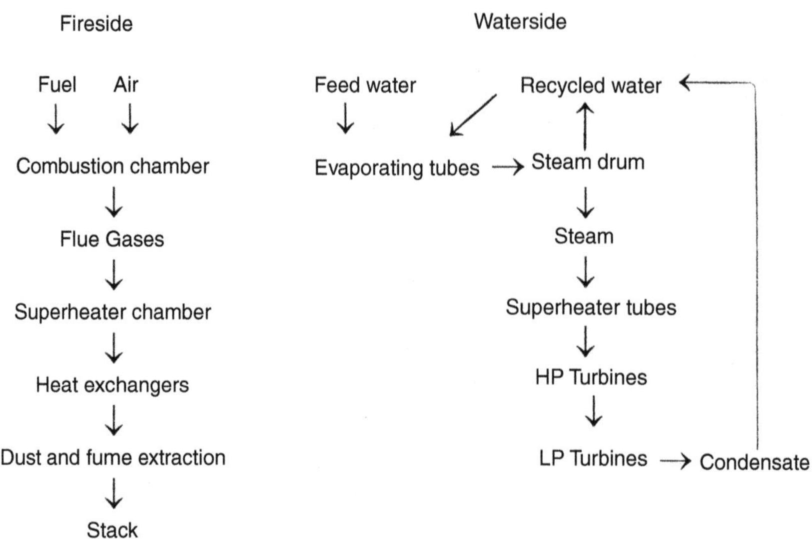

FIGURE 26.1
Progression of fuel, flue gases, water and steam through a fossil fuel recirculating boiler system.

26.1.1 Passivation of Iron in Superheated Water and Steam

The Pourbaix diagram for iron at 25°C and 101 MPa, given in Figure 3.2, does not apply at elevated temperatures and pressures because the parameters $E^{\ominus}$ and pH and the Nernst equation are all temperature dependent. In principle, it is possible to construct diagrams for conditions prevailing in steam raising boilers but there is insufficient information on standard electrode potentials for reactions of iron in superheated water. However, Figure 3.2 suggests that the metal can be passivated with a film of magnetite, Fe_3O_4.

In practice, the metal can be passivated in mildly alkaline deaerated water at temperatures <~220°C by $Fe(OH)_2$:

$$Fe + 2H_2O \rightarrow Fe(OH)_2 + 2H^+ + 2e^- \tag{26.1}$$

and at temperatures >~220°C by magnetite forming as a conversion coat on the iron surface:

$$3Fe + 4H_2O \rightarrow Fe_3O_4 + 8H^+ + 8e^- \tag{26.2}$$

Any pre-existing $Fe(OH)_2$ is oxidized to magnetite by the reaction:

$$3Fe(OH)_2 = Fe_3O_4 + 2H_2O + 2H^+ + 2e^- \tag{26.3}$$

Corresponding reactions for oxidation by dry steam are:

$$Fe + 2H_2O\ (\text{steam}) \rightarrow Fe(OH)_2 + H_2\ (\text{gas}) \tag{26.4}$$

and

$$3Fe + 4H_2O\ (steam) \rightarrow Fe_3O_4 + 4H_2\ (gas) \tag{26.5}$$

At temperatures above 570°C, magnetite is undermined by wüstite, FeO formed at the iron/magnetite interface by reversal of the eutectoid reaction in the iron–oxygen system-described in Section 7.3.2:

$$Fe_3O_4 + Fe \rightarrow 4FeO \tag{26.6}$$

This sets a practical limit well below 570°C as the temperature at which plain carbon or low alloy steels can be contemplated for use in watertube boilers.

The magnetite film is strongly coherent and adherent and its thickness increases from ~0.025 mm at start up to ~0.25 mm after ~10 years operation. Its resistance to superheated water and dry steam is so good that plain carbon or low alloy steels are among the best materials to resist waterside corrosion at temperatures ≤~550°C, provided that the steel substrate is meticulously prepared as described in Section 26.1.2 and the boiler feed water is treated appropriately as explained in Section 26.1.3.

26.1.2 Surface Preparation of Steel for Passivation

The condition of the steel substrate is critically important because the initial magnetite film will not adhere properly to surfaces that are not scrupulously clean. Semi-finished steel products are usually prepared by the manufacturer to remove millscale, lubricants and rust using the standard methods described in Section 6.1.2, but it is virtually impossible to prevent some rusting and contamination during on-site storage and fabrication, and it is common practice to clean the internal surfaces of new systems after assembly. In this operation the system is *boiled out* with a hot moderately strong alkaline solution to disperse oils, grease and detritus and flushed out. It is then refilled with a hot strong acidic solution, for example, inhibited sulfuric acid to remove rust and scale, leaving bright surfaces. The acid is flushed out with nitrogen, residual acid is neutralized with alkaline solution containing an oxygen scavenger, for example, hydrazine and the system is dried and sealed pending commissioning.

To preserve the established surface condition of an existing boiler that is temporarily out of service, it must be drained, dried completely and sealed, preferably inserting trays of a dehydrating agent such as quicklime. The surface condition is also at risk if the load is temporarily reduced due to the introduction of air through faults or in reserve makeup water.

26.1.3 Treatment and Control of Boiler Water

The treatment and control of boiler water has the following objectives:

1. Control of pH to values for which magnetite is stable at operating temperatures
2. Removal of dissolved oxygen to protect the magnetite film
3. Elimination of the precursors of calcareous deposits that interfere with heat transfer
4. Monitoring for contaminants
5. Use of inhibitors

26.1.3.1 Control of pH

Sodium hydroxide or ammonia are added to the feed water to adjust the pH to values for which the magnetite film is stable. Typical values are 10.5–11.0 for high pressure (10 MPa) boilers and 11.0–11.5 for low pressure (4 MPa) boilers, measured at ambient temperature. It is advisable to monitor and control these values accurately because magnetite can be destabilized by insufficiency or excess of alkali. Another hazard introduced by excess alkali is increased potential for stress-corrosion cracking under residual or thermally induced stresses that can occur if the pH is raised locally by evaporation of water, for example, at weld defects or in isolated porous deposits as explained in Section 5.1.4. Although the risk is small in carefully designed all-welded construction, it cannot be entirely disregarded.

26.1.3.2 Removal of Dissolved Oxygen

Dissolved oxygen significantly reduces the protection afforded by the magnetite layer stimulating corrosion in a characteristic pitting morphology. Views on the responsible mechanisms are controversial. A recurring theme is based on the well-known conversion of magnetite by oxygen to hydrous haematite, which is more voluminous and less compact:

$$4Fe_3O_4 + O_2 + 6nH_2O \rightarrow 6Fe_2O_3 \cdot nH_2O \qquad (26.7)$$

and the effects are variously attributed to local weaknesses induced in the passive film or to the creation of local differential aeration cells. The sensitivity to oxygen content increases with increasing pressure and there is general consensus that the maximum permissible oxygen concentrations are of the orders $<3 \times 10^{-8}$ mass fraction for older boilers operating at <4 MPa and $<7 \times 10^{-9}$ mass fraction for modern boilers operating at >10 MPa.

It is standard practice to reduce the oxygen contents of raw waters to the required values in two stages. The water is first physically purged of dissolved oxygen by spraying it countercurrent to a stream of live steam and the final traces are removed chemically by scavenging, preferably with hydrazine, N_2H_4:

$$N_2H_4 + O_2 \rightarrow N_2 + H_2O \qquad (26.8)$$

Scavenging with hydrazine has the advantage that the normal reaction products are innocuous and volatile, leaving no residual solutes in the water. It must be used with discrimination because slow thermal decomposition of residual unreacted hydrazine can yield ammonia:

$$2N_2H_4 \rightarrow 2NH_3 + N_2 \qquad (26.9)$$

which can cause corrosion of any copper alloys in the system, such as tubes in the condenser. An alternative oxygen scavenger, sodium sulfite, is out of favour because a side reaction can yield sulfide which can stimulate the corrosion of iron and any nickel alloys in the system:

$$4Na_2SO_3 \rightarrow 3Na_2SO_4 + Na_2S \qquad (26.10)$$

Carbon dioxide is also undesirable in boiler feed water apparently because of its effect in lowering pH, but it is removed together with oxygen during steam purging.

26.1.3.3 *Elimination of the Precursors of Calcareous Scale*

Calcium and magnesium bicarbonates are present in supersaturated solution in hard natural waters and before they are used for boiler feed water, the calcium and magnesium cations are removed by precipitation with lime and soda ash or by ion exchange resins to avoid fouling heat transfer surfaces with calcareous scale:

$$Ca(HCO_3)_2 \rightarrow CaCO_3 + CO_2 + H_2O \tag{26.11}$$

$$Mg(HCO_3)_2 \rightarrow MgCO_3 + CO_2 + H_2O \tag{26.12}$$

26.1.3.4 *Monitoring for Contaminants*

It is advisable to monitor the system for deleterious species introduced inadvertently. Examples are chlorides from leaks in condensers and traces of copper ions dissolved in the cuprosolvent soft feed water flowing over copper alloys in the system. Chlorides are depassivating and traces of copper can establish galvanic cells indirectly stimulated by replacement reaction as described in Section 4.1.3.5. A simple estimate indicates that a copper ion concentration as low as 2×10^{-8} mol dm^{-3} in a large boiler evaporating water at a rate of 0.1 m^3 s^{-1} can theoretically deliver as much as 6 g of copper per day into the system. The remedies are to identify and eliminate the sources of these species.

26.1.3.5 *Use of Inhibitors*

Empirical applications of inhibitors include phosphates, nitrates and amines added to boiler water to ameliorate various aspects of corrosion or stress-corrosion cracking. It is difficult to generalize information obtained for the effectiveness and *modus operandi* of different inhibitors, because it cannot easily be transferred between boilers that differ fundamentally in construction, firing, temperature, pressure, size and quality of feed water. The use of inhibitors is best approached on the basis of local experience.

26.1.4 Alloys for High Temperature Superheater Tubes

For the reasons given earlier, plain carbon steels are unsuitable for superheater tubes operating at the highest temperatures in large high pressure boilers, that is 450–650°C. The principal limitation of plain carbon steels for steam temperatures in the range 450–570°C is insufficient hot strength and this is remedied by selecting a suitable low alloy steel, for example, iron–1.25 mass % chromium–0.5 mass % molybdenum–0.1 mass % carbon.

The limitations for steam temperatures >570°C are more serious. Besides the loss of strength, creep resistance becomes a significant factor and plain carbon and low alloy steels lack corrosion resistance on both the waterside and the fireside due to the formation of wüstite at the magnetite/metal interface, as explained earlier. Suitable materials for these conditions are austenitic stainless steels selected on the basis of experience from the AA 300 series, specified in Table 8.3 and characterized in Section 8.4.

26.2 Fireside Corrosion Control

The fireside of a boiler comprises the combustion chamber, external surfaces of evaporation tubes, and superheating tubes and downstream ancillary equipment including heat exchangers and facilities for dust removal from flue gases. According to circumstances and location, explained later, various surfaces may be exposed to one or another of three different kinds of aggressive environments:

1. Gaseous combustion products including volatile species, excess air or unburned fuel gas
2. Molten deposits of oxides and/or salts
3. Aqueous acidic condensates

These environments are derived from the fuel/air mixtures and are first approached by considering the nature of fuels.

26.2.1 Fuels

26.2.1.1 Coals and Oils

Table 26.1 gives typical ranges for results of routine analyses of significant components of coals and oils. The carbon/hydrogen ratios correspond with the dominant combustible materials, that is carbon in coals and hydrocarbons of high molar mass in fuel oils. The nitrogen and sulfur contents, generally in the range 0.5–4.0 mass %, are derived from the biological origins of fossil fuels. The difference between the ash contents of coals and oils or gases is because coals remain in the source rocks, retaining detritus entrapped during formation whereas oils and gases migrate from the source rocks to traps from which they are recovered. Gases, liquid and pulverized solid fuels burn in suspension in the combustion chamber and the finely divided ash residues that they produce are carried through the system in the flue gas. Ash from coals fired on grates forms clinker which is periodically removed.

Certain common impurities in fuels that are not necessarily reported in formal analyses can have severe effects on the corrosion potential of flue gases out of all proportion to the small quantities present. Species of particular concern include precursors of sodium oxide, Na_2O, sodium sulfate, Na_2SO_4, sodium chloride, NaCl and vanadium pentoxide, V_2O_5, as explained later in Section 26.2.2.

The low ash contents of oils might suggest that they would be less liable than coals to contribute these species, but oils can contain metals chelated in porphyrins and certain

TABLE 26.1

Composition Ranges for Significant Components of Some Coals and Fuel Oils

	Analysis Mass %						
Fuel	**Carbon**	**Hydrogen**	**Oxygen**	**Nitrogen**	**Sulfur**	**Ash**	**Volatile Fraction Mass %**
Hard coals	87–89	3–4	2.0	1.0–1.5	1.0	3.0–4.0	11.2
Bituminous coals	70–87	4–5	2.5	1.0–1.5	1.0	2.0–7.0	18.8
Medium fuel oils	84–85	11–12	N/A	0.8	1.4–3.5	0.03	N/A
Heavy fuel oils	84–88	10–12	N/A	0.8	0.4–3.8	0.1	N/A

other organic nitrogenous complexes associated with their biological origins. The chelated complexes are oliophilic and persist in solution not only during the geological processes that convert the biological matter to oils but also through oil refinery operations. They carry 0–0.04 mass % of vanadium in the oil with certain other metals and deliver them as oxides on combustion. They are the primary source of vanadium in oils especially from Venezuela. Oils can also carry as much as 0.05 mass % of sodium in emulsified solutions of sodium chloride, either as residue from the crude oil or derived from contamination in refinery processing or transport over sea.

26.2.1.2 Natural Gas

Natural gas is a rich hydrocarbon gas, mainly methane, with a typical composition given in Table 26.2. It can be used as boiler fuel for generating electricity without many of the disadvantages of coal and oil, but it is a premium fuel not universally available and usually most profitably exploited for distribution as a public utility.

26.2.2 Fireside Environments

26.2.2.1 Combustion Chamber Atmospheres

The gaseous products of combustion vary widely, according to the fuel, air/fuel ratio and the aerodynamics of the combustion chamber. Table 26.3 gives some representative values. The principal components are carbon dioxide and water vapour. Complete combustion requires an excess of oxygen over the stoichiometric quantities. Conversely, carbon monoxide is present if combustion is incomplete. These products are accompanied by the nitrogen content of the air, which is essentially a diluent. Nearly all fossil fuels contribute oxides of sulfur which are precursors of sulfate deposits in hot zones and of acidic condensates in cooler zones.

26.2.2.2 Ionic Liquid Environments

Mixtures of oxides and/or salts derived from inorganic impurities in the fuel, notably Na_2O, Na_2SO_4, NaCl and V_2O_5, can stimulate catastrophic corrosion if they can condense as liquid ionic phases on heated metal surfaces, as explained in Section 3.3.3.7. The

TABLE 26.2
Composition of Natural Gas

Component	Methane, CH_4	Ethane, C_2H_6	Propane, C_3H_8	Butane, C_4H_{10}	Nitrogen, N_2
Volume %	86–90	3–5	0.5–1.5	0–0.5	1–7

TABLE 26.3
Typical Compositions of Gaseous Combustion Products

	Volume %					
Fuel	CO_2	H_2O	O_2	SO_2	SO_3	N_2
Hard coal	12	7	<6	0.03–0.3	<0.01	Balance
Fuel oil	13	11	<6	0.05–0.3	<0.01	Balance

TABLE 26.4

Melting Points of Some Components of Ionic Deposits

Species	Melting Point°C
NaCl	801
Na_2SO_4	884
$NaCl/Na_2SO_4$ eutectic	625
Na_2O	920
V_2O_5	675
Low mp. compounds in the Na_2O/V_2O_5 system	570–630

determining factors are the compositions of the precursors and the temperatures required to melt phases derived from them. Table 26.4 gives the examples for the $NaCl/Na_2SO_4$ system which includes a eutectic with a melting point of 625°C and for the Na_2O/V_2O_5 system which exhibits series of compounds with melting points in the range 570–630°C. Certain other impurities, notably alkali metals, sulfates and chlorides and abnormal combustion conditions can generate other ionic liquids with low melting points. In normal circumstances, liquid phases derived from combustion of boiler fuels do not form on surfaces at temperatures below about 570°C, but it is prudent to be aware that they can form at higher temperatures.

26.2.3 Corrosion Control of Principal Components

26.2.3.1 Water Evaporation Tubes

In normal operation, carbon steel water evaporation tubes are protected by a passive film, because the copious flow of water maintains fireside surface temperatures within a range in which the magnetite/iron interface is stable. They are also unaffected by the usual inorganic impurities in fuels because the temperatures are too low for ionic liquid phases to condense.

Abnormal conditions that interfere with the water flow, such as steam blanketing or internal deposits induce local overheating that can raise the wall temperature into the range in which the protective magnetite is undermined by wüstite. The remedy is usually to rectify faults in design or operation.

26.2.4 Superheater Tubes

Mild steel and low alloy steel superheater tubes resist oxidation at temperatures up to about 550°C because they are also protected by a stable magnetite film, but for tubes operating at the highest temperatures, 570–650°C, a magnetite film would not provide adequate protection because it would be undermined by wüstite, as explained in Section 26.1.1. Therefore, they are manufactured from oxidation-resistant alloys, usually austenitic stainless steels in the AISI 300 series, specified in Table 8.3.

26.2.4.1 Combustion Chamber

The function of the combustion chamber is to contain and utilize the burning fuel. The flame reaches a temperature >1500°C and besides heating the water evaporating tubes, it also heats the walls of the combustion chamber and associated steel components by radiation and impingement to a degree determined by the aerodynamics of the combustion chamber. These surfaces are not water-cooled and therefore reach high temperatures that expose them to degradation by wüstite formation and liquid deposits. Complications due to ash are erosion–corrosion due to abrasion of protective oxides by dust and deposits of porous clinker that can hold ionic liquids. Vulnerable surfaces can be protected by cast iron heat sinks and refractory materials secured by studs as used in other contexts, for example, metallurgical furnaces.

26.2.4.2 Downstream Ancillary Equipment

Residual heat is recovered from the flue gases emerging from the combustion chamber and associated superheater systems by passing them in sequence through a heat exchanger to preheat air for combustion and an economizer to preheat the feed water. They then enter electrostatic precipitation equipment to remove dust before discharge.

To maximize thermal efficiency and meet the demands of public authorities for minimal environmental impact of emissions, the flue gases are discharged at temperatures that are below the acid dew point so that oxidizing acidic liquors condense in ancillary plants at such low temperatures. A representative value of the dew point for condensation of sulfuric acid from typical sulfur trioxide concentration in flue gas is 150°C and hydrochloric acid may also form at lower temperatures. Steel is rapidly attacked by the condensate at a rate that depends on temperature. The quantity of condensate increases but the kinetics are progressively slower as the temperature falls, so that the corrosion rate passes through a maximum that occurs at about 100°C. This presents difficult problems in material selection for ancillary equipment, including not only heat exchangers and electrostatic precipitators, but also the stack and associated fans. Mild steel can be used where the temperature is above the dew point and where condensate does not form, but materials more resistant to sulfuric acid must be used elsewhere. These can include a weather resisting steel with the nominal composition, 1 mass %–Cr–0.5 mass % Cu–0.1 mass % P, selected copper alloys, and some nonmetallic materials, including glass, resin coatings and masonry for stacks. Measures that can be taken to limit the effect include controlling excess air to the minimum needed for efficient combustion and injecting small quantities of ammonia, amines or magnesium oxide to neutralize the acid condensates. In pulverized coal fired boilers, basic oxides derived from ash in the flue gases can assist in neutralizing sulfur dioxide.

26.2.4.3 Care of Fireside Surfaces during Idle Periods

If a boiler is taken out of service without proper precautions, deliquescent deposits on fireside surfaces absorb water on cooling, stimulating corrosion of the underlying substrate. For a short idle period, the onset of corrosion can be delayed by purging the system of flue gases with an inert gas and isolating it to retain sensible heat and prevent ingress of ambient air. For longer periods, the deposits are removed by steam or water applied at high pressure and the system is washed with a basic solution, drained, purged and dried.

27

Some Corrosion Issues in Nuclear Engineering

27.1 Overview

The nuclear domain has been part of corrosion since the 1950s. Before the discovery of nuclear fission in the late 1930s, uranium was an obscure element with few uses. Although radiation had been a subject of research since the days of the Curies and of Becquerel, the concept of nuclear materials was unknown. This changed with the advent of the Manhattan Project. After the end of World War II, nuclear energy, produced by nuclear fission, was promoted as a potentially cheap source of almost limitless power.

This was the impetus to understanding engineering materials in the nuclear environment. In the first generation of designs for nuclear power stations it was realized that the power plants would be built with a projected service life. The expense of development and operation necessitated that this would require facilities to be in operation for 20, 30 or even 40 years. The case for nuclear power was predicated on economic production of electricity, so cost-effective solutions for materials issues was a priority. Thus by the 1950s, considerable effort had been carried out in countries such as the United States, the United Kingdom and France to understand the causes of degradation of metals under irradiation, whether by physical effects such as creep or by chemical means such as corrosion.

Twenty years ago, when the first edition of this book, *Corrosion Science and Technology* was written, a chapter on the nuclear aspects of corrosion would have focused exclusively on the corrosion issues inherent in operating nuclear plants, many of which had been in operation for 20 or 30 years. There has been much change since those times. Nuclear power fell out of favour because of the problems surrounding radioactive waste and the increasing economic cost per unit of electricity which was partly bound up with the decommissioning costs of power plants coming to the end of their lives. After this, the attention of politicians and the public was drawn to the use of carbon fuels and the projected harm which fossil fuels have on climate change. Briefly, nuclear power came into fashion as a nil carbon footprint source of power. Since then, in many countries government subsidies have combined to boost renewable energy sources such as wind power, wave power and biofuels. An inherent weakness of most of these energy sources is that they are not continuous in operation. The demise of large scale storage media, such as the Regenesys project has magnified this and nuclear power has been promoted to be a solution to this, albeit a contentious one.

At the present time several companies are in the process of designing the so-called fourth-generation power plants. This is particularly active in those countries which have preserved their research base in nuclear engineering, particularly France. Parallel to this are renewed activities on developing economic fusion power and ongoing concerns with long-term nuclear wastes. At the end of the second decade of the twenty-first century, several Western countries are considering future energy needs and whether to commit to new nuclear projects. The whole situation is flavored with ecological concerns and geopolitical affairs, such as dependence on gas imports from potentially hostile countries. Another development is that nowadays, unlike the 1960s, before a contract is placed for a new facility, the whole means for control of nuclear contamination must be designed into the commissioned project by the constructor; this is normally overseen by the national nuclear regulator of the country concerned.

Therefore, the theme of corrosion of nuclear materials covers a broad area, including corrosion in conventional plants, future materials for new plants and handling of waste. This chapter discusses certain of these aspects in greater detail.

27.2 Fusion

Several attempts have been made to create a fusion reactor capable of producing power economically, but so far no fusion process is self-sustaining. The great advantage of fusion power is that there are hypothetically no radioactive fusion products. The challenge is two-fold. First, materials need to be able to hold a continuous plasma and second, there needs to be efficient thermal harvesting of the fusion power. The principal international project over the last 30 years has been the European Union's Joint European Torus (JET). The lessons learned from the JET project during this time are being used to design a new generation fusion reactor, the International Thermonuclear Experimental Reactor (ITER). ITER is a $20 billion consortium made up of the European Union, the United States, China, Russia, Japan, India and South Korea.

The development of materials for fusion power is likely to be led by the ITER project for the foreseeable future. The core of the reactor is a steel vacuum vessel housing nine Tokamak plasmas and is hermetically sealed. The lining of the vessel has to be constructed from a plasma resisting material with good nuclear properties. For example, the ITER design proposes to use beryllium as the plasma facing material (PFM). This was also used on the JET project when it was relined with beryllium to replace graphite.

Beryllium has never been used before in such quantity in civil power generation. It has a low atomic number which is desirable for this purpose and is a neutron reflector. The disadvantage is that it has a relatively low melting point of 1278°C. The corrosion of beryllium is covered in some depth in Chapter 18. The principal agents for corrosion are chloride, fluoride and sulfate ions. The extreme toxicity of beryllium means that corrosion of this material in service is a potential health hazard.

A major issue with fusion power is transfer of heat from the plasma to generate electricity. For primary transfer from the plasma a thermal conveyor is required. As an example, the ITER scheme is planned to have a primary cooling circuit, but is not designed for power production. That is planned for the next phase provisionally called DEMO after 2035. This is thus the domain of ongoing research and is a likely area for future material development.

27.3 Fission

Existing corrosion issues in nuclear reactors relate to systems employing some rarer metals in unusual environments that are required to sustain and control fission of the

TABLE 27.1

Some Nuclear Reactors in Current Use

Reactor Type	Moderator	Coolant	Fuel
Pressurized water reactor (PWR)	Water	Water	Enriched UO_2
Boiling water reactor (BWR)	Water	Water	Enriched UO_2
Pressurized heavy water reactor (CANDU)	Heavy water	Heavy water	Natural UO_2
Advanced gas-cooled reactor (AGR)	Graphite	CO_2	Enriched UO_2

^{235}U content of uranium and to arrangements that deliver the heat of fission to generate electrical power.

Fission is induced by capture of slow (low energy) neutrons, *n*, yielding isotopes of lighter elements and emitting fast fission neutrons, *n*, in events of the general form:

$$n(\text{slow}) + {}^{235}U \rightarrow {}^{236}U \rightarrow {}^{92}Kr + {}^{141}Ba + 3n(\text{fission}) \quad (27.1)$$

The fission neutrons are recycled to maintain a steady state chain reaction, but natural uranium contains only 0.72 mass % ^{235}U and the kinetic energy of fission neutrons released is so high that most of them evade capture. The problem is overcome by enriching the ^{235}U content of natural uranium to 4% or 5%, as described in Chapter 19 and placing the nuclear fuel within a confined space, the *reactor core*, filled with a *neutron moderator* to reduce the kinetic energy of fission neutrons. Neutron moderators are materials with low neutron capture cross sections to which neutrons can transfer kinetic energy by elastic collisions. They include water, deuterium, hafnium, boron and silver alloys, selected as appropriate to reactor designs.

The heat generated is extracted from the reactor core by a closed loop coolant and used to raise steam to drive turbines generating electricity. Some types of reactors in current use are listed in Table 27.1. Many are based on the use of water as both moderator and coolant. Two representative designs in general use, pressurized water reactors (PWR) and boiling water reactors (BWR), are briefly described later on in the chapter.

27.3.1 Pressurized Water Reactors (PWRs)

27.3.1.1 The Reactor Core

The reactor core is a pressure vessel containing the nuclear fuel, circulating water as both moderator and coolant, and control rods. The fuel is sintered pellets of uranium dioxide enriched in ^{235}U, assembled as *fuel rods*, inside sealed tubes of a hafnium-free zirconium alloy selected because of its good mechanical properties, outstanding corrosion resistance, and low neutron capture cross section. The fuel rods are grouped in bundles immersed in the water in the reactor core. Control rods are of materials with high neutron capture cross sections, notably hafnium.

27.3.1.2 The Steam Generator

27.3.1.2.1 General Arrangement

Figure 27.1 illustrates a well-established design for a PWR steam generator. It comprises two circuits, a closed primary circuit delivering heat from the reactor core through a heat transfer surface to a secondary circuit in which steam is generated. The main body is a

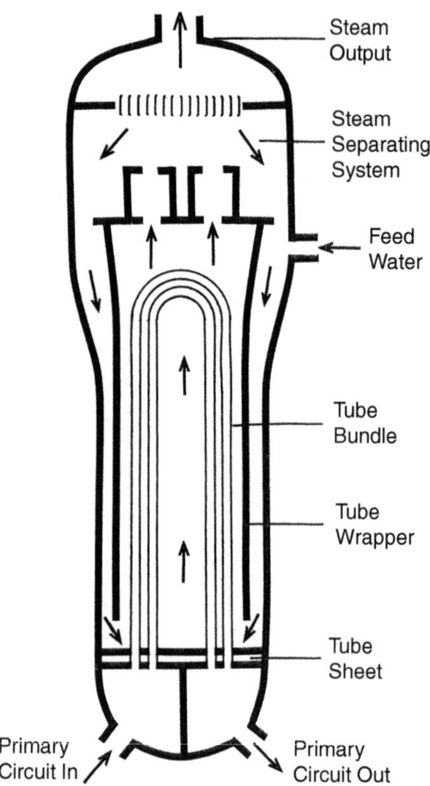

FIGURE 27.1
Schematic diagram of recirculating PWR steam generator.

vertical low alloy steel shell with a safety margin to contain pressures in equilibrium with liquid water at temperatures up to ~347°C (that is, 15 MPa). The heat transfer surface is provided by multiple tubes, typically 22.5 mm in diameter, in an inverted U configuration. For clarity, the figure illustrates only two tubes, but the actual number exceeds 100 and is designated the *tube bundle*. They are welded to a horizontal plate near the bottom of the shell called the *tube sheet*, through which they communicate with the primary water circuit. The tube sheet is of steel, overlaid on the primary side with a nickel alloy (for example, Hastelloy specified in Table 11.1), deposited by welding. The space below is divided by a *partition* plate into two sections, forming a collective inlet and outlet for all of the tubes. The tube bundle is contained within a steel cylindrical *tube bundle wrapper*, to direct water in the secondary circuit over the full extent of the bundle. A steam separating system above the tube bundle extracts steam for delivery to turbines and returns water for recirculation.

An essential feature of the tube bundle is omitted from Figure 27.1 for clarity in illustrating water paths. A series of horizontal flat steel plates, ~19 mm thick, are provided at intervals along the bundle to support the tubes individually and to locate them in the correct positions. In plan view, the plates are of trefoil or quadrefoil form with holes for the tubes with annular clearances of ~0.4 mm, forming crevices that in adverse circumstances can often induce a corrosion-related phenomenon, *denting*, to be described in Section 27.3.1.6.2.

TABLE 27.2
Alloys Used for PWR Steam Generator Tubes

	Composition, Mass %							
Alloy	**Ni**	**Cr**	**Fe**	**C**	**Mn**	**Si**	**Cu**	**Others**
Inconel 600	>76	14–17	6–10	<0.15	<1	<0.5	<0.5	–
Inconel 690	>58	27–31	7–11	<0.05	<0.5	<0.5	<0.5	–
Incalloy 800	30–35	19–23	Balance	<0.10	<1.5	<1.5	<0.75	0.15–0.6 Ti, 0.15–0.6 Al
AISI 316	10–14	16–18	Balance	<0.08	<2.0	<2.0	–	2–3 Mo

Note: Co – all < 0.10; S – all < 0.015.

27.3.1.2.2 Tube Materials

The compositions and condition of materials for steam generator tubes are particularly important to avoid leaking between primary and secondary circuits. Besides general corrosion resistance, the materials selected must be resistant to local effects that can perforate the tube walls, notably pitting and environmentally sensitive cracking. This limits selection to a range of alloys, including those in the nickel–chromium–iron system listed in Table 27.2. The austenitic stainless steels, AISI 304, AISI 316 and AISI 347, are usually reliable in chloride-free aqueous media but they are vulnerable to stress-corrosion cracking in hot solutions of chlorides and oxygen, for example, acquired from leaks in the condenser system. This problem has been addressed by using Inconel 600, Inconel 690 or Incalloy 800 because they have good resistance to stress-corrosion cracking combined with exceptional resistance to superheated water and steam. Even so, to further reduce susceptibility to stress-corrosion cracking, residual internal stresses introduced during manufacture must be relieved. These stresses are imposed during manufacture by tube straightening, forming the U bends and mechanically preparing surfaces. A typical stress-relief anneal is 10–15 hours at ~700°C.

Carbon and cobalt are normal impurities in nickel, nickel based alloys and stainless steels. The carbon content of the nickel alloys and austenitic stainless steels must be low enough to avert intergranular corrosion by sensitization, as described for stainless steels in Section 8.3.3.4. Some authorities advocate a range of carbon contents with an upper limit to avoid sensitization and a lower limit to yield sufficient carbide precipitation to inhibit grain growth by pinning the grain boundaries. The cobalt contents must be as low as possible to minimize the production of the long lived isotope, ^{60}Co, by irradiation of traces of corrosion product circulated through the reactor in the primary circuit.

27.3.1.2.3 Primary Water Circuit

In its function as coolant, the water in the primary circuit is heated in the reactor core to a temperature of about 315°C, delivered to the steam generator where it is cooled to about 215°C and returned to the reactor core for reheating.

27.3.1.2.4 Secondary Water Circuit

The water in the secondary circuit circulates down the annulus between the shell and the tube wrapper to the bottom of the generator and up over the tube bundle inside the wrapper, generating a two-phase mixture of steam and water. A steam separation system in the top of the shell separates the steam phase and delivers it through a nozzle at the top of the generator to supply the turbines. The water phase is mixed with feed water to replenish the evaporated fraction and recirculated as indicated by the arrow in Figure 27.1.

The two-phase mixture typically contains ~25% steam, so that the recirculation ratio is about 4:1. The exhaust steam is condensed and the condensate is recirculated as feed water to the steam generator, thereby completing the secondary circuit. The integrity of the condenser must be assured to avert contamination of the condensate with cooling water so that condenser tubing must be made of a highly corrosion-resistant material, for example, aluminium bronzes or titanium.

27.3.1.3 Water Treatment

27.3.1.3.1 Primary Circuit

The water in the primary circuit is very pure. In particular, the oxygen content is reduced to the lowest practicable value and controlled with hydrazine as a scavenger and excess hydrogen under pressure to suppress oxygen formation by decomposition of the water. Other additives must be compatible with the radioactive environment through which the coolant is recirculated. In its function as neutron moderator, it contains an appropriate concentration of boric acid to regulate the rate of fission and the pH is buffered by lithium hydroxide. Table 27.3 gives a representative analysis.

27.3.1.3.2 Secondary Circuit

As with conventional boilers, PWR steam generators rely on magnetite films to protect steel in contact with superheated water and high pressure steam and the need to control the pH of feed water to maintain it. Further requirements are to eliminate impurities that can concentrate by evaporation in crevices yielding aggressive species, for example, acid chlorides and sulfates and to prevent buildup of sludge in regions of local restricted flow.

A former approach was to balance pH and phosphate additions to precipitate residual hardness constituents and maintain sufficient alkalinity in the water yet suppress the concentration of alkali in crevices by evaporation. The approach is largely superseded by full-flow on-line purification of the condensate combined with an *all volatile treatment*, AVT, to reduce the quantity of solids that can be deposited.

A typical full-flow purification plant to treat condensates with high pH values comprises a deep bed of mixed anion and cation exchange resins. The deep bed also functions as a filter removing corrosion product particulates. The resins are periodically removed for regeneration using sulfuric acid for the cation resin and sodium hydroxide solution for the anion resin and replaced. If the cation resin is used in the ammonium form, ammonia is retained in the condensate but a high degree of regeneration is required to produce water of the required quality. Conversely, if it is used in the hydrogen form a much lower degree of regeneration suffices but it incurs the disadvantage of removing ammonia used for AVT. The AVT additions are ammonia or a volatile amine to control pH and hydrazine to act as an oxygen scavenger. Table 27.3 gives a representative analysis of secondary water after purification and AVT treatment.

TABLE 27.3

Representative Compositions of Water in PWR Steam Generators

		Mass Fraction				
Circuit	**pH**	**Oxygen**	**Hydrazine**	**Boron**	**Lithium**	**Chloride**
Primary	4.5–10	$<10^{-7}$	1.5×10^{-7}	4.4×10^{-3}	10^{-6}	1.5×10^{-6}
Secondary	8.8–9.2	$<5 \times 10^{-9}$	10^{-8}	–	–	–

27.3.1.4 Corrosion and Corrosion Control

Different control strategies are used in the primary and secondary water circuits of a pressurized water reactor. The two systems exhibit characteristic corrosion phenomena which require bespoke corrosion control methods. These are discussed in the next sections.

27.3.1.5 Corrosion in the Primary Water Circuit

27.3.1.5.1 General Corrosion

General corrosion of the tube bores and tube sheet in the primary coolant with the composition given in Table 27.2 is small but significant in respect of the radiation dose delivered to steam generator operatives by the recirculated burden of long-lived isotopes accumulated as irradiated corrosion products. It is usually described by the *release rate* of corrosion product. Laboratory tests in borated water indicate that Inconel 690 and Incalloy 800 are more resistant to general corrosion than Inconel 600, an advantage attributed to the higher chromium contents.

27.3.1.5.2 Environmental Sensitive Cracking

Stress-corrosion cracks initiated on the primary side are uncommon but some have been observed on occasion. They occur at the apices and base tangents to the U bends of tubes with the smallest radii, suggesting that they are associated with hoop stresses, perhaps induced by small inward movements of the U tube legs. Interpretation of the observed phenomena in terms of environmental factors is inconclusive since the circuit is sealed, composition of the water is carefully controlled and no specific agent for stress-corrosion cracking has been identified. Some authorities suggest that cracking initiated on the primary side could occur even in pure water.

27.3.1.6 Corrosion in the Secondary Water Circuit

27.3.1.6.1 General Corrosion

To ensure formation of a reliable protective magnetite film, steel surfaces must be cleaned of rust and fabrication lubricants before commissioning as fossil fuel boilers. With this proviso, the corrosion resistance of open surfaces of steel and of tube alloys is generally satisfactory in feed water purified by ion exchange resins and treated using AVT.

Potential corrosion problems are associated with crevices, contaminants and sludge which are manifest as effects denoted by the industry terms, *denting, tube/tube sheet crevice corrosion* and *tube wall thinning*. These effects were formerly related to phosphate water treatments, faulty AVT or leaking condensers. They are not necessarily current in existing plants where they have been addressed by abandoning phosphate treatments, correcting AVT procedures and reviewing system integrity, for example, by retubing condensers with titanium. Historically these problems formed part of the learning curve but still serve to illustrate what could happen when vigilance is relaxed.

27.3.1.6.2 Denting

Denting is a phenomenon that can occur in any PWR steam generator. It denotes constriction of tubes under the pressure exerted by voluminous corrosion product generated by rapid corrosion of the steel in the crevice between the tubes and the support plates described in Section 27.3.1.2. Figure 27.2 illustrates the phenomenon. It is particularly associated with AVT water treatment and systems in which chloride contamination can be

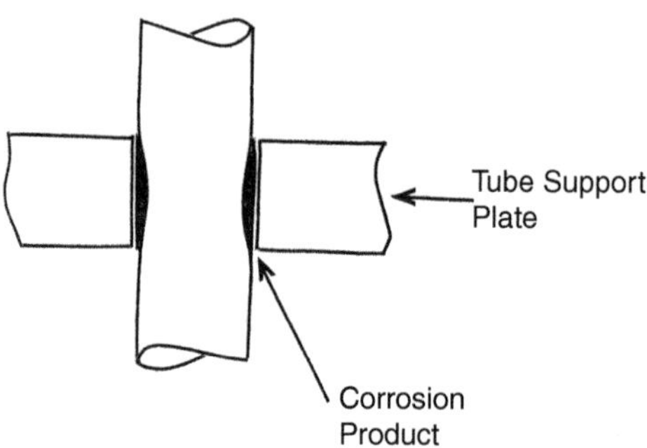

FIGURE 27.2
Denting corrosion in tube support plate.

introduced in condensers cooled by seawater or brackish water. The effect is not crevice corrosion in the usual sense because the medium is free from oxygen. The accelerated corrosion can be attributed to a thermohydraulic mechanism in boiling crevices that concentrates and hydrolyses contaminants introduced during service, yielding aggressive species, notably acid chlorides. Concentration of solutes at heat transfer surfaces is inherent in evaporation and it is noteworthy that local high concentrations of chloride can be found in dented locations. This view is supported by simulated denting by acid chloride media reproduced in the laboratory. Treatment with sodium secondary phosphate at a low concentration that does not conflict with the AVT concept can inhibit it to some degree.

Denting affects both the tubes and the support plates. Distortion of the tubes is not usually symmetrical and introduces stresses that can initiate cracks. Distortion and fracture of the support plates can restrict the flow of water and disturb local support of the tube bundle. More recently, the problem has been addressed by the apparently obvious solution of replacing steel support plates with corrosion-resistant Inconel 600 grids.

27.3.1.6.3 Tube/Tube Sheet Crevice Corrosion

Corrosion in crevices between Inconel 600 tubes and steel tube sheets is superficially similar to denting in that the effect is associated with boiling in a crevice, but there are substantial differences. The attack is intergranular corrosion of the Inconel 600 tubes and it is associated with crevice deposits containing phosphates. Circumstantial evidence suggests that the deposits are derived from current or former phosphate water treatments. The mechanism is uncertain but it is consistent with the view that constant evaporation of boiler water drawn into the crevices concentrates primary and secondary phosphates from dilute solution until the concentration is sufficient to produce free phosphoric acid by the equilibria:

$$2NaH_2PO_4 = Na_2HPO_4 + H_3PO_4 \tag{27.2}$$

$$3Na_2\,HPO_4 = 2Na_3PO_4 + H_3PO_4 \tag{27.3}$$

27.3.1.6.4 Tube Wall Thinning

Tube wall thinning or wastage is manifest as general corrosion of the surfaces of tubes immersed in sludge accumulated on the tube sheet at the bottom of the generator. The depth of sludge could be as much as 30 cm after a few years of operation. It is associated with phosphate water treatment and is much less prevalent with the replacement of phosphate treatment by AVT. The mechanism appears to be similar to that causing tube/tube sheet crevice corrosion. Deep sludge restricts the flow of water so that constant evaporation of boiler water drawn into the sludge concentrates soluble phosphates producing free phosphoric acid by Equations 27.2 and 27.3.

27.3.1.6.5 Environmental Sensitive Cracking

One of the main reasons for preferring Inconel 600 and Inconel 690 to austenitic stainless steels and Incalloy 800 for tubing in steam generators is their much better resistance to stress-corrosion cracking in media contaminated by chlorides. It is usually attributed to their higher nickel contents, evident in Table 27.2. However, sensitized Inconel 600, that is with carbides precipitated by excessive carbon content or faulty heat treatment is not entirely immune to stress-corrosion cracking in hot aqueous chloride media. The remedy lies in rectifying condenser leaks, specifications and reviewing heat-treatment procedures.

Some instances of stress-corrosion cracking in Inconel 600 and Inconel 690 tubes have been attributed to alkali as the specific agent. Potential sources cited are decomposition of sodium phosphates, species admitted through condenser leaks or accidental release of sodium hydroxide from ion exchange resins. Concentrations needed to initiate cracking can be developed by evaporation of boiling feed water in locations where flow is restricted. Remedies include abandoning phosphate treatments in favour of AVT, relieving fabrication stresses, rectifying leaks and attending to regeneration of ion exchange resins, as appropriate.

The flow and boiling action causes the tubes to vibrate. Vibration against the tube support plates can cause fretting and cracking by the synergistic action of corrosion wear and fatigue. The role of wear is to interfere with the formation of passive films. Stress raisers that can initiate fatigue cracks include wall thinning, fretting or local corrosion damage.

27.3.2 Boiling Water Reactors (BWRs)

Boiling water reactors also use sintered uranium dioxide enriched in ^{235}U as fuel, assembled inside sealed tubes of a hafnium-free zirconium alloy with water as both moderator and coolant but, unlike pressurized water reactors, the water is allowed to boil in the reactor core, generating steam which is fed directly to the turbines. The exhaust steam is condensed and returned to the reactor core, completing a single circuit. The water is maintained at a pressure of about 7.5 MPa so that it boils in the core at about 285°C. This is considerably lower than that of a typical pressurized water reactor, which operates at around twice this pressure and 60°C hotter than for a BWR. The reactor and turbine are enclosed by a steel radiation containment shell.

Although many of the materials used are those that are also used in pressurized water reactors and procedures in water chemistry have similarities in both designs, there are important differences in applications for corrosion control which are due to the fact that the same circuit is used for both the reactor core and for turbine operation. The most obvious of these is that the water chemistry contains no moderator material, such as soluble borate species. Thus control of the reactor relies on control rods and these must impart a more homogeneous neutron flux than for pressurized water reactors; because the water is allowed to boil and the vapour reduces moderating efficiency, these rods must be inserted from beneath the reactor.

Another difference is that since the water from the reactor core is irradiated and transferred to the generating equipment, radiation can accumulate in areas where occupational exposure of workers can take place, for example, in maintenance. The particular species of concern is cobalt-60, ^{60}Co, which can account for 80%–90% of radiation exposure. Thus in boiling water reactors, corrosion control has an added dimension, that of controlling buildup of corrosion products that could constitute potential radiation fields. Two techniques are commonly employed in current usage to prevent accumulation of radioactive contamination by corrosion products, these are *noble metal chemical addition* and *zinc addition*. These are discussed in more detail below. These treatments are combined with the advanced chemical water conditioning using all volatile treatment (AVT) and hydrogen injection methods to remove oxygen as described in Section 27.3.1.3.

The principal concern in BWR operation is environmentally assisted corrosion, particularly stress-corrosion cracking. This is intergranular in nature and is often referred to as intergranular stress-corrosion cracking (IGSCC). This is described for PWRs in Section 27.3.1.2. Principles of alloy selection and usage are similar for BWRs. The use of low carbon alloys and high nickel alloys has reduced thermal sensitization of alloys in service to IGSCC. The effect of surface cold-working in the fabrication of materials means that IGSCC remains a risk to be considered in aging plants.

In other designs, reducing buildup of sludge and precipitates in circuitry is an aid to eliminating stress-corrosion cracking. In boiling water reactors this is tempered by the increased need to control radiation fields. Improvements in water purity and conditioning by the methods mentioned previously has reduced the potential for sludge buildup. It must be stressed, however, that boiling water reactors are not optimized for corrosion control using exactly the same strategies as for PWRs, but mitigation of IGSCC by oxygen removal by hydrogen water conditioning (HWC) and noble metal chemical addition (NMCA) is the main priority.

27.3.2.1 Noble Metal Chemical Addition

This involves adding proprietary amounts of usually platinum or rhodium chemicals to the reactor coolant so as to deposit a thin catalytic layer of precious metal on wetting surfaces. This deposit catalyses recombination reactions with oxidants such as oxygen or peroxide reactions to inhibit precipitation buildup on surfaces. There is also a variant of this called *on-line noble chemistry* (OLNC), which is designed to inject noble metal concentrations when the plant is close to full operating power.

As mentioned previously, BWR water chemistry also involves treatments such as zinc addition designed to reduce the buildup of radioactive corrosion products.

27.3.2.2 Zinc Addition

Zinc oxide is added to reactor coolant to reduce the buildup of radiation fields arising from ^{60}Co accumulation in films on primary piping structures. This reduces dosage in areas such as the drywell leading to a reduction in shutdowns. Zinc oxide is believed to inhibit corrosion in nickel alloys and stainless steels and also to replace cobalt ions in spinel films found on stainless steels. A modern variant is *depleted zinc oxide* addition (DZO) in which natural zinc oxide is depleted to less than 1% ^{64}Zn (natural zinc is 48% ^{64}Zn) is used to eliminate secondary radiation from irradiated ^{64}Zn used in the process.

Other recent developments have included the application of monitoring methods, particularly electrochemical corrosion potential (ECP) techniques to forewarn of the conditions required to stimulate buildup of solid radioactive corrosion products. This complements existing strategies such as radiation monitoring and similar methods in the future are likely to be the focus of development and automation control.

Other issues concerning corrosion control in BWRs concern the containment. Corrosion has occurred in the past at locations where secondary condensation from leaks from sources of water outside the primary circuit can accumulate. This includes, for example, sand beds which cushion the reactor in the drywell, housing for monitoring equipment and neutron filters. Progressive improvements in design have now often eliminated the earlier causes of such corrosion. The mention of the containment shell in this context raises the question of external forces. The reactors breached during the Fukushima earthquake and tsunami in Japan on March 11, 2011, were of the boiling water reactor type. Although no design could have reasonably withstood such forces, it has focused much attention on the structural integrity of reactors, particularly those of BWR design.

27.3.3 Breakaway Oxidation

27.3.3.1 Breakaway Oxidation of Zirconium Alloys by Steam

The ZrO_2 film formed in water free from aggressive ions at temperatures below 300°C is normally coherent and adherent to the metal surface and reaches a virtually limiting thickness of a few nanometers forming a self-healing passive film. If the water coolant is lost or obstructed so that the temperature of the fuel assemblies rises above 300°C, the zirconium cladding can react vigorously with residual water or steam, yielding hydrogen by the Reaction 17.4 given in Chapter 17. In a loss-of-coolant accident in a damaged nuclear reactor, hydrogen embrittlement accelerates the degradation of the zirconium alloy cladding of the fuel rods exposed to high temperature steam:

$$Zr + 2H_2O = ZrO_2 + 2H_2 \tag{17.4}$$

This reaction was the precursor of a hydrogen explosion inside the reactor building of the Three Mile Island nuclear power plant (USA) in March 27, 1979. A similar reaction occurred in the boiling water reactors of the Fukushima nuclear power station (Japan) after cooling was interrupted by related earthquake and tsunami events in the disaster of March 11, 2011, venting hydrogen gas into the maintenance halls of these reactors generating explosive mixtures of hydrogen with air which detonated, severely damaging installations and associated buildings.

To avoid more explosions, many pressurized water reactor (PWR) containment buildings, have a catalyst-based recombinator installed to convert hydrogen and oxygen into water at the ambient temperature before an explosive mixture can form.

27.3.3.2 Breakaway Oxidation in Steels

Early generations of nuclear plants often used carbon dioxide as the coolant. After extensive tests it was determined that carbon steels were satisfactory up to 400°C. It was found, however, after around three years of operation that bolts made of mild steel were prone to failing due to breakaway corrosion. The effect was also observed in low chromium steels at higher operating temperatures.

It was found that interaction of the oxide film on the metal, though protective of the steel, was somewhat porous to carbon dioxide and over time a thin carbon deposit is laid down at grain boundaries in the oxide. For low chromium steels there first has to be precipitation of surface chromium carbide, $Cr_{23}C_7$ before the above process can happen as for carbon steels, so breakaway oxidation occurs later.

The remedy was to adjust the operating conditions in terms of temperature, atmosphere and humidity so as to reduce the rate of oxidation before breakaway. Since the conditions to eliminate such oxidation can shorten the lifetime of the graphite moderator, conditions were optimized to balance the life limiting criteria of each component.

27.3.4 Radiation-Induced Creep and Oxidation

A well-known physical effect of radiation on materials is *radiation-induced creep.* When irradiated a material incurs damage; the major effect of this is the creation of *Frenkel pairs.* This is a pair of defects consisting of an interstitial atom and a vacancy. The impingent irradiation dislocates an atom from a regular lattice space leaving a vacancy behind; the displaced atom occupies an extra space in the lattice as an interstitial atom. As mentioned in Section 2.3.5.4 for oxides, they are not normally present in un-irradiated oxides because of the large size of the anion. Also they do not theoretically affect the conduction band of a protective oxide, although they can contribute to ionic conductance. The intensity of irradiation is measured by a property called displacements per atom (DPA). Over the lifetime of a reactor component, the number of displacements per atom can be very high, approaching 30 DPA, and this has significant effects on the physical properties of certain alloys.

Alloys such as zirconium which are hexagonally close packed suffer from radiation-induced creep. This is because the Frenkel pairs are high energy and after formation try to minimize energy. To do this it turns out that the interstitials are preferentially adsorbed in time on the basal plane of the crystal lattice resulting in a movement of mass along the *c*-axis. Texture imparted during manufacture of cladding material means that the crystal structure in the cladding material is partly aligned along one direction. Over years of irradiation there is significant creep along the cladding. Another effect of Frenkel pairs is that the mechanical properties change, including tensile strength, residual stress and fracture toughness. Taken together, aged components can suffer sagging or induced stress due to displacement and this needs to be taken into account in considering corrosion control of power plants, especially when they are beyond their designed life.

Finally, it should be noted that the possibility of radiation-induced creep is a feature of neutron irradiation fluxes of certain materials exposed to fission processes. This has been studied extensively since it was discovered in the early days of research into civilian power plants. It is also true that neutron fluxes are also an inherent feature of fusion processes. Thus in the design and construction of a viable commercial fusion reactor such effects and their impact on corrosion control will need to be taken into account.

27.4 Nuclear Propulsion

Small nuclear fission reactors are used in propulsion of naval vessels, usually submarines. These propulsion units produce only a few hundred megawatts, compared to typically one to two thousand megawatts for a civil nuclear power station. Nuclear propulsion units are pressurized water reactors (PWR) so many of the inherent corrosion issues described in Section 27.3.1 are applicable to nuclear propulsion units. The difference is the much higher power density; the fuel is usually composed of highly enriched uranium (HEU) in a metallic alloy, for example, uranium–zirconium. The oxidation characteristics of this and similar alloys, the so-called 'stainless' uranium alloys, are given in Section 19.4.4. Reactors are designed to operate for fairly long periods without maintenance and there are certain modifications to PWR design, for example, the presence of a neutron shield to protect steel components from the high irradiation flux from the reactor core.

27.5 Regulated Materials

Nuclear materials such as highly enriched uranium are under the jurisdiction of international nonproliferation treaties. The International Atomic Energy Authority (IAEA) requires accountability for all such material. Stocks of metal if oxidized represent a change in the mass to be reported. Thus such stocks are kept in storage and monitored periodically, including their corrosion status. Therefore, corrosion of such materials is not normally an issue.

27.6 Containment, Decommissioning and Disposal

One of the more challenging subjects in nuclear engineering is the treatment of spent material and redundant equipment after operation has ceased. The problem of spent effluent goes back to the 1940s when process enrichment of natural uranium was carried out at Hanford, Washington (United States). From this time, 200 million liters of high level effluent has been stored on this site. The original single walled tanks to contain it have experienced leaks due to corrosion and now the effluent is housed in double walled steel tanks and monitored every six months. Thus the problem of long-term containment of radioactive material has a long history. A decommissioned plant also requires safe deconstruction and disposal. Spent nuclear fuel has high levels of radioactivity; usually they are placed in cooling ponds for several years. After reprocessing, high level waste is vitrified and stored in high integrity containers (HIC) often stainless-steel canisters. Low level waste is usually stored in carbon steel drums. Drums and canisters are placed in the depository and backfilled with either soil or cement.

In the United States, a state of the art repository was designed for Yucca Mountain in Nevada. As yet there has been very little research into alloys which can be used for very long storage facilities. In this regard, the Yucca Mountain facility represents an opportunity to test the suitability of materials for corrosion control in containment of nuclear waste. The site, however, is presently halted in construction due to political and economic issues. Other countries have experienced similar situations. The development of corrosion-resistant materials for long-term storage of nuclear waste must be regarded as an ongoing project. Critical to this is commissioning a fully implemented treatment of ultra-long-term mechanistic models to predict corrosion rates under storage conditions over many decades. The kinetic treatment of chromium alloys in Section 30.2.3 represents the type of fundamental model required to underpin such a model for one material, stainless steels. In this case, the factors that containment vessels would experience in terms of both interior (irradiation) and external (repository and geochemical) environments need to be taken into account over extremely long time frames.

27.7 Summary

In this chapter, some aspects of the corrosion of current and future nuclear materials have been addressed. In the space available here only the briefest overview of some corrosion characteristics could be addressed. There are, however, specialist texts particularly on the nuclear engineering aspects of currently operational civil nuclear power plants. There are also specialist scientific journals such as the *Journal of Nuclear Materials,* which includes reports on corrosion matters. For an in-depth treatment of nuclear corrosion issues, the reader is advised to consult such resources for more details. Nuclear engineering constitutes a subject in its own right and the associated materials applications within it are themselves a separate discipline.

Further Reading

Taylor, G. F. and Dautovich, D. P., Materials performance in CANDU nuclear steam generators, *Nuclear Technology,* 55, 30, 1981 (ANS Second International Conference, St Petersburg, Florida, October 6–10, 1980).

28

Oilfield Corrosion

28.1 Overview

Corrosion control is a primary requirement in the oil industry. Oil breaches are a problem to the environment and a hazard to health. Safety considerations also impel operators to maintain high levels of corrosion protection. The costs of oil contamination and cleanup can run into billions of dollars. Examples are the *Deepwater Horizon* (2012) and the *Exxon Valdez* (1989) and *Amoco Cadiz* (1987) disasters. While these were not directly corrosion related, the Prudhoe Bay spillage (2006) was caused by corrosion and including rebuilding and improvement expenditure, lost revenue, fines and other settlements it accounted for about half a billion dollars in extra costs. The variable economics of oil production means that profitability is strongly dependent on the oil price. In times when the oil price is low, the profit of running existing oilfields can become marginal or even negative. Thus the operating costs of such enterprises, including corrosion control costs, must be minimized without compromising plant integrity.

Certain features of oilfield exploration and production lend themselves to bespoke corrosion control strategies. One of these is the sheer scale of facilities; this makes monitoring, inspection and servicing regimes very expensive due to the amount of manpower and capital required. Another factor is the remote locations of many oilfields which make supply and habitation support for operations demanding; in the present day, this includes installations in marine locations such as the Gulf of Mexico and the North Sea, and terrestrial environments such as the Arctic and desert locations in the Middle East. Such inhospitable locations have been a challenge to the oil industry since at least the end of the nineteenth century; to understand this, the reader only has to look at the historical development of, for example, the desert-based oilfields in Persia and jungle-based oilfields in Burma and Indonesia. For this reason, corrosion control strategies have continuously evolved over the last hundred years, especially with respect to ferrous alloys, particularly carbon steels which are the main materials used to build practical and reliable structures in the field of oil extraction. Fresh impetus in optimizing materials selection and life-cycle optimization was created by an extension of oil exploration into new theatres, such as deep sea and polar theatres in the 1960s and 1970s.

A newer challenge to the oil industry is the desire to exploit reserves at ever deeper depths. At these depths within the earth's crust, latent temperatures are higher, up to around 250°C and well pressures are higher. High pressures mean that the activity of corrosive gases, especially carbon dioxide and hydrogen sulfide are greater and thus the exposure to steel components and other engineering structural assets is more severe. Optimization of materials selection for corrosion control in these environments is an ongoing issue.

28.1.1 Terminology

In the oil industry, like many other industrial sectors, there is a body of in-house terminology to describe items such as the different sectors of the industry, the chemical makeup of types of oil sources and production and maintenance equipment.

28.1.1.1 Sectors

Activities in the oil industry are usually divided into three sectors. These are the upstream operations, midstream operations and downstream operations. Upstream operation includes all exploration and extraction as well as primary processing at the wellhead. Midstream operations include transport of crude oil from the oilfield whether by oil supertankers, pipeline or rail. The corrosion of some of these items of interest is covered elsewhere in this book, particularly in Section 25.2 on corrosion control in ships. Downstream operations include all processes whereby crude oil is converted into petrochemical products. The downstream sector is thus largely the domain of chemical engineers. Upstream operations include all activities in the oilfield environment and control of corrosion in oilfields comes under this heading.

28.1.1.2 Oil Types

Different oilfields around the world produce crude oils with a different range of compositions. Each oil reservoir has a distinct chemistry. The primary classification is to divide oil types into heavy crude and light crude. Heavy crude contains a larger proportion of higher boiling hydrocarbons and produces a greater proportion of lower value products such as fuel oil and bitumen when refined. Light crude has a greater fraction of kerosene and is a more valuable chemical feedstock. Consequently, light crude oil generally commands a higher market price. A typical heavy crude oil is Saudi crude, whereas Brent crude from the North Sea oilfields is a representative light crude oil.

Another classification of oil types which is of great importance concerning corrosion control in oilfield applications is the distinction between sweet and sour sources. This describes the degree of hydrogen sulfide present in the oil as it is extracted. If the hydrogen sulfide content is below the NACE international specification (MR0177) the oil is classified as sweet in terms of corrosion control; if it is above this value it is classified as a sour service environment. The specification is for fluids containing water and H_2S at a pressure of at least 0.4 MPa with a hydrogen sulfide partial pressure of greater than 0.0003 MPa.

28.1.1.3 Location Descriptions and Equipment

The oil and gas industry has its own nomenclature for different locations within the oilfield assemblies and also for certain pieces of equipment. The term *downhole* describes locations underground from the oil reservoir to the surface. Various features of the downhole assemblies are the *drillstring*, consisting of the *bottomhole assembly* and the drillpipe, as well as the *wellbore* which includes the drilled hole including the steel *casing*.

The term *wellhead* refers to the structures at the top of the downhole and includes the blowout preventer, a large valve to prevent uncontrolled release of pressure. Other installations are either surface structures (such as the *production string*, designed to protect casing and linings from physical damage or corrosion) or, in the case of marine locations, offshore platforms and subsea structures.

Certain hardware in the oil industry has its own terminology. For example, during shutdowns of oil pipelines an instrument called a pig is used to clear debris from the interior surfaces of pipelines. This fills the interior of the pipe and is moved along between sections to scour the metal surface. It is particularly important in removing biological material that could be a source of microbial corrosion. Occasionally a type of pig is used which has instrumented measuring devices and can record aspects of the structural integrity of the pipe interior and environmental parameters such as temperature, pH and salinity. This is usually called a 'smart' pig.

28.2 Oilfield Chemistry

Materials used in oilfield applications typically have to be ability to operate at temperatures of 180 to 250°C and pressures in the range 100 to 200 atmospheres (1500 to 3000 psi) in a bottomhole environment. While crude oil itself is not highly corrosive, other components contained within it are. The principal agent is water contained within the oil. Before it is piped to the surface, crude oil is typically a multiphase medium containing water with high levels of dissolved gases and chloride ions as well as an abrasive solid suspended phase derived from drilling mud.

The most common corrosion problem encountered in upstream environments is carbon dioxide corrosion. During the diagenesis of the organic matter in sedimentary rocks to form oil, the oxygen bearing part of that organic content tends to be converted into carbon dioxide. Carbon dioxide is an acidic gas; the relationship between pH and partial pressure of carbon dioxide is given by Equation 2.24.

$$\text{pH} = -0.5\log(K_H K_1 p_{CO_2} + K_w) \tag{2.24}$$

The constants are $K_H = 4.67 \times 10^{-2}$ mol dm^{-3} atm^{-1} $K_1 = \text{x}\ 3.81 \times 10^{-7}$ mol dm^{-3} and $K_w = 4.60 \times 10^{-15}$ mol^2 dm^{-6} at 15°C. As the value of p_{CO_2} increases the pH drops. For example, using a partial pressure of only 0.01 atmospheres from Equation 2.24, the pH under these conditions is 4.7. In the presence of saturated water hardness (calcium carbonate) the pH is modified by the buffering action of the hardness. The equation for this is given in Chapter 2 as

$$\text{pH} = 5.94 - \tfrac{2}{3}\log p_{CO_2} \tag{2.23}$$

For a partial pressure of p_{CO_2} equal to 0.01 atmospheres, pH = 7.3. For intermediate water hardness, the pH will be between 4.7 and 7.3. Thus the environment is acidic for even quite low partial pressures of carbon dioxide. As the partial pressure of carbon dioxide increases to the higher levels found in many oil wells this range moves progressively towards the acidic region of pH.

Many of the materials encountered in oilfield corrosion are carbon steels and the presence of high levels of carbon dioxide adds new complexities to corrosion control of these alloys. In the presence of levels of carbon dioxide in excess of atmospheric levels, the mineral siderite becomes a corrosion product. Siderite, or iron(II) carbonate, $FeCO_3$, begins to form at partial pressures of carbon dioxide over about 0.1% of atmospheric pressure (101 Pa). This modifies the Pourbaix diagram for the iron–water system, Figure 3.2; in this case, siderite covers part of the predominant area of magnetite, Fe_3O_4 with siderite, $FeCO_3$. A detailed treatment of the iron–water–carbon dioxide system can be found in Chapter 7 of *The Geochemistry of Natural Waters*, Third Edition, by J. I. Drever, cited in Further Reading in this chapter. In that text the system in question is labelled Fe–O–H_2O–CO_2.

In the Pourbaix diagram for the iron–water system the predominance area for magnetite signifies a passive region of the diagram; this is shown in Figure 3.3. The new phase siderite is also protective above pH 6.8 to pH 7. Below this range, the carbonate scale is regarded as having poor cohesion to the metal and is not protective. As discussed previously, only low levels of carbon dioxide are needed to lower the pH sufficiently to be the case.

The second chemical agent in oilfield chemistry which can have a major impact on corrosion control in the oil industry is hydrogen sulfide. Hydrogen sulfide is an extremely toxic gas, approximately three times more toxic than hydrogen cyanide. It normally has a poor solubility in water, except under pressure because of the poor dissociation constant of dissolved H_2S. In oil wells classified as sour the sulfur content is greater than 0.0003 MPa (3 matm) and most of this is present as dissolved hydrogen sulfide. Like carbon dioxide, hydrogen sulfide produces an acid solution. Also like carbon dioxide the presence of a partial pressure of hydrogen sulfide modifies the Pourbaix diagram for the iron–water system, Figure 3.2. A detailed treatment of the iron–water–hydrogen sulfide system can also be found as for the iron–water–carbon dioxide system in Chapter 7 of *The Geochemistry of Natural Waters*. In that text, the system in question is labelled Fe–O–H_2O–H_2S. The main difference in oilfield corrosion to the geochemical environment is that in a natural system pyrites, FeS_2 is the stable phase. Pyrites is formed from the monosulfide marcasite, FeS which is the kinetically favoured product which slowly transforms to FeS_2. In oilfield corrosion the normal sulfide product is the kinetic product FeS in sour service environments, which does not have the geological timescale to transform to pyrites. Apart from this the stability domains are similar.

28.2.1 Chemical Modelling

Several tools have been developed to model oilfield chemistry with the objective of predicting corrosion issues. These are generally based on physical chemical principles and features such as multiphase interactions combined with a practical appreciation for the observable details of corrosion. They are thus the domain of specialists. The first widely applied model was due to De Waard in the 1980s for carbon dioxide corrosion. This was later developed by Pots into the HYDROCOR Code used by Royal Dutch Shell. The other major oil companies each have their own proprietary models, for example the CASSANDRA model developed at British Petroleum.

The issues concerning such corrosion models are highlighted in Section 30.6.3 on prediction. Such models combine such components as thermodynamic equilibria, transport and flow phenomena with multiphase flow models and electrochemical treatments of corrosion cells. The materials and geometrical aspects are abstracted. The problem of such models is twofold; first, they are not mechanistically based since a lot of the mechanisms are not fully understood under all circumstances. Second, a common feature of chemical systems outside the laboratory is that they are often not in equilibrium (they are metastable). Even if thermodynamics dictates a chemical change (in this case corrosion) can happen, it does not necessarily mean that it will happen. Under such circumstances empirical or semi-empirical terms are often included to tailor models to give an approximation to the real phenomena seen in service.

Such models are an important step to reliable prediction of corrosion conditions in the field; in particular they can give extra value to field data and also to tests and trials. However, to be of optimal importance they need consistent support in terms of future development, both in more realistic descriptions of the real environment and in the provision of good values for parameters which these models require.

28.3 Materials Issues

28.3.1 Marine Applications

The structures involved in oil production employ a range of metals and alloys. In marine oil exploitation the ubiquitous presence of seawater environments is a primary consideration. Materials selection and corrosion control for marine structures are described elsewhere within this book, in Section 25.3. The section on sour service environments in Chapter 25 is generally applicable to non-marine systems.

28.3.2 Corrosion-Resistant Alloys

The bottomhole environment is a most aggressive one in the oilfield service environment. The trend in oil exploration is to exploit reserves at greater depths. The challenge for the industry in materials selection is to use alloys that can safely resist higher temperatures and pressures in the presence of CO_2, H_2S and chloride without maintenance or being vulnerable to phenomena such as stress corrosion. The materials must be able to have a service life commensurate with the likely duration of production. A popular class of alloys developed to do this is known as the *corrosion-resistant alloys* (CRAs). These alloys are ferrous alloys with a minimum of at least 25% chromium and molybdenum. Some general properties of these alloys are given in Table 28.1. There has been considerable effort in testing, trials and improving materials selection criteria for these alloys according to the different chemical environments they might experience. Material selection is done according to specific types of corrosion requirements and optimization of mechanical performance. For example, the duplex alloys, 22Cr and 25Cr are cold-worked and have good resistance against both carbon dioxide and chloride corrosion, but are vulnerable to stress-related cracking in sour service environments.

TABLE 28.1

Corrosion-Resistant Alloys for the Oil Industry

	Typical Composition, Mass %							
Alloy	**Ni**	**Cr**	**Fe**	**C**	**Mn**	**Mo**	**N**	**Others**
Duplex 22Cr	5	22	Balance	0.1	1	3	0.1	–
Super Duplex 25Cr	7	25	Balance	0.1	1	4	0.3	–
Alloy 28	31	27	Balance	0.01	1	3.5	–	1.0 Cu
Alloy 825	42	22	Balance	0.03	0.5	3	–	0.9 Ti, 2.0 Cu
Alloy 2550	50	25	Balance	0.03	–	6	–	1.2 Ti
Alloy 625	Balance	22	2	0.05	0.2	9	–	3.5 Nb
C-276	Balance	15	6	0.01	–	16	–	2 Co, 3.5 W

A key part of understanding the performance of CRAs especially in sour service environments is the behaviour with carbon dioxide, hydrogen sulfide and these two gases in combination. The mechanism of attack by hydrogen sulfide on chromium alloys is a variant of that by thiosulfate. The corrosion mechanism outlined in Section 30.2.5 for thiosulfate and stainless steels is based on the fact that passivation of chromium alloys is a kinetic effect and is circumvented by ionic chemical species containing elements with a lower electronegativity than oxygen, such as chlorine, bromine and sulfur that can temporarily promote oxidation to chromium(VI) species.

Thiosulfate does this by adsorbing on the surface electrostatically like other anions but possessing a sulfide type group, $^{-}S-SO_{3^-}$ that can participate with reactions on the metal surface. Hydrogen sulfide does this via direct dissociation

$$H_2S = H^+ + HS^- \tag{28.1}$$

The hydrogen sulfide ion HS^- acts in a similar way to the thiosulfate ion in initiating localized attack at the surface. This can in turn lead to either pitting or general corrosion depending on circumstances. It was shown in Section 30.2.3.3 that 15.7 mass % of chromium confers reasonable passivation characteristics on the surface oxide film of stainless steels, but that the degree of resistance to corrosion should continue to improve until 31.8 mass % is reached. CRAs occupy this range and while chromium contents of around 15.7% might be suitable for domestic use, bottomhole sour environments with high temperature, high H_2S and CO_2 pressures, low pH and high chloride activity require higher corrosion performance. Typically, chromium content of at least 22 mass % or higher of chromium is required.

28.3.3 Oilfield Corrosion Control Phenomena

A number of common issues in upstream corrosion control are briefly summarized as follows.

28.3.3.1 Carbon Dioxide Corrosion

As described previously, carbon dioxide stimulated corrosion is a principal risk for loss of structural integrity. It is, however, a more complicated phenomenon than described simply by the oil chemistry as described in Section 28.3.1. The flow characteristics also need to be considered to understand the effect of the abrasive medium within the pipe. Under different circumstances, the morphology of CO_2 corrosion can be pitting induced by the flow regime, general area corrosion or mesa-type corrosion (i.e., areas of siderite deposits surrounded by denuded areas). A rule of thumb which has been sometimes applied to carbon steels is that where the partial pressure of carbon dioxide is greater than two atmospheres, corrosion should occur, while below one half atmosphere it is less likely. For the intervening range 0.5 to 2 atmospheres partial pressure of carbon dioxide it is a strong consideration in corrosion control.

28.3.3.2 Wet H_2S Cracking

At relatively low temperatures, carbon steels are susceptible to stress failure in oilfield sour liquors. This is often called *wet sulfide cracking*. This is discussed in more detail in Section 5.1.4 of this book.

28.3.3.3 Top of Line Corrosion

In horizontal carbon steel pipes for gas interiors corrosion is often found preferentially along the line representing the highest point of the interior of the pipe. This is associated with condensate in the pipe. Corrosion is caused either by carbon dioxide or acetic acid from decomposed organic matter in the condensate. Applying an inhibitor, as described in Section 28.4 to protect the steelwork, coats this area of the pipe less thickly than other areas and top of line corrosion can be a feature of deficient inhibitor concentration at the uppermost zone of the pipe.

28.3.3.4 Drill Mud

The physical detritus transported during drilling operations, *drill mud,* is itself a source of corrosion when in the presence of oxygen. To minimize this, an oxygen scavenger is often added to reduce corrosion further downstream.

28.3.3.5 Microbial Corrosion

A feature of oilfield corrosion is that it can be accentuated by microbial corrosion. Several of the bacteria present are autotrophic, that is they can synthesize all their organic requirements from carbon dioxide. It is important to periodically clear potential microbiological contamination away from the interior of pipework using pigs during maintenance.

28.3.3.6 Cathodic Protection

The principles of cathodic protection are given in Chapter 20. Many of the principles of cathodic protection outlined are generally applicable to oilfield corrosion. For marine applications, cathodic protection methods are discussed in Sections 20.3.2 and 25.3.1.

28.3.3.7 Corrosion under Insulation

Corrosion under insulation (CUI) is a potential problem in the oil industry as it is throughout processing planta in a number of industries. The real issue is integrity of the overlying insulation and moisture ingress. Corrosion control is via regular replacement of insulation before its minimum life cycle in the prevailing ambient conditions.

28.4 Inhibitors

A standard method of corrosion control for protection of the metal substrates such as steel pipework, from carbon dioxide and sour service corrosion, is with the use of inhibitors. In the oil industry these are normally film forming inhibitors. They act as a physical barrier and prevent contact between the corrosive medium and the metal surface. The inhibitors used are based on chemical species such as quaternary ammonium or imidazole species that are polar enough to be absorbed onto the internal surface of the pipe even when there is some corrosion deposit. They have a hydrophobic portion which has an affinity with the hydrocarbon phase in the oil. The inhibitor thus coats the interior of the pipe and protects the metal surface but ensures flow along the pipe. However, this protection can only last a finite time and must be replenished by further application of inhibitor. There are two ways of doing this, by batch application or by continuous application.

28.4.1 Batch Application

This is the process whereby inhibitor is added periodically by injection. It is simple to establish, but comes at the cost of some downtime. It requires manpower to administer and the amount of inhibitor that must be used is greater than with continuous application. This is because it is necessary to ensure that sufficient inhibitor will last to provide adequate protection until the next batch can be administered; therefore, a degree of excess must always be added depending on the frequency of applications.

28.4.2 Continuous Application

Continuous application of inhibitors has the benefit that application is much more evenly administered. Thus there is an environmental advantage as less inhibitor is used. It also needs less manpower and requires less downtime than for batch application. The big disadvantage is the complexity of the automated equipment that must be installed and the high capital cost of commissioning such a system.

28.5 Summary

Because of the economic importance of the oil industry, the field of corrosion and corrosion control within it represents a whole subject in its own right. The brief treatment within this chapter should, therefore, be viewed as an introductory text. For a more comprehensive source, the reader is encouraged to consult more advanced texts such as that by Craig, cited in Further Reading below.

Further Reading

Craig, B. D., *Oilfield Metallurgy and Corrosion*, Fourth Edition, MetCorr, 2004.

Drever, J. I., *The Geochemistry of Natural Waters*, Third Edition, Prentice Hall, New Jersey, 1997, Chap. 7, pages 144 and 148.

29

Principles of Corrosion Testing

This brief summary is not concerned with comprehensive but essential minutiae of the procedures for performing corrosion tests. To do so would entail tedious repetition of material that is well reviewed elsewhere, for example, in Shreir's book and ASTM publications, quoted in Further Reading. Rather it considers the principles, limitations and applications of testing as a means of assuring consistency of existing corrosion-resistant metal products and for developing improved versions.

Tests are required to provide information validating the suitability of metal products for corrosion-resistant service. The information may be needed for either of the following purposes:

1. To meet acceptance standards for supply and procurement of commercial products.
2. To provide information for the development of alloys and of surface coatings.

29.1 Accelerated Tests

Acceptance standards for most properties and characteristics apply only to the material and are based on fail-safe criteria. They can be specified in terms of the results of straightforward mechanical and physical measurements on test pieces submitted for evaluation. In contrast, standards for resistance to corrosion apply to properties of a *system* comprising the material and its environment and are time-dependent so that they must be based on safe-life criteria.

There is an immediate difficulty in defining the environmental component of a corroding system for testing because a given metal product may serve in a variety of different environments which may be imprecisely known and variable. Only the corrosion process itself accurately represents the environment and time scale and it provides the best opportunity to acquire information to examine the validity of tests. This requires systematic records and cooperation between producers and users.

To be timely, acceptance tests are artificially accelerated to yield information on the quality of current production and performed in controlled environments to permit interpretation and extrapolation to service time scales. Tests can be accelerated by exposing test pieces to test environments that are qualitatively similar to those anticipated in the field but more aggressive to produce failure or indicate survival in a much shorter time scale. Various tests are available and three common examples are:

1. Tests on flat panels in cabinets with arrangements to produce atmospheres with high controlled humidities.
2. Tests in intermittent salt sprays with drying during the intervening periods.
3. Tests in ultraviolet light to accelerate degradation of paint coatings on flat metal test pieces.

The application of anodic potentials to test pieces has been proposed as the basis for accelerated tests but in practice, it distorts the relevant reactions to such an extent that the results are inconsistent, unreliable and bear little relation to natural corrosion processes. The principle has been abandoned.

Evaluation of the results of the tests and interpretation in terms of service in the field is inevitably subjective but can be made quantitative by appropriate measurements. General corrosion can be quoted as either the rate of metal loss or the gain in mass of corrosion products for test pieces withdrawn from exposure at predetermined intervals. Pitting can be quoted as the population density of pits observed under a low power microscope and as the statistical distribution of pit depth determined by the travel of the microscope between the metal surfaces. Assessment of the degradation of painted metal surfaces is inevitably subjective. Visual comparisons can be made against sets of standards indexed to represent progressive degrees of blistering, rust, peeling, chalking and whatever other defects are appropriate from time to time. These comparisons can be interpreted in terms of water penetration, absorption of iron ions, retained millscale and breakdown of binding media.

Accelerated tests introduce distortion because they can introduce mechanisms that do not prevail in the field and do not allow for the time-dependency of corrosion or for random variation of an environment. In fact, the greater is the acceleration and the better is the control exercised over a corrosion test the less likely it is to represent service in the field. Nevertheless, provided that the limitations are appreciated they are useful in providing bench marks for quality assurance and maintaining consistency of metal products. Producers and users must agree on test parameters and duration and on techniques for evaluation, preferably on an industry-wide basis.

29.2 Exposure Tests

Tests in which test pieces are exposed to the natural environment can provide reliable direct evidence to supplement and compare with accelerated tests. An obvious disadvantage is that they are so long term that by the time that they yield significant information, the materials that they represent have been in service for many years and may not always be typical of current production or user requirements.

Tests in the atmosphere are usually conducted on multiple flat panels supported on racks constructed from an insulating material. Samples of metals are usually used in the condition supplied by the manufacturer and those for testing paint systems may be left as supplied or given prior, whichever represents the conditions in which they will be coated. All of the test pieces are set at an angle in the same orientation so that they are equally exposed to rain or snow, but shed it immediately so that they dry rapidly when it stops. In temperate climates, it is conventional to define four kinds of sites for the tests, that is, urban, industrial, rural and coastal, terms that are self-explanatory. The orientation of the racks of test pieces must take account of the traverse of the sun and directions of prevailing winds, Thus, a full test of a range of materials requires several sites, each with multiple racks of test pieces.

Test pieces buried in soils produce results that are required to select materials for underground utilities. They are of only local significance because of the very wide range of natural soil compositions and structures and of the effects of human activities in industrial, urban and rural sites. The information required may be either the relative responses of different materials in the same soil or the responses of the same material in different soils. The samples must take account of orientation to allow runoff of soil waters and return soil overburden must be restored to its original physical condition.

Exposure of small test samples by immersion in the sea yields results that must be interpreted with caution because they cannot represent the behaviour of larger structures in such a complex dynamic conducting system, as explained in Chapter 25.

29.3 Pilot Tests

Pilot tests are used to evaluate or compare new materials selected for modifications to existing plant or new constructions. In chemical or similar plants, the samples may be temporary replacement of existing parts by equivalent parts made of proposed new materials or coupons suitably disposed to represent critical features, with facilities to remove them periodically for examination. Paint and other applied coatings can be pilot tested by replacing small areas of paint coatings on existing metal structures with new systems or formulations to select the most appropriate coatings for future repainting.

29.4 Stress-Enhanced Corrosion Tests

29.4.1 Tests for SCC

The essence of tests for stress-corrosion cracking is to load a sample in crack opening mode and expose it to a solution of the specific agent. The load may be applied as constant stress or as constant strain. Many tests are available, but the most convenient and probably the most common test is the U bend test, in which a load is applied in bending mode at constant strain. The sample is prepared as a U shape with holes at the tops of the legs through which a bolt with a nut is passed to apply the load. Initial internal stress in the sample is relieved after forming by minimal heat treatment and a test load is applied by tightening the bolt. If the heat treatment cannot be tolerated, an alternative arrangement is a flat sample clamped in a C jig and loaded at its center by a screw normal to one face. It is essential that the nut and bolt or C jig and screw are made from the same or a closely similar material to the sample to avert galvanic effects that could invalidate the test. Racks of test pieces loaded to prescribed values in an appropriate range are immersed in prescribed concentrations of the specific agent under test and examined at intervals for failure manifest as cracks at the outside of the bend and relaxation of the applied load.

29.4.2 Corrosion Fatigue Tests

Corrosion fatigue tests are usually carried out on standard Wöhler rotating bend machines in which a rapidly rotating cylindrical sample is loaded in cantilever by a dead weight and life to fracture is measured by the number of stress reversals recorded on a revolution counter. An aqueous or other liquid corroding medium can be applied by drip, by wick or by application in a soft plastic tube sealed around the sample.

29.4.3 Tests for Erosion–Corrosion Cavitation and Impingement

Tests that assess the resistance of metals to stress-related effects due to relative motion between a metal and an aqueous environment, such as erosion–corrosion, cavitation and impingement are difficult to standardize and are best conceived in relation to the applications envisioned. Many devices for such tests have been reported and their designs are tributes to the ingenuity of the innovators. Examples include test pieces in the form of discs spinning in aqueous media to induce erosion–corrosion, test pieces forced to vibrate at resonant frequencies and controlled amplitudes in static aqueous media to induce cavitation, water streaming through Venturi tubes also to induce cavitation, water jets directed at flat test pieces to attack them by impingement and water streaming through test pieces in the form of pipe fittings with constrictions or abrupt changes of direction to produce various effects associated with plumbing.

29.5 Tests for Resistance to Thermal Oxidation

29.5.1 Thermogravimetric Measurements

Thermogravimetric balances are available commercially for following the progress of chemical reactions from mass changes of solid or liquids exposed to vacuum or a controlled gas phase. They are sensitive to small mass gains of the order of milligrams and are designed to accommodate samples with a mass of a few grams. They can be applied to determine the initial rates of oxidation of metals in controlled atmospheres and are used to compare oxidation of the same metal in different atmospheres or to compare the oxidation of different metals in the same atmosphere. Although thermogravimetric balances are sometimes included in a list of corrosion tests, their role is limited by the following factors:

1. They accept only small samples.
2. They are restricted to short-term isothermal tests and cannot follow time-dependant changes in oxidation laws or temperature fluctuations.

29.5.2 Exposure and Pilot Tests

Accelerated tests are out of the question because changes in environment or temperature change the nature of the oxide and its relation to the metal surface on which protection depends. The best approach is the use of long-term exposure or pilot tests in which test pieces are exposed to the environments that the material will experience in the field. The procedures to be followed are usually self-evident and comprise supporting test pieces in positions corresponding to those envisioned for application of the material. There are usually no galvanic effects with which to be concerned. Evaluation can be by correlating mass changes of test pieces withdrawn at prescribed intervals with visual observations of the condition of the scale and with the profiles of temperature and environmental changes.

An interesting example of pilot tests is used by gas turbine manufacturers to evaluate alloys and surface treatments for turbine blades and associated hot components. The problem is to design engines to operate at ever greater temperatures. There is too much at stake to risk using new highly complex alloys without the most thorough realistic evaluation. It is clearly impossible to use an airborne engine for pilot tests and the solution is to simulate actual service using a *demonstrator engine*, which is a stationary test rig comprising the front section of an existing engine delivering a flame in which test pieces of experimental candidate materials are exposed on a rotating carrier. To meet competition and to develop improved materials in time for airline re-equipment dates, empirical evaluation in demonstrator engines takes precedence over longer term evaluation of the underlying fundamental principles. As one materials engineer of a gas turbine manufacturer expressed it, 'My job is to prove new materials and fly them. It is for others to assess the factors that contribute to improved performance'.

Further Reading

Shreir, L. L., *Corrosion and Corrosion Control*, John Wiley, New York, 1985.
American Society for Testing Materials Standards (ASTM), 2806.

30

Prediction of Corrosion Failures

30.1 Overview

In one way, the philosophy of this the last chapter in the third edition of *Corrosion Science and Technology*, departs slightly from the original concept of the first edition, but in a second way it remains true to the original vision. That edition was created in the late 1990s to be the template of a course at U.S. masters' level, but also to double as a reference text. As such, the content is instructive covering the aspects of corrosion from the best current knowledge. Prediction of corrosion failures is different because premature failure – which is what prediction methods of corrosion failures are trying to avoid – is often a result of deficiencies in the knowledge of the materials system or its application in service or its environment. In this way this chapter is at least in part addressing faulty application of the principles outlined in the chapters of this and other similar texts on corrosion control.

In another way, however, this chapter remains true to the original vision of *Corrosion Science and Technology*. The reader is advised at this point to briefly review the first part of Chapter 1, 'Overview of Corrosion and Protection Strategies', especially Sections 1.1 and 1.4. Central to corrosion prediction, apart from the issue of faulty application of corrosion control, are the principles of corrosion being a systems characteristic and that best practice for corrosion control strategies involves a series of informed judgements of a range of control options. These are based on projected corrosion behaviour under ordinary circumstances within the constraints of the particular engineering application; the latter involves an implicit prediction of future corrosion performance in service.

What this chapter does not set out to do is to recapitulate the different types of corrosion, the tenets of good corrosion design – minimizing crevices, differential aeration or bimetallic couples and other such considerations, nor to review current best recipes for inhibitors. There are many good texts which do this well. The reader should be able to address many design issues by referring to individual chapters throughout this book concerning corrosion control in particular industries and also chapters outlining corrosion resistance of relevant metals.

The nature of engineering design is often that the general methods are known and that a new design is an optimization of best practice for a new application; best practice is enshrined in various standards and protocols which are understood by design leads – chemical engineers, mechanical engineers, civil engineers and the like. Thus teaching texts which treat corrosion control by design as if the corrosion engineer is solving these problems afresh do not fully represent the nature of corrosion control. The parallel is with computer science in that undergraduate courses begin with teaching students to write virgin code, but the reality of the professional programmer is that the most common task is software development or modification of existing workable code.

This chapter, therefore, seeks to address some of the issues that arise from trying to predict corrosion in service. This is an inexact subject and some of the comments must come down to the opinion and judgement of practitioners. However, it draws on three topics which are needed to inform these judgements, namely (i) the nature of the engineering application and systems interplay, (ii) knowledge of the fundamental mechanisms to demonstrate that the means of life limitation are correct and (iii) evidence for both of these to enable review and decision making. It also recognizes that there are three different aspects to corrosion prediction. These are (i) predicting the observable phenomena of corrosion, which is based on mechanism, (ii) predicting the anticipated rate of corrosion, that is the kinetic aspects which are based on physics and chemistry, especially electrochemistry and (iii) predicting the actual corrosion in service which is based on the engineering aspects of the subject.

In many ways the most important feature of corrosion is the last of these – to anticipate the typical performance of a structure, a component or artifact in service, and then to quantify the risk of it failing to achieve this. Unfortunately, this is the hardest task in all of corrosion science and is almost never discussed outside proprietary methods used in certain industries or with certain alloys. Tools that can be used to construct prediction methods represent an eclectic mix of approaches and several of these tools are described in this section. For a particular application, it is necessary to select the tools required and then to link them in a way to give confidence in the prediction of corrosion performance.

The situation is complicated by the fact that certain terms advanced as pertaining to predicting material performance are often used in an imprecise and sometimes semi-interchangeable manner, especially in academic publications with indirect interest in corrosion, for example, surface physics. There are parallels with other subjects, such as environmental hydrogeology in which monitoring methods have been advanced without adequate thought for the reliability of information for decision making. The most important of these terms are prediction, diagnosis, failure mode, mechanism, risk and technique. The following parts of this section define these terms with relevance to corrosion prediction methods.

Prediction: The word *prediction* comes from the Latin, *pre-dicere*, 'to say beforehand'. The word prediction does not imply veracity in the quality of prediction. Different circumstances dictate the level of confidence required. For example, prediction of typical corrosion performance and minimum service life can be used to put conditions on a contractor to replace components which fail before a set number of years. At the other extreme are attempts to predict corrosion behaviour from first principles, for example, case critical structures which cannot undergo inspection, for instance safety switches in sealed modules. In this case, the task is usually to underwrite a guaranteed minimum lifetime for which *every* unit will function when required.

Diagnosis: The word *diagnosis* is from the Greek, *dia-* meaning through, and *-gnosis*, meaning knowledge. Diagnosis is a tool in failure analysis, but it is only indirectly related to corrosion prediction. This is because in general, either a component is not designed to fail, or else the usual route of degradation is assumed at the design stage. Also since most methods involve permanent removal of a component from service, diagnostics is generally restricted to investigations *after* a failure has occurred. Key issues with the diagnostic treatments of failures are that often it is dealing *de facto* with the atypical situation. The strategies used in translating diagnostic evidence into systems for the prediction of future corrosion failures are very complicated, as they involve transmogrification of the chance of the observed failure with the statistical risk of it reoccurring.

Failure Mode: The word *mode*, from the Latin, *modulus*, signifies the means by which a material fails. There is often confusion between the terms *mechanism* and *failure mode*. In older literature, for example, *Corrosion Engineering*, by Fontana and Greene, cited at the end of this chapter in further reading, corrosion is characterized into eight different 'types' of corrosion where each form of corrosion could be said to be a different form of failure via an environmentally induced process. The modern view is that this method of organization is via failure modes, for example, filoform corrosion or stress-corrosion cracking. Interestingly, in this type of treatment, modes of failure are elsewhere sometimes referred to as mechanisms or models, for example, Fontana and Greene's classical diagram of chloride-induced pitting in the text mentioned previously, could be seen as a failure mode. This is because in essence it is a descriptive portrayal and *classifies* a process which does not comply with the strict definition of mechanism.

Mechanism: This is from the Greek, *mecha*, meaning contrivance. Mechanism has two meanings, one philosophical in nature, that is the notion that all processes within the physical world conform to the laws of physics and chemistry. The other meaning is that a particular process can be envisaged as if it is composed of components which fit together to predict all the essential observations from first principles as a *direct and inescapable consequence* of the components of the mechanism. Put simply, it implies that corrosion conforms to the Newtonian idea of the clockwork universe and the mechanism as described needs merely to be activated (the metaphor is to wind up the clockwork) to predict the physicochemical aspects of the system.

Thus, for a theory or a model of corrosion to be properly described as a mechanism, it *must* be able to predict all the key associated phenomena. Paramount among these is the species-specific nature of many corrosion processes. In general corrosion, the species conditions are adequately described by Pourbaix diagrams (see Section 3.1.5) but in much local corrosion, such as pitting or stress-induced processes there is no universally accepted mechanism which predicts the species-specific nature of these phenomena. This includes examples such as why stainless steels are pitted by chloride and bromide ions but not fluoride ions but commercial aluminium is pitted by all three ions. Another example is why thermal oxidation of uranium occurs at low temperatures even though the 'protective' surface oxide is close packed.

In these cases, all that can be done is that a mechanism is constructed (see the original Greek root of the word *mechanism* meaning contrivance) and then allowed to act on the description of the system to predict observable phenomena without imposing any constraints arising from individual measurements. Application of the mechanism to the real system should produce a *self-consistent* prediction of a set of observable and testable phenomena, the majority of which are well established in the literature as characteristic of that particular type of corrosion. The validity of a particular mechanism lies with its ability to predict most or all of these observations. It is thus at odds with the aims of certain techniques, which try to 'prove' a particular 'mechanism' by claiming to find crucial evidence. A true mechanism does not rely on one technique alone for validation.

As mentioned, there is at present no universally accepted mechanism for any major corrosion process. Accordingly any particular mechanism or partial mechanism must usually be constructed to address the particular questions that need to be answered concerning corrosion performance. Thus, a particular mechanism can be a proprietary treatment according to circumstances. This is addressed in more detail in Section 30.2 and the approaches to three mechanisms developed by the authors of this current work, that of stainless-steel passivation, of halide-induced pitting of aluminium alloys and of uranium oxidation are given therein.

Risk: The work *risk* is from the French, *risques*, meaning danger. Modern engineering and health and safety legislation imposes constraints on corrosion protection which are more than purely economic in nature. Central to this is the notion of the risk assessment. Secondary issues include consideration of environmental consequences (for example, through the international standard ISO14001) of breaches of containment and change control, that is consideration of destructively synergistic interactions between activities such as structural integrity, maintenance regime and operation when a process or facility is modified. Consideration of the role of corrosion in this relies on a degree of prediction of potential faults. This is often highly regulated and depends on tried and tested standards for materials in service. However, at any one time there are always emerging technologies and materials for which the future behaviour of artifacts cannot always rely on previous experience.

Technique: Originally, the word *technique* derives from the Latin for craft and was used to describe the method of application of a medium in the arts, for example, oils or watercolour. From this the word has come to mean a method of analysis for scientific research. Often technique-based work is engaged in a search for a productive application and in surface physics, corrosion mechanisms are sometimes cited as uses for certain techniques, especially that a certain technique might provide 'proof' for a particular mechanism; it should be noted, however, that the evidence of a mechanism should rely on its consistency with the whole body of evidence, including general in-service behaviour and also the fundamental processes of physics and chemistry. For research into corrosion mechanisms and prediction, the role of the technique is often better applied to measurements of parameters to quantify the rate of processes rather than to act as primary proof for the existence of the processes themselves.

30.1.1 Engineering Performance and Economics

It can be seen that terms which are often used in corrosion prediction can have different meanings, depending on the context in which they are used. It must be emphasized, however, that the whole field of materials science is driven by the application of such materials to practical engineering demands. Thus corrosion is governed by the *realpolitik* of practicality, service design, reliability and cost. The last two factors in particular have an important bearing on whether a particular material or formulation has a future. Inherent in this is a measure of prediction of performance versus cost and implied with this is a prediction of performance in service measured against design criteria. Concerning metallic materials in the chemical domain, this implies satisfactory prediction and risk assessment of items for corrosion resistance.

The overriding objective is to achieve maximum stated performance at the minimum cost. There is thus a trade-off between the inherent corrosion risk of the metal and its cost, including installation and maintenance and the exact solution to a particular engineering need requiring special consideration. This is accomplished with access to information from several sources to provide the best judgement and often in practice involves the pragmatic application of a safety factor to guarantee the correct performance. Of principal importance is the historical performance of a material in service. This is primary information and is the most reliable information concerning actual corrosion performance. It is reinforced by other secondary information; this comes in several guises. These and various emerging tools which can be used to guide material selection and application to corrosion in service are the subject of much debate. Only the briefest overview of the various factors involved in its application to corrosion prediction can be given and correct prediction of material performance ultimately comes down to an informed judgement.

30.1.2 Laboratory Tests and Standards

As mentioned in the previous section, the best clue to corrosion performance of a material is through the experience of how it has performed in the past. However, prior to first use in a new situation extensive laboratory corrosion tests are usually necessary. These are then used to make a decision that the material in question is suitable for a particular corrosion environment. This involves some treatment of the quantification results of tests and of service returns. The basic factors involving statistical interpretation of such information are outlined in Section 30.4. The general discussion as to design and interpretation of tests is given in Chapter 29.

By their nature, most laboratory tests are an abstracted and standardized extrapolation of what is meant to happen in real life. Owing to the paradox that to predict in a timely manner means trying to gain advanced knowledge, most tests are exaggerated approximations of what is thought to happen in real life. The statistics of such a set of circumstances are outlined in Section 30.4. What is clear, however, is that for any test to have validity then the corrosion processes which are operating in the test environment must be the ones happening during the real corrosion process the test is designed to replicate.

To translate the test processes with the real system there must be confidence that the corrosion mechanism is the same in each case; in this case, hypothetically, the difference between the two systems is given by the external parameters – for example, temperature, chemical activity of corroding species, oxygen potential and time – but in practice this is rarely straightforward. An often quoted analogy is the effect of temperature on a chicken's egg; at 37°C the result is a live chick, but at 100°C it is a boiled egg. At 37°C the mechanism is biological, but at 100°C thermal denaturing of the chemical structure of the egg occurs and the product is different. Thus, in this case temperature cannot be used as an accelerating vector for a test. The same is true with many corrosion processes; the full mechanism is not known with certainty, therefore, extrapolation from ad hoc tests for predictive purposes must be treated with caution.

30.1.3 Redesign, Misdesign and Rectification

One feature of corrosion prediction which must be considered is the concept that in the design stage a structure is not designed to corrode prematurely. Also when designing a metal component it is not always envisioned that it will undergo future in-service modifications or change of use. Increasingly though in the procurement phase of complex projects, the whole life cycle of the item in question is considered using system engineering methods. It is important that the corrosion engineer is familiar with such concepts and can raise corrosion-induced life-cycle issues as part of the system assessment. The impact of corrosion on systems assessments is outlined in the next section.

The problems arising from potential misdesign are often overlooked when commissioning equipment. It must be emphasized that nothing is designed to fail prematurely. Thus, when an unforeseen corrosion problem arises one of six things has happened. These are

1. The wrong material has been used.
2. The supplied material was defective.
3. The material has been used for the wrong environment.
4. Installation has not been performed adequately.
5. Good maintenance has not been carried out frequently enough or to the correct standards.
6. The failure is at least in part due to misuse or changing use.

The common theme of all these causes is human factors. Human factors research constitutes a subject in its own right within systems engineering. It has been argued that the role of people in defective systems can even be seen as a source of 'secondary mechanisms' which can be considered to be as important as the effect of the usual corrosion failure modes. While the nomenclature used in other texts may be different especially the term mechanism, the highlighting of the critical nature of the human element for corrosion control is often overlooked, both in applications in corrosion control and in corrosion teaching.

The influence of the human element can be illustrated by the many case histories that are available in the literature or in textbooks. For example, causes of corrosion such as differential aeration cells may arise from misdesign of joints which introduce crevices at the join. Another example is that of not installing the outtake of a tank at the lowest point so as to eliminate regions of stagnation at the bottom of a tank. What is not always appreciated is that a well engineered design that has proved suitable in previous installations may prove unsuitable in circumstances where although the specific corrosion control characteristics are similar, the scale or general location may play a role. Two case histories illustrate this fact.

Case 1: A bridge manufacturer experienced a corrosion problem in equipment used to clean and filter recycled cutting fluid. Corrosion was in the form of perforations of a steel tank which collected cutting fluid prior to filtration. The cutting fluid was aqueous in nature and was pH buffered using a proprietary aqueous solution which was regularly dosed during maintenance. The manufacturer of the equipment had experienced no corrosion problems with supplied equipment to other customers of cutting gear. There were also no obvious differences in use by the operator from other users.

The main difference of this facility to all others was that the cutting bed was designed to handle larger structures. The cutting bed was of a larger surface area to other installations, and although the cutting frequency of each piece by robot cutting equipment was similar, the sheer size of the bed meant that the flux of carbon dioxide from the air which the cutting fluid experienced was sufficient to exhaust the buffering capacity of the fluid which resulted in particles of carbonate precipitating in the fluid and caused local differential aeration cells to form under them.

The result was indirectly due to misdesign. Whereas the original design by the equipment supplier was sufficient for all previous installations, the possible effects of an increased size of plant had not been foreseen.

Case 2: A supermarket chain experienced persistent corrosion-induced failure of pipework for refrigeration circuits of chiller cabinets at one particular store. The store in question was in a port location. Twenty-five other stores in the region had experienced no faults, including other stores in coastal locations. Pipework with a design life of 30 years had failed in 4 years. The pipes were of carbon steel and were lagged with PVC foam and wrapped with PVC tape. The failed pipework was located between two buildings – the main supermarket building and a compressor building – and comprised two pipes, a lower one taking refrigerant to the supermarket chiller cabinets and an upper one returning the expanded and cooled refrigerant for recompression. The integrity of the insulation of these pipes was poor compared to the lagging of the pipework for the refrigeration circuits, for example, on the supermarket roof.

The direct cause of failure was due to the large presence of sea birds in the location. The position of the upper pipe provided an extra safe perch from the ships. The PVC wrapping of the upper pipe was breached by birds pecking at it, allowing salt-laden air from the wharf side to soak through the foam insulation. The fact that the cold return pipe was the upper pipe meant that this pipe, when breached, was susceptible to external condensation from the warm pipe beneath. This was a design fault in the original design but had not become a cause for failure in other sites because the PVC wrap outer coating had not been breached.

These two cases illustrate two important points. The first one is that a perfectly good design may become unsuitable simply because of scale. In this case, the human factor is because the corrosion control of the system has not been revisited to assess the impact of scaling up; other factors in scaling up can be the degree of maintenance required or accessibility. In the case of the supermarket, the failure was not in the specific environment that the corroding pipework was in but in not appreciating how the wider systems characteristics were perturbed. Both cases illustrate the necessity of systems assessment in corrosion control of the design for new installations.

30.1.3.1 Quality Control, Quality Management, Certification and Other Data

In assessing the causes of misdesign, there are certain activities of good practice which can provide some measure of control over unforeseen events. The central philosophy is that of continuous review. The general tools are based on the frequent review and the Deming cycle. The Deming cycle was originally devised in the United States, but it found real use in Japan for quality control. The cycle of continuous review (Plan-Do-Act-Review) has been used in many forms, including manufacturing (ISO 9001) for quality management and environmental management (ISO 14001). Many other developments in application of continuous review processes have been advanced such as total quality management (TQM) and six sigma (6σ); in some circles there is some criticism of such tools as a bar to creative development of new ideas. However, they do represent a body of evidence in manufacturing operations which is a source of information for corrosion control. Taken together with certification methods for process change and the paper trails associated with quality control of manufactured items, such as route cards, there is often a considerable body of evidence which can be used by the corrosion engineer to predict future corrosion failures from occurring due to misdesign.

The importance of communication between the corrosion engineer and the owners of information on quality is clear. For optimized quality control, quality managers need to know what information is available within their organization. Equally important is that quality control managers understand that corrosion accounts for around 3% of all costs (Hoar Report). Sales returns, certification of modifications, maintenance records, health and safety reviews and environmental licensing can all provide information to provide evidence for corrosion prediction, either in product or plant so as to avoid misdesign or else in protocols to include the role of corrosion in misdesign or redesign. It is important that the professional corrosion engineer maintains a clear sight of as much secondary information as possible.

30.1.3.2 Knowledge-Based Systems

In corrosion texts there are many case histories which illustrate where corrosion control has been overlooked. The preceding discussions have illustrated the role which human error or misjudgement can have in contributing to materials failures by corrosion. Elimination of human error at whatever stage can be difficult as an oversight is *de facto* something which can only be dealt with by taking decisions in hindsight, for example, in design, procurement, modification or maintenance. The main defense is often to rely on the existing best available information and best practice standards (International Standards such as ISO and also industry standards) for materials specifications, implementation and usage. However, recent attempts have been made at devising methods of implementing corrosion control procedures using knowledge-based systems, whereby the decision processes are made at least in part automatically by algorithmic methods, artificial intelligence or neural networks. The rapid expansion of computing power in the last 30 years has enabled the processing of large amounts of data and attempts at modelling aspects of decision making in corrosion control have been attempted. Sadly, at present, very little of this activity has been successful enough to filter through to practice. Central to this is the difficulty in the fundamental prediction of the behaviour of a system under all eventualities so that a knowledge-based system can operate sufficiently well enough so that decisions can be acted upon by engineering managers. Specific points are:

1. Compared to certain other systems where cause and effect are well established, corrosion issues can exhibit latency. It must be remembered that apart from gold, all metals are unstable in air and thus the system is a metastable one. In simple terms, whereas it can often be assumed that corrosion might occur, this does not mean that it will occur. Because of this assumptions are often made using human judgements which err on the side of caution. For example, when dealing with corrosion under insulation (CUI), it is usually prudent to assume that corrosion will occur after a certain time and standard practice is to replace components in a pre-planned schedule all at once to minimize downtime; piecemeal replacement via a knowledge-based system of say, vulnerable points such as joints or exposed corners, is at present not economic. Thus while, for example, in manufacturing, machining processes can be automated and the mechanical properties of artifacts throughout their life can be projected in terms of flaw density and applied stress – for example, in fatigue or fracture mechanics, it is much more difficult to do this in predicting corrosion occurrences. This is because the full mechanisms for corrosion are not known, the propensity to corrode depends implicitly on the microenvironment – which gives rise to local variations in the electrochemical characteristics, and projections for lifetime of a structure or component do not take into account very well the statistical spread of in-service results.
2. *Systems integration*: At lower levels of the system, knowledge-based operations are sometimes used. For example, a component may be designed to be optimally protected in isolation but in combination with other components or with modifications with changes of use inter-relation of different parts of the system becomes unmanageable. The situation here is not that a knowledge-based system is inappropriate, but rather that simplifying the assessments involved in integrating different components of a system requires human judgement.
3. *Management*: Prediction of corrosion requires the right quality of information. Unfortunately, this information often comes from many sources, and for

several of these sources such as manufacturing, sales returns, design and the like, it is usually only of secondary importance for the supplier of the information although it is of primary importance to the corrosion engineer. The existence of a knowledge-based system for corrosion prediction, therefore, requires commitment from senior managers to be effectively applied to eliminating causes. Ownership needs to be assigned at a senior level.

4. *Optimization*: Under certain circumstances, knowledge-based constructs can be used to optimize predicting the corrosion performance of an entire system from design to operation in terms of fitness for purpose and cost. The common element in such systems is usually that there is one ubiquitous parameter which controls whether corrosion should occur or not and this parameter acts as a transferable vector for all parts of the system. This cuts down on potential latency in the system and minimizes hysteresis of behaviour. Appropriate vectors are those which are not based on mass transfer or chemical transformation of state. Good examples are those of electrolyte conductivity including humidity, pH and electrical (redox) potential. For the first two examples controlling moisture or conductivity renders the electrochemical processes inactive and for the last two the control of these parameters can be accomplished by, for example, application of corrosion inhibitors.

Another good example of a system which can be controlled by optimization of a single parameter by knowledge-based methods is the application of Impressed Current Cathodic Protection (ICCP) to marine vessels. In this case, the parameter is the electrical potential of the impressed surface as a function of space in the dielectric medium. Increasingly with naval vessels the stealth signature of ICCP systems is important in minimizing detection of ships. The design and operation of a fully stealth optimized ICCP system would also need to take into account variations in the conductivity of water bodies with differing salinities than Standard Mean Ocean Water (SMOW) – for example, the Baltic Sea which has low salinity. When taken together with physical scale modelling techniques such as Dimension and Conductivity Scaling (DACS), plus finite element methods of current impressment, solution chemical treatments to model chalking of anodic hull sites, and electrical field strength monitoring under sail, it represents a system amenable to a knowledge-based approach. The eventual objective is a tunable ICCP system which can be varied according to environment.

To summarize, the application of knowledge-based approaches is still in its infancy. It does represent a potentially valuable method to partially eliminate those causes of corrosion that are due to human error, either in the design stage, during operation or due to inadequate maintenance. For certain relatively simple systems, automatic prediction of the corrosion characteristics during service life can be achieved.

30.1.3.3 Competences

In the discussions contained within this chapter, the central theme of corrosion prediction is predicated on best practice and a knowledge-driven approach. Implicit in the latter is that the level of competence in decision making at each stage of development in the prediction is maintained at a high level. The competences required encompass a myriad of different subjects within the realm of science and engineering, and it is of critical importance that deficiencies in knowledge are rectified by either training or by appropriate experience acquired on the job.

It is important that such experience of professional persons engaged in aspects of corrosion prediction is continually assessed and documented. Some competences, such as metallographic sample preparation and interpretation, can take years of practice, but do not necessarily have attached qualifications. As the observation and detection of corrosion symptoms is primarily an on-site field discipline, it cannot be learned in a classroom. Other aspects can be learned in a teaching environment, but the tuition needs to be given by suitably trained and accredited personnel with experience of corrosion issues, their monitoring and *cost effective* remedy in service. Good examples of the latter are these various accredited courses offered by the National Association of Corrosion Engineers (NACE).

The situation is best illustrated by three hypothetical examples of perceived mid-career engineers with interests relating to predicting corrosion in design or operation.

Individual 1 is a chartered mechanical engineer with ten years' experience in a general manufacturing capability. After a degree in mechanical engineering, this individual completed a graduate training course as a manufacturing engineer. This individual encounters corrosion and corrosion control matters as part of their general activities in product design and manufacturing. He or she has a good knowledge of the cost aspects of engineering and can apply these to corrosion control and also understands how aspects of production, such as machining or cleaning components can affect lifetime performance of artifacts. There is also familiarity of the operation of quality methods, international standards and certification. They are also able to pick up the principles of good practice, such as knowing how to minimize bimetallic corrosion cells and avoid debris traps and crevices.

The self-perceived weakness of this individual is that they are unsure that what they believe is the specialist expertise required for corrosion assessment and prediction, especially in the chemical domain. However, the underlying skills in mathematics and physics of a mechanical engineering degree, combined with the undergraduate classes in materials science included in most such courses means that the individual has the capacity to pick up the chemical and environmental aspects of corrosion if suitable material is at hand.

NOTE: In the original commission for the First Edition of this book *Corrosion Science and Technology*, the authors envisioned a number of potential readers. One of these was the type of individual who was a seasoned engineer with an unfamiliarity with chemistry. The technical ability of a Chartered Engineer to acquire corrosion expertise from a suitable text was recognized by the authors at the time. *Individual 1: Competence Effectiveness Rating 7.5/10.*

Individual 2 acquired technical proficiency through an apprenticeship scheme after leaving school. They are now in a medium-level production role, in charge of production implementation of a shop floor facility. Through a long career in industry they have acquired a practical knowledge of engineering which includes workshop practice and good housekeeping in a production environment. The latter includes familiarity with quality control reporting methods and dealing with service returns. He or she has also been involved with production trials and quality tests for corrosion resistance of new products such as salt-spray cabinets.

They have a sound knowledge of using this information to highlight defects in the product and, therefore, have a sense of assessing the macroscopic features of corrosion in a practical way through a knowledge of materials quality and defects, the features imparted by cleaning and joining of artifacts and elimination of problems at the source by monitoring the training requirements of production staff. This individual knows that one of the best resources for prediction of product performance at the manufacturing stage is the vigilance of shop floor staff to report back any unusual occurrences. Thus this individual operates and champions a cost-effective incentive scheme for shop floor employees who are encouraged to highlight quality issues related to corrosion performance.

The above discussion describes a production manager with on the job training concerning corrosion prediction in a shop floor environment. It should also be recognized that similar arguments are applicable to other similar employees. For example, if the former apprentice mentioned here had instead of production experience, pursued a career in the maintenance of structures in the field, the situation would be similar. In this case, he or she would not have the production-centered corrosion prediction experience of the previous example, but would have equivalent experience and skills in observing and pre-empting corrosion issues in the field. They are particularly familiar with the practical aspects of avoiding differential aeration cells, bimetallic couples, the effect of stress (in SCC and corrosion-fatigue), the role of flow-induced corrosion (such as cavitation–erosion in pump impellers) and its converse (quiescence) and also in reducing microbially induced corrosion. They are also aware of the potential of stray current corrosion.

This illustrates the importance in corrosion prediction of often engaging with a community of different people who can each bring their own expertise and experiences to obtain the best corrosion prediction methods for any situation. It is a feature of many activities that many roles can have a working knowledge of corrosion prediction without knowing it; a good example of the latter is a building or facility manager who understands the processes within his or her building. *Individual 2: Competence Effectiveness Rating 6.5/10.*

Individual 3: The third individual described in terms of competence for corrosion prediction methods is at first sight the most qualified for the role. However, it is important to realize that it is often the case that hidden competences are often more important than apparent qualification. Individuals 1 and 2 described above exhibit many of these hidden competences – for example, design and commissioning experience in the case of the mechanical engineer, and shop floor or on-site field experience, and workshop and quality implementation experience in the case of the former apprentice.

Individual 3 represents a contrasting case. This person has a PhD from a large materials science research group from a university with a large publication output in corrosion science. This is on top of a physics degree specializing in instrumentation. The PhD project utilized this individual's expertise in building and operating apparatus to develop potential new materials, such as ultra-hard novel surface coatings or matrix composites to maximize mechanical properties and then to assess the corrosion characteristics of such products in a range of laboratory environments. Novel coatings were prepared by variations in the formulation of standard methods of coating applications.

This individual then spent several years at a research and development facility in a large manufacturing company characterizing materials properties for potential new products. This person has been a co-author of several papers on characterization of potential new material systems utilizing both instrumented laboratory electrochemical methods and sophisticated surface physics techniques, such as atomic force microscopy (AFM), scanning tunneling microscopy (STM), focused ion beam milling (FIB), micro-indentation and electrochemical impedance spectroscopy (EIS), the latter mainly being used to investigate the integrity of paint and conversion coatings. They also have considerable experience in applying these techniques to standard test procedures for production quality control.

While this individual has quite a lot of experience in the corrosion field and has a PhD in materials science, their knowledge is quite specialized in nature and is deficient in many regards.

Specific weaknesses are as follows:

1. *A lack of engineering culture*: The ethos of fitness for purpose and time and cost-effectiveness are second nature to chartered engineers, especially when concerned with underwriting engineering structures or manufactured products. For scientists whose main focus is on research and development in a laboratory environment, the technology transfer aspects of scaling laboratory-sized investigations to the real problem is not always readily apparent. The aspects of such cultural implications are discussed in more detail in Section 30.1.4.3.
2. A lack of detailed knowledge of shop floor operations, on-site review protocols and remedies. For laboratory-based individuals, the knowledge of the practical aspects of looking for the symptoms which could indicate future corrosion failures in the field is not necessarily apparent.
3. *Technique centeredness*: In most organizations, many different techniques have been tried in establishing evaluation methods for materials performance. Commonly, while new techniques can show promise, it is often the case that in a commercial environment they do not live up to expectations and are under-utilized. For example, the authors of this work know of at least one manufacturing company where an atomic force microscope suite was commissioned for quality control purposes but is idle due to the fragility of the probes used. This type of situation can sometimes lead to situations where a scientific facility has purchased and commissioned a specialist technique together with a qualified operator but cannot justify running costs. In this situation, the temptation of the specialist is to find applications for an underused resource and the financial importance of corrosion prediction makes it a ready target.
4. *Poor cost-effectiveness*: In an academic or quasi-research environment, the measure of performance is often at least in part by peer-reviewed journal papers. This usually requires a degree of novelty and an involved programme of work to produce such output. The use of instrumentation-based techniques in this way is an expensive business and normally research-centered facilities are not very cost effective at providing good corrosion prediction. This is dealt with more fully in Section 30.1.4.3.
5. The difficulty of translating instrumentation techniques to real situations (parameterization). A potential use of technique-based measurements in corrosion prediction is to provide information to corrosion models, which are then used to describe a time-based description of what happens in service. Corrosion models for prediction are discussed in Section 30.5. Non-empirical models can be fitted to data from field experience. However, there is often a desire to provide more fundamentally derived models based on chemical and physical principles. This is discussed in detail in Section 30.2 on mechanisms. Even the best mechanistically derived model relies on reliable parameters to feed into the model so that the most accurate prediction can be obtained. It is notoriously difficult to provide good values for such parameters so most models rely on existing data from the 1920s to the 1970s. This includes parameters such as exchange current densities and thermodynamic parameters for phases – for example, for Pourbaix diagrams and alloy phases and general transport phenomena.

6. *A lack of systems thinking*: Corrosion is a systems phenomenon. To best be able to apply all the available information, a broad approach and a general awareness of many factors is needed. In this regard, the specialist, especially from a research environment, can be at a disadvantage.

From the above arguments in assessing Individual 3 for competence in predicting corrosion, a score in relation to the other individuals can be given.

Note although the rating assessment for this type of individual's competence might appear low, it only describes the *competence effectiveness*, that is the relative ability of such individuals to contribute to a process whereby the future performance of the system in terms of its corrosion behaviour can be predicted. What can be seen is that the knowledge and experience required in predicting corrosion behaviour to a level whereby it can be acted on is very involved and requires inputs from various parties. *Individual 3: Competence Effectiveness Rating 3.5/10.*

Aside from the discussion of expertise needed to make reliable predictions are the concepts of risk and liability. Risk is a function of the truthfulness of the corrosion prediction as well as other things, but liability costs (such as insurance premiums for costs to a third party, for example, to a customer for remediation of available corrosion) involves a factor for professional 'esteem', that is the external qualifications and representations of an individual such as membership in professional bodies. In this case, individuals with qualifications such as Chartered Engineer are of especial importance.

30.1.4 Systems Assessment

Systems engineering is a relatively new discipline. It relies on a concept called *systems thinking*. Systems thinking was an idea which originally started in the late 1950s and progressively developed until the present day and was designed to deal with complicated problems governed by applied stimuli. At its heart, it treats any problem as a system comprised of set of parts with inter-relational dependencies. The individual parts can be both physical, for example, a menu of materials options for design, quantifiable such as cost, abstract such as human factors, or conditional (voluntary or involuntary) such as interruption of a supply chain (see Section 30.1.4.4). The system must have a goal which can be optimized by correct application of systems principles.

Often the capture of systems requirements, especially the correct goal to be obtained is critical. Two examples from history illustrate this in more detail.

Case 1: The Battlecruiser The battlecruiser was developed at the start of the twentieth century to answer the question as how to maximize the combat effectiveness of surface fleets. The theory was thus; the battleship was the main weapon of the (then) modern major naval powers, Great Britain, France, the United States, Russia and Japan. They had the heaviest armorment and the most armor, but were the slowest ships and also required the greatest support from land. Potential conflict with other world powers also required numbers of battleships to be grouped as battle fleets to counter enemy battle fleets. The problem for those powers with colonial empires was how to maximize the effectiveness of surface fleets. Around the time of the World War I, this was solved by the battlecruiser. A battlecruiser had the armorment of a battleship, usually with 14 or 15 inch guns, but combined with light armor. They could outrange smaller ships, but had the mobility to avoid enemy battleships unless protected by their own battleships. Their higher speed also meant that battlecruisers could be dispatched more quickly to deal with enemy threats around the world. Thus they could project sea power to a greater surface extent than slower ships.

However, in World War II, battlecruisers proved very vulnerable to air power. The first incidence was three days after Pearl Harbor, on December 10, 1941, when the British battlecruiser *Repulse* together with the battleship *Prince of Wales* was sunk by air attack from the Japanese carrier fleet near present day Malaysia. The Japanese navy had approached the problem of naval presence in another way than other countries. The question they were interested in was different from that of the Western powers. This was how to maximize the projection of sea power. They realized that the key concept was projection and this could only be maximized using air power as it projected a greater area of attack; battleships and battlecruisers were mainly used in a defensive capacity to protect the fleets of aircraft carriers. The Japanese goal for the question of optimizing naval power was how to maximize the projection of sea power, whereas the other powers answered the question of how to maximize the combat effectiveness of surface fleets. From a systems perspective, the wrong systems requirements were in place.

Case 2: The Battalion Carré System The preceding example is from sea-based military history, but solution of systems problems can also involve organizational matters and a good example of this is the development of modern command structures. It can be said that the beginnings of organizational management derive from this. Prior to 1800, the emphasis on all European armies was on maximum combat effectiveness of individual units. The emphasis was on obtaining the most efficient firepower per unit and conflict was based on lines of infantry three men deep all of which could fire. The systems requirement was combat efficiency. Higher organization was via the regiment and each regimental colonel required orders from the commanding general. It took considerable coordination to arrange units so that units could mutually support each other and this made tactics rather static.

With the advent of the French Revolution, France instituted conscription and the army was organized by Lazare Carnot, the father of Sadi Carnot, the thermodynamicist, to supply and equip it. The revolutionary armies thus had a superiority of manpower and although much of it was not trained to professional standards it had high morale due to revolutionary ardor.

The system requirement was to force a resolution whereby superiority in numbers would be maximized. To this end, the Battalion Carré system was introduced. For this, a new permanent structure was set up which was comprised of formations 25–30 miles apart (Corps) marching on parallel routes, each of which was given enough resources to maintain itself in position against the enemy until reinforcements arrived. The responsibility for decisions was given to the commander of the new formation (the Corps) who could act independently and use force of numbers at key points while his adversary was waiting for orders. Combat efficiency was not the systems requirement. The above example is an illustration of where the correct solution to a system requirement was an organizational one. For corrosion prediction of engineering systems this is manifest, for example, by effective monitoring or maintenance programmes.

The question is how is such reasoning at all relevant to corrosion prediction? The answer is that in predicting corrosion performance, we are satisfying a systems requirement which is controlled by a subsystem–materials–environment interaction. Depending on the system, some requirements are universal, such as cost, availability and physical function, but others are different in that they are system specific. As an example, consider the four chapters in this book on aviation (Chapter 21), automobiles (Chapter 22), food (Chapter 23) and buildings (Chapter 24). In this case, the systems requirements are shown in Table 30.1. These systems requirements are now described in more detail.

TABLE 30.1
Some Systems Requirements Concerning Corrosion in Different Applications

Application	System Requirement	Notable Issues
Aviation	Cost	Availability
	Continuing function during life cycle	Maintainability & reliability
	Airworthiness	Safety certification procedures
Automobiles	Cost	Availability
	Continuing function during life cycle	Maintainability & reliability
	Marketability	Aesthetics
Food	Cost	Availability
	Continuing function during life cycle	Maintainability & reliability
	Health and hygiene	Containment & contamination
Buildings	Cost	Availability
	Continuing function during life cycle	Maintainability & reliability
	Structural compatibility	Building lifetime options

Cost: Materials selection is driven by cost of the means of construction and the materials used are governed by a range of other cost factors. For example, the way that modern retail buildings are designed takes into account the extra retail sales generated by fast construction times and the materials selected accommodate this. As mentioned in Chapter 1, corrosion control is ultimately cost driven. In any systems assessment, the cost of materials and their maintenance is paramount. Factors such as cost discounting of future outlay play a part; for example, if two designs have the same materials costs but for one the cost of materials is upfront (it is maintenance free) and for the other some of the cost is deferred – such as periodic repainting – then the latter is preferred because the cost deferment is amortized by inflation. In this regard the availability of maintenance materials *throughout the life of the material* also needs to be considered as issues such as discontinuation of, for example, paint systems by suppliers may risk incurring future costs.

Continuing function during life cycle: In any design, the paramount requirement of the material is that it must continue to carry out its engineering function for the lifetime of the product. Life-cycle time is thus inherently a system requirement. Beyond this, the maintenance and monitoring for corrosion control should impact as little as possible on the function in terms of downtime. Lack of access for monitoring or maintenance is important as are designs of joints and geometric traps. This must be factored in when predicting corrosion performance.

Airworthiness: For the aviation industry, a critical systems requirement is that materials in service must satisfy the complicated legislation laid down by the international aviation authorities and implemented by the various domestic aviation authorities. This requires certification to very high safety protocols which are difficult to achieve for new materials and processes. For this reason, there are industry-specific protocols for corrosion control strategies, lifetime prediction and planned redundancy.

Marketability: In modern automobiles, corrosion is no longer simply a safety characteristic; this is already satisfied in developed countries as they usually employ annual roadworthiness tests where defective parts are replaced if necessary. At the stage that a vehicle ceases to be usable due to corrosion it has already lost almost all of its value before it becomes unroadworthy. So, even though paint protection systems are designed for corrosion control, their main worth is in maintaining the resale value of purchased vehicles and customer loyalty for future sales. For similar reasons, manifestation of corrosion in trim such as filiform corrosion which would be trivial in other applications are important in customer perception of automobiles. Simply put, *aesthetics* is a systems characteristic in this case and corrosion control strategies must take this into account. For this reason, quality control of procedures such as surface finishes during manufacture is very highly controlled.

Health and hygiene: When it comes to food (and also for medical applications), the systems requirement for corrosion control is to minimize contamination. Materials used must be non-toxic. Corrosion deposits are potential sources of pathogen-containing harmful agents, bacteria, fungi and the like, and cannot be tolerated. Materials used in food production and preparation must also be amenable to cleaning procedures designed to eliminate health hazards.

Structural compatibility: An extra systems requirement within control in the building industry is that corrosion control usually must be compatible with existing building practices. Factors include the range of main building materials (concrete, brick, wood, steel frame), the very long life of certain buildings and the wide scope of the environment, incorporating building use, climate, location, soil mechanics and geology. One feature of corrosion control in buildings is that the long lifetime of a large proportion of buildings means that over the lifetime of a building best practice in material use and corrosion control is likely to change over the life cycle of the structure. Thus, materials requirements must satisfy the likelihood of in-service modifications.

The above discussions illustrate various system requirements which have an impact on the procedures adopted for corrosion control. It should be noted, however, that a full systems treatment often involves more inputs than this, especially at the design stage.

30.1.4.1 Systems Engineering Tools

For procurement of major engineering products, there are complicated systems engineering tools which are tailored to optimize the competing demands of purchasing the most cost-effective, well-designed and fabricated equipment available. The most advanced of these are used for defense procurement, for example, by the United States Department of Defense or the United Kingdom's Ministry of Defence. U.S. Defense Procurement is governed by the U.S. Defense Acquisition System – DoD5000. This is based on life-cycle systems assessment procedures and is a general protocol for acquisition of equipment from very large items (ships, aircraft and armored fighting vehicles) to combat equipment for individual personnel. The life-cycle assessment is based on five stages known as the Defense Acquisition Process. These are shown in Table 30.2

Corrosion control is considered throughout and is laid down as a requirement in any acquisition process. As part of the logistics footprint, the UK Ministry of Defence operates a similar life-cycle assessment denoted by the name CADMID. This is shown in Table 30.3. CADMID is the acronym for the stages in the process.

Both processes have various defined points in the assessment where decisions can be made to halt or modify the acquisition and encompass considerations such as availability, reliability and maintainability, obsolescence, supply chain management and future in-service modification.

TABLE 30.2

United States Department of Defense DoD5000 Acquisitions Process

Step 1	Concept refinement
Step 2	Technology development
Step 3	Systems development and demonstration
Step 4	Production and development
Step 5	Operations and support

TABLE 30.3

United Kingdom Ministry of Defence Acquisition Process

Step 1	Concept
Step 2	Assessment
Step 3	Demonstration
Step 4	Manufacture
Step 5	In-service
Step 6	Disposal

Other developed countries have similar acquisition protocols for military procurement and various industries employ systems procurement methods, albeit usually in a simpler manner.

30.1.4.2 Design Hierarchy

Within a design there are often effective hierarchies. One concerns scale – with component being the smallest, subsystem assembly being the mid-size intermediate and systems level the highest level. Interaction at each level, for example, for materials compatibility for mixed metal systems, needs to be considered, first component to component, then subsystem to subsystem and finally in a holistic whole system analysis.

Another hierarchy that often arises is in where materials considerations are addressed. In practice, design often starts with the question of what the engineering function is to be and only later is the problem of what to make it out of addressed. Occurrences have included new processes in chemical industries that work well at laboratory scale in glass vessels that then present problems during pilot plant scale-up and become uneconomic at plant scale due to corrosion problems. This is a case of the research chemist designing a process, chemical and mechanical engineers then creating a means of putting it into production and a corrosion scientist inheriting the problem of making sure that the plant can last long enough to be a viable option. Obviously, for this to be successful there must be ongoing discussions in the design process from the beginning which involves the corrosion scientist advising the other parties into materials selection.

The final design hierarchy is based on materials properties. When constructing methods to predict likely corrosion performance there is an inherent hierarchy in materials factors; this is:

1. Materials Utilization
2. Corrosion Process
3. Failure Mode

The hierarchy is thus the *failure mode*, which is the final symptom of the corrosion process and is the stage at which the design criteria are breached. This might be mechanical failure, for example, in corrosion-fatigue, rupture of joints due to corrosion products or a chemical failure such as contamination of food. However, these are all a feature of the *corrosion process* which is itself is a feature of the *utilization of the material* for a particular environment including the material selection, its microstructure, engineering geometry, chemistry of its surroundings, applied stress and control measures such as coatings or inhibitors. Predictive models, therefore, have system restraints which define the failure criteria including via the failure mode (hierarchy level 3 as above) and decision criteria via the first step (materials utilization) to satisfy the life-cycle requirements.

30.1.4.3 The OODA Loop and Refining of Options for Corrosion Prediction

The tools required for a systematic analysis of corrosion processes, engineering and customer requirements is very complicated. It must be always borne in mind that above all, the reason that prediction of corrosion performance is carried out is to produce some sort of risk based judgement criteria under economic constraints. While it is very difficult to construct a robust holistic system to predict corrosion performance, it is much easier to systematically review tools that have been proposed so as to select or eliminate them from the range of options.

One way to do this is via a decision process called the OODA loop. OODA is the acronym for the stages in the process, as given below. The OODA loop was originally devised by John Boyd to be a decision method for air-to-air combat, but has since been used in many applications in business and engineering. The objective of the OODA loop is to make the optimum decision in systems where available information is changing; in the original application, decisions were made by a pilot in air-to-air combat with an adversary. The output of the decision loop was to make a *better* decision *faster* than the adversary when available information such as the trajectory of the adversary was constantly changing. To an extent, the purpose is to try and anticipate information. Thus there is a predictive element to the process and this can be applied to corrosion control.

At its core, the OODA loop is a feedback loop. This is made of four different parts. These are:

1. **O**bserve
2. **O**rientate
3. **D**ecide
4. **A**ct

This has similarities with the Plan-Do-Act-Review feedback loop of the Deming cycle. The difference is in the second stage 'Orientate'. In this stage, there is an appraisal of all available information using five criteria. In the original OODA loop as applied to air-to-air combat these were:

1. Previous Experience
2. New Information
3. Genetic Heritage

4. Cultural Traditions
5. Analysis and Synthesis

When applied to other fields, items 1, 2 and 5 are unchanged, but items 3 and 4 can be modified. For example, when considering corrosion these five elements can be:

1. Previous Experience
2. New Information
3. Human Factors
4. Subject Culture
5. Analysis and Synthesis

Note the term *Previous Experience* includes generally accepted information on materials selection in terms of cost and performance. It thus incorporates a whole set of information on best practice from various standards organizations. Human factors include such information as human accessibility for maintenance and inspection. For the layperson trying to understand corrosion control information purposes, the most illustrative item is Subject Culture. There have been many remedies proposed for corrosion control and also as methods for increasing the understanding of the fundamentals of corrosion. The impact of subject culture is best demonstrated by two examples. The first example is a hypothetical incentive scheme for inspection of components during manufacture. The second example is one of a range of techniques which have been proposed as beneficial to corrosion prediction, focused ion beam milling (FIB).

30.1.4.3.1 Incentive Scheme

Consider a scheme concerning a product whereby a manufacturer offers regular maintenance and servicing at low-margin cost; this includes annual servicing where consumable parts are regularly replaced by service engineers who are encouraged to report corrosion problems to production facilities. The results are shared with production and design engineers and professional corrosion engineering expertise is brought in to understand on-site issues when required. The scheme has support from senior management to ensure that it is linked into company quality management procedures and commercial value in terms of customer goodwill and manufacturing/service improvements are captured. The cultural tradition is that of the engineer in the field, backed by the engineering function of the organization.

A good way to assess the value of such schemes is by a simple scoring system. Table 30.4 illustrates this for the present example.

TABLE 30.4
Systems Assessment of an Incentive Scheme for Corrosion Prediction

Topic	Score	Notes
Cost	5/5	Can be accomplished using existing resources
Feasibility	5/5	Little investment needed
Relevance	4/5	Related to product in service
Value Capture	3/5	Customer goodwill and company reputation
Ease of Modification	3/5	Scheme is easy to change in the light of accruing data from the scheme such as common faults

Consider another scheme from a laboratory environment.

30.1.4.3.2 Focused Ion Beam Milling

A contract research organization has a facility for focused ion beam (FIB) milling. This is an expensive facility requiring special apparatus; the technique 'mills' surface layers over a small scale by ablation with an ion beam to give a depth profile of surface material. This technique has found greatest use in the electronics industry, but can be used for depth characterization of surface films. The technique requires dedicated cost outlay in purchasing a facility as well as a dedicated operator and service support. To be cost-effective, each instrument needs to be used as much as possible and focused ion beam milling has been proposed for a range of subjects including for understanding of corrosion control.

The contract research organization also has a partnership with a further education provider to enable it to apply for government research grants and industrial funding. The university's interest is in producing research output and using a specialist technique for characterization is a low risk way to produce results for publication. In terms of corrosion prediction, however, such technique-based work can present certain problems; the first is that it is not always clear how characterization data can be used for estimating how long an artifact can last. This is because predictive methods are in essence a numeration exercise and characterizing the surface of a material cannot give a definitive measure of how fast a process might occur unless certain properties, for example, film resistance can be measured. Vague assurances that such techniques can help elucidate corrosion mechanisms must be taken with caution. Elucidation of mechanisms for prediction is described in Section 30.2.

Using the scoring method used previously in Table 30.4 for systems assessment, we get a value of the relevance of focused ion beam milling for corrosion prediction of metals in service in this case. This is given in Table 30.5.

In terms of subject culture, for a surface physicist these scores would be much higher than for a corrosion scientist. For this reason, much technique-based work on corrosion must be treated with caution by those with an interest in predicting corrosion events and their frequency during service life. This argument is fairly general for many surface techniques. Because metals are electrically conducting they can sometimes provide a specious application to justify work for technique development. This is not to say that surface techniques do not have a part to play. Microstructural characterization of bulk phases relies heavily on surface analysis of metallographic samples and this is critical to understanding mechanical properties of alloys.

The OODA loop offers a tool to provide a sanity check the validity of different options for corrosion prediction. Some examples that are of doubtful value are as follows:

1. Any form of electrochemically stimulated initiation of corrosion – for example, pre-pitting of samples in a laboratory. Depassivation is not normally an electrochemical process.

TABLE 30.5

Systems Assessment of a Laboratory Scheme for Corrosion Prediction

Topic	Score	Notes
Cost	1/5	High equipment cost and dedicated staff required
Feasibility	1/5	High initial capital outlay and low throughput
Relevance	2/5	Not service conditions; Laboratory method
Value Capture	1/5	Largely research and development
Ease of Modification	1/5	Difficult as method is fixed

2. Many surface instrumentations in which analysis cannot be performed *in situ*. Corrosion processes are the inter-relation between a material and its environment; techniques which require separation of the material from its environment or treatment must be treated with care.
3. Mass balance calculations based on closed systems. Attempts have been made to compose speciation models for modelling the chemical species using conservation of mass. As such, these are thermodynamic equilibrium models and must be treated as inherently unreliable. The problems attendant in modelling corrosion processes this way are addressed in more detail in Section 30.5.

NOTE: The practical engineer may come across publications in the scientific literature from time to time in which different methods are sometimes proposed to be suitable tools in the understanding and prediction of corrosion for corrosion control. The reader should be aware that – as with other scientific subjects – government funding of projects at, for example, universities, is often oversubscribed and thus to gain funding some claims may sometimes be overstated. Thus the reader is advised to consider the relevance, especially of each piece of technique-based work, on methods of corrosion prediction on its own merits and not necessarily to take all claims at face value.

30.1.4.4 Availability, Reliability and Maintainability

One topic which needs to be addressed separately in any systems approach to corrosion control over the life cycle of its use is the continuity of the resources required for corrosion control. In systems engineering, this subject is known under several names, but most commonly as Availability, Reliability and Maintainability (ARM). As such it includes areas such as supply chain management but also topics such as the economics of maintenance and monitoring, the criticality of component failures and contract obligations in underwriting continuing performance. Much of this is not directly in the domain of the corrosion engineer, but certain areas of interest require technical input during the lifetime of the product. One of these is when a resource is discontinued, for example, a paint formulation or inhibitor; in this case, replacement options must be considered and testing undertaken to underwrite a change in process. It is prudent during the design stage of an engineering project that future availability, reliability and maintainability issues are addressed from the outset.

30.2 Mechanisms

30.2.1 General Mechanistic Principles

It has often been stated in corrosion texts that to be able to understand corrosion processes accurately a knowledge of mechanism is required. The main purpose of doing so is two-fold. First, if a corrosion mechanism is known then the validity of any prediction is increased as the failure mode is demonstrated and the processes behind it are understood. Second, any remediation methods to be considered can be applied with some confidence in how they act.

The definition of the words mechanism and failure mode were given in Section 30.1. Constructing a mechanism from first principles is a particularly demanding exercise.

A full corrosion mechanism is given in Section 30.2.3 for stainless steel. The essential points of developing a mechanism are thus: First, a mechanism is a *hypothesis* that is tested by available evidence. The hypothesis is then available to be modified if required. For corrosion processes much available evidence is already known – that various metals are susceptible to attack under certain chemical environments and stress regimes. A proposed mechanism should not conflict with any previously known observations concerning metals behaviour nor should it be at odds with the basic processes of chemistry and physics.

30.2.2 Depassivation and Electrochemistry

Corrosion processes are comprised of potentially three processes. These are the stages of initiation of corrosion, propagation of corrosion and – in principle – repassivation. For some metals, the onset of corrosion is a step which can take many years and is difficult to predict. It usually involves some sort of depassivation mechanism, most of which are poorly understood. Propagation is always an electrochemical phenomenon except for dry oxidation, with an anodic region and a cathodic region; the mechanism is simply that of an electrochemical cell. Repassivation is normally associated with self-healing processes, for example, when a cathodic protection is applied to an unprotected structure. For repassivation to be effective there must be intimate contact between the passivation layer and the metal. The special case of chromium and chromium alloys is discussed in Section 30.2.3.

30.2.2.1 Depassivation

The starting point for understanding the mechanism of depassivation of most metals is via the relevant Pourbaix diagram for each metal. Most metals, with the notable exception of chromium alloys, exhibit passivity due to a stable protective layer of oxide or other surface film. The stability of such phases is governed by the thermodynamic stability of the film under certain pH and redox conditions. As described in Chapter 3, this is illustrated by the application of Pourbaix diagrams to understand passivity; the acceptance of Pourbaix diagrams in this regard as describing the conditions of passivity predicates that the general conditions which control the solubility of *bulk phases* determine the passivity of metals. This is the starting point for construction of any mechanism concerning initiation of corrosion processes. The initiation reactions are *not* electrochemical in nature.

In constructing any mechanism, one is hampered by the fact that once corrosion has been observed, the original site of initiation – that is the place where depassivation started has been destroyed. Thus mechanisms of depassivation must be inferred from the hypothetical passivation layer over a metal, including special regard for the microstructure of the material and the possible inorganic phases overlying it. This approach is demonstrated in Section 30.2.4, the pitting initiation of aluminium.

30.2.2.2 Implied Consequences of the Tafel and Butler–Volmer Equations

Predictions of corrosion rate during the electrochemical propagation are normally carried out using the Butler–Volmer equation or its simplified form, the Tafel equation. The Tafel equation was originally an experimentally deduced expression, but was given a theoretical foundation through the derivation of the Butler–Volmer equation as derived in Chapter 3. It must be realized that this derivation is based on a model of electrode processes and that this itself constitutes a mechanism for the transfer of charge to or from the electrode to create ions.

The mechanism of electrode processes has been extensively studied and is a complex and demanding subject, involving solvation of ions, absorption of ions and solvent molecules on the electrode surface, quantum mechanics of electron transfer and the distortion of species in the applied electric field. Such processes have been described in detail in advanced electrochemical texts beyond the scope of this book. Certain mechanisms of inhibitors interfere with these electrode processes and are manifest by the reduction of the exchange current density of the electrode to a minimal value.

30.2.3 Stainless Steels

In this section, an example of a fully derived chemical mechanism is presented. This is the pitting of chromium alloys by certain ions, particularly halides. The basic precepts in this mechanism were summarized in Chapter 8, Section 8.3.2, but the full mechanism is described here. The reader is advised that the chemistry contained within it is rather involved, but as is stated in Section 30.2.1, since any mechanism is a hypothesis, for it to be tested to the highest degree all aspects of such a theory should be given so as to invite critique.

It is well known that the chemistry of metallic chromium in aqueous solutions is anomalous. It is a fact that the thermodynamic stability of chromium, as defined by the chromium Pourbaix diagram shown in Figure 8.12, does not adequately describe the chemical behaviour of chromium or chromium containing alloys in contact with aqueous media. Thus, chromium imparts a chemically passive character to ferrous alloys but they are susceptible to localized attack in the form of pitting corrosion in chloride, bromide and thiosulfate environments.

In this section, the fundamental chemistry is investigated of the molecular interactions of chromium-bearing iron surfaces in aqueous solution and in halide solutions to construct a chemical model for their dissolution (depassivation) from first principles. Specifically included are two hypotheses, which were introduced previously in Chapter 8; first, kinetic inertia of the trivalent state of chromium (Cr^{III}) suggests dissolution of a protective film on such alloys ought to be kinetically controlled. Second, mixed species such as the chlorochromate ion and the similar bromochromate and thiosulfochromate ions are considered. From these considerations, a new and complete mechanism for the slow kinetics of dissolution of chromium alloys and their catalytic depassivation by chloride, bromide and thiosulfate ions has been elucidated. It is also shown that several phenomena of the passivity of chromium alloys to attack are a direct consequence of the fundamental chemistry of chromium.

30.2.3.1 General Features of Pitting of Chromium Alloys

The characteristic features of pitting corrosion of stainless steels are given in Section 8.3.2.2. Iron–chromium alloys with a minimum content of 13% chromium have outstanding corrosion resistance due to their ability to establish and maintain a passive surface condition in a wide range of aqueous media. Additional components can enhance passivating characteristics but cannot confer them without chromium. The purpose of the present approach is to examine the chemical basis of this passive state with particular reference to its limitations.

The treatment here is conducted from a different perspective than much of the work on the corrosion of stainless steels. The approach used is to take the classic texts on inorganic and physical chemistry of a level typical of undergraduate university teaching at a British chemistry bachelor's degree or U.S. master's degree and to take the fundamental chemistry contained within these resources to build up a model to describe the chemical behaviour of stainless steels. Since the authors of this current text desired that the evidence presented here is that which is universally accepted by the chemistry community, the selected texts were deliberately chosen to be those already in print by 1970 and as far as possible to conform to the description of the 'standard' texts used to teach generations of chemistry students. The chemistry references used in building the mechanism were:

1. F. A. Cotton and G. Wilkinson, *Advanced Inorganic Chemistry*, Second Edition, Longmans, London, 1966
2. J. W. Mellor, *A Treatise on Inorganic and Theoretical Chemistry*, Second Edition, Longmans, London, 1953
3. L. Pauling, *The Nature of the Chemical Bond*, Cornell University Press, New York, 1966
4. J. O'M. Bockris and A. K. N. Reddy, *Modern Electrochemistry*, Plenum Press, New York, 1970
5. L. Sillen and A. E. Martell, *Stability Constants*, The Chemical Society, London, 1964
6. S. Glasstone and D. Lewis, *Elements of Physical Chemistry*, Second Edition, Macmillan & Co., London, 1960
7. R. A. Robinson and R. H. Stokes, *Electrolyte Solutions*, Second Edition, Butterworths, London, 1959

The following discussions of the chemistry underpinning the observable phenomena of the passivation and activation of the corrosion processes for stainless steels thus depend only on standard physics and chemistry. Two principles of traditional discourses on the physical chemistry of corrosion processes were used as a starting point. These are as follows.

The first topic is that of an electrode process of a metal plate immersed in water. As mentioned previously, this is manifest in the physical reality behind the tools of the Tafel equation and the Butler–Volmer equation. The implicit assumption underpinning the process is that the physics of electrostatic interactions hold true in all circumstances. In the work described here, the mathematical treatment of electrostatically contact adsorbed ions on a metal plate is used to extend the treatment to species-specific considerations. The treatment of contact adsorbed ions is given in its complete form in Bockris and Reddy, Volume 2, Chapter 7 on the electrified interface. A simplified treatment is given in Section 30.2.3.6 of this work.

The second topic is that the formation and maintenance of the passive state depends on external environmental factors which influence the thermodynamic stability of species as would normally be described by a Pourbaix diagram. The important influences are the redox potential applied by (cathodic) reactions at the metal surface, the pH and the influence of the so-called depassivating ions, Cl^-, Br^- and thiosulfate $S_2O_3^{2-}$, if present.

The influences of these factors on passivation are known from anodic polarization characteristics for iron–chromium alloys determined by potentiostatic measurements conducted in oxygen-free media. Some representative characteristics are illustrated schematically in Table 8.4 for acidic media in which critical values for the applied electrical potential E and anodic current density, i are defined as follows: *rest potential*, $E_R = -0.2$ V (SHE), *passivating potential*, $E_P = +0.0$ V (SHE), *breakdown potential*, $E_B = +1.3$ V (SHE), *critical current density*, $i_{CRIT} = 10^{-3}$ A cm^{-2} and *passive current density*, $i_{CRIT} = 10^{-5}$ A cm^{-2} where SHE is the Standard Hydrogen Electrode. In chloride media there is a *pitting potential*, E_{PP}. $E_{PP} = +0.5$ V (SHE) in 0.01 M Cl^-. This defines four ranges; An *active range*, $E < E_P$, where general metal dissolution occurs as ions in low oxidation states, a *passive range*, $E_P < E < E_B$, in acids, or $E_R < E < E_B$, in neutral media where dissolution does not occur to significant degree, a *transpassive range*, $E > E_B$, where general dissolution occurs as ions in high oxidation states and a *pitting region*, $E_{PP} < E < E_B$, where local dissolution at potentials in the normal passive range is stimulated by Cl^-, Br^- or thiosulfate ions. These are illustrated in the corrosion velocity diagrams in Figures 8.6 to 8.10.

The passive surface on stainless steels is assumed to be conferred by an oxide film present on the metal surface immersed in an aqueous medium. The film is indeterminate and can be vanishingly thin, but to assess its kinetic stability, some reasonable working assumptions are needed as follows.

1. The relative proportion of iron and chromium in the film is similar to that in the metal.
2. Chromium species dominate the chemical properties of the film because it is the passivating agent.
3. Cations within the film tend to coordinate with oxygen ions in arrangements related to corresponding bulk phase, for example, Cr^{III} aspires to octahedral coordination.
4. Cations at the free surface coordinate with water and hydroxyl ions.

These working assumptions are taken in the next section to construct a species-specific mechanism for iron–chromium alloys from first principles to demonstrate the causes of passivity for these alloys.

30.2.3.2 The Chemical Approach to Passivity

The first stage in constructing a chemical model to explain why chromium imparts passivity to ferrous alloys is to understand the fundamental chemistry of chromium. This is summarized as follows.

30.2.3.2.1 The Formal Oxidation States for Chromium

As is well known, the common oxidation states are Cr^{II}, Cr^{III} and Cr^{VI}. Both Cr^{II} and Cr^{III} form soluble cations and hydrous/hydroxide phases. Cr^{VI} species exist in highly oxidizing media as the chromate anions CrO_4^{2-}, and $HCrO_4^-$ Equilibria between these phases in the pH range 0 to 14 for activities of soluble species 10^{-6} mol dm^{-3} are displayed in Figure 8.12, extracted from Pourbaix's classic schematic potential–pH diagram for the chromium–water system. Although it applies strictly to elemental chromium, it also provides a chemical context within which to explore the behaviour of chromium as an alloying element in iron–chromium alloys. The notable feature of Figure 8.12 is that over much of the acidic region of the diagram, the thermodynamically stable species are soluble ions and thus a protective oxide film should not form. In the light of the obvious fact that iron–chromium alloys are corrosion resistant in aerated acidic media (see Figure 8.6 for reference), simple thermodynamic stability cannot account for this.

30.2.3.2.2 *Passivity: The Resistance of Cr^{III} to Dissolution*

Passivity on chromium or on alloys on which it is a component is associated with the Cr^{III} oxidation state that resists dissolution from the substrate by ligand exchange reactions, for example:

$$-[Cr^{3+} - OH^-]_{substrate} + 6H_2O = -[OH^-]_{substrate} + Cr(H_2O)_6{}^{3+} \quad (30.1)$$

The reaction is virtually arrested by the exceptionally large kinetic inertia that inhibits ligand exchange reactions of Cr^{III} in general. The magnitude of the inertia is illustrated in Table 8.5, which shows that the rate constant for ligand exchange in the chromium complex $Cr(H_2O)_6{}^{3+}$ is less than the constants for most other $M(H_2O)_6{}^{n+}$ complexes by a factor of at least 10^{-9}. The reason for this kinetic inertia is due to a well-established chemical phenomenon of transition metal cations called *ligand field theory*. The particulars of this theory, which are beyond the scope of this text, can be found in Cotton and Wilkinson referenced above, but the important fact for this treatment is that octahedrally coordinated cations with three *d*-electrons (d^3) are extremely inert due to a property called ligand field stabilization energy.

The easy passivation of iron–chromium alloys shows that Cr^{III} can impart similar kinetics to films on alloys in which chromium is a significant component. This argument can be extended to estimate the kinetic inertia due to ligand field stabilization of oxide anions in the passive layer which share two Cr^{III} octahedrals. These calculations shows that effective complete kinetic stability is conferred on the film when the chromium content is around 13%–15% by mass, and is consistent with the amount of chromium required as an alloying element to passivate. This is calculated in Section 30.2.3.3.

30.2.3.2.3 *The Active Range: Depassivation by Reduction of Cr^{III} to Cr^{II}*

Ligand exchange reactions for Cr^{II} are very fast as shown in Table 8.5 and when present they generally facilitate the solution of Cr^{III} in chromium chemistry. Equilibrium between the Cr^{3+} and Cr^{2+} ions is given by

$$Cr^{3+} + e^- = Cr^{2+} \quad E^{\ominus} = 0.41 \text{ V SHE} \quad (30.2)$$

The activity of Cr^{3+} in the system rises progressively with the cathodic polarization and the onset of the active range can be attributed to the introduction of sufficient Cr^{2+} ions to outweigh the kinetic inertia of ligand exchange on Cr^{III} ions by offering an alternative fast dissolution route. This releases Cr^{2+} ions which are continuously replenished to restore the equilibrium given in Equation 30.2 until the film is effectively consumed. The Cr^{2+} cations are a transient species because they are then oxidized by water to Cr^{3+}:

$$Cr^{2+}{}_{(aq)} + H_2O_{(1)} = Cr^{3+}{}_{(aq)} + OH^-{}_{(aq)} + \tfrac{1}{2}H_{2(g)} \quad (30.3)$$

Applying the Nernst equation at 25°C to Equation 30.2:

$$E/V\ (SHE) = -0.41 - 0.0591 \log(a_{Cr}{}^{2+}/a_{Cr}{}^{3+}) \quad (30.4)$$

This yields values for the $a_{Cr}{}^{2+}/a_{Cr}{}^{3+}$ ratio = 10^{-7} at E = +0.0 V (SHE). Applying the rule that Cr^{3+} slows the kinetics by a factor of 10^{-9}, as shown in Table 8.5, the relative rate of dissolution via Cr^{2+} at this potential is 10^2. This is consistent with critical parameters determined from anodic polarization characteristics. For example, using representative values for an AISI 304 in 0.05 M sulfuric acid, given in Table 8.4, the ratio of active to passive current densities at the passivating potential is

$$i_{CRIT}/i_{PASSIVE} = (10^{-3})/(10^{-5}) = 10^2 \qquad (30.5)$$

The difference in standard electrode potential for the reaction in Equation 30.2 and the transition and the actual potential at which the active range starts (at the passivating potential E_P) is illustrated by comparing the values in Table 8.4 with the chromium–water system in Figure 8.5 at pH 1.2, commensurate with a solution comprising 0.05 M sulfuric acid.

30.2.3.2.4 The Transpassive Range – Oxidation of Cr^{III} to Cr^{VI}

If the metal surface is anodically polarized to the breakdown potential in aqueous media, Cr^{III} at the surface of the passive oxide film is oxidized *in situ* to chromate, changing the environment of chromium from the octahedral coordination to the tetrahedral coordination of the Cr^{VI} species. The reaction is not ligand replacement, so it is not inhibited by the kinetic inertia associated with solution of Cr^{III} and depassivation by general dissolution of the film ensues.

30.2.3.3 Calculation of the Percentage of Chromium to Passivate Stainless Steels

The next step in the mechanism is to determine how much chromium is required to passivate iron–chromium alloys. To do this, it can be assumed that the passive oxide which protects the metal is close packed and the metallic composition within it corresponds to the composition of the alloy. The passive film is vanishingly thin but it can be assumed that the oxidation states of ions within the oxide phase are those which are stable under normal circumstances, that is between the oxygen and hydrogen lines in the Pourbaix diagram. These are Fe^{2+}, Fe^{3+} and Cr^{3+}. The calculation of the amount of chromium that will passivate an iron–chromium alloy only requires this information, plus the fact that Cr^{3+} cations are only ever found in an octahedral (i.e., six-coordinate) environment which are the result of ligand field stabilization energy for the ion. The exact structure of the passive film is not required.

Nevertheless, it is illustrative to visualize a hypothetical structure which conforms to this set of circumstances. Luckily, there is a generic structure which approximates to a close-packed oxide phase which can accommodate iron and chromium in these oxidation states. This is the *2, 3 spinel* phase. Spinel phases are described in Section 2.3.3.4.

The argument for this is as follows; iron and chromium metal both potentially react with water (i.e., without the need for oxygen) to liberate hydrogen as does Cr^{2+}, therefore, they can be oxidized in the absence of air. Only species in the domain of stability for water, as given by Section 3.1.5.3, need be considered. Thus for the iron and chromium diagrams, Figure 3.2 and Figure 8.12, iron and chromium species such as $HCrO_4^-$ and FeO_4^{2-}, which require a potential for formation greater than can be provided by oxygen can be excluded. From this, it can be seen that for iron and chromium the only possible oxidation states in the film are as Fe^{II}/Fe^{III} and Cr^{III} species under normal conditions.

Consider the hypothetical juxtaposition of a fresh alloy surface with aerated water. The activity of water in dilute aqueous solution is essentially unity. The solubility of oxygen is, however, very limited; at 25°C, the solubility is about 3×10^{-4} mol dm^{-3}. The formation of the first atomic layers of the oxide occur extremely quickly via mechanisms such as quantum tunneling via Cabrerra–Mott theory as discussed in Section 3.3.2; initially the surface of the metal mainly oxidizes by reaction with adjacent water molecules to Fe^{II} and Cr^{III} for the reasons above, for example

$$Fe + H_2O = Fe^{2+} + O^{2-} + H_2 \quad (30.6)$$

$$2Cr + 3H_2O = 2Cr^{3+} + 3O^{2-} + 3H_2 \quad (30.7)$$

These ions are accommodated in an oxide lattice which forms the passive film; this layer has maximum density for electrostatic reasons and is coherent with the metal surface so that atoms need to tunnel the smallest distance. Furthermore, as it is very thin, it is likely to be epitaxially compatible with the alloy surface which is of cubic morphology. Both of these criteria are fulfilled by a close-packed cubic oxide layer with holes filled by cations which hold the close-packed lattice together.

The common divalent oxidation states of many of the first row transition metals such as Fe^{2+}, Mn^{2+} and Ni^{2+} and trivalent states Cr^{3+}, Mn^{3+} and Fe^{3+} are easily accommodated in this close-packed type of oxide structure as evidenced by the following minerals which exist as either spinels, inverted spinels or disordered spinels; $FeCr_2O_4$, $NiCr_2O_4$, $NiFe_2O_4$, $MnCr_2O_4$, $MnFe_2O_4$, Fe_3O_4 and Mn_3O_4. In the last two cases, the phase is comprised of two oxidation states, that is $Fe^{II}Fe^{III}{}_2O_4$ and $Mn^{II}Mn^{III}{}_2O_4$. All these phases have +2 cations and +3 cations in the ratio of 1:2 within the same oxide lattice. Several of these phases are isomorphous, that is they can accommodate foreign ions as long as charge is conserved.

These spinel phases possess the cubic close-packed lattice of oxide (O^{2-}) anions required in a face-centered arrangement with four anions in each unit cell. This means that the anions in a unit cell have eight charges so the cations within it must also have eight charges. To balance this, each unit cell has three cations with a total charge of eight. The formula of the whole unit cell is, therefore, $M^{2+}(M^{3+})_2O_4$. For a spinel oxide on an iron–chromium alloy, the chromium cations must be Cr^{3+}, but iron cations can be in either the +2 or +3 oxidation state. There is insufficient chromium from the alloy in most commonly used stainless steels to form the pure chromite spinel, $FeCr_2O_4$ so there is substitution of Fe^{3+} for Cr^{3+} ions. Thus the model for the passive oxide protecting the metal is $Fe^{2+}(Fe^{3+}{}_{1-x},Cr^{3+}{}_x)_2O_4$ where x is 1.5 times the mole fraction of chromium in the alloy. For most purposes, the mole fraction of iron and chromium can be taken as similar to the mass fraction due to the similar molecular masses of iron and chromium.

As was mentioned previously, it is a feature of the chemistry of chromium that due to the ligand field of the d^3 state, kinetically inert Cr^{3+} ions are only ever found existing in 6-coordinate octahedral environments. Thus in any 2–3 spinel, the Cr^{3+} ions are located in the octahedral interstices in the oxide lattice. Each such ion within the bulk crystal of this lattice is adjacent to 6 oxide ions and each oxide ion is adjacent to 3 octahedral ions. It can be shown that each kinetically slow Cr^{3+} ion in the bulk phase of a spinel lattice is, therefore, linked to 12 other octahedral cations via oxide bridges. These octahedra may contain either Cr^{3+} or Fe^{3+} ions. This is shown in Figure 30.1.

From this the proportion of chromium to passivate stainless steels can now be determined. As stated previously in Section 30.2.3.2, the slow kinetics of Cr^{3+} means that the half-life of breaking a Cr–O bond is reduced by a factor of 10^{-9} compared with other cations. For this reason, the half-life of exchange for the Cr–O bond is around 3.5 days as shown by the value of 10^5 s in Table 8.5. A half-life is a manifestation of any bond having the energy at any instant to dissociate. For this reason, if there is an identical coincident bond within a lattice which simultaneously requires dissociation from a Cr^{3+} ion, then the net half-life of dissociation is reduced by another 10^9 times as both bonds must be severed at the same time. In this case, the half-life of dissociation is

$$3.5 \text{ days} \times 10^9 = 3.5 \times 10^9 = 9{,}600{,}000 \text{ years} \quad (30.8)$$

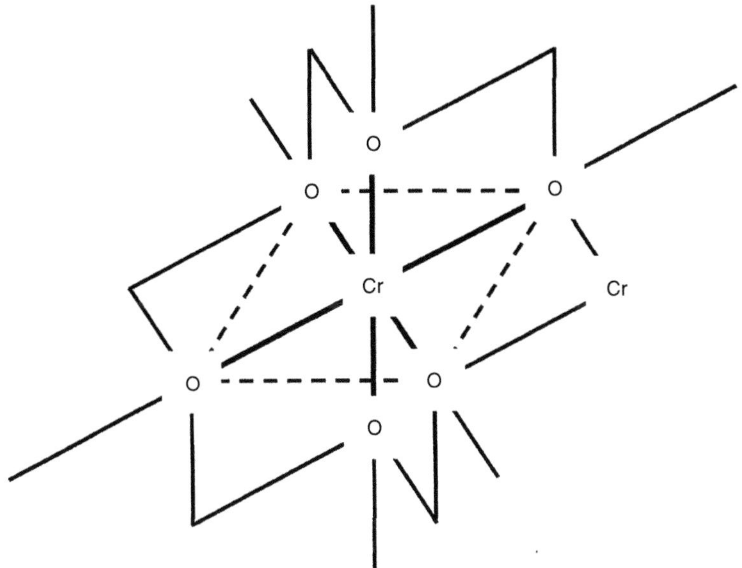

FIGURE 30.1
Schematic representation of the linkage of Cr^{3+} cations in the kinetically inert film. Note Cr^{3+} ions are 6 coordinate and have 12 potential links to other Cr^{3+} octahedrals.

Since the human life span is only of the order of 70 years, the phase appears to be completely stable, although this is a purely kinetic effect. It means that the material is technically metastable to dissolution and does not correlate with thermodynamic tools such as Pourbaix diagrams, especially the chromium–water diagram. It is known that the chromium diagram does not predict the stainless character of iron–chromium alloys so this indicates that the corrosion resistance of these alloys is due to this kinetic effect. The proportion of chromium required to achieve this can be calculated as follows.

If all Cr^{3+} ions in the lattice are on average attached via common oxide anions to other Cr^{3+}octahedrals, then the dissolution of the oxide is slowed. This is because it is unavoidable to destroy the oxide lattice without breaking kinetically inert bonds. From Equation 30.8, if all Cr^{3+} ions in the lattice are on average attached via common oxide anions to two other Cr^{3+}octahedrals, then the dissolution of the oxide should be so slow as to be imperceptible. Dissolution can only occur at the surface of the film and at these sites the octahedral cations have lower coordination connecting them to the matrix of the passive layer. The three types of octahedral surface sites that can exist are as follows.

Face site	Five bonds to other sites in the lattice	One free bond
Edge site	Four bonds to other sites in the lattice	Two free bonds
Corner site	Three bonds to other sites in the lattice	Three free bonds

The free bonds are usually occupied by water molecules absorbed onto the outer surface of the passive film. It is shown in Figure 30.1 that each octahedral cation in the bulk phase is linked via oxide ions to 12 other octahedra; this rule can be extended to surface sites.

Face site	Five bonds to other sites in the lattice	Linked to 10 octahedra
Edge site	Four bonds to other sites in the lattice	Linked to 8 octahedra
Corner site	Three bonds to other sites in the lattice	Linked to 6 octahedra

From this, the mole fraction of chromium required in the oxide to have (i) each octahedral linked once and (ii) linked twice to other chromium octahedra gives the proportion of chromium required to (i) start to passivate the film and (ii) complete passivation of the film.

Site	Linked Once	Linked Twice
Face site	1/10 (10 atomic %)	2/10 (20 atomic %)
Edge site	1/8 (12.5 atomic %)	2/8 (25 atomic %)
Corner site	1/6 (16.7 atomic %)	1/3 (33 atomic %)

Allowing for the fact that the molecular mass of chromium is 51.996 daltons and iron is 55.847 daltons, this gives, in terms of the mass of chromium in the alloy:

Site	Linked Once	Linked Twice
Face site	9.4 mass %	18.9 mass %
Edge site	11.7 mass %	23.7 mass %
Corner site	15.7 mass %	31.8 mass %

Thus, even the least vulnerable sites (face sites) are not kinetically protected until nearly 10% chromium is added to the alloy. Thereafter, increasing the chromium content to around 16% confers significant kinetic protection to the most vulnerable sites. Complete immunity begins to be conferred after 19% chromium. The percentages required for passivation are in very good agreement with that observed for iron–chromium alloys. It is interesting to note at this point, by referring to the polarization characteristics of AISI 304 Steel, as given in Table 8.4 and illustrated in Figure 8.5, that even when in the passive range, this alloy does actually still exhibit a small passive current density of 10^{-5} A cm^{-2}. This gives strong support to the idea that the passivity of stainless steels is a kinetic one and even though greatly retarded the alloy does still exhibit a corrosion rate under the apparently passive conditions, albeit very small.

Note the above argument does not require the assumption of a spinel structure; as shown in Figure 30.1, it depends only on the fact that the coordination of Cr^{3+} ions is octahedral and that the proportion of chromium in the oxide is the same as in the alloy.

30.2.3.4 Depassivation by Replacement Ligands

Chloride, bromide and thiosulfate ions depassivate iron–chromium alloys at potentials below the breakdown potential by reducing the anodic polarization (thus the electrochemical potential) required to raise the oxidation state of Cr^{III} to Cr^{VI}, thereby promoting the following sequence of reactions initiated at the surface of the film. They otherwise follow the well-established rules for complexes in solution:

1. Replacement of water/hydroxyl ions coordinated with Cr^{III} at the surface of the substrate with $X^{-}_{(aq)}$
2. Oxidation of the Cr^{III} to Cr^{VI} *in situ* and its realignment into tetrahedral coordination
3. Separation of the oxidation product from the substrate and dissolution
4. Reduction of Cr^{VI} to Cr^{III} and recovery of $X^{-}_{(aq)}$ in the aqueous environment

In summary, this yields

$$-Cr^{III}{}_{surface} \rightarrow Cr^{III}{}_{(aq)} \tag{30.9}$$

which is simply the dissolution of Cr^{III} from the film surface.

An ion must fulfill the following requirements to participate in the reactions described above; (i) it must be accepted as a ligand in a direct complex with Cr^{III}, (ii) it must be accepted as a replacement ligand in the chromate ion forming a soluble chlorochromate-type species, CrO_3X^{n-}, but also hydrolyze in the presence of excess water to re-release the species, X^{n-} for further catalysis and (iii) it must reduce the potential required to oxidize Cr^{III} to Cr^{VI} so that it occurs in aerated solution. Candidates for X^{n-} include all those ions which could hypothetically substitute into the Cr^{III} coordination sphere. These include fluoride, F^-, chloride, Cl^-, bromide, Br^-, iodide, I^-, sulfate, SO_4^{2-}, nitrate, NO_3^-, bicarbonate, HCO_3^-, dihydrogenphosphate, $H_2PO_4^-$ and thiosulfate, $S_2O_4^{2-}$ ions.

The relative ease of oxidation of the species Cr–X to CrO_3X, where X, is F, O, Cl, Br or I, can be estimated using a Born–Haber cycle as follows. The energy of oxidation is related to the work done on changing the oxidation state from Cr^{III} to Cr^{VI} and a factor in this is the differing ionic character of the Cr–X bond which influences the relative charge on the central Cr ion. Ionic characters of the various Cr–X bonds can be calculated from Pauling's scale of electronegativity using Table 2.8, from which the relative hypothetical increase of charge on the chromium during oxidation, that is the apparent ionization energy, versus that of the hypothetical adsorbed surface water, Cr–H_2O, can be determined. This is shown in Table 30.6. *e* is the electronic charge; for purely ionic character, the residual charge on Cr^{III} is $+3e$.

The residual charge in the hypothetical ionization of the Cr^{3+} to Cr^{6+} step in the Born–Haber calculation can be determined by considering the ionic character of the oxide interactions between Cr^{3+} and oxide. This determined by Pauling's method and the degree of ionic character is calculated to be 60%, so the residual charge on each chromium cation is $+3e$ times 0.6 equals 1.80. Since the adsorbed surface water, Cr–H_2O, is bonded through the oxygen atom, the electronegativity difference of the Cr–H_2O bond is identical to the oxide lattice. For this reason, the hypothetical residual charge for the Cr–H_2O entity can also be assigned a value of 1.80; that is the hypothetical charge of the cation is $Cr^{1.80+}$. From Table 30.6, the hypothetical residual charge can be calculated for the other species. This is shown in Table 30.7 together with the hypothetical chromium(VI) charge in the ionization step of the Born–Haber cycle.

TABLE 30.6

Calculated Difference in the Residual Charge on Chromium for Halide Ions Compared with Adsorbed Water Molecules

Species	Charge Difference	Ionic Character of Bond
Cr–H_2O	$+0.00e$	60% ionic
Cr–F	$+0.17e$	76% ionic
Cr–Cl	$-0.20e$	39% ionic
Cr–Br	$-0.29e$	30% ionic
Cr–I	$-0.41e$	18% ionic

Note: e is the electronic charge.

TABLE 30.7

Hypothetical Residual Charge on Chromium for Halide Ions Compared with Adsorbed Water Molecules

Species	Hypothetical Charge on Cr^{III}	Hypothetical Charge on Cr^{VI}
$Cr–H_2O$	$Cr^{1.80+}$	$Cr^{4.80+}$
Cr–F	$Cr^{1.97+}$	$Cr^{4.97+}$
Cr–Cl	$Cr^{1.60+}$	$Cr^{4.60+}$
Cr–Br	$Cr^{1.51+}$	$Cr^{4.51+}$
Cr–I	$Cr^{1.39+}$	$Cr^{4.39+}$

This argument using Born–Haber calculations is a hypothetical one; however, the influence of electronegativity on the oxidation potential of cations is a real effect. This is described in detail in Section 30.2.3.5. The results of this treatment as given in Table 30.7 distinguish between four kinds of ligands. These are:

1. Cl^- and Br^- are less electronegative and *reduce* the energy to raise the oxidation state of Cr^{III} to Cr^{VI}, therefore, they can enable the attack of the passive oxide film at potentials below the normal breakdown potential.
2. I^- in principle also reduces the energy to raise the oxidation state, but at potentials relevant to the present context its potency is diminished by conversion to iodine at the surface.
3. F^- *increases* the energy required to raise the oxidation state of Cr^{III} to Cr^{VI} and therefore does not pre-empt the normal transpassive range above the breakdown potential.
4. Ligands with oxygen, for example, sulfate, OSO_3^{2-} and dihydrogenphosphate, $OPO(OH)_2^-$, do not significantly change the molecular environment from that of the oxide and have little effect. This is because replacing $Cr–H_2O$ with, for example, adsorbed sulfate ion, $Cr–OSO_3^{2-}$, does not change the electronegativity environment around the cation. It is still surrounded by six Cr–O bonds.
5. The thiosulfate ion is a special case because ligand substitution into the Cr^{III} environment can be theoretically either via the terminal sulfur atom or else via oxygen atoms in a similar manner to that with sulfate. This represents a special case and as shall be described in the next section, the behaviour of thiosulfate with Cr^{III} provides strong evidence that the mechanism outlined here is substantially correct.

30.2.3.5 The Fundamental Evidence

The mechanism outlined in the previous section relies on two hypothetical properties of the system. These are (i) that changing the electronegativity of the metal-bonding atom in the ligand changes the ease of oxidation of the metal and (ii) in the case of chromium, the +III state can be oxidized to a chromato-species. If these two properties can be shown to be true, then the mechanism outlined above can be shown to be a most likely process for depassivation. The simplest system to illustrate the first property of the system is to review the redox chemistry of the solvated +2 and +3 ions of the first transition series (titanium to copper) with different ligands.

This is to compare

$$M(X)_6^{3+} + e^- = M(X)_6^{2+} \tag{30.10}$$

with

$$M(Y)_6^{3+} + e^- = M(Y)_6^{2+} \tag{30.11}$$

where M = Ti, V, Cr, Mn, Fe, Co, Ni, or Cu and X and Y are ligands of the same charge, but of different electronegativity. The redox process does not change the geometry of the complex. Since corrosion mechanisms almost always occur in aqueous media, for relevance, ligand X should be water and also the medium of interest should be aqueous solution. For direct comparison, ligand Y should solvate the metal cation in a similar way to water, that is via an sp^3 lone pair arising from hybridization of 2s, $2p_x$, $2p_y$ and $2p_z$ orbitals; the obvious candidates for this are to compare the corresponding hexamine ions with the hexaquo ions.

$$M(H_2O)_6^{3+} + e^- = M(H_2O)_6^{2+} \tag{30.12}$$

and

$$M(NH_3)_6^{3+} + e^- = M(NH_3)_6^{2+} \tag{30.13}$$

The bonding interactions of water and ammonia are very similar. The structure of water is shown in Figure 2.3. In an inorganic complex, such as in Equation 30.12, water acts as a ligand by using one of the lone pairs (described in Section 2.2.2) to solvate the cation and gives an arrangement similar to that shown in Figure 2.5 for magnesium, which is in a six-coordinate octahedral environment. For ammonia, the orbital used is the same as that for water in Figure 2.3, that is a lone pair sp^3 orbital derived from 2*s* and 2*p* atomic orbitals. The difference is that the other lone pair is now a bonding orbital with a third hydrogen atom to give the NH_3 molecule. The important fact is that in comparing the ease of oxidation of

$$M(H_2O)_6^{3+} + e^- = M(H_2O)_6^{2+} \tag{30.12}$$

and

$$M(NH_3)_6^{3+} + e^- = M(NH_3)_6^{2+} \tag{30.13}$$

the bonding orbitals are identical and the major difference is the ionic character of the M−O and M−N bonds.

In practice, at least one of the oxidation states in Equations 30.12 and 30.13 has to be kinetically slow enough not to undergo ligand exchange, otherwise the amino complexes will hydrolyze in solution because of the excess of water as solvent. Certain of the first row transition metals are not suitable comparisons for this due to instability of either the +2 or +3 oxidation states (for example, Mn^{3+}) or do not form purely octahedral complexes (for example, Cr^{2+}, Ni^{3+}). Fortunately, however, the cobalt system offers sufficient stability in water to compare the general effect of changing ligands with different electronegativities on the redox chemistry. For cobalt, the ease of oxidation of the aquo and amine complexes in aqueous solution can be compared.

$$Co(H_2O)_6^{3+} + e^- = Co(H_2O)_6^{2+} \qquad E^{\ominus} = 1.84 \text{ V (SHE)} \tag{30.14}$$

and

$$Co(NH_3)_6^{3+} + e^- = Co(NH_3)_6^{2+} \qquad E^{\ominus} = 0.11 \text{ V (SHE)} \tag{30.15}$$

Therefore, changing the ligand from H_2O to NH_3 reduces the potential required to oxidize the central metal by 1.73 V. This is greater than the domain of stability of water and also even greater than the standard electrode potential of the Cr^{III}/Cr^{VI} redox reaction as given by

$$Cr_2O_7^{2-} + 14H^+ + 6e^- = Cr^{3+} + 7H_2O \qquad E^{\ominus} = 1.33 \text{ V (SHE)} \tag{30.16}$$

under standard species activities. The dramatic effect of changing electronegativity of the ligand (either O or N bonded) on the ease of oxidation is self-evident. The role of electronegativity in stabilizing the +3 state in the $Co(NH_3)_6^{3+}$ ion compared to the $Co(H_2O)_6^{3+}$ ion can be evaluated by calculating the charge on the metal due to partial covalent bonding with the ligands using Pauling's method. This is similar to the calculations done previously for the Cr^{III}/Cr^{VI} oxidation in this chapter and given in Table 30.5.

For the cobalt system, this can be illustrated in a simplified manner as follows. Using the values for electronegativity on the Pauling scale of 3.5 for oxygen, 3.0 for nitrogen and 1.8 for cobalt, the ionic character of hypothetical single bonds between Co–O and Co–N can be calculated from Table 2.8. The electronegativity difference of Co–O is 1.7, equivalent to a partial ionic character of 51% and for Co–N it is 1.2, which is equivalent to 30% partial ionic character. If the bonds were completely ionic then the central cation would have a charge of $+2e$ for Co^{2+} and of $+3e$ for Co^{3+}. This is reduced by the partial covalent character of the Co–O and Co–N bonds by $0.49e$ and $0.70e$ (i.e., 51% and 30% ionic character), respectively, per single bond order. It can be seen, therefore, that the charge density on the nitrogen-bound cobalt cations in the hexamino complexes is less than that of the equivalent hexaquo complexes. This is manifest by the greater stability of the +3 state in the hexamino complex than in the hexaquo complex as shown by the redox potentials.

The net change in charge is given in Table 30.8. In practice, as illustrated by Pauling's treatment of the $Co(NH_3)_6^{3+}$ ion in his classic text, *The Nature of the Chemical Bond*, cited at the end of this chapter, the central cation of a complex tries to achieve electroneutrality

TABLE 30.8

Relation between the Residual Charge on Cobalt Ions and Electrode Potential

Species	Hypothetical Charge on Cation	Electrode Potential (SHE)
$Co(H_2O)_6^{2+}$	$-0.94e$	+1.84 V
$Co(H_2O)_6^{3+}$	$+0.24e$	
$Co(NH_3)_6^{2+}$	$-2.2e$	+0.11 V
$Co(NH_3)_6^{3+}$	$-1.2e$	

Note: SHE is the standard hydrogen electrode.

by conforming to the ionic character to place zero charge on the central cation. The values in Table 30.8 represent an indication of the electrical work needed to accomplish this over the natural electronegativity of the elements. The most notable feature of Table 30.8 is the similar charge density for the $Co(H_2O)_6^{2+}$ and $Co(NH_3)_6^{3+}$ ions, which are the normal stable ions in aqueous solution.

The above evidence demonstrates the general effect of decreasing electronegativity of the atoms bound to cationic complexes on their redox stability. When applied to chromium chemistry, this discussion predicts that decreasing the electronegativity of ligands coordinated to chromium ions should promote the ease of oxidation of the chromium species. For halides, the intermediate is a halo-chromate species, CrO_3X^-, $X = F$, Cl, Br or I which when created will eventually be hydrolyzed by excess water in solution. Thus it is impossible to isolate the transient species and proof of the mechanism must be achieved by indirect methods. This is done by considering another ion, thiosulfate, which is known to pit stainless-steel alloys and which in this case can act in a similar way to halide ions.

The thiosulfate ion, $S_2O_3^{2-}$, is structurally similar to the sulfate ion SO_4^{2-}, but with one oxygen atom substituted for a sulfur atom. Both ions are tetrahedral and are similarly sized with a mean negative charge on each terminal atom of 0.5*e*. The salts of sulfate anions with divalent and trivalent solvated metal cations are numerous and many of these cations also form salts with thiosulfate ions. Examples from within the first transition series include salts with manganese, iron, cobalt, nickel and copper ions. However, for $Cr(H_2O)_6^{3+}$, whereas the sulfate salt is known and can be crystalized from aqueous solution, the thiosulfate salt has proved impossible to prepare. This is remarkable since thiosulfate ions are often isomorphous with sulfate ions because of the similar geometry and size within the lattice.

Furthermore, if instead of an aqueous solution containing $Cr(H_2O)_6^{3+}$ ions, an aqueous solution of chromammine, $Cr(NH_3)_6^{3+}$ is used to prepare salts, then thiosulfate salts of the form $[Cr(NH_3)_{6-n}(H_2O)_n{}^{3+}]_2[S_2O_3{}^{2-}]_3 \cdot xH_2O$ are formed where n = 1 or 2. Ions of the type $[Cr(NH_3)_{6-n}(H_2O)_n{}^{3+}]$ represent partial substitution of water for ammonia ligands in the coordination sphere of $Cr(NH_3)_6^{3+}$, which suggests that slow hydrolysis of this ion is a competing reaction during precipitation of the thiosulfate salt. However, it is self-evident that when the Cr^{3+} ion is bonded to nitrogen-bearing ligands such as ammine ligands, the thiosulfate ion remains unreacted, but when bound to water a reaction between thiosulfate and Cr^{3+} ions occurs. As a comparison, in the case of the sulfate ion a number of salts can be made which all generally conform to the type $[Cr(NH_3)_{6-n}(H_2O)_n{}^{3+}]_2[SO_4{}^{2-}]_3 \cdot xH_2O$ in an analogous manner to the thiosulfate species except that *n* can be any number from 0 to 6.

In comparing the difference in the chemistry of these salts, it is necessary to look at the essential difference between the sulfate and thiosulfate ions. This is the terminal group involving oxygen and sulfur, $^-O-SO_3^-$, in sulfate and $^-O-SO_3^-$ in thiosulfate. Now as mentioned previously, the thiosulfate ion is the only other ion specifically listed as a pitting ion for stainless steels apart from chloride and bromide ions. Using the premise that the mechanism described for chloride and bromide ions is correct, a similar reaction can be written for the Cr^{III}–thiosulfate reaction.

To recap the general mechanism; this is:

1. Replacement of water/hydroxyl ions coordinated with Cr^{III} at the surface of the substrate
2. Oxidation of the Cr^{III} to Cr^{VI} *in situ* and its realignment into tetrahedral coordination

3. Separation of the oxidation product from the substrate and dissolution
4. Reduction of Cr^{VI} to Cr^{III} and recovery of $X^-_{(aq)}$ in the aqueous environment

Step 1 is similar to the reaction with the other pitting ions, chloride and bromide

$$-[Cr^{3+}–OH^-]_{substrate} + Cl^- = -[Cr^{3+}–Cl^-]_{substrate} \quad (30.17)$$

$$-[Cr^{3+}–OH^-]_{substrate} + Br^- = -[Cr^{3+}–Br^-]_{substrate} \quad (30.18)$$

that is

$$-[Cr^{3+} - OH^-]_{substrate} + S_2O_3^{2-} = -[Cr^{3+} - S - SO_3^{2-}]_{substrate} \quad (30.19)$$

Step 2 is possible because sulfur has an electronegativity of 2.5 on the Pauling scale; this is lower than either Cl or Br and thus oxidation of the result of reaction of Equation 30.19 should be favourable for the reasons discussed above. However, whereas for chloride or bromide ions, well-known species are the result, CrO_3Cl^- and CrO_3Br^-, in the case of thiosulfate the analogous species requires transfer of the terminal sulfur from the thiosulfate ion to the Cr^{3+} inner complex sphere. As the sulfide entity $-S^-$ is a poor leaving group, this is probably accompanied by an S_N2-like substitution reaction at the central sulfur atom in the thiosulfate ion, that is

$$-[Cr^{3+} - S - SO_3^{2-}]_{substrate} + OH^- = -[Cr^{3+} - S^- \cdots SO_3^- \cdots OH^-]_{substrate}$$

$$= -[Cr^{3+} - S^-]_{substrate} + HSO_4^- \quad (30.20)$$

$$-[Cr^{3+} - S^-]_{substrate} + HSO_4^- = -[Cr^{3+} - SH]_{substrate} + SO_4^{2-} \quad (30.21)$$

and Cr^{III} is oxidized to Cr^{VI} in a manner identical to the other pitting ions

$$-[Cr^{3+} - SH]_{substrate} \rightarrow Cr^{VI}O_3SH^-_{(aq)} \quad (30.22)$$

The sulfochromate species CrO_3SH^- is then hydrolyzed in solution to the chromate ion by a process akin to that of the chlorochromate and chlorobromate ions. Reactions 30.20, 30.21 and 30.22, represent the oxidation of Cr^{III} by thiosulfate ion at the oxide surface on a metal. This reaction is also in principle possible in solution via the $Cr(H_2O)_6^{3+}$ ion and explains the non-existence of chromium(III) thiosulfate as a salt.

$$Cr(H_2O)_6^{3+}{}_{(aq)} + S_2O_3^{2-}{}_{(aq)} \rightarrow [Cr(H_2O)_5^{3+} - S - SO_3^{2-}]_{(aq)} + H_2O$$

$$[Cr(H_2O)_5^{3+} = S - SO_3^{2-}]_{(aq)} + OH^- \rightarrow [Cr^{3+} - S^- \cdots SO_3^- \cdots OH^-]_{(aq)} \quad (30.23)$$

$$= [Cr(H_2O)_5^{3+} - S^-]_{(aq)} + HSO_4^-{}_{(aq)} \quad (30.24)$$

$$[Cr(H_2O)_5^{3+} - S^-]_{(aq)} + HSO_4^-{}_{(aq)} = [Cr(H_2O)_5^{3+} - SH]_{(aq)} + SO_4^{2-}{}_{(aq)} \quad (30.25)$$

$$[Cr(H_2O)_5^{3+} - SH]_{(aq)} \rightarrow Cr^{VI}O_3SH^-_{(aq)} \quad (30.26)$$

It is consistent with the fact that Cr^{3+} does not form a thiosulfate salt from aqueous solution. However, for the analogous reaction series via the $Cr(NH_3)_6^{3+}$ ion the reaction akin to the reaction in Equation 30.26, that is

$$[Cr(NH_3)_5^{3+} - SH]_{(aq)} \rightarrow Cr^{VI}(N - species)_3SH^-_{(aq)} \tag{30.27}$$

cannot occur because the +VI oxidation state for chromium is unknown for nitrogen ligands. Thus, the presence of ammonia ligands blocks the oxidation reaction of Cr^{III} with thiosulfate and thus salts such as $[Cr(NH_3)_{6-n}(H_2O)_n^{3+}]_2[SO_4^{2-}]_3 \cdot xH_2O$ are precipitated from solution.

The arguments set out in this section present strong evidence that effectively proves that depassivation of the kinetically inert Cr^{III} film on stainless steel by chloride, bromide and thiosulfate occurs via an oxidative mechanism involving Cr^{VI} species. It has been shown that the increased ease of oxidation is a direct consequence of the lower electronegativity of chlorine, bromine and sulfur compared to oxygen, leading to the conclusion that reactions such as this can be predicted from basic chemistry. It has also been shown that the three ions which react in this way are exactly those three ions which can attack the surface of stainless steels by pitting. Finally, if the oxidative mechanism is blocked – for example, by substitution of nitrogen ligands within the co-sphere of Cr^{III} so as to preclude oxidation to a stable Cr^{VI} species redox reactions cannot occur. This body of evidence gives sound background to the mechanism proposed in this work.

30.2.3.6 Dependence of the Initial Reaction on Species Activity

The initial depassivation reaction depends on the interaction of two species. These are (a) the pitting ion, that is chloride, bromide or thiosulfate and (b) the surface activity of chromium on the protective film. The formation of the CrO_3X^{n-} (X = chloride, bromide, thiosulfate) intermediate requires a direct interaction between surface Cr^{III} ions and X^- ions, that is contact absorption of the depassivating ion, and to sustain a detectable dissolution rate, X^- must cover a sufficient fraction of the surface. Bockris and Reddy derived an equation from first principles relating the surface charge density, q_M to the fraction of the surface covered, θ and the bulk activity in solution of the pitting ion, a_{X^-}; this can be simplified to the following form

$$Aq_M = -\text{constant} - \ln a_{X^-} + \ln\theta - \ln(1-\theta) + B\theta^{3/2} \tag{30.28}$$

where A and B are constants and the surface charge density, q_M, is proportional to the electrode potential, E. Assuming that reactions involving substitution of the pitting ions, that is Equations 30.17, 30.18 and 30.19 obey the normal substitution kinetics in that the rate of reaction is proportional to the concentration of the substituting ion at the surface, for an arbitrarily significant reaction rate (i.e., when the corrosion rate reaches a certain value) there will be a threshold value of θ. At this threshold value, the potential equals the pitting potential, E_{PP}. As at the pitting potential θ is a fixed value and the last three terms of Equation 30.28 are constant. This yields a relation between the pitting potential and the activity of the pitting ion in solution.

$$E_{PP} = \text{constant} - \Phi \ln a_{X^-} \tag{30.29}$$

Φ is a collection of constants. Thus pitting potentials should be an approximate logarithmic function of the activity of the depassivating species in solution as is observed. This means that if the solution concentration of the depassivating species is raised from 0.001 to 0.01 mol dm^{-3}, for example, the pitting potential will drop by the same amount as, for example, when the bulk concentration is raised from 0.01 to 0.1 mol dm^{-3} or alternatively from 0.1 to 1.0 mol dm^{-3}. Eventually, a concentration is reached when the pitting potential is lowered sufficiently to eliminate the passive range altogether.

The dependence of the passivity of stainless steels on the metal composition is more difficult to calculate from first principles and to do so requires certain assumptions. The calculation of the proportion of chromium to passivate the oxide has already been covered; the role of molybdenum is discussed in Section 30.2.3.7.

30.2.3.7 Other Phenomena

Secondary symptoms of corrosion of stainless steels are as follows (a) mechanical activation, (b) the stochastic nature of pitting, (c) enrichment and acidification of the liquid within the pit and (d) the role of certain alloying elements, particularly molybdenum and nitrogen in arresting the depassivation of stainless steels. The first of these, mechanical activation, is important. The next two, stochastic episodes in pitting and acidification, are strictly phenomena of pit propagation and morphology and only of secondary relevance for this work; however, if some of the behaviour of these phenomena can be explained as a logical consequence of the initiation mechanism it adds self-consistency to the whole theory. The last phenomenon listed, that of the effect of certain alloying elements on retarding the initiation process, depends on the interference of these species with the most probable corrosion mechanism. Some of the thoughts advanced in this section do not entail complete proof of the interactions of the species involved, but some general traits can be inferred from standard chemistry. Each of these phenomena is discussed in turn.

30.2.3.7.1 Mechanical Activation

It is well known that the application of stress has a dramatic effect on the corrosion of stainless steels. Both stress-corrosion cracking and corrosion-fatigue demonstrate this synergy and it is illustrative to consider the effect of stress on the chemical mechanism outlined in this work. As calculated previously, the proportion of chromium needed to passivate the alloy is at least 15.7 mass %, which is sufficient to ensure that all cations in the passive film are linked to each other through kinetically hindered Cr–O–Cr linkages.

When a cyclic stress is applied to stimulate corrosion fatigue, persistent slip bands (PSBs) form within the metal which slide back and forth over each other. Since the passive oxide film is very thin, this stress is carried into the film. Thus, planes of atoms in the film slide over each other repeatedly because of the applied cyclic stress. Taking a particular Cr–O–Cr linkage, it can be seen that in the situation where this undergoes slippage perpendicular to the Cr–O–Cr linkage, then passivity is potentially reduced unless the underlying cation is also a Cr^{3+} ion.

That is if on cyclic slippage

$$\text{Cr–O–Cr linkage} \leftrightarrow \text{Cr–O–Cr linkage}$$

no activation occurs. But if the situation is thus

$$\text{Cr–O–Cr linkage} \leftrightarrow \text{Cr–O–Fe linkage}$$

then mechanical activation has been achieved as the Cr–O–Fe is not kinetically hindered as the O–Fe bond is labile. Since chromium is a minor component of the film, the second option described is more likely than the first one so depassivation is aided by the applied cyclic stress. Also, since the amount of activation caused in this way depends on the number of bonds sheared per unit time, the depassivation of stainless steels in corrosion-fatigue should be a function of the rate of applied cyclic stress and not the amplitude of the cycle. This is indeed what is observed. For more information, see Talbot, Martin, Chandler and Sanderson (1982), cited in the Further Reading of this chapter.

30.2.3.7.2 Nuclear Activation

In the same way as mechanical activation occurs through physical shear of Cr–O–Cr linkages, these bonds can be physically broken by bombardment by nuclear particles. In this case, radiation impact causes displacement of atoms in the oxide film. This not only severs Cr–O–Cr linkages, but also, because it creates zones of destruction, increases the number of lower coordination sites, that is the edge and corner sites, around the site of impact.

30.2.3.7.3 Other Alloying Elements

For the full mechanism to be truly general in scope, the role of other alloying elements on the corrosion of stainless steels must be considered. Before this can be done, it is useful to make some more reasonable assumptions about the structure of the hypothetical oxide on the surface of the metal. To recap, as stated previously, the oxide must be coherent with the metal surface so as to present a physical barrier between the metal and the atmosphere; the model envisions this as a +2, +3 spinel containing iron and chromium. However, a typical steel contains more elements than iron and chromium, for example, AISI 304, which has an approximate composition as follows, from Table 8.3.

Fe	70 mass %
Cr	18 mass %
Ni	9 mass %
Mn	2 mass %
Si	1 mass %
C	<0.08 mass %

Silicon forms a refractory oxide, quartz, which is incompatible with spinel structures, carbon oxides are gaseous. Therefore, the passive spinel layer only needs to accommodate the elements iron, chromium, nickel and manganese. Converting from the mass percentage in the alloy, the atomic ratio of this is

Fe	71.4 atomic %
Cr	19.1 atomic %
Ni	8.5 atomic %
Mn	1.0 atomic %

The effect of iron and chromium in the alloy has already been discussed. To make the model as general as possible, all stable oxidation states for the other metals, nickel and manganese at potentials between that of the free metal and the oxygen/water line, need to be addressed. As was mentioned in Section 30.2.3.3, the ions Ni^{2+}, Mn^{2+} and Mn^{3+} can all be accommodated into the spinel lattice. These are the ions corresponding to the oxidation states within the water stability area of the Pourbaix diagrams for nickel and manganese so only these ions need to be considered. Since Cr^{3+} is always in the octahedral coordination, these other ions simply replace Fe^{2+} and Fe^{3+} within the spinel lattice. Thus nickel and manganese have little net effect on the mechanism of passivation for stainless steels as long as the alloy is uniform.

A secondary effect with manganese which is well known is that it can cause depletion of chromium in the alloy around manganese sulfur clusters caused by sulfur content in the steel. When the depletion of chromium is such that it falls locally below about ten atomic percent, the metal is no longer protective. The corrosion susceptibility is due to the local lack of Cr^{3+} ions in the spinel so the effect of manganese is indirect. Sensitization, caused by depletion of chromium around welds by precipitation of chromium from the matrix as $Cr_{23}C_7$, produces a similar effect when the local concentration of chromium drops to below about 10 atomic %.

There are two alloying elements which have a beneficial effect versus halide attack of stainless steels during pitting or crevice corrosion; these are nitrogen and molybdenum. The resistance of alloys is often given by a quantity called the Pitting Resistance Equivalence Number (PREN). For austenitic stainless steels

$$\text{PREN} = \%\text{Cr} + 3.3\% \times \text{Mo} + 16\% \times \text{N} \tag{30.30}$$

where percentages are in terms of mass. A PREN of at least 40 indicates good pitting resistance to halide media. This relationship is entirely empirical, but it demonstrates the potency of molybdenum and nitrogen especially when the above equation is converted to atomic percentages.

$$\text{PREN} = \%\text{Cr} + 6.1 \times \%\text{Mo} + 4.3 \times \%\text{N} \tag{30.31}$$

It can be seen that the potency of both molybdenum and nitrogen in their resistance to pitting corrosion by halide ions is far greater than that of chromium. In the case of nitrogen, the reason for this is clear. In the review of proof for this mechanism in Section 30.2.3.5, the ease of oxidation of Cr^{III} to Cr^{VI} species was discussed with respect to depassivation of stainless steels. In this it was shown that there was conclusive evidence that an internal reaction occurred between thiosulfate and Cr^{3+} ions and that this gave strong proof that depassivation at ambient redox potentials by chloride, bromide and thiosulfate ions was because of an electronegativity effect. As part of this proof the lack of reactions of hexamino chromium complexes were discussed with respect to the inability of Cr^{3+} to oxidize to Cr^{VI} species when coordinated to nitrogen atoms. The incorporation of nitrogen atoms in the passive film is presumably as N^{3-} nitride anions in place of O^{2-} ions, counterbalanced with some Fe^{2+} ions converted to Fe^{3+} so as to maintain charge balance. The Cr^{3+} ions coordinated to nitrogen anions cannot oxidize to Cr(VI) species. This inhibits the mechanism for halide ions to depassivate the surface film by oxidation to the halochromate species.

The role of molybdenum is different. Molybdenum chemistry is dominated by the Mo^{VI} oxidation state. In this state, molybdenum in an oxygen environment forms polymolybdates in which soluble ions have at least seven metal atoms in an anionic complex. A feature of these species is that they form encapsulating heteropoly species with many foreign ions including Cr^{3+}, all of which are surrounded by MoO_6 octahedra. In the case of Cr^{3+} in solution, the free species is $CrMo_6O_{18}(OH)_6^{3-}$ with the chromium octahedra surrounded on six faces by molybdenum octahedra. Since molybdates are incompatible with the spinel structure, it is likely that molybdenum forms structures with exposed corner Cr^{3+} octahedra so as to form a partial structure like $CrMo_6O_{18}(OH)_6^{3-}$ to encapsulate these corner chromium sites. This blocks these sites for exchange with the depassivation ions, chloride bromide and thiosulfate.

30.2.3.7.4 Pitting

Up to this point, the mechanism has not addressed the phenomenological aspects of pitting. As described in Chapter 8, the attack of stainless steels by species such as chloride, bromide and thiosulfate is most often encountered in the form of pitting, that is local attack where most of the metal is unaffected but concerted attack is at certain regions. The locations attacked can be associated with inclusions or grain boundaries or else with various procedures which induce stress such as work hardening. The ions chloride, bromide and thiosulfate are often referred to as pitting ions.

The depassivation mechanism differentiates between three geometries of surface chromium cations. These are corner sites (with three bonds to the spinel matrix), edge sites (with four bonds to the spinel matrix) and face sites (with five bonds to the spinel matrix). For the reasons described, at a chromium content typical of stainless steels, corner sites are most vulnerable to depassivation by dissolution with pitting ions. The spinel is a cubic lattice and the presence of corner, edge and face sites gives rise to features called steps, kinks and terraces on the surface of the film.

The role of such surface morphology on the ease of dissolution and removal of corner, edge and/or face sites has been studied extensively from a mechanistic perspective. The chemistry of such surface reaction mechanisms is very well established, especially in geochemistry where natural dissolution and precipitation kinetics of minerals is important. This has direct relevance to depassivation of metals since corrosion usually happens in a field environment. More information can be found, for example, in Chapter 11 of *The Geochemistry of Natural Waters, Surface and Groundwater Environments,* Third Edition, by James I Drever (1997) cited in Further Reading. The detailed mechanism of dissolution of minerals in this way is a complex subject, but the main application to understanding corrosion is that the areas of a surface which is disrupted – for example, at a grain boundary, tend to have step and kink formations which can be eaten away by removing corner sites. The act of eating away corner sites creates more underlying corner sites that are then also eaten away causing a pit to initiate.

Another feature of surface mechanistics of crystal dissolution is that if a crystal lattice has a defect, for example, a screw dislocation, then these features can be self-sustaining leading to local dissolution and depassivation. Since such defects are indicative of defects in the underlying metal – for example, from work hardening or inclusions, pitting can also be a feature of disruption in the metal surface.

30.2.3.7.5 Secondary Solution Reactions and Stochastic Behaviour

Two phenomena associated with pitting of stainless steels, especially in chloride media, have been extensively studied with a view to understanding the mechanism involved. These are acidification of the electrolytic environment within an active pit and the appearance of transient 'metastable pits' which then self-repassivate, leading to stochastic behaviour of pit sustainability.

The characteristics behind acidification of the pH of the electrolyte have been described in great detail by Szklarska-Smialowska (see Further Reading). It has been shown that the pH within the pit can drop to quite low values and it is often said that the process is autocatalytic. That is the increase of acid stimulates the anodic reaction which then produces a more acidic environment. This has always been a difficult proposition to explain because the chromium Pourbaix diagram does not predict the corrosion behaviour of stainless steels, especially concerning the low pH region. For example, stainless steel can be used to contain nitric acid.

The stochastic nature of pitting of stainless steels was first modelled by Williams and co-workers in the 1970s. Much work since then has been carried out by Burstein and others on the experimental aspects of stochastic pitting. A stochastic system is one where randomness prevails. The mathematical tools for this type of process rely on such mathematical constructs as Markov chains and the statistics involved has been applied to many fields. The inherent problem with such an approach is not the analysis of such a process, but rather why it happens instead of a deterministic system. In essence, it is a question of scale. Deterministic processes are governed by statistical thermodynamics and the role of randomness is reflected in the partition of energy between molecular states. The extremely small scale of molecules and the large number of them means that random partitions of molecular properties are averaged out to conform to steady bulk properties. Thus the key question with the apparent stochastic behaviour of pitting of stainless steels is what causes such an effect, rather than the deterministic population resulting from statistical thermodynamics.

Let us consider the corrosion products of pitting of stainless steels in chloride media by the mechanism described in this chapter. The product is the chlorochromate ion, CrO_3Cl^- and this is formed at the surface of the passive film at a potential below that of chromate in the Pourbaix diagram. This enables the film to dissolve releasing chromium as CrO_3Cl^- ions and also Fe^{2+} and Fe^{3+} ions from the spinel. The chemistry of chlorochromate is very well known both in the solid, for example, as Peligot's salt, CrO_3Cl^-, and in aqueous solution. In dilute solution, the chlorochromate ion hydrolyses to chromate by the following reaction

$$CrO_3Cl^- + H_2O = HCrO_4^- + Cl^- + H^+ \qquad (30.32)$$

This reaction is favoured by an excess of water and in dilute solutions proceeds entirely towards the right-hand side of the equation. Concerning the corrosion mechanism, this reaction has three important effects. These are

1. The active species for pitting corrosion, the chlorochromate ion, CrO_3Cl^- is destroyed liberating chloride ions. These chloride ions are free to reabsorb on the surface of the oxide film to further depassivate the surface. The chlorochromate ion is a transient species and is therefore not detected as part of the process.
2. The hydrolysis of the chlorochromate ion consumes water and also produces hydrogen ions. Therefore, unless there is free diffusion, the amount of water in a pit will tend to decrease and thus concentrate the electrolyte species present. Over and above this, the hydrogen ion activity will rise, therefore, the pH will drop.
3. As was stated previously, the chlorochromate ion is formed at electrode potentials below that of the chromate ion. Therefore, once the ion is hydrolyzed the resulting chromate ion is an oxidizing species with respect to the solution.

The dissolution of the spinel lattice is a source of Fe^{2+} ions. These react with chromate ions produced by hydrolysis of the chlorochromate ion via the following reaction

$$3Fe^{2+} + HCrO_4^- + 8H_2O = 3Fe(OH)_{3(s)} + Cr(OH)_{3(s)} + 5H^+ \quad (30.33)$$

In practice, the solid hydroxide phase is probably a mixed hydroxide, $3Fe(OH)_3 \cdot Cr(OH)_{3(s)}$. The reaction also produces hydrogen ions which contribute to the acidity of the pit. The solid mixed hydroxide is precipitated from solution and forms the solid corrosion products over a corrosion pit. Equation 30.31 raises two interesting questions. First, although the reactants are both aqueous ions they have opposite charges and thus, for the anodic reaction, $3Fe^{2+}$ has a positive transport number but the $HCrO_4^-$ ion has an effective negative transport number, that is it has to migrate against the potential field. This means that changing the applied potential field in an artificial way, say via potentiometry has an effect in changing the relative rate of transfer of $3Fe^{2+}$ and $HCrO_4^-$ into solution, and thus influences the reaction of Equation 30.33. For this reason, applying the results of potentiometric studies on pit morphology of stainless steels to real systems should be treated with some caution. The suitability or otherwise of potentiometric techniques to predict actual in-service corrosion behaviour because of these effects needs to be reviewed.

The second question arises from the kinetic inertia of the Cr^{3+} ion. Equation 30.33 is an oxidation–reduction reaction whereby one of the products is a solid phase containing the Cr^{3+} ion. This phase forms a product from solution which overlies the anodic site, and encloses and traps an electrolyte, which for reasons given previously consumes water and generates hydrogen ions. Assuming that the corrosion product restricts the access to the corroding pit the pit electrolyte will become concentrated over time.

Ordinarily, a reaction such as Equation 30.33 which produces a solid product from solution requires nucleation and this is followed by a precipitation reaction in dynamic exchange with a smaller dissolution back reaction to give a more thermodynamically favoured phase than the initial product. This also allows the corrosion product to continuously form in a way so as to minimize capillary pressure in the trapped electrolyte. However, with the reaction of Fe^{2+} and $HCrO_4^-$ this back reaction is not possible due to the slow substitutional exchange of the Cr^{3+} ion in the product. Thus when the last stage of the reaction 30.33 occurs – presumably by an inner sphere reaction between iron and chromium, any initial structure in the solid product is frozen into place. In essence, this means that the structure of the first formed corrosion product is stochastic and irregular, both in terms of structure and by extension also in terms of mechanical properties. The initial corrosion product is, therefore, not well formed. However, the depassivation of the spinel structure also produces Fe^{3+} ions which do not undergo any oxidation–reduction reactions, but precipitate as ferric hydroxide.

$$Fe^{3+} + 3H_2O = 3Fe(OH)_{3(s)} + 6H^+ \quad (30.34)$$

This is a slow reaction but is presumably nucleated on the hydroxide phase formed in reaction 30.32. It is thought that this secondary precipitation of ferric hydroxide strengthens the original corrosion products formed by reaction 30.32.

From Equations 30.32 and 30.33, it is clear that these reactions consume water. Therefore, the initially formed stochastic corrosion film is placed under increasing hydrostatic stress as the pit grows and the water within it is consumed. Mechanical rupture of such a film is responsible for the apparent behaviour of 'metastable' pits before the reaction of Equation 30.34 can strengthen the hydroxide phase. As they rupture the process is reactivated.

30.2.3.8 Mechanism Summary

The preceding sections give an example of a complete species-specific mechanism which predicts all the generally observed characteristics of pitting corrosion of stainless steels. From the premise that the passivity of stainless steels is a kinetic phenomenon because of the well-known inertia of Cr^{3+} chemistry, the following observed features of stainless-steel corrosion have all been predicted *as a natural consequence* of this hypothesis.

1. The minimum proportion of chromium in the alloy to confer some passivity is 10% by mass. Where depletion of chromium occurs, for example, around manganese sulfide inclusions or by precipitation of chromium carbides, corrosion can occur when the local chromium content is below 10%.
2. The passivity of this alloy improves until around 31% by mass of chromium has been added.
3. Pitting is expected to be the main mode of attack because areas of disruption and discontinuity of the overlying film are most vulnerable. This is due to the presence of more corner and edge sites in the passivating film.
4. Chloride, bromide and thiosulfate ions are the only common pitting ions under ambient conditions. Hypochlorite is often considered a pitting ion, but commonly, solutions of hypochlorite bleach made from chlorine contain chloride ions via the reaction

$$Cl_2 + H_2O = Cl^- + ClO^- \tag{30.35}$$

   so hypochlorite pitting is caused by chloride ions. The hydrogen sulfide ion, HS^- is also a pitting ion for similar reasons to the thiosulfate ion, but is usually only encountered in very highly alkaline media, above pH 12. However, in sour service environments in oilfield corrosion, attack occurs due to the high partial pressure of hydrogen sulfide and can be potentially damaging. This is discussed in more detail in Chapter 28.
5. Fluoride, iodide, nitrate, sulfate and bicarbonate are not pitting ions. The prediction of fluoride as not being a pitting ion is especially important. For many other metals where chloride is a pitting ion, for example, aluminium and beryllium, fluoride is also a pitting ion. Because of the high electronegativity of fluorine, for stainless steels the correct behaviour for fluoride is predicted. This means that the theory that the lack of pitting observed with fluoride ions on stainless steels is due to solvation of the anion is not correct.
6. In Section 3.2.4.3, the two main traditional theories of conventional, that is, thermodynamic passivity, were discussed. These are the film theory, originated by Faraday, and the adsorption theory, advanced by Koloturkin and Uhlig. The current theory presented here for stainless steels has common features with the film theory in that it considers depassivation to be due to exterior dissolution of a passive film. It also has features of the adsorption theory in that the basic mechanism involves substitution of the pitting ion for a water molecule on the surface coordinated to a Cr^{3+} in the surface film. The main difference here is that there is a problem with both the film theory and adsorption theory, namely in that if the depassivating ion can interact strongly enough to cause disruption, then a case usually needs to be made as to how it can then disengage from this interaction

to initiate the pit for growth. In the current theory, disengagement of the pitting ion, for example, chloride is made only *after* it has left the surface as the oxidized halochromate ion so a contrived disengagement step does not need to be required.

7. Molybdenum should improve corrosion resistance. The chemistry of molybdenum has been reviewed and it has been proposed that molybdates cap vulnerable sites on the passive surface. This chemistry is unique to molybdenum and tungsten, however, tungsten forms a more complicated structure – the Keggin structure – with the Cr^{3+} ion $CrW_{12}O_{40}^{5-}$. which requires twice the metal as for the is $CrMo_6O_{18}(OH)_6^{3-}$ ion. Taking into account that tungsten has twice the relative mass and is also more expensive then molybdenum, then only the effect of the later needs to be considered.
8. Nitrogen should improve corrosion resistance. The mechanism predicts, as is observed, that nitrogen as an alloying element has a beneficial effect on the corrosion resistance of stainless steels. This is because with nitrogen ligands Cr^{3+} is difficult to oxidize to hexavalent chromium.
9. Other alloying elements should not have a significant effect. The alloying elements in nickel and manganese are easily incorporated into the oxide film, the cubic 2,3 spinel structure without detriment to the passivating characteristics of the film.
10. Stochastic behaviour for pitting should be expected. In previous work on pitting of stainless steels, the random behaviour of the onset of pitting has been extensively characterized. Since the general randomness of molecular interactions are normally governed by statistical mechanics principles, which when summed over all available states usually yields steady, that is, deterministic properties, this has been difficult to rationalize. The present theory predicts that stochastic behaviour of the first stages of pitting should be observed because of the kinetic inertia of the corrosion product and the lack of a back reaction for dissolution of the product.
11. Acidification of pits is predicted as a consequence of the hydrolysis reactions within the pit. The reaction is catalyzed by pitting ions being re-released due to hydrolysis of the halochromate-type product in solution. The hypothesis that the pitting reaction is autocatalyzed by hydrogen ions is unclear, but probably incorrect.
12. A log-linear relationship of pitting ion activity with pitting potential is predicted. The variation of pitting potential with pitting ion activity is a result of the assumption that pitting ions are electrostatically adsorbed on the surface of the passive film.
13. The effect of stress on the kinetics of depassivation is observed. The effect of the application of stress, especially on the corrosion rate, can be predicted. Imposition of an applied stress stimulates dissolution as measured by the transient current observed under such processes. The magnitude of the transient current is a function of the rate of change of displacement.
14. The mechanism presented is a kinetic treatment to explain the passivity of stainless steels. The dissolution of the film is slowed under conditions where the Cr^{3+} ion should be soluble according to the chromium Pourbaix diagram. These conditions correspond to acidic media. Polarization characteristics are shown in Table 8.4 for the alloy AISI 304 in 0.05 M sulfuric acid at pH 1.2. The critical current density $i_{PASSIVE}$ is 1×10^{-6} A cm^{-2} and as illustrated in Figure 8.5, the corrosion rate

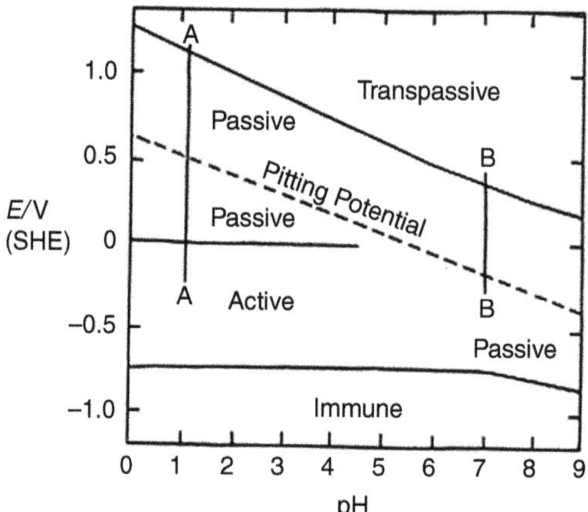

FIGURE 30.2
Pseudo-Pourbaix diagram for the chromium–water system. Note from Equation 30.8, the stability of these domains is not indefinitely stable but possesses a mean lifetime of 9,600,000 years.

of these alloys is still measurable even in the passive range. This lends support to this theory of passivation by kinetic inertia since the dissolution rate of the film hardly proceeds under conditions where a protective film is not thermodynamically stable.

15. A pseudo-Pourbaix diagram for stainless-steel alloys can be constructed. Taking the evidence above, the premise that for alloys containing sufficient chromium (around 13 mass % or greater so that face and edge sites plus a proportion of corner sites are protected), a domain can be added for the kinetically passive Cr^{III} state. Domains of activity still apply for the areas where Cr^{2+}, $HCrO_4^-$ and CrO_4^{2-} are the main species. This is shown in Figure 30.2. The lines A–A and B–B refer to the corrosion velocity diagrams given in Figure 8.6 for acid media and Figure 8.7 for neutral aqueous media. It can be seen that the active–passive regions for these diagrams fit well with that of the pseudo-Pourbaix diagram in Figure 30.2. Adding depassivating media such as chloride ions lowers the boundary in Figure 30.2 by a log-linear relationship and this is reflected by the corrosion velocity diagrams conforming to lines A–A and B–B illustrating the behaviour shown in Figure 8.8 and Figure 8.9.

Historical Observations: The mechanism outlined in Section 30.2.3 for stainless steels represents a complete theoretical treatment of the chemical processes behind the corrosion behaviour of such alloys. The failure previously to apply the basic chemistry of chromium to the system but to treat it as principally, in terms of thermodynamic passivity aided by studies using various potentiometric and surface analysis techniques has delayed the development of the subject. One is reminded of the 'Great Nernstian Hiatus' as described by Bockris and Reddy in which dynamic electrochemistry was hidebound for 30 years due to the success of classical Nernstian electro-thermodynamics.

The root cause of this arises from the suggested theory of Fontana and Greene in their classic 1967 book *Corrosion Engineering* (cited in Further Reading at the end of this chapter), concerning crevice corrosion and initiation. In it they proposed a schematic which they described as a mechanism, to explain why corrosion starts and a growing corrosion pit forms or crevice corrosion starts. They intimated that the mechanism was applicable to more than one metal. Unfortunately, their proposed initial stage of corrosion relies on the fact that hydroxide ions are less mobile that hydroxide ions (page 43 of their book *Corrosion Engineering*). This is actually untrue and the reverse is actually the case as demonstrated by Table 2.6 of this book. Hydroxide ions that travel through aqueous solutions via the Grotthus mechanism as shown in Figure 2.6 and thus do not need to physically be moved through the liquid and are highly mobile so the Fontana and Greene assumption is wrong on this account.

The error was compounded by Fontana and Greene in their description of pitting as an 'autocatalytic process' when what they probably meant was that corrosion pits became more robust in propagation, that is less likely to stall, as they grew. The general acceptance of this theory combined with the progressively greater access to new surface analysis techniques (especially post-AFM techniques) and electrochemical (potentiometric) techniques has produced over the last 50 years a climate in which considerable activity in corrosion research of an instrumentational nature has searched for evidence of this so-called mechanism. The goal of professional instrumentationalists has sometimes been to demonstrate that certain techniques can offer conclusive proof and this has unfortunately obscured the problems inherent in the theory. In this, it can be perceived that there is a certain element which reminds one of the activities of ancient alchemists in their search for the Philosphers' Stone and its ability to turn metals into gold. If the reader is to take only a few ideas away from this discussion, they should be

1. The corrosion mechanisms of metals are all species specific to each metal.
2. The general rules of physics and chemistry are generally applicable.
3. A mechanism is likely to be so complicated that any one technique cannot 'prove' it. The proof is by self-consistency with all existing phenomena.
4. Procedures which deliberately change the chemical processes, for example, electrochemical pre-pitting of samples or ion-beam milling of oxide films are not representative of field conditions and should not be regarded as 'safe' methods for investigation.
5. Such mechanisms have enough complexity to make accurate mathematical modelling of them next to impossible.

30.2.4 Pitting of Commercial Grade Aluminium

Like stainless steels, aluminium alloys can be susceptible to pitting by chloride ions. However, unlike stainless steels, aluminium alloys are also pitted by fluoride ions. In developing predictive methods for the corrosion behaviour of these alloys, a specific mechanism must be created which is different from that of stainless steels. As pitting is a local phenomenon, special importance must be given to the microstructure surrounding incubation sites.

Unlike for stainless steels where depassivation is a kinetic effect, the aluminium–water diagram in Figure 3.4 does accurately describe the corrosion characteristics of pure aluminium. Thus the passive phase under normal circumstances, Al_2O_3 (i.e., corundum) is protective. To be able to predict the corrosion behaviour of pitting it is necessary to understand the morphology of initiation sites.

In Section 9.1.1, the secondary phases which strengthen commercial grade aluminium, for example, AA 1100, are given. These include the intermetallic phases such as Fe_3SiAl_{12} and $FeAl_3$. These inclusions are known for being initiation sites for pitting under certain circumstances in neutral chloride and fluoride electrolytes. The bulk of the metal surrounding these nucleation sites can be thought of as principally aluminium grains overlain with a protective alumina film.

30.2.4.1 Initiation

The question arises in predicting the corrosion behaviour of pitting of aluminium as to what is the reaction that initiates attack? In situations such as this, the requisite mechanism must be inferred indirectly because the nature of the first step in establishing the resulting corrosion cell is that the initiation site is destroyed. All that can be done is to understand the microstructure of the metal and the chemical interactions of it with the depassivating agent (i.e., the pitting ion, principally chloride or fluoride ions). It must be realized, however, that the mechanism must be different from that of stainless steels because in the latter case fluoride ions do not cause pitting.

The principal initiation sites are associated with the phases Fe_3SiAl_{12} and $FeAl_3$. As both iron and aluminium are electropositive metals, an oxide film will be formed over these inclusions. Consider the environment of an $FeAl_3$ inclusion. In the presence of an electrolyte in contact with the metal, there must be a layer of a hydroxide phase. It is this phase which interacts with the pitting ion. The adsorption of pitting ions, Cl^- or F^- onto the outer surface of the hydroxide film proceeds in a similar manner to the first step in any pitting interaction. Thus as in Equation 30.28, the relationship between surface coverage of chloride or fluoride ions, θ, is given by

$$Aq_M = -\text{constant} - \ln a_{X^-} + \ln \theta - \ln(1-\theta) + B\theta^{3/2} \tag{30.28}$$

which gives the log-linear relationship with pitting ion activity as shown previously

$$E_{PP} = \text{constant} - \Phi \ln a_{X^-} \tag{30.29}$$

The nature of the interaction of the pitting ion with the surface is thus bound up with the phase stability of hydroxides in the presence of halide ions. The hydroxide phase over $FeAl_3$ contains both Fe^{3+} and Al^{3+} ions, in the proportion of the inclusion stoichiometry, that is 1 to 3. Both these cations form amorphous hydroxide phases of the same valency so the chemistry of this layer is likely to reflect an amorphous mixture of the two. The difference of the chemistry around the inclusion to that of the surrounding metal grains is due to the presence of iron in the hydroxide phase.

The ferric ion has an affinity for fluoride and chloride ions. The interaction of Fe^{3+} with fluoride ions is very strongly favoured, the association constants for FeF^{2+} being around 10^5 mol dm^{-3}. Chloride ions will also form direct bonds with ferric ions, the yellow-brown hexahydrate of ferric chloride actually being a complex with Fe–Cl bonds, *trans*-$[FeCl_2(H_2O)_4]Cl \cdot 2H_2O$. Iron hydroxide is digested slowly by chloride solutions by a process called dialysation. For freshly precipitated ferric hydroxide, this process takes about five months in concentrated brine. The general process is that the solid phase adsorbs negative chloride or fluoride ions into the hydroxide matrix by establishing Fe–Cl or Fe–F bonds. This draws negative charge into the matrix and eventually this stimulates dispersal of the solid into negatively charged particles which repel each other in the form of a colloid.

This reaction is the likely initiation step of depassivating the film over an $FeAl_3$ inclusion. As the reaction does not change, the bulk concentration of the chloride or fluoride ions in the electrolyte, it is a quasi-steady-state reaction so the initiation reaction should proceed at a constant pace for a set ion activity. However, once this has occurred the exposed intermetallic phase is noble to the bulk metal so it will act as a cathode and the exposed aluminium grains are rendered anodic.

30.2.4.2 Propagation Kinetics

Depassivation creates an electrochemical couple. The available area for the cathode (the inclusion) is of fixed area, but the area of the anode grows as a function of the rate of corrosion. However, the anodic and cathodic currents are equal. Thus the total cathodic current, which is the surface area multiplied by the exchange current density for the inclusion, is constant and balances the mean current per unit area on the anodic site. As the rate of corrosion depth with time, dx/dt is proportional to the anodic current per unit area of anode; this means that if there was no resistance to the system, the anodic corrosion current density would be

$$i = At^{-2} \tag{30.36}$$

to balance the total cathodic current. However, as the corrosion pit grows the mean distance between anode and cathode gets larger and thus the resistance of the system – film resistance and solution resistance also gets larger. The resistivity of the film and electrolyte are unchanged so the resistance of the system is proportional to the distance between anode and cathode, that is

$$R = Bx \tag{30.37}$$

But from Ohm's law

$$V = iR \tag{30.38}$$

so

$$V = At^{-2} \cdot Bx \tag{30.39}$$

but x, the distance between the anode and cathode, is a function of the integral of the corrosion process. This is because the cathode (the inclusion) is immobile and as corrosion progresses the anode is progressively further away. It is difficult to predict the exact dependence of x with time, t, because it depends on geometry; for example, the relative contributions of film and electrolyte resistance have difference dependencies on pit radius. It has been found experimentally that the pitting of commercial aluminium by chloride ions follows a cube root law

$$\frac{da}{dt} = Ft^{-3} \tag{30.40}$$

where a, is the mean pit depth and F is an arbitrary constant. By comparison with Equation 30.39, this shows that the kinetics of the system can be described by equations composed of a term based on Equation 30.36 and also a term based on Equations 30.37 to 30.39 in which the total resistance of the system is proportional to $1/t$. This implies that aluminium pitting functions as a simple linear system and that resistivity of a phase is constant, but the effective distance between anode and cathode is also a function of $1/t$.

It is not possible to present a more theoretical predictive model for this type of kinetics at least in part because certain physicochemical parameters are not measurable. For example, the cathodic exchange current density required is that *on the inclusion, $FeAl_3$,* this is different from that measured on sample of the metal, which would be the cathodic exchange current density across the whole surface.

It must be stressed that a semi-empirical treatment of the kinetics of pitting as done here is for a self-sustaining electrochemical cell. In circumstances where the cell is not self-sustaining, for example, with pure aluminium in the absence of pitting ions, the situation is different, for example, if the metal is forced to initiate artificially by electrochemical pre-pitting, the kinetics can be treated as two first-order reactions. The forward reaction is anodic dissolution. This is constant since the anodic and cathodic sites are both aluminium metal grains and are not constrained in area. The backward reaction is the natural re-establishment of the passive film and this is also first order as it is essentially a reaction between the metal and the solution, both of which have constant composition.

For this type of situation, the variation of depth with time is of the form

$$\frac{dx}{dt} = K_{\text{anodic}}\left(1 - \exp(-bt)\right) \tag{30.41}$$

K_{anodic} is the corrosion velocity after pre-pitting. Eventually, the exp[$-bt$] term $\rightarrow$ 1 so the corrosion velocity drops to zero indicating repassivation. Thus stimulated pitting has completely different kinetics to the natural unstimulated phenomena. For this reason, pre-pitting of corrosion samples as a predictive method of corrosion assessment for real scenarios should not be used in any circumstances. The observed kinetic equation (30.40) also shows that there is no repassivation for the natural case.

30.2.4.3 Summary

The semi-empirical treatment of aluminium pitting described predicts the follow features.

1. Around iron–aluminium inclusions, fluoride and chloride ions should induce pitting. Phases are consistent with the process of dialysation of the overlying hydroxide layer.
2. The depassivation kinetics should be steady state. This is because it is a reaction between the surface oxide film and an electrolyte which is not changed by the process.
3. The propagation kinetics should be a form of power law. This is because the cathodic site is a fixed area so the current for that site is fixed. This current is balanced by the anodic current which is dispersed across the growing anodic site. As the anodic site grows, the distance between anode and cathode gets larger and this adds a further dependence on the reaction rate.

It should be noted that the study of the mechanistic treatment of pitting of aluminium is not as well developed as that for stainless steels, especially with regard to the species-specific elements of the system. The main difference is that pitting processes of aluminium and similar metals, for example, beryllium are started by the breach of a thermodynamically stable barrier, whereas for stainless steel there is no initiation.

30.2.5 Low Temperature Thermal Oxidation

In Section 3.3, the thermodynamics and kinetics of dry oxidation were discussed. In Section 3.3.3, the growth of thick films on metals at elevated temperatures was interpreted with the help of Wagner's theory of oxidation. The oxidation kinetics is a function of the disorder within the protective oxide layer, especially interstitial ions and vacancies. The electrical conductivity of the matrix is critical so as to provide an electron population of the conduction band. Because of this, the rate of oxidation is very dependent on impurities and the chemical valency of ions of alloying elements as this strongly affects the number of electrons in the conduction band. The latter is governed by Hauffe's rules.

For uranium and its alloys, the approach of Wagner theory does not predict the corrosion behaviour of the metal. The corrosion of uranium and its alloys is covered in Chapter 19. Uranium thermally oxidizes rapidly in air at a relatively low temperature. The oxidation is considerable at temperatures in excess of 120°C. The reaction is retarded in the presence of carbon dioxide. The oxidation rate is also reduced by alloying elements such as titanium, molybdenum, tantalum, niobium, chromium and nickel. These elements produce ions of different valencies so Hauffe's rules are not obeyed.

As described in Chapter 19, the normal oxide on uranium, UO_2 is often extremely non-stoichiometric with excess oxygen, but most of the anions, O^{2-}, are close packed in a fluorite structure. It is thus very surprising that oxide growth is via apparent transport of O^{2-} through the lattice. The oxidation mechanism of uranium is thus not understood with any degree of confidence, although because of the well-known affinity of uranium (VI), as uranyl, UO_2^{2+} for carbonate ions, for example, $UO_2(CO_3)_2^{2-}$, and in the U–O–H_2O–CO_2 system, there is probably a role for surface uranium(VI) species in the inhibition by carbon dioxide.

The greatest activity in the field of uranium oxide has been in x-ray determinations of its complex structure. Comparatively little effort has been expended in understanding the transport mechanisms of oxygen anions through the matrix. It is certain though that some of the features that have been proposed in the structure of UO_2, such as the Willis, 2,2,2 structure cannot function as passive diffusing entities because of their size. Much work remains to be done before a comprehensive predictive model of low temperature thermal oxidation of uranium alloys can be created.

30.3 Physical and Biological Vectors

In most of the predictive treatments of corrosion mechanisms described in this chapter, the effect of the biological world has been largely ignored. The role of the physical domain, particularly stress, is covered in Chapter 5. In the prediction of corrosion problems, both of these vectors are usually treated separately to the normal rules of corrosion initiation.

30.3.1 The Intervention of Stress

The effect of stress in corrosion processes is often concerned with the morphology of cracking during the failure stage, such as if it is transgranular or intergranular, whether the stress is static, as in stress-corrosion cracking (SCC) or cyclic as in corrosion-fatigue. Systems are usually considered as characteristic of the metal, for example, aluminium or carbon steels in susceptible environments. Considerable thinking has been advanced concerning the mechanisms involved. The basic mechanisms from the mechanical perspective are described in Sections 5.1.2.3, 5.1.3.4 and 5.1.4.2 for SCC of aluminium, stainless steels and carbon steels and a general mechanism is covered in Section 5.2.2 for corrosion-fatigue. What is clear is that the environmental element is often unclear, especially why certain alloys are susceptible to certain media. In terms of prediction, the conditions required to avoid stress-corrosion cracking or corrosion-fatigue are generally well known and are critical in the design stage. However, if, due to misdesign, a structure or component is susceptible to cracking or fatigue, it is normally very difficult to predict a service life. The exception is for nonferrous metals without an endurance limit, where a prescribed design life of say 10^7or 10^8 cycles can be applied.

30.3.2 Biological Vectors

The impact of biological factors on corrosion and corrosion control is quite difficult to quantify. The effect of biological material is to alter the intimate environment of a metal. This can be by biofouling of vessels, by growth of plants in crevices which can lead to differential aeration cells, by the action of fungi in aviation fuel tanks or else due to by-products of animal residues, for example, of amines or ammonia that can degrade copper alloys. However, the greatest attention concerning biologically induced corrosion is usually that of microbial-induced corrosion (MIC). This includes algae, single cell fungi such as yeasts and microscopic animal species but is most commonly concerned with the effect of contaminating bacteria. Bacterial corrosion in the food industry is briefly outlined in Section 23.1.2.1.

30.3.2.1 Microbial-Induced Corrosion

The action of microbial-induced corrosion can under certain situations be very serious and cause the failure of very large installations. For example, the global cost of biodegradation of buried pipelines due to microbial corrosion is of the order of billions of dollars per annum. Microbial corrosion can also be a product of biofouling, in that in nature there is often a film of microorganisms covering other substrates and established macrospecies such as aquatic weed that can act as a habitat for bacteria and a venue for microbial-induced corrosion.

Microbiological ecology comprises more than half of all species on earth. These organisms have been in existence for millions of years and have evolved to utilize resources in the natural ecosystem in which they live. Corrosion only became an issue when man started to employ metal artifacts, roughly 6000 years ago. Thus, microbiologically induced corrosion is simply the natural world's reaction to man-made perturbation of the natural order.

It might seem strange that MIC is discussed under the heading of corrosion prediction. However, it must be realized that to be able to understand the risk of corrosion by MIC in a particular context a microbiological determination first needs to be carried out. It is a specialist subject, but once this has been done then it is often possible to predict whether MIC may be a factor. This is because the fundamental state is the degree of sterility or biological contamination. Damage to metals arising from microbial-induced corrosion is usually via the alteration of the local chemical environment as a by-product of respiration, either by acidification or a redox reaction. It is convenient to classify microbial corrosion according to the species involved.

In this work, some of the older nomenclature is used; this is to provide continuity with other corrosion texts. In 2000, some of the organisms active in MIC were reclassified which resulted in changes of name. For example, a new group of bacteria was created called *Acidithiobacillus* which includes such organisms previously named thiobacillus thioxidans and thiobacillus ferrooxidans. The new classification reflects the features of these organisms in that they are acid loving sulfur bacteria. However, in corrosion, the older classification is normally used for microbially induced corrosion.

The first class of organisms is thiobacillus (for example, thiobacillus thioxidans and thiobacillus ferrooxidans as mentioned above) a class of sulfate-reducing bacteria (SRB). These microbes are fairly commonly found in soil or water. They are classified as chemolithotrophic organisms in that they can utilize carbon dioxide in synthesis, but derive the energy for this from oxidation of sulfur compounds to give sulfuric acid. Corrosion is due to acid attack. These bacteria flourish under acid conditions and the effect of thiobacilli can be detected by the low pH of the metal environment. Corrosion by this means has been found in sewer pipes, manhole covers and gas mains. Damage can be at the top of the interior of the pipe, called *top of line* corrosion. Because of the ubiquity of thiobacilli in the environment, corrosion control is usually either by eliminating sources of sulfur or by cathodic protection because elimination of bacteria is rarely feasible.

Another form of acid producing SRB is ferrobacilli. This type of bacterium is commonly encountered in the environment where the natural mineral iron pyrites, FeS_2 is exposed to natural waters. This includes runoff from mine wastes and other areas of extraction. Since the electrolytes contain Fe^{3+} ions from the oxidation of pyrites there is a deposit of alkaline iron(III) sulfate.

Cellulose decomposing bacteria are prevalent in situations where there is a high level of paper or wood which can degrade in a waterlogged anaerobic environment. This is most common in facilities such as landfill sites where there is a large amount of paper and cloth waste material. The action of bacteria on cellulose produces a mixture of fatty acids, especially acetic acid and n-propanoic acid (butyric acid). It is important to realize that the same type of decomposition can occur where cellulose material contacts metal substrates such as rag-based insulating material on older structures.

Corrosion can also be an issue in enclosed spaces at elevated temperatures due to thermophilic bacteria. Such bacteria are incubated typically at temperatures between 40°C and 55°C and include microbes such as the clostridium family. Such organisms usually only present themselves as a corrosion hazard in warm environments.

Prediction of microbiologically induced corrosion requires specialist microbiological analysis. It depends on the type of organisms likely to be encountered. It is generally managed by good housekeeping, regular monitoring and an understanding of the type of physical traps which can act as a location for microbial colonies. Concerning the latter, it is handy to realize that these often require a certain range of humidity or water activity and a degree of quiescence.

30.3.2.2 Ecology

In considering the effect of the biological world on metals in the environment, it can be useful to consider the wider biological ecosystem. In many cases the chemical environment of a metal can change throughout the year when submersed in nature. Though materials selection is often conducted using standard materials which are designed for a purpose, it is handy for the corrosion engineer to check the potential risks for corrosion across each season to realize the greatest threat. Examples include:

1. *Jetties*: The buildup of waterweed or other biofouling in the spring and summer can cause problems when cut. Rotting vegetation can create areas of deaeration where respiration of microbial matter can be significant.
2. *Buildup of leaf mould and other detritus during bad weather events*: In temperate zones, the annual shedding of leaves from deciduous plants produces high amounts of organic matter in the fall or autumn. This can create fluxes of organic acids from cellulose decomposing bacteria at certain times of the year. This occurs under anoxic conditions so corrosion by seasonal deaeration can also be a problem.
3. *Eutrophication*: In surface water environments where there is significant runoff of nutrients, for example, nitrate from intensive agriculture or phosphate from urban drains, this can cause a process known as *eutrophication* in which there is an excess of growth of algae in the spring. Under these conditions there can be an explosion of such species, called an algal bloom which rapidly exhausts nutrients leading to the death and decay of the algae. This can create temporary anoxic conditions high in organic material which can overwhelm filtration plants leading to localized corrosion. The detritus can provide an ideal habitat for microbial-induced corrosion.
4. *Limnology*: The chemistry of lakes, ponds and reservoirs is often very stratified in the summer and winter. The bottom of the water body is usually composed of cool water which is deoxygenated. The upper strata of the water body is where photosynthesis takes place but the lower reaches of it are anoxic because of detritus filtering to the bottom of the water body. The gradient of redox potential is often very sharp – of the order of only a few centimeters and is accompanied by a steep thermal gradient called a thermocline. The anoxic conditions are persistent because reducing conditions converts iron and manganese in sediment to the reduced forms of Fe^{2+} and Mn^{2+} ions.

   This stratification is less distinct in spring and fall or autumn because mixing occurs during the storm season and thus metal artifacts in such a carbon steels can be vulnerable to corrosion, particularly in the summer or winter. Inspection of the iron Pourbaix diagram, Figure 3.2, illustrates that under neutral conditions anoxic environments are vulnerable to corrosion because in these conditions, the metal is unprotective.

In practice, many of the issues which arise are handled by good materials selection and the standard use of metals used in such circumstances is usually based on years of experience of what actually works in service. It is, however, always a good policy at the design stage to briefly review the variations in seasonal ecology and the attendant chemistry throughout the life cycle of the component or structure in service to see if unusual circumstances might arise.

30.3.2.3 Environmental Impact Assessments

In recent years, environmental legislation in many countries has become increasingly comprehensive. This has been brought together in a series of pan-national protocols under the International Standards Organization (ISO). The protocols for environmental management are covered by the series of standards beginning with ISO 14000 and the means of carrying out environmental protection methods is covered by ISO 14001 which dictates procedures for *Environment Management Systems* (EMS). These apply to organizations which carry out any activity which may harm the environment, and includes manufacturing companies, service industries, local civic authorities and government bodies.

As part of an EMS, an assessment must be made on the potential damage which any activity might cause. This is called an Environmental Impact Assessment (EIA). A specific duty of an EMS is covered in the definition 3.2.7 of ISO 14000, Prevention of Pollution directive. An environmental impact assessment is also required when there is any man-made change to the environment, either physical as in new construction, or chemical, as in a change in reagents used and which might be exposed to the wider ecosystem through spillage or disposal. An EMS also covers organizational and cultural issues which might expose an organization to inflicting future environmental damage.

The corrosion engineer will be familiar with the principles of maintaining metallic structures so as provide containment for materials that are harmful to the environment. Containment is of course the sole purpose of pipelines, although it is usually regarded that the main point of maintaining such integrity is to transfer a product or a resource to where it is needed. At present, most EMS's of organizations do not specifically address the role of corrosion control in reducing potential environmental damage in specific environmental impact assessments. This is presumably because most environmental impact assessments are carried out by those whose competencies are largely within the biological field, such as ecology or environmental science. Though there is often a familiarity with the tools of carrying out risk analyses, such personnel are normally unfamiliar with the principles of corrosion or corrosion control.

This situation is likely to change over the next few years. High profile spillages generate bad publicity for large corporations, such as in the oil industry. This is described in more detail in Chapter 28. The interplay between the systems analysis of risk of an accident, the environmental damage caused and the cost of cleanup is likely to become more integrated. The duty of bringing together the impact assessment for an EMS via ISO 14001 together with the understanding of corrosion processes, corrosion control and prediction of degradation of plant and infrastructure will increasingly become a feature of environmental engineering in the future.

30.4 Statistics

The nature of acquiring data for the assessment and prediction of corrosion processes, especially in service, is a very inexact one. There is a natural variation in the statistical population of any series of test results. Corrosion tests are often analyzed using methods of statistical analysis similar to those used in engineering failure analyses. For a detailed treatment of engineering failure analysis methods, the reader is advised to refer to one of the many good texts on this subject. The following sections present an outline of some of the points of interest in predicting corrosion behaviour from tests and service returns.

30.4.1 Trials and Tests

Test trials are a common feature of predicting corrosion performance. Test procedures are described in more depth in Chapter 29. When assessing the results of trials, it is advisable to bear in mind the following statistical factors.

1. *Statistical validity of populations*: When testing for corrosion resistance the presence or absence of corrosion within a set of samples does not mean that corrosion will not occur. Often the labour intensiveness of analyses for laboratory tests is prohibitive, thus gaining enough samples for a population to be statistically relevant can be a problem; this is especially so if there are a large range of parameters of concern, such as, for example, proposed length of service, alloy specification, humidity, surface finish, temperature or presence or absence of a protective coating such as a lacquer or geometry. The application of the Taguchi method in this case would indicate the need for several hundred test samples for the results of a test to be statistically significant. In practice, by using data from service returns, together with experience of tests and applying materials standards, the number of samples needed can usually be less, but it must always be realized that this is a compromise with purely statistical methods.
2. *Microstructure*: One factor that can be difficult for large structures when conducting statistical analysis of laboratory tests is that the amount of material under test is often not sufficient to account for the variations of microstructure. For example, if a trial is composed of 200 coupons of area 400 cm^2 each (20 cm $\times$ 20 cm), then the total test area of all samples is only 8 m^2. If the size of the artifact being underwritten for corrosion is larger than this it can be seen that this may be too small to reproduce the entire range of microstructural features that might be precursors to initiation of corrosion. If all the samples are machined from the same stock, then the chance of reproducing the full range of metallurgical forms which could act as initiation sites for corrosion is even less. For this reason for prediction of corrosion performance from coupon-based trials, it is imperative that the microstructural aspects of the material are addressed in a quantitative manner. The full methods for quantitative microstructural analysis are given in *The Quantitative Description of the Microstructure of Materials*, by Kurzydłowski and Ralph, listed in Further Reading at the end of this chapter.

   One consequence of the statistical variation in trials, whether from micrographic or other effects, is that there may occur a false negative result. In a collection of samples in a trial there is a finite possibility that even though the material could be vulnerable to corrosion that the conditions required do not occur on any sample.

   The consideration of sample size and relevance is important in any investigation concerning corrosion prediction. Consider the following hypothetical circumstance. A new coating is vulnerable to micrographic features where a high enough local density of inclusions and surrounding defects can cause microscopic areas of decohesion. This is because there is risk of failure since a previously applied conversion coating might not adequately cover these areas. If the likelihood of these areas is not common, a series of trials using test coupons may not produce any corrosion during testing.

   The current argument is also applicable to laboratory instrumentation techniques involved in corrosion research. This is because the samples usually required are small. It is often good practice to prefer techniques that can have a high sample throughput or use larger samples rather than on more complicated methods.
3. *Artifacts*: The nature of a test piece can influence the results of a series of tests. For example test blanks have edges. Edge effects can come from a range of factors such as mechanical stresses in corners or from contamination, especially with

chloride ions from handling. For this reason, tests in which corrosion is disproportionately reported close to the edge or side of a test piece should be treated with care.

For similar reasons, any corrosion detected preferentially around areas where a specimen is supported much also be viewed with caution. It is a matter of using practical judgement as to where recorded corrosion on a test piece is found.

4. *Acceleration vectors*: If a metal artifact is designed for a life cycle of many years, it can be impracticable to test it for corrosion resistance using real-time methods. In this case, acceleration vectors are often used. Tests are often carried out at elevated temperature or relative humidity from that encountered in service. Alternatively, a more aggressive chemical environment may be used. If the test is designed to deliberately overstress the corrosion resistance of the metal so that it fails, then it is known as an elephant test. The metaphor is that the overstress ensures failure in the same way as an item stepped on by an elephant will be broken by the sheer force applied.

The difficulty arises when an accelerated test is used to extrapolate service life. In the past, various empirical rules have been applied to accelerated tests. Probably the most common one of these is known as the Arrhenius relationship which describes the temperature dependence of a reaction

$$\frac{d[Product]}{dt} = A(\exp[-E^{\ddagger}/\boldsymbol{R}T]) \tag{30.42}$$

where [*Product*] is the result of any reaction, $\boldsymbol{R}$ is the universal gas constant, T is the temperature in Kelvin and A is the pre-exponential factor. The term $E^{\ddagger}$ is the *activation energy*. Svante Arrhenius first proposed Equation 30.42 as representing the temperature dependence of chemical reactions when working on the optical isomerization of sucrose in water. This is a special type of reaction in that the reactant (optically active sucrose) has the same free energy as the product (a mixture of left and right handed sucrose). In general, the Arrhenius relationship for the temperature dependence of a reaction relies on the matrix in which the reaction occurs (in the above case the solvent water) being unchanged and also that the speed of the reaction is controlled by only one principal step, called the *rate determining step*.

It should be obvious that if a reaction rate is controlled by a rate determining step, then it is a prerequisite that for the Arrhenius relationship to be applicable the chemical mechanism must remain the same for all temperatures. Referring to Section 30.2 on corrosion mechanisms, especially Section 30.2.3 on stainless steels, it can be seen that corrosion mechanisms can involve many steps so it is often not safe to apply the Arrhenius relationship to accelerated corrosion tests without significant caveats.

One empirical 'law' which is often encountered is the so-called 10°C rule. This states that for every 10°C temperature rise the reaction rate doubles; when applied to accelerated tests this is translated into a rule of thumb that raising the temperature speeds up the corrosion process so that the apparent service life can be simulated with a shorter-timed test at elevated temperatures.

The 10°C rule comes from testing of electrical insulation of cables in the early twentieth century. This insulation was made of a form of natural rubber called gutta-percha. This polymeric material degrades by a reduction of elasticity as it ages due to a loss of double bonds in the polymer chain. This is a form of perishing. The task for these tests was to discover the maximum current that could be carried by such cables; for a set voltage the temperature of the cable is proportional to the current flowing (as the electrical resistance of the wire is constant).

It was found that as the current was increased, the temperature of the wire and insulation was increased and that the rate of perishing increased at a rate of doubling for every 10°C rise in temperature. In this case, the process is the destruction of C=C double bonds and the 10°C rule results from the particular activation energy of this bond. While the 10°C rule worked in this case, it is not valid to apply it unquestionably to every situation. Thus, the 10°C rule should not be taken as a general rule for more complicated processes such as corrosion assessment.

30.4.1.1 Weibull Distribution and Truncated Data

In the statistical treatment of corrosion tests, methods are often used that are common practice in engineering failure analysis. In failure analysis, data is presented – either from in-service returns, or from test trials – in which the date that a specimen has failed is recorded. This type of data is classified as binary truncated data. It is binary data because the results are presented as only one of two conditions; either a specimen has passed the failure criteria or it has not. It is truncated data because when a specimen is recorded as having failed it has actually failed at some time between the current time and the last observation.

The most common statistical distribution applied to engineering failure is the Weibull distribution. This is useful because a predicted service life can be calculated with a relatively small number of samples and it is applicable to binary truncated data. Thus it has been extensively applied to corrosion tests and trials and to a lesser extent corrosion failures in service. The Weibull distribution must, however, be treated with some caution as it presupposes that there is only one mode of failure and it has a fixed statistical probability. For a simple corrosion test this may have some validity, but it is an oversimplification of real corrosion problems. In practice, the Weibull distribution is used when it happens to work well; in other cases semi-empirical distributions such as a log-linear or other distributions can be tried.

30.4.1.2 Elephant Tests, Burn-In and Burn-Out

An elephant test is one which is designed to test a specimen to destruction; the metaphor is that a hypothetical elephant has stepped on it. In one way this is a predictive tool. If the specimen or component only fails under conditions that will never be seen in service then it can be often predicted that the component will last for the designed service life. In corrosion, elephant tests include tests involving overaggressive media or excessive temperature. Usually, it is not possible to use such tests in any calculation of life cycle as the failure modes will be different from that encountered in service.

Manufactured items will often exhibit failure probabilities which are a reflection of either the manufacturing process or else of secondary materials fabrication. In this case, phenomena such as burn-in or burn-out can be encountered. Burn-in is where the failure rate of new components in service is quite high and in later service the failure rate slows to a steady value. Burn-in often arises from manufacturing defects; these fail quickly, and for items such as incandescent bulbs a quick test after manufacture such as running them for 10 minutes can eliminate defective stock being sent to the customer. In corrosion, burn-in affects products with, for example, corrosion-resistant coatings. Where there is a blemish in the coating, the item can be predicted to fail early in the life cycle.

Burn-out is a phenomenon where after a relatively steady failure rate, the number of failures rises continuously once a certain point in the life cycle has passed. In corrosion burn-out, it can be a feature of joints and pipe seals. In this case, once some play has occurred in a joint, seal abrasion proceeds which tends to either remove protective measures from the metal (coatings) or creates debris which can promote an electrolyte to be in contact with the metal. In burn-out, after a certain time, the amount of failure progressively increases and become a life limiting issue.

To summarize, the application of statistical treatment methods is essential in optimizing corrosion prediction. However, this is far from straightforward and to do it adequately requires a thorough knowledge of manufacturing processes, microstructure, mechanistic principles and chemistry. It is always preferable to rely as much as possible on in-service data rather than that generated via artificial laboratory tests.

30.4.2 In-Service Considerations

Service returns are the most reliable data for statistical analysis. For maximum usefulness, a full record should be kept. This includes a record of production methods, including batch number, the days when manufacturing operations were conducted, any observations as to unusual effects seen during manufacture such as discolourations, issues with cleaning and joining and general metal quality. A regular schedule of maintenance regimes for products in the field should be logged including when and where inspected, changes and repairs made, changes in environment or weather conditions and observations of wear and tear should be done.

Units that fail in service should be examined and the nature, degree and type of corrosion observed should be recorded quantitatively. Where possible, regular appraisal of new data should be carried out with reference to the existing database and service standards to tease out service returns that are atypical of normal service.

30.5 Kinetics, Modelling and Prediction

The *raison d'être* of corrosion prediction is to be able to anticipate whether corrosion will occur, its location and how long it will take before it becomes a cause for concern. With the exception of gold, all metal systems in a terrestrial environment are metastable; thus the prediction of corrosion processes is in essence a kinetic problem, but with the proviso that if the metal is well matched to its surroundings the kinetic process will be close to zero. To this end, there are several approaches to quantitatively evaluate the metal and its interaction with its environment. The more traditional of these can be described as physico-electrical models. The more recent schemes are chemical models.

30.5.1 Electrical Models

The simplest way to predict the rate at which corrosion occurs is to measure the current carried. This is because through Faraday's law the amount of corrosion in terms of metal loss is proportional to the charge expended. The current–voltage characteristics of a corroding metal system are normally readily available and there is often no need to understand the exact mechanism of corrosion. The corrosion process can be divided into two parts, initiation and corrosion.

30.5.1.1 Initiation

A technique which has recently received much attention is *electric impedance spectroscopy* (EIS). This uses an alternating current (AC) of variable frequency to determine the electrical resistance and capacitance of an electrode process; the resistance can be divided into two components, namely the resistance of an electrolyte and the resistance of any oxide film on the metal. The use of AC current methods is well established in conductimetric measurements of electrolytes, where DC methods would electrolyze a solution. In EIS measurements the reason for using AC methods is similar; by reversing the direction of the potential, polarization effects (i.e., distortion of the system) are eliminated. The mathematical treatment of charges in an oscillating field in real systems is very involved, especially the dependence on frequency. For example, even in the simplest system – a soluble symmetrical electrolyte such as sodium chloride in water – a frequency dependence of the molar conductance (the real part of impedance) was predicted as long ago as 1929. This effect, the Debye–Falkenhagen effect arises from oscillation of ions at high frequency.

Despite this electrical impendence spectroscopy has shown some promise in measuring film resistance and film capacitance for treated metal surfaces such as those with coatings or appreciable oxide thicknesses. The exact means of analysis is usually performed with specialist software and is beyond the scope of this text, but it involves two plots. The first, a *Nyquist plot*, is a representation of the real and imaginary parts of the impedance; the second a *Bode plot* is an impedance-phase angle-frequency plot.

When the electrical resistance and therefore, impedance of a film overlying a metal is high the sample is a near insulator. However, if during aging, the film degrades and its resistance declines this can be determined using the EIS method. It is, therefore, applicable for systems such as paints and lacquers which can depolymerize and become waterlogged thus initiating underlying corrosion. From this it is possible to estimate the initiation time of certain corrosion processes.

30.5.1.2 Propagation

Once initiation has started, an electrochemical cell becomes active. The corrosion rate then can make use of a corrosion–velocity diagram. These are sometimes called *Evans diagrams* after Ulick R. Evans and are described in Section 3.2.3.1. An example of their use for the calculation of the corrosion rate of iron in aerated water is given in Example 3 from Chapter 3. By calculating the anodic current density the corrosion rate during propagation can be determined.

Two geometric considerations alter the situation. This is first where the relative area of the anode and cathode change during corrosion, and second where differential aeration is involved. There are various treatments for the effect of a fixed cathode, for example, the approach given in Section 30.2.4.2. Where both cathode and anode vary in morphology, the mathematics of modelling such a system is very complicated. Differential aeration kinetics can be handled by judicious application of corrosion–velocity diagrams. For example, Figure 3.15 shows the situation for the corrosion of iron in aerated water. This is explored further in Example 3 from Chapter 3. In this example, it should be expected that the effect of a low oxygen potential should be described through the Nernst equation for the oxygen reduction reaction; indeed this equation includes a term $1/(a_{O_2})^{1/2}$. The system is not as simple as this, however, due to polarization kinetics of the anodic reaction and its dependence on oxygen potential. This is discussed in Section 3.2.3.2.

Despite this, corrosion models derived from treatments based on electrode physics do have several advantages. The first one is that many of the parameters required to quantify such models are readily measurable. The second advantage is the output parameter (current) is directly proportional to the amount of corrosion. Third, in order to model corrosion processes in this way the geometry of the system can be conveniently abstracted which simplifies calculations.

30.5.2 Chemical Modelling

30.5.2.1 General Points

In recent years, there has been considerable effort in developing corrosion models which are based on a chemical description of corrosion processes. Most commonly, these are models of the chemical transport of corrosion in crevices. Parallel to this are other tools in assessing the propensity of a protected system to experience corrosion if a chemical species is exhausted. The latter approach uses concepts such as buffer capacity for pH, redox potential or degree of water hardness, but does not seek to provide an accurate chemical description of corrosion processes.

30.5.2.2 Simple Tools

In corrosion assessments, the use of potential–pH (Pourbaix) diagrams is commonplace. These diagrams show for each metal, the conditions in terms of redox potential (referenced to the standard hydrogen electrode) and pH at which the metal is either thermodynamically immune to corrosion or else passivated. To ensure that the chemical environment in which a metal is placed remains favourable for corrosion control, the chemical environment can be controlled so that either pH or redox potential do not vary greatly even when a chemical agent that is acidic, alkaline, oxidizing or reducing is added.

The resistance to change in this way is determined by a factor called the *buffer capacity* of the system. For pH change this is given by the relationship

$$\beta = -\frac{dC_{\text{acid}}}{dpH} \tag{30.43}$$

where C_{acid} is the amount of acid added to perturb the pH of the chemical system. There is an equivalent function for the redox buffer capacity. This can be illustrated in the case of a pH buffer with a monoprotic buffering agent HA. This can dissociate thus

$$HA = H^+ + A^- \tag{30.44}$$

The buffer capacity can be derived through mass balance calculations. The buffer capacity, β, is given by the formula

$$\beta = 2.303\left(\frac{K_w}{[H^+]} + [H^+] + \frac{MK_{HA}[H^+]}{(K_{HA} + [H^+]^2)}\right) \tag{30.45}$$

The full derivation is given in J. N. Butler's *Ionic Equilibrium: A Mathematical Approach*, cited in Further Reading at the end of this chapter. The terms for $[H^+]$ and K_w are the same as in Equation 2.23. The two new terms are the dissociation constant K_{HA} of reaction 30.44 and the total concentration, M, of the buffering chemical HA.

$$M = [HA] + [A^-] \tag{30.46}$$

The two most important features of Equation 30.43 are thus: first, the buffering capacity is optimum at the pH which equates to $-\log K_{HA}$, that is when the concentration $[HA] = [A^-]$. Second, the buffering capacity, β, is almost proportional to the amount of buffering material, M, present. This is true whether the buffer is a pH buffer or a redox buffer. For this reason where the proportion, M, of a chemical buffering agent is very small the buffering capacity of it is very small. For this reason, the effect of dissolved oxygen should not ever be factored into any mass balance determinations concerning the corrosion environment. Nevertheless, if the buffer capacity for pH and redox potential is known for a corrosion environment, then the resilience of these parameters to change is also known. From this, the resistance of a metal to corrosion in its environment can be estimated. A pH or redox buffer can be naturally occurring, for example, carbonate or phosphate in the case of pH in biofluids or anoxic sediment in the case of redox. It can also be due to artificially added chemical agents that are designed to keep pH or redox at the desired levels.

Another tool which can be used to provide a level of prediction to natural systems is the Langelier index. This is a measure of the degree of water hardness and the propensity to form a hard scale. As described in Section 3.2.6.1 calcium carbonate is a natural cathodic inhibitor. In hard waters, a natural scale of this mineral can be laid down on the surface of the metal. This impedes the cathodic reaction of oxygen.

The Langelier index gives the thermodynamic indicator that the system is oversaturated with calcium carbonate. It is defined as

$$LI = pH - pH_s \tag{30.47}$$

LI is the Langelier index. pH is the pH of the solution exposed to the metal and pH_s is the pH at which the solution would be saturated with calcium carbonate. The later can be calculated from Equations 2.16, 2.18 and 2.28 if the calcium concentration and alkalinity of the solution are known. This is somewhat cumbersome, so in practice, it is usually done by referring to data in tabular form. A Langelier index of zero means that the solution is in equilibrium with calcium carbonate. If it is positive this means that it is supersaturated and a scale of calcareous deposit might be expected to form. In this case the metal has some protection from corrosion.

30.5.2.3 More Complex Models

In recent years, various models have been proposed to describe corrosion processes from a physical chemical perspective. One of the earliest of these was by Oldfield and Sutton in 1978. They created a crevice corrosion model for stainless steels. This envisaged four stages in the crevice model. These are:

1. Reduction of oxygen in the crevice due to consumption by the cathodic reaction
2. Increase of acidity in the crevice electrolyte due to hydrolysis of cations
3. Breakdown of the passive film on the surface at a critical value of pH
4. Continuation of corrosion in the crevice causing further hydrolysis and acidification

In essence, it was a development of the idea put forward by Fontana and Greene, referenced in Section 30.2.3. The apparent success of the model was that it was supposedly able to predict acidification within pits and the concentration of pitting ions within the liquor. Further developments of the core model by several different sets of co-workers has focused on applying computational tools that were developed from chemical engineering or geochemical models on aqueous chemistry and transport. Examples of such tools are packages such as PHREEQC for advection, speciation and reaction kinetics, available from the U.S. Geological Survey and the developments of WATEQ which was originally composed by Truesdell and Jones in 1974, cited in Further Reading. Application of these and other tools yield superficially plausible models for the chemical equilibria and species transport within a corroding crevice.

Such models should be treated with caution. To illustrate this, consider the precipitation processes inherent in calcite precipitation, as discussed previously with respect to the Langelier index. In natural solutions which are supersaturated with calcium carbonate, precipitation is actually rather difficult. This is because the precipitated mineral (calcite) is anisotropic and has comparatively few sites for crystal growth. In natural systems such as rivers and lakes and also domestic water supplies, the growth of the mineral calcite is slow; in practice it requires the presence of some magnesium to precipitate a more disordered product known as low magnesium dolomite which contains 4% magnesium. Even here, the situation is complicated by the inhibiting effect of organic material such as biofilms in poisoning nucleation sites.

This complexity of real systems must be taken into account when considering the corrosion in certain environments. In Chapter 25 on marine environments, the difference in corrosion behaviour of materials in seawater and in laboratory demonstrations is highlighted. Thus although crevice corrosion models as described above often have plausible stages, the actual situation of these processes is almost certainly more complicated than the model takes into account.

To illustrate this it is convenient to consider that there are other chemical systems in materials science which have long been thought of as autocatalytic but have been recently been shown not to be. One example is the polymerization of epoxy-amines which was demonstrated to be due to differential solvation. This is given in greater detail by Talbot, cited in Further Reading at the end of this chapter.

Of special note is the third stage of the model by Oldfield and Sutton. In the light of comprehensive mechanistic treatments such as described in Section 30.3.2, it can be seen that the assumption that the breakdown of the passive film occurs at a certain pH is not strictly valid. This is because the depassivation step is a kinetic effect and the chromium Pourbaix diagram does not predict the pH dependency of the phases present. Unfortunately, it must be concluded that the present predictive validity of chemical corrosion models is not sufficient at the moment for them to be used in any other capacity than primarily as academic research exercises.

30.6 Epilogue

The ultimate goal of corrosion and corrosion control is to be able to predict corrosion processes with certainty so as to eliminate unplanned corrosion events. This chapter has tried to cover some of the modern developments in corrosion prediction. This includes systems approaches and fundamental mechanistics as well as updated versions of tried and tested methods to critically evaluate materials–environment interactions so as to determine the risk of corrosion. It must be remembered in all of this that the art of corrosion control is in essence an economic one and is driven by the demands of possibility. Underneath everything, corrosion and corrosion control are essentially field disciplines. For all their merits, perfect but expensive solutions are not the answer.

At this point, it is instructive to ask two questions. The first one is: What is the state of play concerning the information required to improve our understanding of corrosion processes? The second question is: How can this be harnessed to inform better decisions in design, implementation and operation of engineering applications?

By the 1970s, many of the basic tools that are currently used to make decisions in corrosion control were already in existence. These include Pourbaix diagrams, corrosion velocity diagrams, alloy thermodynamics, metallography and the fundamentals of fracture mechanics and micromechanics. The great development since then has been in the development, commissioning and application of surface physics techniques driven at least in part by the suitability of metallic materials as samples due to their conducting nature and robustness in vacuum chambers. The ever-increasing cost of these new techniques will, however, at some point impact on their usefulness and they will become increasingly the domain of the instrumentation physicist and of 'big' science. It could be said that the dominance of technique based work in academic corrosion research over the last 30 years has some similarities with the 'Great Nernstian Hiatus' described by Bockris and Reddy which happened in electrochemistry from 1910 to 1940. This was a period where the kinetic understanding of electrode processes was retarded due to the oversuccess of the thermodynamic approach of the Nernst equation. In the present day, a case can be made that the understanding of corrosion *process* has been dominated by study of corrosion *characterization* and has arisen at least in part because of the ease of use of certain surface techniques that can be carried out in a laboratory.

Fortunately, there is a wealth of legacy data available, both from metals in service and from techniques to move the subject of corrosion, its prediction and corrosion control forward. The two great challenges in the present day in understanding corrosion are the species-specific interactions and the overarching systems treatment of corrosion. Both of these areas of interest present special difficulties, the most obvious one being that the practitioner has to be comfortable with extreme complexity but at the same time be able to distil this into practical solutions for metals in real environments. This duality is not simple. One is reminded though of the adage *'May you live in interesting times'*. If the corrosion scientist or engineer can embrace these challenges then the elimination of waste through corrosion can reach new levels of efficiency. That is good for us all.

Further Reading

Bockris, J. O'M. and Reddy, A. K. N., *Modern Electrochemistry*, Plenum Press, New York, 1970.

Burstein, G. T., Pistorius, P. C. and Mattin, S. P., The nucleation and growth of corrosion pits on stainless steel, *Corrosion Science*, 35, 57, 1993.

Butler, J. N., *Ionic Equilibrium: A Mathematical Approach*, Addison-Wesley, Reading, Massachusetts, 1964.

Cotton, F. A. and Wilkinson, G., *Advanced Inorganic Chemistry*, Second Edition, Longmans, London, 1966.

Drever, J. I., *The Geochemistry of Natural Waters*, Third Edition, Prentice Hall, New Jersey, 1997.

Fontana, M. G. and Greene, N. D., *Corrosion Engineering*, Second Edition, McGraw-Hill, New York, 1967.

Glasstone, S. and Lewis, D., *Elements of Physical Chemistry*, Second Edition, Macmillan & Company, London, 1960.

Kurzydłowski, J. K. and Ralph, B., *The Quantitative Description of the Microstructure of Materials*, CRC Press, Boca Raton, Florida, 1995.

Mellor, J. W., *A Treatise on Inorganic and Theoretical Chemistry*, Second Edition, Longmans, London, 1953.

Oldfield, J. W. and Sutton, W. H., Crevice corrosion of stainless steels. A mathematical model, *British Corrosion Journal*, 13, 3, 1978.

Pauling, L., *The Nature of the Chemical Bond*, Cornell University Press, New York, 1966.

Pourbaix, M., *An Atlas of Electrochemical Equilibria*, Cebelcor, Brussels, 1965.

Robinson, R. A. and Stokes, R. H., *Electrolyte Solutions*, Second Edition, Butterworths, London, 1959.

Sillen, L. and Martell, A. E., *Stability Constants*, The Chemical Society, London, 1964.

Szklarska-Smialowska, Z., *Pitting Corrosion of Metals*, NACE International, Houston, Texas, 1986.

Talbot, D. E. J., Martin, J. W., Chandler, C. and Sanderson, M. I., Assessment of crack initiation in corrosion fatigue by oscilloscope display of corrosion current transients, *Metals Technology*, 9, 130, 1982.

Talbot, J. D. R., The kinetics of the epoxy amine cure reaction from a solvation perspective, *Journal of Polymer Chemistry Part A*, 42, 3579, 2004.

Truesdell, A. H. and Jones, B. F., WATEQ – A computer program for calculating chemical equilibria of natural waters, *Journal of Research, USGS*, v2, 233, 1974.

Index

B

C

D

E

F

G

I

O

P

T

Milton Keynes UK
Ingram Content Group UK Ltd.
UKHW052028071024
449327UK00027B/2473